LONDON MATHEMATICAL SOCIETY LECTURE NOTE SERIES

Managing Editor: Professor Endre Süli, Mathematical Institute, University of Oxford, Woodstock Road, Oxford OX2 6GG, United Kingdom

The titles below are available from booksellers, or from Cambridge University Press at www.cambridge.org/mathematics

367 Random matrices: High dimensional phenomena, G. BLOWER
368 Geometry of Riemann surfaces, F.P. GARDINER, G. GONZÁLEZ-DIEZ & C. KOUROUNIOTIS (eds)
369 Epidemics and rumours in complex networks, M. DRAIEF & L. MASSOULIÉ
370 Theory of p-adic distributions, S. ALBEVERIO, A.YU. KHRENNIKOV & V.M. SHELKOVICH
371 Conformal fractals, F. PRZYTYCKI & M. URBAŃSKI
372 Moonshine: The first quarter century and beyond, J. LEPOWSKY, J. MCKAY & M.P. TUITE (eds)
373 Smoothness, regularity and complete intersection, J. MAJADAS & A. G. RODICIO
374 Geometric analysis of hyperbolic differential equations: An introduction, S. ALINHAC
375 Triangulated categories, T. HOLM, P. JØRGENSEN & R. ROUQUIER (eds)
376 Permutation patterns, S. LINTON, N. RUŠKUC & V. VATTER (eds)
377 An introduction to Galois cohomology and its applications, G. BERHUY
378 Probability and mathematical genetics, N. H. BINGHAM & C. M. GOLDIE (eds)
379 Finite and algorithmic model theory, J. ESPARZA, C. MICHAUX & C. STEINHORN (eds)
380 Real and complex singularities, M. MANOEL, M.C. ROMERO FUSTER & C.T.C WALL (eds)
381 Symmetries and integrability of difference equations, D. LEVI, P. OLVER, Z. THOMOVA & P. WINTERNITZ (eds)
382 Forcing with random variables and proof complexity, J. KRAJÍČEK
383 Motivic integration and its interactions with model theory and non-Archimedean geometry I, R. CLUCKERS, J. NICAISE & J. SEBAG (eds)
384 Motivic integration and its interactions with model theory and non-Archimedean geometry II, R. CLUCKERS, J. NICAISE & J. SEBAG (eds)
385 Entropy of hidden Markov processes and connections to dynamical systems, B. MARCUS, K. PETERSEN & T. WEISSMAN (eds)
386 Independence-friendly logic, A.L. MANN, G. SANDU & M. SEVENSTER
387 Groups St Andrews 2009 in Bath I, C.M. CAMPBELL *et al* (eds)
388 Groups St Andrews 2009 in Bath II, C.M. CAMPBELL *et al* (eds)
389 Random fields on the sphere, D. MARINUCCI & G. PECCATI
390 Localization in periodic potentials, D.E. PELINOVSKY
391 Fusion systems in algebra and topology, M. ASCHBACHER, R. KESSAR & B. OLIVER
392 Surveys in combinatorics 2011, R. CHAPMAN (ed)
393 Non-abelian fundamental groups and Iwasawa theory, J. COATES *et al* (eds)
394 Variational problems in differential geometry, R. BIELAWSKI, K. HOUSTON & M. SPEIGHT (eds)
395 How groups grow, A. MANN
396 Arithmetic differential operators over the p-adic integers, C.C. RALPH & S.R. SIMANCA
397 Hyperbolic geometry and applications in quantum chaos and cosmology, J. BOLTE & F. STEINER (eds)
398 Mathematical models in contact mechanics, M. SOFONEA & A. MATEI
399 Circuit double cover of graphs, C.-Q. ZHANG
400 Dense sphere packings: a blueprint for formal proofs, T. HALES
401 A double Hall algebra approach to affine quantum Schur–Weyl theory, B. DENG, J. DU & Q. FU
402 Mathematical aspects of fluid mechanics, J.C. ROBINSON, J.L. RODRIGO & W. SADOWSKI (eds)
403 Foundations of computational mathematics, Budapest 2011, F. CUCKER, T. KRICK, A. PINKUS & A. SZANTO (eds)
404 Operator methods for boundary value problems, S. HASSI, H.S.V. DE SNOO & F.H. SZAFRANIEC (eds)
405 Torsors, étale homotopy and applications to rational points, A.N. SKOROBOGATOV (ed)
406 Appalachian set theory, J. CUMMINGS & E. SCHIMMERLING (eds)
407 The maximal subgroups of the low-dimensional finite classical groups, J.N. BRAY, D.F. HOLT & C.M. RONEY-DOUGAL
408 Complexity science: the Warwick master's course, R. BALL, V. KOLOKOLTSOV & R.S. MACKAY (eds)
409 Surveys in combinatorics 2013, S.R. BLACKBURN, S. GERKE & M. WILDON (eds)
410 Representation theory and harmonic analysis of wreath products of finite groups, T. CECCHERINI-SILBERSTEIN, F. SCARABOTTI & F. TOLLI
411 Moduli spaces, L. BRAMBILA-PAZ, O. GARCÍA-PRADA, P. NEWSTEAD & R.P. THOMAS (eds)
412 Automorphisms and equivalence relations in topological dynamics, D.B. ELLIS & R. ELLIS
413 Optimal transportation, Y. OLLIVIER, H. PAJOT & C. VILLANI (eds)
414 Automorphic forms and Galois representations I, F. DIAMOND, P.L. KASSAEI & M. KIM (eds)
415 Automorphic forms and Galois representations II, F. DIAMOND, P.L. KASSAEI & M. KIM (eds)
416 Reversibility in dynamics and group theory, A.G. O'FARRELL & I. SHORT
417 Recent advances in algebraic geometry, C.D. HACON, M. MUSTAŢĂ & M. POPA (eds)
418 The Bloch–Kato conjecture for the Riemann zeta function, J. COATES, A. RAGHURAM, A. SAIKIA & R. SUJATHA (eds)

419 The Cauchy problem for non-Lipschitz semi-linear parabolic partial differential equations, J.C. MEYER & D.J. NEEDHAM
420 Arithmetic and geometry, L. DIEULEFAIT *et al* (eds)
421 O-minimality and Diophantine geometry, G.O. JONES & A.J. WILKIE (eds)
422 Groups St Andrews 2013, C.M. CAMPBELL *et al* (eds)
423 Inequalities for graph eigenvalues, Z. STANIĆ
424 Surveys in combinatorics 2015, A. CZUMAJ *et al* (eds)
425 Geometry, topology and dynamics in negative curvature, C.S. ARAVINDA, F.T. FARRELL & J.-F. LAFONT (eds)
426 Lectures on the theory of water waves, T. BRIDGES, M. GROVES & D. NICHOLLS (eds)
427 Recent advances in Hodge theory, M. KERR & G. PEARLSTEIN (eds)
428 Geometry in a Fréchet context, C.T.J. DODSON, G. GALANIS & E. VASSILIOU
429 Sheaves and functions modulo p, L. TAELMAN
430 Recent progress in the theory of the Euler and Navier–Stokes equations, J.C. ROBINSON, J.L. RODRIGO, W. SADOWSKI & A. VIDAL-LÓPEZ (eds)
431 Harmonic and subharmonic function theory on the real hyperbolic ball, M. STOLL
432 Topics in graph automorphisms and reconstruction (2nd Edition), J. LAURI & R. SCAPELLATO
433 Regular and irregular holonomic D-modules, M. KASHIWARA & P. SCHAPIRA
434 Analytic semigroups and semilinear initial boundary value problems (2nd Edition), K. TAIRA
435 Graded rings and graded Grothendieck groups, R. HAZRAT
436 Groups, graphs and random walks, T. CECCHERINI-SILBERSTEIN, M. SALVATORI & E. SAVA-HUSS (eds)
437 Dynamics and analytic number theory, D. BADZIAHIN, A. GORODNIK & N. PEYERIMHOFF (eds)
438 Random walks and heat kernels on graphs, M.T. BARLOW
439 Evolution equations, K. AMMARI & S. GERBI (eds)
440 Surveys in combinatorics 2017, A. CLAESSON *et al* (eds)
441 Polynomials and the mod 2 Steenrod algebra I, G. WALKER & R.M.W. WOOD
442 Polynomials and the mod 2 Steenrod algebra II, G. WALKER & R.M.W. WOOD
443 Asymptotic analysis in general relativity, T. DAUDÉ, D. HÄFNER & J.-P. NICOLAS (eds)
444 Geometric and cohomological group theory, P.H. KROPHOLLER, I.J. LEARY, C. MARTÍNEZ-PÉREZ & B.E.A. NUCINKIS (eds)
445 Introduction to hidden semi-Markov models, J. VAN DER HOEK & R.J. ELLIOTT
446 Advances in two-dimensional homotopy and combinatorial group theory, W. METZLER & S. ROSEBROCK (eds)
447 New directions in locally compact groups, P.-E. CAPRACE & N. MONOD (eds)
448 Synthetic differential topology, M.C. BUNGE, F. GAGO & A.M. SAN LUIS
449 Permutation groups and cartesian decompositions, C.E. PRAEGER & C. SCHNEIDER
450 Partial differential equations arising from physics and geometry, M. BEN AYED *et al* (eds)
451 Topological methods in group theory, N. BROADDUS, M. DAVIS, J.-F. LAFONT & I. ORTIZ (eds)
452 Partial differential equations in fluid mechanics, C.L. FEFFERMAN, J.C. ROBINSON & J.L. RODRIGO (eds)
453 Stochastic stability of differential equations in abstract spaces, K. LIU
454 Beyond hyperbolicity, M. HAGEN, R. WEBB & H. WILTON (eds)
455 Groups St Andrews 2017 in Birmingham, C.M. CAMPBELL *et al* (eds)
456 Surveys in combinatorics 2019, A. LO, R. MYCROFT, G. PERARNAU & A. TREGLOWN (eds)
457 Shimura varieties, T. HAINES & M. HARRIS (eds)
458 Integrable systems and algebraic geometry I, R. DONAGI & T. SHASKA (eds)
459 Integrable systems and algebraic geometry II, R. DONAGI & T. SHASKA (eds)
460 Wigner-type theorems for Hilbert Grassmannians, M. PANKOV
461 Analysis and geometry on graphs and manifolds, M. KELLER, D. LENZ & R.K. WOJCIECHOWSKI
462 Zeta and L-functions of varieties and motives, B. KAHN
463 Differential geometry in the large, O. DEARRICOTT *et al* (eds)
464 Lectures on orthogonal polynomials and special functions, H.S. COHL & M.E.H. ISMAIL (eds)
465 Constrained Willmore surfaces, Á.C. QUINTINO
466 Invariance of modules under automorphisms of their envelopes and covers, A.K. SRIVASTAVA, A. TUGANBAEV & P.A. GUIL ASENSIO
467 The genesis of the Langlands program, J. MUELLER & F. SHAHIDI
468 (Co)end calculus, F. LOREGIAN
469 Computational cryptography, J.W. BOS & M. STAM (eds)
470 Surveys in combinatorics 2021, K.K. DABROWSKI *et al* (eds)
471 Matrix analysis and entrywise positivity preservers, A. KHARE
472 Facets of algebraic geometry I, P. ALUFFI *et al* (eds)
473 Facets of algebraic geometry II, P. ALUFFI *et al* (eds)
474 Equivariant topology and derived algebra, S. BALCHIN, D. BARNES, M. KEDZIOREK & M. SZYMIK (eds)
475 Effective results and methods for Diophantine equations over finitely generated domains, J.-H. EVERTSE & K. GYŐRY
476 An indefinite excursion in operator theory, A. GHEONDEA
477 Elliptic Regularity Theory by Approximation Methods, E.A. PIMENTEL

London Mathematical Society Lecture Note Series: 478

Recent Developments in Algebraic Geometry

To Miles Reid for his 70th Birthday

Edited by

HAMID ABBAN
Loughborough University

GAVIN BROWN
University of Warwick

ALEXANDER KASPRZYK
University of Nottingham

SHIGEFUMI MORI
Kyoto University

CAMBRIDGE
UNIVERSITY PRESS

University Printing House, Cambridge CB2 8BS, United Kingdom

One Liberty Plaza, 20th Floor, New York, NY 10006, USA

477 Williamstown Road, Port Melbourne, VIC 3207, Australia

314–321, 3rd Floor, Plot 3, Splendor Forum, Jasola District Centre,
New Delhi – 110025, India

103 Penang Road, #05–06/07, Visioncrest Commercial, Singapore 238467

Cambridge University Press is part of the University of Cambridge.

It furthers the University's mission by disseminating knowledge in the pursuit of education, learning, and research at the highest international levels of excellence.

www.cambridge.org
Information on this title: www.cambridge.org/9781009180856
DOI: 10.1017/9781009180849

First published 2022

A catalogue record for this publication is available from the British Library.

Library of Congress Cataloging-in-Publication Data
Names: Abban, Hamid, 1983– author. | Brown, Gavin, 1968– author. |
Kasprzyk, Alexander, 1980– author. | Mori, Shigefumi, author.
Title: Recent developments in algebraic geometry : to Miles Reid for his
70th birthday / edited by Hamid Abban, Loughborough University, Gavin
Brown, University of Warwick, Alexander Kasprzyk, University of
Nottingham, Shigefumi Mori, Kyoto University.
Description: Cambridge, United Kingdom ; New York, NY, USA : Cambridge
University Press, 2022. | Includes bibliographical references and index.
Identifiers: LCCN 2022009684 | ISBN 9781009180856 (paperback) |
ISBN 9781009180849 (ebook)
Subjects: LCSH: Reid, Miles (Miles A.) – Influence. | Geometry, Algebraic.
Classification: LCC QA564 .A198 2022 | DDC 516.3/5–dc23/eng20220716
LC record available at https://lccn.loc.gov/2022009684

ISBN 978-1-009-18085-6 Paperback

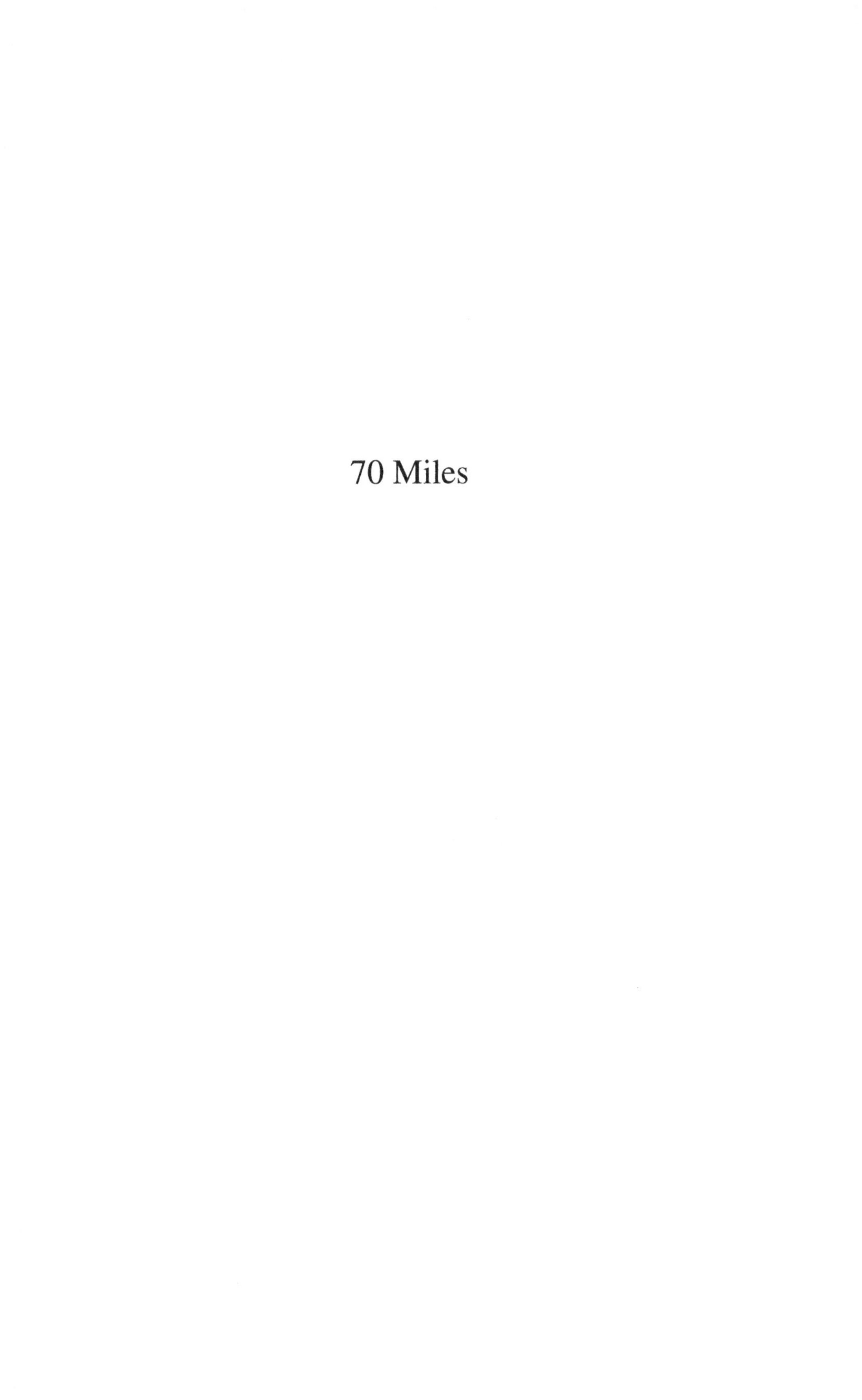

70 Miles

Contents

Contributors

Hamid Abban
Department of Mathematical Sciences, Loughborough University

John Armstrong
Department of Mathematics, King's College London

Fedor Bogomolov
Courant Institute of Mathematical Sciences, New York University

Christian Böhning
Mathematics Institute, University of Warwick

Gavin Brown
Mathematics Institute, University of Warwick

Meng Chen
School of Mathematical Sciences, Fudan University

Hailong Dao
Department of Mathematics, University of Kansas

Igor V. Dolgachev
Department of Mathematics, University of Michigan

Tom Ducat
Mathematical Sciences, Durham University

Mark Gross
The Department of Pure Mathematics and Mathematical Statistics, University of Cambridge

Paul Hacking
Department of Mathematics and Statistics, University of Massachusetts

Yong Hu
School of Mathematics, Korea Institute for Advanced Study

Alexander Kasprzyk
School of Mathematical Sciences, University of Nottingham

Yujiro Kawamata
Graduate School of Mathematical Sciences, University of Tokyo
Department of Mathematical Sciences, The Korea Advanced Institute of Science and Technology
National Center for Theoretical Sciences, Mathematics Division, National Taiwan University

Sean Keel
Department of Mathematics, 1 University Station Austin

János Kollár
Princeton University

Shigefumi Mori
Kyoto University Institute for Advanced Study, Kyoto University

Matteo Penegini
Università degli Studi di Genova, Dipartimento di Matematica

Alena Pirutka
Courant Institute of Mathematical Sciences, New York University

Yuri Prokhorov
Steklov Mathematical Institute of Russian Academy of Sciences
Department of Algebra, Moscow Lomonosov University
National Research University Higher School of Economics

N. I. Shepherd-Barron
Department of Mathematics, King's College London

Evgeny Shinder
School of Mathematics and Statistics, University of Sheffield
National Research University Higher School of Economics

Bernd Siebert
Department of Mathematics, 1 University Station Austin

Claire Voisin
Collège de France

Happy Birthday

Hamid Abban, Gavin Brown, Alexander Kasprzyk and Shigefumi Mori

It is our great pleasure and honour to say Happy Birthday to our friend Miles Reid, for your 70th birthday and indeed a few subsequent ones; it takes a long time to make a big pot, as they say, perhaps especially when there are so many potters.

These 11 papers by 20 authors give some idea of the wide range of subject areas, people and countries you have visited and influenced. It would have been easy to fill a book several times over with papers from your friends, and we regret only that we had to stop at some point while there were still so many to ask.

It may be traditional for the introduction of a Festschrift to survey the mathematics where the maestro made their major contributions, but fortunately we are spared that exercise by your own regular surveys [Rei87b, Rei87, Rei00, Rei02a, Rei02b], from the tendencious to the congressional and from the young to the old. There is also your own webpage, which knows nothing of personal data protection and reveals that in the time it has taken to compile this volume your family has grown by several grandchildren.

Nevertheless, to run through a few of the blockbusters, we first learned about canonical singularities from [Rei80], and minimal models became reality in [Rei83a, Rei83b]. Shortly afterwards 'Reid's fantasy' was revealed in [Rei87a]. Several papers on the McKay correspondence include your first coauthored paper [IR96] and another joint paper whose arXiv title 'Mukai implies McKay' is more memorable than the one the journal preferred [BKR01].

Scattered throughout are various long-running obsessions. There are always surfaces, from the smallest invariants [Rei78], to the nonnormal [Rei94], to the positively characteristic [Rei19]. And lurking nearby are codimension 4 Gorenstein rings, projection and its converse unprojection [PR04], and their applications to constructing varieties and maps [CPR00, BKR12, BR13] culminating in the general structure theory [Rei15]. Since the latter paper promises to raise more questions than it answers, perhaps we can guess some of the things you will be doing in the coming decade.

With your eighth decade well under way, it is time to say:

Happy 70th Birthday, Miles!

References

[BKR01] Tom Bridgeland, Alastair King, and Miles Reid. The McKay correspondence as an equivalence of derived categories. *J. Amer. Math. Soc.*, 14(3):535–554, 2001.

[BKR12] Gavin Brown, Michael Kerber, and Miles Reid. Fano 3-folds in codimension 4, Tom and Jerry. Part I. *Compos. Math.*, 148(4):1171–1194, 2012.

[BR13] Gavin Brown and Miles Reid. Diptych varieties, I. *Proc. Lond. Math. Soc. (3)*, 107(6):1353–1394, 2013.

[CPR00] Alessio Corti, Aleksandr Pukhlikov, and Miles Reid. Fano 3-fold hypersurfaces. In *Explicit birational geometry of 3-folds*, volume 281 of *London Math. Soc. Lecture Note Ser.*, pages 175–258. Cambridge Univ. Press, Cambridge, 2000.

[IR96] Yukari Ito and Miles Reid. The McKay correspondence for finite subgroups of $\mathrm{SL}(3, \mathbf{C})$. In *Higher-dimensional complex varieties (Trento, 1994)*, pages 221–240. de Gruyter, Berlin, 1996.

[PR04] Stavros Argyrios Papadakis and Miles Reid. Kustin-Miller unprojection without complexes. *J. Algebraic Geom.*, 13(3):563–577, 2004.

[Rei78] Miles Reid. Surfaces with $p_g = 0$, $K^2 = 1$. *J. Fac. Sci. Univ. Tokyo Sect. IA Math.*, 25(1):75–92, 1978.

[Rei80] Miles Reid. Canonical 3-folds. In *Journées de Géometrie Algébrique d'Angers, Juillet 1979/Algebraic Geometry, Angers, 1979*, pages 273–310. Sijthoff & Noordhoff, Alphen aan den Rijn—Germantown, Md., 1980.

[Rei83a] Miles Reid. Decomposition of toric morphisms. *Arithmetic and geometry, Vol. II*, volume 36 of *Progr. Math.*, pages 395–418. Birkhäuser Boston, Boston, MA, 1983.

[Rei83b] Miles Reid. Minimal models of canonical 3-folds. In *Algebraic varieties and analytic varieties (Tokyo, 1981)*, volume 1 of *Adv. Stud. Pure Math.*, pages 131–180. North-Holland, Amsterdam, 1983.

[Rei87a] Miles Reid. The moduli space of 3-folds with $K = 0$ may nevertheless be irreducible. *Math. Ann.*, 278(1-4):329–334, 1987.

[Rei87b] Miles Reid. Tendencious survey of 3-folds. In *Algebraic geometry, Bowdoin, 1985 (Brunswick, Maine, 1985)*, volume 46 of *Proc.*

Sympos. Pure Math., pages 333–344. Amer. Math. Soc., Providence, RI, 1987.

[Rei87] Miles Reid. Young person's guide to canonical singularities. In *Algebraic geometry, Bowdoin, 1985 (Brunswick, Maine, 1985)*, volume 46 of *Proc. Sympos. Pure Math.*, pages 345–414. Amer. Math. Soc., Providence, RI, 1987.

[Rei94] Miles Reid. Nonnormal del Pezzo surfaces. *Publ. Res. Inst. Math. Sci.*, 30(5):695–727, 1994.

[Rei00] Miles Reid. Twenty-five years of 3-folds—an old person's view. In *Explicit birational geometry of 3-folds*, volume 281 of *London Math. Soc. Lecture Note Ser.*, pages 313–343. Cambridge Univ. Press, Cambridge, 2000.

[Rei02a] Miles Reid. La correspondance de McKay. Number 276, pages 53–72. 2002. Séminaire Bourbaki, Vol. 1999/2000.

[Rei02b] Miles Reid. Update on 3-folds. In *Proceedings of the International Congress of Mathematicians, Vol. II (Beijing, 2002)*, pages 513–524. Higher Ed. Press, Beijing, 2002.

[Rei15] Miles Reid. Gorenstein in codimension 4: the general structure theory. In *Algebraic geometry in east Asia—Taipei 2011*, volume 65 of *Adv. Stud. Pure Math.*, pages 201–227. Math. Soc. Japan, Tokyo, 2015.

[Rei19] Miles Reid. The Tate Oort group scheme $\mathbb{TO}_p$. *Proc. Steklov Inst. Math.*, 307:245–266, 2019.

On Stable Cohomology of Central Extensions of Elementary Abelian Groups

Fedor Bogomolov, Christian Böhning and Alena Pirutka

We feel honoured to dedicate this article to our friend and colleague Miles Reid on the occasion of his 70th birthday.

Abstract

We consider that kernels of inflation maps associated with extraspecial p-groups in stable group cohomology are generated by their degree-2 components. This turns out to be true if the prime is large enough compared to the rank of the elementary abelian quotient, but false in general.

1 Introduction and Statement of Results

Throughout k will be an algebraically closed field of characteristic $l \geq 0$ and p will be a prime number assumed to be different from l if l is positive. Let G be a finite p-group. One defines the stable cohomology $H^*_{\mathrm{s},k}(G, \mathbb{Z}/p) = H^*_{\mathrm{s}}(G, \mathbb{Z}/p)$ in the following way (this does depend on k, but we suppress it from the notation when there is no risk of confusion): for a finite-dimensional generically free linear G-representation V, let $V^L \subset V$ be the open subset where G acts freely. Then the ideal $I_{G,\mathrm{unstable}}$ in the group cohomology ring $H^*(G, \mathbb{Z}/p)$ is defined to be, equivalently, the kernel of the natural homomorphism:

$$H^*(G, \mathbb{Z}/p) \longrightarrow H^*(\mathrm{Gal}(k(V/G)), \mathbb{Z}/p) \tag{1.1}$$

or, more geometrically, the kernel of

$$H^*(G, \mathbb{Z}/p) \longrightarrow \varinjlim_{U \subset V^L/G} H^*_{\text{ét}}(U, \mathbb{Z}/p),$$

where U runs over all nonempty Zariski open subsets of V^L/G.

Definition 1.1. We define $H^*_{\mathrm{s}}(G, \mathbb{Z}/p)$ as

$$H^*_{\mathrm{s}}(G, \mathbb{Z}/p) = H^*(G, \mathbb{Z}/p)/I_{G,\mathrm{unstable}}$$

A priori, this seems to depend on the choice of V, but really does not [Bog07, Theorem 6.8]. We often identify $H^*_{\mathrm{s}}(G, \mathbb{Z}/p)$ with its image in $H^*(\mathrm{Gal}(k(V/G)), \mathbb{Z}/p)$.

$H^*_{\mathrm{s}}(G, \mathbb{Z}/p)$ is contravariant in the group G: if $\varphi\colon G' \to G$ is a group homomorphism, V' and V are generically free G' and G-representations with a dominant intertwining map $\Phi\colon V' \to V$ (meaning $\Phi(g'v') = \varphi(g')\Phi(v')$ for all $g' \in G'$, $v' \in V'$), and $U' \subset V'$ and $U \subset V$ are nonempty G', G-invariant open subsets with $\Phi(U') \subset U$, then the diagram

$$\begin{array}{ccc} BG' & \longrightarrow & BG \\ \uparrow & & \uparrow \\ U'/G' & \longrightarrow & U/G \end{array}$$

induces a homomorphism $H^*_{\mathrm{s}}(G, \mathbb{Z}/p) \to H^*_{\mathrm{s}}(G', \mathbb{Z}/p)$ independent of all choices. For a subgroup $H = G'$ of G and φ the inclusion, we call the induced homomorphism $H^*_{\mathrm{s}}(G, \mathbb{Z}/p) \to H^*_{\mathrm{s}}(H, \mathbb{Z}/p)$ restriction or inflation. We also sometimes say that a class in $H^*_{\mathrm{s}}(G', \mathbb{Z}/p)$ is induced from G when it is in the image of $H^*_{\mathrm{s}}(G, \mathbb{Z}/p) \to H^*_{\mathrm{s}}(G', \mathbb{Z}/p)$ and if this map is implied unambiguously by the context.

Definition 1.2. Put $L = k(V/G)$. The unramified group cohomology

$$H^*_{\mathrm{nr}}(G, \mathbb{Z}/p) \subset H^*_{\mathrm{s}}(G, \mathbb{Z}/p)$$

is defined as the intersection, inside $H^*(L, \mathbb{Z}/p)$, of $H^*_{\mathrm{s}}(G, \mathbb{Z}/p)$ and $H^*_{\mathrm{nr}}(L, \mathbb{Z}/p)$; here, as usual, $H^*_{\mathrm{nr}}(L, \mathbb{Z}/p)$ are those classes that are in the kernel of all residue maps associated with divisorial valuations of L, that is, those corresponding to a prime divisor on some normal model of L.

In this article, we study a rather special class of groups.

Definition 1.3. For a prime p, an extraspecial p-group G is a p-group such that its center $Z(G)$ is cyclic of order p and $G/Z(G)$ is a non-trivial elementary abelian group.

This differs a bit from the arguably most common definition using the Frattini subgroup [Suz86, 4., Section 4, Definition 4.14], but it is equivalent to it by [Suz86, 4., Section 4, 4.16].

Thus, each extraspecial p-group sits in an exact sequence

$$1 \longrightarrow Z \longrightarrow G \xrightarrow{\pi} E \longrightarrow 1, \tag{1.2}$$

where $Z \simeq \mathbb{Z}/p$ is the center of the group G and $E \simeq (\mathbb{Z}/p)^n$ is an elementary abelian. Moreover, the skew form given by taking the commutator of lifts of elements in E,

$$\omega\colon E \times E \to Z,$$
$$(x, y) \mapsto [\tilde{x}, \tilde{y}],$$

must be a symplectic form if G is extraspecial. Hence $n = 2m$, and the order of G is of the form p^{1+2m} for some positive integer m. One can be much more precise and prove that, for each given order p^{1+2m} there are precisely two extraspecial p-groups of that given order, up to isomorphism [Hup67, III, Sections 13, 13.7 and 13.8] or [Gor07, Chapter 5, 5.], but we do not need this detailed structure theory. We want to study the kernel of the 'inflation map'

$$K^G = \ker\left(\pi^*\colon H^*_{\mathrm{s}}(E, \mathbb{Z}/p) \to H^*_{\mathrm{s}}(G, \mathbb{Z}/p)\right).$$

This is a graded ideal in the graded ring $H^*_{\mathrm{s}}(E, \mathbb{Z}/p)$ (graded by cohomological degree). It is natural to expect that this should be, in general, generated by its degree-2 component or, even more precisely, by the class $\omega \in \mathrm{Hom}(\Lambda^2 E, \mathbb{Z}/p) = H^2_{\mathrm{s}}(E, \mathbb{Z}/p)$ given by the extension; see also Formula (3.1) in Section 3 for the description of the stable cohomology of abelian groups. In fact, Tezuka and Yagita in [TY11] studied a very similar problem in Section 9, p. 4492 and the following; see especially the problems they have mentioned on p. 4494, top and bottom, concerning what they cannot yet prove. Indeed, the expectation above is false in general (this is similar to the situation in ordinary group cohomology where conjectures that kernels of inflation maps associated with central extensions should always be the expected ones are false as well; see [Rus92, Proposition 9]). We show:

Theorem 1.4. *Let G be an extraspecial p-group of order p^{1+2m} as above. Then, provided $p > m$, the ideal K^G is generated by $\omega \in K^G_2$.*

Note that $\omega \in K^G_2$ always because it is the image, in $H^2_{\mathrm{s}}(E, \mathbb{Z}/p)$, of the class $\tilde{\omega} \in H^2(E, \mathbb{Z}/p)$, giving the central extension G, which vanishes when pulled back to $H^2(G, \mathbb{Z}/p)$ (the induced central extension of G has a section).

On the other hand:

Theorem 1.5. *Take* $k = \mathbb{C}$. *If* G_0 *is the extraspecial* 2*-group of order* 2^{1+6} *that is the preimage of the diagonal matrices* $\mathrm{diag}(\pm 1, \dots, \pm 1)$ *in* $\mathrm{SO}_7(k)$ *under the natural covering map*

$$\mathrm{Spin}_7(k) \to \mathrm{SO}_7(k),$$

then K^{G_0} *is* not *generated by its degree-*2 *piece* $K_2^{G_0} = \langle \omega \rangle$.

This does not seem to be related to the fact that $p = 2$ is a special prime; we believe similar examples could very likely be given for every other prime p as well, as will become apparent from the construction in the proof.

Remark 1.6. Theorems 1.4 and 1.5 should be seen in the following context, which provided us with motivation for this work.

(1) As pointed out in [BT12, Theorem 11], the Bloch–Kato conjecture (Voevodsky's theorem) implies that, letting $\Gamma = \mathrm{Gal}(k(V/G))$ as above, and denoting

$$\Gamma^a = \Gamma/[\Gamma, \Gamma], \quad \Gamma^c = \Gamma/[\Gamma, [\Gamma, \Gamma]],$$

the natural map $H^*(\Gamma^a, \mathbb{Z}/p) \to H^*(\Gamma, \mathbb{Z}/p)$ is surjective, and its kernel K^{Γ^a} coincides with the kernel of $H^*(\Gamma^a, \mathbb{Z}/p) \to H^*(\Gamma^c, \mathbb{Z}/p)$ and is generated by its degree 2 component $K_2^{\Gamma^a}$ (note that since $[\Gamma, [\Gamma, \Gamma]] \subset [\Gamma, \Gamma]$, there is a natural homomorphism $\Gamma^c \to \Gamma^a$ giving $H^*(\Gamma^a, \mathbb{Z}/p) \to H^*(\Gamma^c, \mathbb{Z}/p)$); this follows not obviously from a spectral sequence argument, but in any case directly from the Bloch–Kato conjecture since for $L = k(V/G)$,

$$H^n(\Gamma^a, \mathbb{Z}/p) \simeq (L^* \otimes_{\mathbb{Z}} \cdots \otimes_{\mathbb{Z}} L^*)/p, \quad H^n(\Gamma, \mathbb{Z}/p) \simeq K_n(L)/p,$$

and the Milnor K-group $K_n(L)$ is a quotient of $L^* \otimes_{\mathbb{Z}} \cdots \otimes_{\mathbb{Z}} L^*$ by the nth graded piece of the ideal generated by the Steinberg relations in degree 2. Thus, whereas on the full profinite level, kernels of inflation maps are generated in degree 2, this property is not inherited by finite quotients of the full Galois group, that is, finite central extensions of finite abelian groups.

(2) The consequence of the Bloch–Kato conjecture in (1) shows the importance to understand central extensions of abelian groups for the computation of stable and unramified cohomology. In [BT17], after formula (1.2), the authors mention that for a finite central extension $0 \to \mathbb{Z}/p \to G^c \to G^a \to 0$ of an abelian group G^a, the kernel

of $H^*_{\mathrm{s}}(G^a, \mathbb{Z}/p) \to H^*_{\mathrm{s}}(G^c, \mathbb{Z}/p)$ *contains* the ideal generated by the kernel of $H^2_{\mathrm{s}}(G^a, \mathbb{Z}/p) \to H^2_{\mathrm{s}}(G^c, \mathbb{Z}/p)$; a preliminary version had *is equal to* instead of *contains*, and the present article shows that the containment can be strict contrary to what was expected.

2 Some Linear Algebra

We establish some results concerning the exterior algebra of a symplectic vector space over a field of any characteristic. Most of this is contained in [Bou05, Chapter VIII, Section 13, 3., pp. 203–210], but since the standing assumption there, Chapter VIII, is to work over a field of characteristic 0 where we are interested in the case of a base field of finite characteristic, it is necessary to point out in detail which statements go through unchanged and which ones require adaptation.

Let $\mathbb{F}$ be any field and V be a finite-dimensional $\mathbb{F}$-vector space of even dimension $n = 2m$. Suppose that V is symplectic, that is, endowed with a nondegenerate alternating bilinear form Ψ. Let $\mathrm{Sp}_{2m}(V, \Psi) = \mathrm{Sp}_{2m}$ be the corresponding symplectic group. From, for example [EKM08, Proposition 1.8], it follows that V is isometric to an orthogonal direct sum of m hyperbolic planes; in other words, there exists a symplectic basis

$$(e_1, \dots, e_m, e_{-m}, \dots, e_{-1}),$$

with $\Psi(e_i, e_j) = 0$ unless $i = -j$ when $\Psi(e_i, e_{-i}) = 1$. This is a statement entirely independent of the characteristic of $\mathbb{F}$, in particular, which also holds in characteristic two (the form is then at the same time alternating and symmetric). Let V^* be the dual vector space to V and (e_i^*) be the basis dual to the basis (e_i). We identify the alternating form Ψ with an element $\Gamma^* \in \Lambda^2 V^*$. Then

$$\Gamma^* = -\sum_{i=1}^{m} e_i^* \wedge e_{-i}^*.$$

Via the isomorphism $V \to V^*$ given by Ψ, the form Ψ induces a symplectic form Ψ^* on V^*. Identifying Ψ^* with an element Γ in $\Lambda^2 V$,

$$\Gamma = \sum_{i=1}^{m} e_i \wedge e_{-i}.$$

One also denotes by $X_- \colon \Lambda^* V \to \Lambda^* V$ the endomorphism induced by a left exterior product with Γ and by $X_+ \colon \Lambda^* V \to \Lambda^* V$ the endomorphism given by a left interior product (contraction) with $-\Gamma^*$; more precisely,

$$X_+(v_1 \wedge \cdots \wedge v_r)$$
$$= \sum_{1 \le i<j \le r} (-\Psi)(v_i, v_j)(-1)^{i+j} v_1 \wedge \cdots \wedge \hat{v}_i \wedge \cdots \wedge \hat{v}_j \wedge \cdots \wedge v_r.$$

Moreover, let $H\colon \Lambda^* V \to \Lambda^* V$ be the endomorphism that is multiplication by $(m-r)$ on $\Lambda^r V$ for $0 \le r \le 2m$. Then, as in [Bou05, p. 207, Example 19], it follows that

$$[X_+, X_-] = -H, \quad [H, X_+] = 2X_+, \quad [H, X_-] = -2X_- \tag{2.1}$$

so the vector subspace of $\mathrm{End}(\Lambda^* V)$ generated by X_+, X_-, H is a Lie subalgebra isomorphic to $\mathfrak{sl}(2, \mathbb{F})$. Moreover, for the action of $\mathfrak{sl}(2, \mathbb{F})$ on $\Lambda^* V$, the subspace $\Lambda^r V$ is the subspace of elements of weight $m - r$.

In the following proposition and its proof, we make the conventions that for integers $i < 0$, $\Lambda^i V := 0$ and for binomial coefficients and positive integers n, $\binom{n}{i} := 0$.

Proposition 2.1. *Put $E_r = (\Lambda^r V) \cap \ker X_+$, the 'primitive elements' in $\Lambda^r V$. If $p = \mathrm{char}\,\mathbb{F} > \dim V/2 = m$ or $\mathrm{char}\,\mathbb{F} = 0$, then*

(1) for $r \le m-1$, the restriction of X_- to $\Lambda^r V$ is injective;
(2) for $r \ge m-1$, the restriction of X_- to $\Lambda^r V$ induces a surjection from $\Lambda^r V$ onto $\Lambda^{r+2} V$;
(3) for $r \le m$,

$$\Lambda^r V = E_r \oplus X_-(\Lambda^{r-2} V).$$

Moreover, E_r coincides with the submodule $F_r \subset \Lambda^r V$ defined as the span of all 'completely reducible' r-vectors $v_1 \wedge \cdots \wedge v_r$ such that $\langle v_1, \ldots, v_r \rangle$ is a totally isotropic subspace of V. Here completely reducible means simply a pure wedge product of the above form $v_1 \wedge \cdots \wedge v_r$.

Proof The proof is based on the following observations.

(I) Let E be any $\mathfrak{sl}(2, \mathbb{F})$-module and ϵ be a primitive element, by which we mean, as usual, $X_+(\epsilon) = 0$, and ϵ is an eigenvector for some $\lambda \in \mathbb{F}$ for H. Then, as long as ν is an integer such that $1 \le \nu < p$, it does make sense to define

$$\epsilon_\nu = \frac{(-1)^\nu}{\nu} X_-^\nu \epsilon, \quad \epsilon_0 = \epsilon, \quad \epsilon_{-1} = 0.$$

Then a straightforward computation with the relations (2.1), done in [Bou05, Chapter VIII, Section 1, 2., Proposition 1], shows that

$$H\epsilon_\nu = (\lambda - 2\bar{\nu})\epsilon_\nu, \qquad X_-\epsilon_\nu = -(\bar{\nu}+1)\epsilon_{\nu+1},$$
$$X_+\epsilon_\nu = (\lambda - \bar{\nu} + 1)\epsilon_{\nu-1}, \tag{2.2}$$

as long as all indices of the occurring ϵ's are $< p$. Here we put a bar on an integer to indicate that we consider it as an element of $\mathbb{F}$ via the natural homomorphism $\mathbb{Z} \to \mathbb{F}$, which for us will however be usually not injective.

(II) If we define F_r as in the statement of the proposition, then obviously $F_r \subset E_r$ and

$$\dim F_r = \binom{2m}{r} - \binom{2m}{r-2}, \quad 0 \le r \le m.$$

This is proven in [DB09, Theorem 1.1] under no assumptions on $p = \operatorname{char}\mathbb{F}$.

For the module $E = \Lambda^* V$, we can thus display the action of the operators H, X_+, X_- schematically in the familiar way:

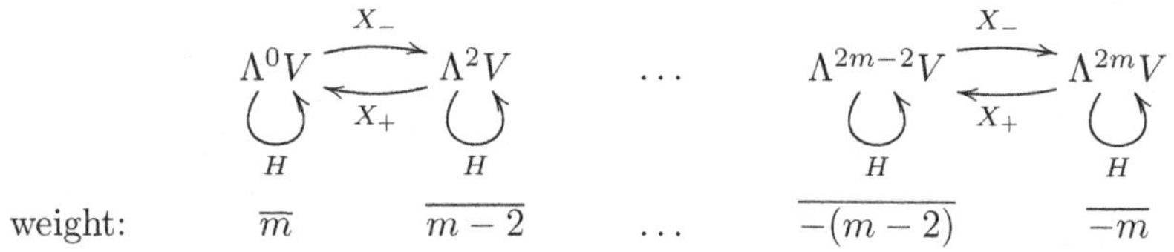

$$\begin{array}{lccccc} & \Lambda^0 V \underset{X_+}{\overset{X_-}{\rightleftarrows}} & \Lambda^2 V & \cdots & \Lambda^{2m-2}V \underset{X_+}{\overset{X_-}{\rightleftarrows}} & \Lambda^{2m}V \\ & \circlearrowleft H & \circlearrowleft H & & \circlearrowleft H & \circlearrowleft H \\ \text{weight:} & \overline{m} & \overline{m-2} & \cdots & \overline{-(m-2)} & \overline{-m} \end{array}$$

and

$$\begin{array}{lccccc} & V \underset{X_+}{\overset{X_-}{\rightleftarrows}} & \Lambda^3 V & \cdots & \Lambda^{2m-3}V \underset{X_+}{\overset{X_-}{\rightleftarrows}} & \Lambda^{2m-1}V \\ & \circlearrowleft H & \circlearrowleft H & & \circlearrowleft H & \circlearrowleft H \\ \text{weight:} & \overline{m-1} & \overline{m-3} & \cdots & \overline{-(m-3)} & \overline{-(m-1)} \end{array}$$

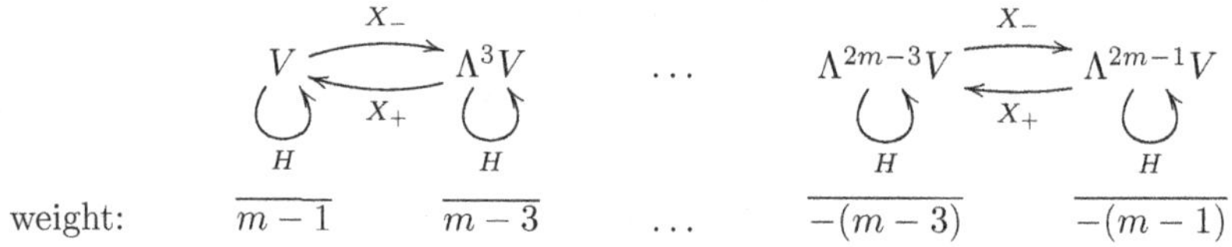

Now we start to use the assumption that $p = \operatorname{char}\mathbb{F} > m$.

If we start with a primitive element $\epsilon = \epsilon_0$ in one of the F_r, $0 \le r \le m$, of weight $\lambda = \overline{m-r}$ in $\{\bar{0}, \dots, \bar{m}\}$, then the ϵ_ν, as in item (I) above, are all defined for $\nu = 0, \dots, m$. Moreover, if μ is the largest integer such that $\epsilon_\mu \ne 0$, then $\mu \le m < p$ and μ can only possibly be equal to m if we start with ϵ_0 in F_0; excluding the latter case for a moment, we can use the third of (2.2) to obtain

$$0 = X_+(\epsilon_{\mu+1}) = (\lambda - \bar{\mu})\epsilon_\mu,$$

where now all indices are still $< p$, and one can only have that $\lambda - \bar{\mu} = 0$ in $\mathbb{F}$ if μ is the unique lift of λ in $\{0, \dots, m\}$. If $\mu = m$ and $\epsilon_0 \in F_0$, the third of (2.2) still shows that all of $\epsilon_0, \epsilon_1, \ldots, \epsilon_m$ must be nonzero (since applying X_+ to any of them the appropriate number of times returns a nonzero multiple of ϵ_0 under the standing assumptions), so putting all this together, we can say that starting from a primitive $\epsilon_0 \in F_r$, $0 \leq r \leq m$, with weight $\lambda = \overline{m-r}$ we get a chain of nonzero vectors

$$\epsilon_0 \underset{X_+}{\overset{X_-}{\rightleftarrows}} \epsilon_1 \quad \cdots \quad \epsilon_{m-r-1} \underset{X_+}{\overset{X_-}{\rightleftarrows}} \epsilon_{m-r}$$

$$\text{weight:} \quad \overline{m-r} \qquad \overline{m-r-2} \quad \cdots \quad \overline{-(m-r-2)} \qquad \overline{-(m-r)}$$

where according to the formulas in (2.2) and since $p > m$, the X_+ and X_- map each of the ϵ's to a nonzero multiple of the subsequent one 'up or down the ladder' as indicated in the previous diagram. In particular, X_+ and X_- induce isomorphisms between the vector subspaces indicated in the following diagram:

$$F_r \underset{X_+}{\overset{X_-}{\rightleftarrows}} X_-(F_r) \quad \cdots \quad X_-^{m-r-1}(F_r) \underset{X_+}{\overset{X_-}{\rightleftarrows}} X_-^{m-r}(F_r)$$

for $0 \leq r \leq m$. In addition, the sum of F_r, $X_-(F_{r-2})$, $X_-^2(F_{r-4}), \ldots$ inside $\Lambda^r V$ (for any $0 \leq r \leq 2m$, noting $F_s = 0$ for $s > m$) is direct: this can be seen by repeatedly applying X_+ and using the third formula of (2.2). Then a dimension count using item (II) at the beginning of the proof yields

$$\Lambda^r V = F_r \oplus X_-(F_{r-2}) \oplus X_-^2(F_{r-4}) \oplus \cdots = F_r \oplus X_-(\Lambda^{r-2} V)$$

for any r, which proves $E_r = F_r$, *(3)* in the statement of the proposition as well as *(1)* and *(2)*. □

We will need one further piece of information concerning $E_r = F_r$, $0 \leq r \leq m$ later.

Theorem 2.2. *Let $\mathbb{F}$ be a field of characteristic p. The* $\mathrm{Sp}(2m, \mathbb{F})$*-module $E_r = F_r$, $1 \leq r \leq m$ is irreducible if*

$$p > m - \frac{r}{2} + 1.$$

More precisely, it is irreducible if and only if p does not divide

$$\prod_{0\le j\le r,\ j\equiv r\,(\mathrm{mod}\,2)} \binom{m-(r+j)/2+1}{(r-j)/2}.$$

Proof This is [PS83, p. 1313, Theorem 2]. □

3 Proofs of Main Results

We start by recalling that for an abelian p-group A,

$$H^*_{\mathrm{s}}(A,\mathbb{Z}/p) \simeq \Lambda^* A^\vee = \Lambda^* H^1(A,\mathbb{Z}/p) = \Lambda^* \mathrm{Hom}(A,\mathbb{Z}/p). \quad (3.1)$$

See, for example [Bog07, Example after Remark 6.10]. We now turn to extraspecial groups G sitting in an exact sequence (1.2) and retain the notation from Section 1.

Definition 3.1. A subgroup $A \subset E$ is called *totally isotropic* if it is a totally isotropic subspace of the $\mathbb{F}_p$-vector space E, that is, the symplectic form ω vanishes identically on A.

Totally isotropic subgroups A can be characterised as the ones such that there exists an abelian subgroup $\tilde{A} \subset G$ of the same p-rank as A mapping onto A via π, where π is the natural surjection $\pi\colon G \to E$ (take for $\tilde{A}$ the subgroup generated by the lifts to G of a minimal set of generators for A).

Proof of Theorem 1.4 For a totally isotropic subgroup A of E, consider the diagram

$$\begin{array}{ccc} H^*_{\mathrm{s}}(E,\mathbb{Z}/p) & \xrightarrow{\pi^*} & H^*_{\mathrm{s}}(G,\mathbb{Z}/p) \\ \downarrow{\scriptstyle r^E_A} & & \downarrow{\scriptstyle r^G_{\tilde{A}}} \\ H^*_{\mathrm{s}}(A,\mathbb{Z}/p) & \xrightarrow{\pi^*} & H^*_{\mathrm{s}}(\tilde{A},\mathbb{Z}/p) \end{array}$$

where the vertical arrows are the restriction maps. From the description of the stable cohomology of abelian groups, one gets that the lower horizontal arrow is injective and, in fact, an isomorphism. In other words, a class $\alpha \in H^r_{\mathrm{s}}(E,\mathbb{Z}/p)$ that is non-trivial on a totally isotropic subgroup is *not* in the kernel of π^*. Applying Proposition 2.1 to the symplectic vector space $V = H^1(E,\mathbb{Z}/p) = E^*$, we see that in order to prove the theorem it suffices to show that every nonzero class $\alpha \in E_r$, $0 \le r \le m$, is non-trivial on some totally isotropic subgroup. Since totally isotropic subgroups are invariant under the action of $\mathrm{Sp}_{2m}(\mathbb{F}_p)$, the classes in E_r that are trivial on all totally isotropic subgroups form an $\mathrm{Sp}_{2m}(\mathbb{F}_p)$ submodule; as E_r is irreducible by Theorem 2.2, this submodule is either reduced to zero or everything. Hence, it suffices to prove that some class $\alpha \in E_r$, for every $0 \le r \le m$, is non-trivial on some totally isotropic subgroups. But this is clear: in the notation introduced at the beginning of Section 2, if we take the totally isotropic subgroup $A = \langle e_1^*, \dots, e_r^* \rangle$, then $\alpha = e_1 \wedge \cdots \wedge e_r$ is non-trivial on it. □

For the proof of Theorem 1.5, we need a few more auxiliary results. Assume $k = \mathbb{C}$ now. As in the statement of that theorem, consider the preimage $G_0 \subset \mathrm{Spin}_7(k)$ of the diagonal matrices with entries ± 1 in $\mathrm{SO}_7(k)$.

Lemma 3.2. *G_0 is an extraspecial 2-group sitting in an exact sequence*

$$0 \longrightarrow \mathbb{Z}/2 \longrightarrow G_0 \longrightarrow (\mathbb{Z}/2)^6 \longrightarrow 0.$$

Proof The existence of such an exact sequence is clear. The point is to verify that $\mathbb{Z}/2$ is the entire center of G_0, and this is done in [Ada96, Chapter 4, Lemma on p. 22]; for this, it is important that 7 coming from $\mathrm{Spin}_7(k)$ is odd: the center of the analogously defined groups for the even Spin groups is larger (the claim that all of them are extraspecial in [Bak02, 5.5, p. 154] is erroneous). □

Lemma 3.3. *Generically free linear $\mathrm{Spin}_7(k)$-quotients $V/\mathrm{Spin}_7(k)$ and generically free linear G_0-quotients W/G_0 are stably birationally equivalent. Here V resp. W are any generically free (finite-dimensional, complex) linear representations of $\mathrm{Spin}_7(k)$ resp. G_0.*

Proof This is proven in [Bog86, Section 3 and in the following], but we include the easy argument for the sake of completeness and give a few more details. Consider the standard representation k^7 of $\mathrm{SO}_7(k)$; via the natural covering map, it is a $\mathrm{Spin}_7(k)$-representation. Inside $k^7 \oplus \cdots \oplus k^7$ (seven times), consider the subvariety R of tuples of vectors $(v_1, \dots, v_7)$ that are mutually orthogonal. This is invariant under the group action, and has the structure of a tower of equivariant vector bundles over any of the summands k^7. Let $P = ke_1 \times \cdots \times ke_7 \subset R$ be the Cartesian product of the lines through the standard basis vectors $e_1, \dots, e_7$. Then P is a $(\mathrm{Spin}_7(k), G_0)$-section of the action and [CTS07, Theorem 3.1] applies; in particular, given any generically free linear $\mathrm{Spin}_7(k)$-representation V, then (a) $(V \oplus R)/\mathrm{Spin}_7(k)$ is stably equivalent to $V/\mathrm{Spin}_7(k)$ since R has the structure of a tower of equivariant vector bundles and one can apply the 'no-name lemma' [CTS07, Theorem 3.6]; (b) in $V \oplus R$ the subvariety $V \times P$ is a $(\mathrm{Spin}_7(k), G_0)$-section, whence $(V \oplus R)/\mathrm{Spin}_7(k)$ is birational to $(V \times P)/G_0$. This concludes the proof. □

Theorem 3.4. *Generically free linear* $\mathrm{Spin}_7(k)$*-quotients are stably rational; in particular, combining this with Lemma 3.3,* G_0 *has trivial unramified cohomology (with finite torsion coefficients).*

Proof The fact that generically free $\mathrm{Spin}_7\,\mathbb{C}$-quotients are stably rational is proven in [Kor00]; see also [CTS07, Section 4.5]. The fact that unramified cohomology of stably rational varieties is trivial is proven, for example, in [CT95, Proposition 4.1.4]. □

Proposition 3.5. *Let* G *be a finite group. Suppose that* $\alpha \in H^*_{\mathrm{s}}(G, \mathbb{Z}/p)$ *is a class whose restriction, for any* $g \in G$*, to* $H^*_{\mathrm{s}}(Z(g), \mathbb{Z}/p)$ *is induced from* $H^*_{\mathrm{s}}(Z(g)/\langle g \rangle, \mathbb{Z}/p)$*; here* $Z(g)$ *is the centraliser of* g *in* G*. Then* α *is unramified.*

Proof This follows from the way residue maps in Galois cohomology are defined; see [GMS03, Chapter II, 7] for the following: for $K = k(V)^G$, V a generically free G-representation, and v a geometric discrete valuation of K, one considers the completion K_v, and the decomposition group Dec_w where w is an extension of v to the separable closure K_s. Then $\mathrm{Gal}(K_v) \simeq \mathrm{Dec}_w \subset \mathrm{Gal}(K)$, and there is a split exact sequence

$$1 \longrightarrow I \longrightarrow \mathrm{Gal}(K_v) \longrightarrow \mathrm{Gal}(\kappa_v) \longrightarrow 1, \tag{3.2}$$

where κ_v is the residue field of v and $I \simeq \hat{\mathbb{Z}}$ is the topologically cyclic inertia subgroup. For a finite constant $\mathrm{Gal}(K)$-module C of order not divisible by $\mathrm{char}(k)$, there is an exact sequence

$$0 \longrightarrow H^i(\mathrm{Gal}(\kappa_v), C) \longrightarrow H^i(\mathrm{Gal}(K_v), C) \xrightarrow{r} H^{i-1}(\mathrm{Gal}(\kappa_v), \mathrm{Hom}(I, C)) \longrightarrow 0,$$

and r is the residue map. The residue of an element $\beta \in H^i(K, C)$ is obtained by restricting to $H^i(K_v, C)$ and afterwards applying r. Now, under the natural map $\mathrm{Gal}(K) \to G$, the topologically cyclic inertia subgroup I will map to a cyclic subgroup of G generated by some element $g \in G$, and (3.2) being a central extension, $\mathrm{Gal}(K_v)$ will map into the centraliser $Z(g) \subset G$. Now, if $C = \mathbb{Z}/p$ and $\alpha \in H^*_{\mathrm{s}}(G, \mathbb{Z}/p)$ is a class whose restriction to $H^*_{\mathrm{s}}(Z(g), \mathbb{Z}/p)$ is induced from $H^*_{\mathrm{s}}(Z(g)/\langle g \rangle, \mathbb{Z}/p)$, then since there is a commutative diagram,

$$\begin{array}{ccccc} \mathrm{Gal}(\kappa_v) & \twoheadleftarrow & \mathrm{Gal}(K_v) & \hookrightarrow & \mathrm{Gal}(K) \\ \downarrow & & \downarrow \psi & & \downarrow \tau \\ Z(g)/\langle g\rangle & \twoheadleftarrow & Z(g) & \hookrightarrow & G \end{array}$$

and a factorisation

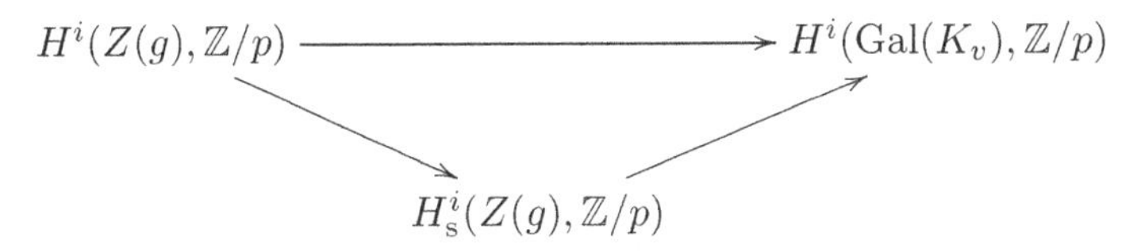

(the latter because $\mathrm{Gal}(K_v)$ sits inside $\mathrm{Gal}(k(V/(Z(g)))) \subset \mathrm{Gal}(K)$, $\mathrm{Gal}(k(V/(Z(g))))$ being the preimage of $Z(g)$ under τ), we get that the restriction of α to the decomposition group comes from $H^i(\mathrm{Gal}(\kappa_v), \mathbb{Z}/p)$; hence, its residue is zero. □

Proof of Theorem 1.5 Let $E = (\mathbb{Z}/2)^6$ and $V = E^*$. On E we choose coordinates $x_1, x_2, x_3, y_1, y_2, y_3 \in E^*$ that form a symplectic basis so that

$$\omega = \sum_{i=1}^{3} x_i \wedge y_i.$$

To prove the theorem, we are going to take the following steps.

(1) We produce a class $\zeta \in H^4_{\mathrm{s}}(E, \mathbb{Z}/2) = \Lambda^4 V$ that is not in the ideal generated by ω. More precisely, we will take

$$\zeta = x_2 \wedge x_3 \wedge y_2 \wedge y_3.$$

(2) We prove that $\pi^*(\zeta) \in H^4_{\mathrm{s}}(G_0, \mathbb{Z}/2)$ is unramified using the criterion given in Proposition 3.5. By Theorem 3.4, we conclude that $\pi^*(\zeta) = 0$ when ζ is a class in the kernel of π^* not in the ideal generated by ω. Indeed, this argument works for any class in $H^4_{\mathrm{s}}(E, \mathbb{Z}/2)$, not just the specific ζ.
(3) We check that the kernel of $\pi^* \colon H^2_{\mathrm{s}}(E, \mathbb{Z}/2) \to H^2_{\mathrm{s}}(G_0, \mathbb{Z}/2)$ is spanned by ω.

As kindly pointed out by the referee, for *(1)*, it suffices to remark that $\omega^2 = 0$, but $\omega \wedge \zeta \neq 0$ so that ζ cannot be a multiple of ω.

Let us now prove *(2)*. This uses another very nice simplification suggested by the referee; our initial proof was more complicated. We prove that $\pi^*(\vartheta) \in H^4_{\mathrm{s}}(G_0, \mathbb{Z}/2)$ is unramified for any class $\vartheta \in H^4_{\mathrm{s}}(E, \mathbb{Z}/2)$, when $H^4_s(E,\mathbb{Z}/2) \to H^4_s(G_0,\mathbb{Z}/2)$ is actually a zero map since G_0 has trivial unramified cohomology.

For an element g in E, we denote $\tilde{g}$ any lift of g to G_0. Denote the image of the centraliser $Z(\tilde{g}) \subset G_0$ in E by S_g. It has the following description:

$$S_g = \langle h \in E \mid \omega(g,h) = 0 \rangle .$$

In other words, it consists of all elements h in E whose preimages in G_0 commute with $\tilde{g}$. In order to show that $\pi^*(\vartheta)$ is unramified in the stable cohomology of G_0, it is thus sufficient, by Proposition 3.5, to show:

(*) For any $\tilde{g} \in G_0$, the restriction of $\pi^*(\vartheta)$ to $H^*_s(Z(\tilde{g}),\mathbb{Z}/2)$ is induced from $H^*_s(S_g/\langle g\rangle,\mathbb{Z}/2)$ via the natural homomorphisms

$$Z(\tilde{g}) \longrightarrow Z(\tilde{g})/\langle\tilde{g}\rangle \longrightarrow S_g/\langle g\rangle.$$

If $g = 0$, this is obvious since ϑ comes from $H^*_s(E,\mathbb{Z}/2)$. Hence, we will assume in the sequel that $g \neq 0$. Without loss of generality, we can also assume that $\dim S_g \geq 4$ since otherwise ϑ restricts to zero on S_g, hence will also be zero on $Z(\tilde{g})$ when (*) is trivially verified. Since g is nonzero and ω is nondegenerate, S_g is a hyperplane, hence has dimension 5. Furthermore, we can assume that $g = e_1$ is the standard basis element, so

$$S_g = \langle e_1, e_2, e_{-2}, e_3, e_{-3}\rangle.$$

Then a direct computation shows

$$\Lambda^4(S_g^*) = (\omega\mid_{S_g} \wedge \Lambda^2(S_g^*)) \oplus \langle x_2 \wedge y_2 \wedge x_3 \wedge y_3\rangle,$$

and the first summand is in the kernel of $H^4_s(S_g,\mathbb{Z}/2) \to H^4_s(Z(\tilde{g}),\mathbb{Z}/2)$, and the second summand is induced from $H^4_s(S_g/\langle g\rangle,\mathbb{Z}/2)$.

Finally, for (*3*), note that there are generically free E- and G_0-representations V_E and V_{G_0} such that, denoting by a superscript L the loci where the actions are free, $(V^L_{G_0})/G_0$ maps dominantly to $(V^L_E)/E$, and the induced field extension $k(V_{G_0})^{G_0} \supset k(V_E)^E$ factors as

$$\begin{array}{ccc} k(V_{G_0})^{G_0} & \xrightarrow{\simeq} & k(\mathcal{S})(t) \\ & & \uparrow \\ & & k(\mathcal{S}) \\ & & \uparrow \\ & & k(V_E)^E, \end{array}$$

where $\mathcal{S}$ is a Severi–Brauer scheme over $k(V_E)^E$ and t is an indeterminate: indeed, one can take for V_E any generically free E-representation, which is at the same time a G_0-representation via the homomorphism $G_0 \to E$, and for V_{G_0}, one takes $W \oplus V_E$, where W is a generically free G_0-representation in which the center $\mathbb{Z}/2$ of G_0 acts non-trivially via multiplication by scalars.

Then the sought-for Severi–Brauer scheme $\mathcal{S}$ is

$$(\mathbb{P}(W) \oplus V_E^L)/G_0 \to (V_E^L)/E$$

over which $(W\backslash\{0\} \oplus V_E^L)/G_0$ is a k^*-principal bundle (Zariski locally trivial). Hence, the tower of fields above. By Amitsur's theorem [GS06, 4.5.1], the kernel of $\mathrm{Br}(k(V_E)^E) \to \mathrm{Br}(k(\mathcal{S}))$ is cyclic generated by the class of $\mathcal{S}$, and $\mathrm{Br}(k(\mathcal{S})) \to \mathrm{Br}(k(\mathcal{S})(t))$ is injective. Since the two-torsion in the Brauer group of the fields Λ involved here is precisely $H^2(\Lambda, \mathbb{Z}/2)$, the definition of stable cohomology given in formula (1.1) shows that the kernel of $\pi^*\colon H^2_{\mathrm{s}}(E, \mathbb{Z}/2) \to H^2_{\mathrm{s}}(G_0, \mathbb{Z}/2)$ is cyclic. Since it contains the non-trivial class ω, it is generated by ω. □

Remark 3.6. As kindly pointed out by the referee, step (*3*) of the above proof can be given a simpler proof based on ideas of multi-linear algebra similar to the ones used in Theorem 1.4, which imply more generally:

Lemma 3.7. *For any extraspecial p-group G, the degree-2 component*

$$K_2^G = \ker\left(H^2_{\mathrm{s}}(E, \mathbb{Z}/p) \to H^2_{\mathrm{s}}(G, \mathbb{Z}/p)\right)$$

is generated by ω.

Proof It suffices to show that if $\eta \in H^2_{\mathrm{s}}(E, \mathbb{Z}/p) = \Lambda^2(E^*)$ is such that it vanishes on each $u \wedge v \in \Lambda^2 E$ for u, v generating an isotropic subspace, then this η is a multiple of ω. Indeed, this is true in any characteristic since the subspace of $\Lambda^2 E$ generated by isotropic planes $u \wedge v$ has codimension 1 (hence, it has orthogonal of dimension 1 generated by ω): indeed, the isotropic planes

$$e_i \wedge e_j, \quad j \neq -i \qquad \text{and}$$
$$(e_i + e_{i+1}) \wedge (e_{-i} - e_{-i-1}), \quad 1 \leq i \leq m-1$$

generate a codimension 1 subspace. (This is also implied by [DB09, 1.1].)

The assertion of this lemma also follows from Amitsur's theorem in a way analogous to the method used in step (*3*) of the proof of Theorem 1.5, but the proof here is clearly much easier. □

Remark 3.8. The phenomenon, on which the proof of Theorem 1.5 is based to a large extent, that X_- as in Proposition 2.1 can fail to be surjective on some $\Lambda^r V$, with $r \geq m-1$ in cases where $p \leq m$, is not related to $p = 2$ but reoccurs for other primes: it is only to do with the fact that $(X_-)^p = 0$ in characteristic p. We therefore strongly suspect that examples of the type given in Theorem 1.5, where K_2^G fails to generate the ideal K^G, for some extraspecial group G, exist for all primes p.

Acknowledgements

For the first author, the study was funded within the framework of the Higher School of Economics (HSE) University Basic Research Program and the Russian Academic Excellence Project 5-100 and by Engineering and Physical Sciences Research Council programme grant EP/M024830. The third author was partially supported by National Science Foundation grant DMS-1601680, The French National Research Agency grant ANR-15-CE40-0002-01, and the Laboratory of Mirror Symmetry National Research University HSE, the Government of the Russian Federation grant ag.\no.\14.641.31.0001.

We would like to thank the anonymous referee for carefully reading the initial manuscript and many helpful suggestions for improvements.

References

[Ada96] J. F. Adams. *Lectures on exceptional Lie groups.* Chicago Lectures in Mathematics. University of Chicago Press, Chicago, IL, 1996. With a foreword by J. Peter May, Edited by Zafer Mahmud and Mamoru Mimura.

[Bak02] Andrew Baker. *Matrix groups: An introduction to Lie group theory.* Springer Undergraduate Mathematics Series. Springer-Verlag London Ltd., London, 2002.

[Bog86] F. A. Bogomolov. Stable rationality of quotient spaces for simply connected groups. *Mat. Sb. (N.S.)*, 130(172)(1):3–17, 128, 1986.

[Bog07] F. A. Bogomolov. Stable cohomology of finite and profinite groups. In *Algebraic groups*, pages 19–49. Universitätsverlag Göttingen, Göttingen, 2007.

[Bou05] Nicolas Bourbaki. *Lie groups and Lie algebras.* Chapters 7–9. Elements of Mathematics (Berlin). Springer-Verlag, Berlin, 2005. Translated from the 1975/82 French originals by Andrew Pressley.

[BT12] Fedor Bogomolov and Yuri Tschinkel. Introduction to birational anabelian geometry. In *Current developments in algebraic geometry*, volume 59 of *Math. Sci. Res. Inst. Publ.*, pages 17–63. Cambridge Univ. Press, Cambridge, 2012.

[BT17] Fedor Bogomolov and Yuri Tschinkel. Universal spaces for unramified Galois cohomology. In *Brauer groups and obstruction problems*, volume 320 of *Progr. Math.*, pages 57–86. Birkhäuser/Springer, Cham, 2017.

[CT95] J. L. Colliot-Thélène. Birational invariants, purity and the Gersten conjecture. In *K-theory and algebraic geometry: connections with quadratic forms and division algebras (Santa Barbara, CA, 1992)*, volume 58 of *Proc. Sympos. Pure Math.*, pages 1–64. Amer. Math. Soc., Providence, RI, 1995.

[CTS07] Jean-Louis Colliot-Thélène and Jean-Jacques Sansuc. The rationality problem for fields of invariants under linear algebraic groups (with special regards to the Brauer group). In *Algebraic groups and homogeneous spaces*, volume 19 of *Tata Inst. Fund. Res. Stud. Math.*, pages 113–186. Tata Inst. Fund. Res., Mumbai, 2007.

[DB09] Bart De Bruyn. Some subspaces of the kth exterior power of a symplectic vector space. *Linear Algebra Appl.*, 430(11-12):3095–3104, 2009.

[EKM08] Richard Elman, Nikita Karpenko and Alexander Merkurjev. *The algebraic and geometric theory of quadratic forms*, volume 56 of *American Mathematical Society Colloquium Publications*. AMS, Providence, RI, 2008.

[GMS03] Skip Garibaldi, Alexander Merkurjev and Jean-Pierre Serre. *Cohomological invariants in Galois cohomology*, volume 28 of *University Lecture Series*. American Mathematical Society, Providence, RI, 2003.

[Gor07] Daniel Gorenstein. *Finite groups*. AMS Chelsea Publishing Co., New York, second edition, 1980 (reprinted 2007).

[GS06] Philippe Gille and Tamás Szamuely. *Central simple algebras and Galois cohomology*, volume 101 of *Cambridge Studies in Advanced Mathematics*. Cambridge University Press, Cambridge, 2006.

[Hup67] B. Huppert. *Endliche Gruppen. I*. Die Grundlehren der Mathematischen Wissenschaften, Band 134. Springer-Verlag, Berlin-New York, 1967.

[Kor00] V. È. Kordonskiĭ. Stable rationality of the group Spin_{10}. *Uspekhi Mat. Nauk*, 55(1(331)):171–172, 2000.

[PS83] A. A. Premet and I. D. Suprunenko. The Weyl modules and the irreducible representations of the symplectic group with the fundamental highest weights. *Comm. Algebra*, 11(12):1309–1342, 1983.

[Rus92] David J. Rusin. Kernels of the restriction and inflation maps in group cohomology. *J. Pure Appl. Algebra*, 79(2):191–204, 1992.

[Suz86] Michio Suzuki. *Group theory. II*, volume 248 of *Grundlehren der Mathematischen Wissenschaften*. Springer-Verlag, New York, 1986. Translated from the Japanese.

[TY11] M. Tezuka and N. Yagita. The image of the map from group cohomology to Galois cohomology. *Trans. Amer. Math. Soc.*, 363(8):4475–4503, 2011.

On Projective 3-Folds of General Type with $p_g = 2$

Meng Chen, Yong Hu and Matteo Penegini

This paper is dedicated to Miles Reid on his seventieth birthday.

Abstract

We classify minimal projective 3-folds of general type with $p_g = 2$ by studying the birationality of their 6-canonical maps.

1 Introduction

Miles was invited by Gang Xiao to visit East China Normal University in 1990 while the first author was a graduate student studying with Xiao. The first author would like to thank Miles for his friendship, constant guidance and encouragement.

This paper is the continuation of [CHP21]. Let X be a minimal 3-fold of general type with at worst $\mathbb{Q}$-factorial terminal singularities. We studied the birational geometry of projective 3-folds of general type with geometric genus $p_g = 1$ and 3 in [CHP21]. Here we consider $p_g(X) = 2$. The main result of this paper, Theorem 1.2, was announced in [CHP21].

Definition 1.1. Let W be a $\mathbb{Q}$-factorial normal projective variety of dimension n. Assume that the two maps $\tau \colon W \dashrightarrow W'$ and $g \colon W' \longrightarrow S$ satisfy the following properties:

(1) W' is a nonsingular projective variety and S is normal projective of dimension $s < n$;

(2) τ is a dominant birational map and g is a fibration.

Then we say that the set

$$\mathcal{F} = \{\hat{F} \subset W | \hat{F} = \tau_*^{-1}(F), F \text{ is a fiber of } g\}$$

forms an $(n-s)$-fold class of W, where $\tau_*^{-1}(\cdot)$ denotes the strict transform. In particular, if $n-s=1$ (resp. $=2$), we call it a *curve class* (resp. a *surface class*). The number $(K_W^{n-s} \cdot \tau_*^{-1}(F))$ (F a general fiber of g) is called *the canonical degree of* $\mathcal{F}$. Such degree is also denoted $\deg_c(\mathcal{F})$.

When $\varphi_{m_0,X}$ is of fiber type (that is $\dim \overline{\varphi_{m_0,X}(X)} < \dim X$), the induced fibration (obtained by taking the Stein factorization of $\varphi_{m_0,X}$) automatically forms either a curve class $\mathcal{C}$ or a surface class $\mathcal{S}$ of X. We also say that X is m_0*-canonically fibered* by a curve class $\mathcal{C}$ (or a surface class $\mathcal{S}$). Note that in our case $m_0 = 1$, so we can simply say *canonically fibered* to state the main result.

Theorem 1.2. Let X be a minimal projective 3-fold of general type with $p_g(X) = 2$. Then one of the following statements is true:

(1) $\varphi_{6,X}$ is birational onto its image;
(2) X is canonically fibered by a $(2,3)$-surface class of canonical degree $\frac{1}{2}$, in which case $\varphi_{6,X}$ is non-birational;
(3) X is canonically fibered by a $(1,2)$-surface class (denote by $\mathcal{C}$ the genus 2 curve class which is naturally induced from $\mathcal{S}$) and one of the following holds:
 (i) $\deg_c(\mathcal{C}) = \frac{2}{3}$;
 (ii) $\deg_c(\mathcal{C}) = \frac{4}{5}$;
 (iii) $P_2(X) = 5$, $\deg_c(\mathcal{C}) = 1$ and $\deg_c(\mathcal{S}) = \frac{1}{2}$;
 (iv) $P_4(X) = 14$, $\deg_c(\mathcal{C}) = 1$ and $\deg_c(\mathcal{S}) = \frac{1}{2}$.

 In this case $\varphi_{6,X}$ is non-birational.
(4) There is an explicit finite set $\mathbb{S}_2$ such that X is canonically fibered by a $(1,2)$-surface class and $\mathbb{B}(X) \in \mathbb{S}_2$, in which case $\varphi_{6,X}$ is non-birational (see 2.6 for the definition of $\mathbb{B}(X)$).

Remark 1.3. The existence of threefolds described in Theorem 1.2(*2*) and (*3*)(i) are provided by the following examples. Denote by X_d a general weighted hypersurface of degree d in the sense of Fletcher [IF00].

(1) The 3-fold $X_{16} \subset \mathbb{P}(1,1,2,3,8)$ has $K^3 = \frac{1}{3}$, $p_g = 2$ and φ_7 is non-birational. This example corresponds to Theorem 1.2(*3*)(i);
(2) The 3-fold $X_{14} \subset \mathbb{P}(1,1,2,2,7)$ has $K^3 = \frac{1}{2}$, $p_g = 2$ and φ_6 is non-birational. This example corresponds to Theorem 1.2(*2*).

Moreover,

(3) We do not know existence results for Theorem 1.2(*3*)(ii)–(iv), nor do we know the existence of any in $\mathbb{S}_2$ (most likely not!).
(4) A complete list of the 263 elements of the set $\mathbb{S}_2$ can be found at:

http://www.dima.unige.it/~penegini/publ.html

We briefly introduce the structure of this paper. Section 2 contains some basic stuff. Section 3 contains the core technical theorems, which will be effectively used to do the classification. These theorems concern 3-folds with $p_g \geq 2$ and canonically fibered by a $(1, 2)$-surface class.

Theorem 1.2 is proved in several steps in 4. Subsection 4.1 takes care of Theorem 1.2(*1*)–(*2*). Theorem 1.2(*3*)(i)–(ii) are proved in 4.2. Subsections 4.3 and 4.4 are devoted to constructing effective numerical constraints on $P_2(X)$, $P_3(X)$, $P_4(X)$, $P_5(X)$ and $P_6(X)$. This repeatedly applies the theorems of section 3 case by case. These constraints on the plurigenera will be used to produce (by computer aided computation) the set $\mathbb{S}_2$ that proves Theorem 1.2(*4*) (subsection 4.5). Finally, in 4.3, Theorem 1.2(*3*)(iii)–(iv) (Propositions 4.12, 4.14) are proved. This also provides more details and insights on the computations in [CHP21].

Notation and conventions. We work over the field $\mathbb{C}$ of complex numbers. A minimal threefold of general type X is a $\mathbb{Q}$-factorial 3-fold with at worst terminal singularities such that the canonical divisor K_X is a nef and big $\mathbb{Q}$-Cartier divisor. Moreover, let $\omega_X = \mathcal{O}_X(K_X)$ be the canonical sheaf. Throughout the paper we use the following symbols.

- $\sim$ denotes linear equivalence or $\mathbb{Q}$-linear equivalence when specified $\sim_{\mathbb{Q}}$
- $\equiv$ denotes numerical equivalence
- $|M_1| \succcurlyeq |M_2|$ (or, equivalently, $|M_2| \preccurlyeq |M_1|$) means, for linear systems $|M_1|$ and $|M_2|$ on a variety, $|M_1| \supseteq |M_2| +$ (fixed effective divisor)
- $Q_1 \geq Q_2$ for $\mathbb{Q}$-divisors means $m(Q_1 - Q_2)$ is an effective Cartier divisor for some positive integer m.

2 Preliminaries

2.1 Set Up

Let X be a minimal projective 3-fold of general type and we assume that $p_g(X) = h^0(X, \mathcal{O}_X(K_X)) \geq 2$. So, we may consider the canonical map $\varphi_1 \colon X \dashrightarrow \mathbb{P}^{p_g(X)-1}$, which is a non-constant rational map.

From the very beginning, we fix an effective Weil divisor $K_1 \sim K_X$. Take successive blow-ups $\pi\colon X' \to X$, which exist by Hironaka's big theorem, such that:

(1) X' is nonsingular and projective;
(2) the moving part of $|K_{X'}|$ is base point free;
(3) the union of supports of both $\pi^*(K_1)$ and exceptional divisors of π is simple normal crossing.

Denote by $\tilde{g}$ the composition $\varphi_1 \circ \pi$. So $\tilde{g}\colon X' \to \Sigma \subseteq \mathbb{P}^{p_g(X)-1}$ is a non-constant morphism by the above assumption. Let $X' \xrightarrow{f} \Gamma \xrightarrow{s} \Sigma$ be the Stein factorization of $\tilde{g}$. We get the following commutative diagram:

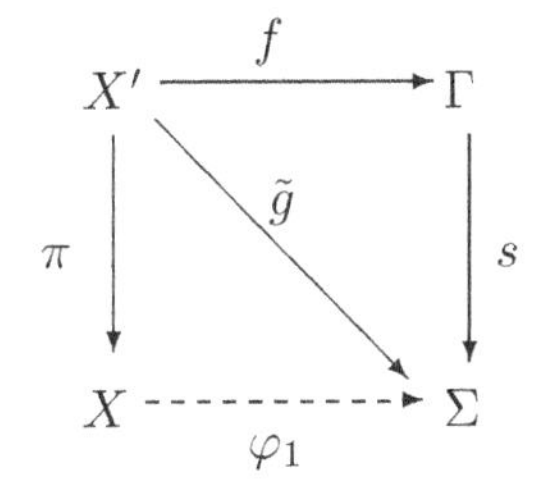

We may write $K_{X'} = \pi^*(K_X) + E_\pi$, where E_π is an effective $\mathbb{Q}$-divisor which is a sum of distinct exceptional divisors with positive rational coefficients. For any positive integer m, we have $\lceil m\pi^*(K_X)\rceil \leq mK_{X'}$. Set $|M| = \mathrm{Mov}\,|K_{X'}|$. Since $h^0(X', M) = h^0(\mathcal{O}_X(K_X))$ and X has at worst terminal singularities, we may also write

$$\pi^*(K_X) \sim_{\mathbb{Q}} M + E',$$

where E' is another effective $\mathbb{Q}$-divisor.

In this paper, we only treat the case with $\dim \overline{\varphi_1(X)} = 1$. We say that X is *canonically fibered by surfaces* with invariants $(c_1^2(F_0), p_g(F_0))$, where F_0 is the minimal model of F via the contraction morphism $\sigma\colon F \to F_0$ and F is a general fiber of f.

We may write $M \equiv \zeta F$ where $\zeta = \deg f_*\mathcal{O}_{X'}(M)$. For any positive integer m, $|M_m|$ denotes the moving part of $|mK_{X'}|$.

Naturally one has $\pi^*(K_X) \sim_{\mathbb{Q}} \zeta F + E'$. In practice, we need such a real number $\mu = \mu(F)$ which is defined to be the supremum of those rational numbers μ' satisfying the following property:

$$\pi^*(K_X) \sim_{\mathbb{Q}} \mu' F + E'_F \tag{2.1}$$

for a certain effective $\mathbb{Q}$-divisor E'_F. Clearly we have $\mu(F) \geq \zeta$.

2.2 Convention

For any linear system $|D|$ of positive dimension on a normal projective variety Z, we may write

$$|D| = \mathrm{Mov}\,|D| + \mathrm{Fix}\,|D|$$

and consider the rational map $\Phi_{|D|} = \Phi_{\mathrm{Mov}\,|D|}$. We say that $|D|$ *is not composed of a pencil if* $\dim \overline{\Phi_{|D|}(Z)} \geq 2$. A *generic irreducible element* of $|D|$ means a general member of $\mathrm{Mov}\,|D|$ when $|D|$ is not composed of a pencil or, otherwise, an irreducible component in a general member of $\mathrm{Mov}\,|D|$. For a nonsingular projective surface S of general type, we say that S *is a* (u,v)*-surface* if $K_{S_0}^2 = u$ and $p_g(S_0) = v$ where S_0 is the minimal model of S.

2.3 Known Inequalities

Pick a general fiber F of f. Assume that $|G|$ is any base point free linear system on F. Denote by C a generic irreducible element of $|G|$. Since $\pi^*(K_X)|_F$ is nef and big, by Kodaira's Lemma, there is a rational number $\beta > 0$ such that $\pi^*(K_X)|_F \geq \beta C$. We assume, from now on, that $\beta = \beta(|G|)$ is the supremum satisfying the above property.

For any integer $m > 0$, we define

$$\xi = \xi(|G|) = (\pi^*(K_X) \cdot C)_{X'}$$
$$\alpha(m) = \alpha_{(|G|)}(m) = (m - 1 - \frac{1}{\mu} - \frac{1}{\beta})\xi$$
$$\alpha_0(m) = \lceil \alpha(m) \rceil .$$

If no confusion is likely to arise, we use the simplified notation ζ, μ, β, ξ and $\alpha(m)$. According to [CC10a, Theorem 2.11], whenever $\alpha(m) > 1$,

$$m\xi \geq \deg(K_C) + \alpha_0(m). \tag{2.2}$$

In particular, if m is sufficiently large so that $\alpha(m) > 1$, (2.2) implies

$$\xi \geq \frac{\deg(K_C)}{1 + \frac{1}{\mu} + \frac{1}{\beta}}.$$

Moreover, by [Che07, (2.1)] one has

$$K_X^3 \geq \mu\beta\xi. \tag{2.3}$$

2.4 Birationality Principle

We refer to [CC10a, 2.7] for the birationality principle. Recall the following concept for point separation.

Definition 2.1. Let $|L|$ be a moving (without fixed part) linear system on a normal projective variety Z. We say that the rational map $\Phi_{|L|}$ *distinguishes sub-varieties* $W_1, W_2 \subset Z$ if, set-theoretically,

$$\overline{\Phi_{|L|}(W_1)} \nsubseteq \overline{\Phi_{|L|}(W_2)} \quad \text{and} \quad \overline{\Phi_{|L|}(W_2)} \nsubseteq \overline{\Phi_{|L|}(W_1)}.$$

We say that $\Phi_{|L|}$ *separates points* $P, Q \in Z$ (for $P, Q \notin \mathrm{Bs}\,|L|$), if $\Phi_{|L|}(P) \neq \Phi_{|L|}(Q)$.

We tacitly and frequently use the following theorem:

Theorem 2.2 ([CC10a, Theorem 2.11]). *Keep the same setting and assumptions as in Subsection 2.1 and Subsection 2.3. Pick a generic irreducible element F of $|M|$. For $m > 0$, assume that the following:*

(1) $|mK_{X'}|$ distinguishes different generic irreducible elements of $|M|$;
(2) $|mK_{X'}||_F$ distinguishes different generic irreducible elements of $|G|$;
(3) $\alpha(m) > 2$.

Then $\varphi_{m,X}$ is birational onto its image.

2.5 A Weak Form of Extension Theorem

Sometimes we use the following theorem which is a special form of Kawamata's extension theorem (see [Kaw99, Theorem A]):

Theorem 2.3 ([CZ16, Theorem 2.4]). *Let Z be a nonsingular projective variety on which D is a smooth divisor such that $K_Z + D \sim_{\mathbb{Q}} A + B$ for an ample $\mathbb{Q}$-divisor A and an effective $\mathbb{Q}$-divisor B and that D is not contained in the support of B. Then the natural homomorphism*

$$H^0(Z, m(K_Z + D)) \longrightarrow H^0(D, mK_D)$$

is surjective for all $m > 1$.

In particular, when Z is of general type and D moves in a base point free linear system, the condition of Theorem 2.3 is automatically satisfied. Taking $Z = X'$, $D = F$ and modulo a process of taking the limit (so we may assume that μ is rational), the following holds

$$|n(\mu + 1)K_{X'}||_F \succcurlyeq |n\mu(K_{X'} + F)||_F = |n\mu K_F|$$

for some sufficiently large and divisible integer n. Noting that

$$n(\mu+1)\pi^*(K_X) \geq M_{n(\mu+1)}$$

and that $|n\mu\sigma^*(K_{F_0})|$ is base point free, we have

$$\pi^*(K_X)|_F \geq \frac{\mu}{\mu+1}\sigma^*(K_{F_0}) \geq \frac{\zeta}{1+\zeta}\sigma^*(K_{F_0}). \tag{2.4}$$

2.6 The Weighted Basket of X

The *weighted basket* $\mathbb{B}(X)$ is the triple $\{B_X, P_2(X), \chi(\mathcal{O}_X)\}$. We use the definitions and notation of [CC10b, §§2–3] such as "basket", "prime packing", "the canonical sequence of a basket", $B^{(n)}$ $(n \geq 0)$, $\chi_m(\mathbb{B}(X))$ $(m \geq 2)$, $K^3(\mathbb{B}(X))$, σ_n $(n \geq 5)$ and so on.

As X is of general type, the vanishing theorem and Reid's Riemann–Roch formula [Rei87] (see also first lines in [CC10b, 4.5]) imply that

$$\chi_m(\mathbb{B}(X)) = P_m(X)$$

for all $m \geq 2$ and $K^3(\mathbb{B}(X)) = K_X^3$. For any $n \geq 0$, $B^{(n)}$ can be expressed in terms of $\chi(\mathcal{O}_X)$, P_2, P_3, $\cdots$, P_{n+1} (see [CC10b, (3.3–14)] for more details), which serves as a powerful tool for our classification.

3 Some Technical Theorems

3.1 Two Restriction Maps on Canonical Class of $(1,2)$-Surfaces

Within this subsection, we always work under the following assumption:

($\pounds$) *Keep the setting in 2.1. Let $m_1 > 1$ be an integer. Assume that $|M_{m_1}|$ is base point free, $d_1 = 1$, $\Gamma \cong \mathbb{P}^1$ and that F is a $(1,2)$-surface. Take $|G| = \mathrm{Mov}\,|K_F|$, which is assumed to be base point free. Let C be a generic irreducible element of $|G|$.*

In this section we prove three technical theorems that relate the numbers β, ξ, μ and α to the linear systems of the definition above. These three theorems will be used systematically in Section 4 together with [CHP21, Props 3.4–3.7].

Theorem 3.1. *Let X be a minimal projective 3-fold of general type with $p_g(X) \geq 2$. Keep assumption ($\pounds$). Let $m_1 > 1$ be an integer. Suppose that $|S_1|$ is any base point free linear system on X' with $h^0(S_1|_F) \geq 2$*

and that, for some integer $j \geq 1$, $M_{m_1} \geq jF + S_1$. *Denote by* C_1 *the generic irreducible element of* $|S_1|_F|$. *Assume that* $|S_1|_F|$ *and* $|G|$ *are not composed of the same pencil. Set* $\tilde{\delta} = (C_1 \cdot C)$.

(1) *When* $\tilde{\delta} \leq 2j$, $\varphi_{n+1,X}$ *is birational for*

$$n = \left\lfloor \frac{1}{\xi(|G|)}(2 - \frac{\tilde{\delta}}{j}) + \frac{m_1}{j} + \frac{1}{\beta} \right\rfloor + 1.$$

(1′) *When* $\tilde{\delta} \leq j$, $\varphi_{n+1,X}$ *is birational for*

$$n = \left\lfloor \frac{2}{\xi(|G|)}(1 - \frac{\tilde{\delta}}{j}) + \frac{2m_1}{j} \right\rfloor + 2.$$

(2) *When* $\tilde{\delta} > 2j$, $\varphi_{n+1,X}$ *is birational for*

$$n = \left\lfloor \frac{2m_1}{\tilde{\delta}} + \frac{1}{\beta} + \frac{1}{\mu} \cdot (1 - \frac{2j}{\tilde{\delta}}) \right\rfloor + 1.$$

(2′) *When* $\tilde{\delta} > 2j$ *and* $S_1|_F$ *is big,* $\varphi_{n+1,X}$ *is birational for*

$$n = \left\lceil \frac{m_1}{j} + \frac{1}{\beta} \right\rceil.$$

(3) *One has*

$$\pi^*(K_X)|_F \geq \frac{j}{j+m_1}\sigma^*(K_{F_0}) + \frac{1}{j+m_1}S_1|_F.$$

(4) *For any integer* $n > \frac{m_1}{j} + \frac{1}{\beta}$ *with* $(n - \frac{m_1}{j} - \frac{1}{\beta})\xi + \frac{\tilde{\delta}}{j} > 1$, *one has*

$$(n+1)\xi \geq \left\lceil (n - \frac{m_1}{j} - \frac{1}{\beta})\xi + \frac{\tilde{\delta}}{j} \right\rceil + 2.$$

Proof Write $m_1\pi^*(K_X) \equiv jF + S_1 + E_{m_1}$, where E_{m_1} is an effective $\mathbb{Q}$-divisor on X'. Set

$$n = \begin{cases} \left\lfloor \frac{1}{\xi(|G|)}(2 - \frac{\tilde{\delta}}{j}) + \frac{m_1}{j} + \frac{1}{\beta} \right\rfloor + 1, & \text{when } \tilde{\delta} \leq 2j; \\ \left\lfloor \frac{2m_1}{\tilde{\delta}} + \frac{1}{\beta} + \frac{1}{\mu} \cdot (1 - \frac{2j}{\tilde{\delta}}) \right\rfloor + 1, & \text{when } \tilde{\delta} > 2j. \end{cases}$$

By the Kawamata–Viehweg vanishing theorem ([Kaw82, Vie82]),

$$\begin{aligned} |(n+1)K_{X'}||_F &\succcurlyeq \left| K_{X'} + \left\lceil n\pi^*(K_X) - \frac{1}{j}E_{m_1} \right\rceil \right| \Big|_F \\ &\succcurlyeq \left| K_F + \left\lceil n\pi^*(K_X)|_F - \frac{1}{j}E_{m_1}|_F \right\rceil \right| \end{aligned}$$

since

$$n\pi^*(K_X) - \frac{1}{j}E_{m_1} - F \equiv (n - \frac{m_1}{j})\pi^*(K_X) + \frac{1}{j}S_1$$

is simple normal crossing (by our assumption), nef and big.

Since $p_g(X) > 0$, one sees that $|(n+1)K_{X'}|$ distinguishes different general F and $|(n+1)K_{X'}||_F$ distinguishes different general C. What we need to do is to investigate the behavior of $|(n+1)K_{X'}||_C$.

Recall that we have

$$\frac{1}{\beta}\pi^*(K_X)|_F \equiv C + H_1 \tag{3.1}$$

where H_1 is a certain effective $\mathbb{Q}$-divisor. Vanishing on F gives

$$\begin{aligned}\left|K_F + \left\lceil n\pi^*(K_X)|_F - \frac{1}{j}E_{m_1}|_F - H_1\right\rceil\right||_C \\ \succcurlyeq \left|K_C + \left\lceil n\pi^*(K_X)|_F - \frac{1}{j}E_{m_1}|_F - H_1\right\rceil\Big|_C\right| \\ = |K_C + \tilde{D}_1|\end{aligned}$$

where $\tilde{D}_1 = \left\lceil n\pi^*(K_X)|_F - \frac{1}{j}E_{m_1}|_F - H_1\right\rceil\Big|_C$ with

$$\deg(\tilde{D}_1) \geq (n - \frac{m_1}{j} - \frac{1}{\beta})\xi + \frac{\tilde{\delta}}{j} > 2.$$

By [Che03, Fact 2.2], we see that $\varphi_{n+1,X}$ is birational, which implies (*1*).

For (*2'*), even if $n = \frac{m_1}{j} + \frac{1}{\beta}$ is integral, the $\mathbb{Q}$-divisor

$$n\pi^*(K_X)|_F - \frac{1}{j}E_{m_1}|_F - H_1 - C$$

is nef and big since $S_1|_F$ is nef and big. Thus we still have $\deg(\tilde{D}_1) > 2$. Hence $\varphi_{\lceil \frac{m_1}{j} + \frac{1}{\beta}\rceil+1,X}$ is birational.

A direct application of the argument in the proof of (*1*) implies that, whenever $\deg(\tilde{D}_1) > 1$, $|K_C + \tilde{D}_1|$ is base point free, so (*4*) follows from $(n+1)\xi \geq \deg(\tilde{D}_1) + 2$.

Finally, modulo a further birational modification, we may and do assume that the linear system $|M'_{3j-1}| = \mathrm{Mov}\,|(3j-1)(K_{X'} + F)|$ is base point free. It is clear that M'_{3j-1} is big. By vanishing and Theorem 2.3

$$\begin{aligned}|(3j + 3m_1)K_{X'}||_F &\succcurlyeq |K_{X'} + M'_{3j-1} + 3S_1 + F||_F \\ &\succcurlyeq |K_F + (3j-1)\sigma^*(K_{F_0}) + 3S_1|_F| \\ &\succcurlyeq |3j\sigma^*(K_{F_0}) + 3S_1|_F|.\end{aligned}$$

Note that we have $| \lfloor (3j+3m_1)\pi^*K_X \rfloor | = \mathrm{Mov}\,|(3j+3m_1)K_{X'}|$. Thus we finish the proof of *(3)*.

Statement *(1')* follows from the similar argument to that for *(1)*. Instead of using the relation (3.1), one may use the statement *(3)*, namely:

$$\frac{j+m_1}{j}\pi^*(K_X)|_F \equiv C + \frac{1}{j}S_1|_F + H''$$

for an effective $\mathbb{Q}$-divisor H'' on F. *(1')* follows by taking

$$n = \left\lfloor \frac{2}{\xi(|G|)}(1-\frac{\tilde{\delta}}{j}) + \frac{2m_1}{j} \right\rfloor + 2.$$

We are left to treat *(2)*. By Kawamata–Viehweg vanishing and (2.1),

$$\begin{aligned} |(n+1)K_{X'}||_F &\succcurlyeq \left| K_{X'} + \left\lceil n\pi^*(K_X) - \frac{2}{\tilde{\delta}}E_{m_1} - \frac{1}{\mu}E'_F \right\rceil \right| \Big|_F \\ &\succcurlyeq \left| K_F + \left\lceil n\pi^*(K_X)|_F - \frac{2}{\tilde{\delta}}E_{m_1}|_F - \frac{1}{\mu}(1-\frac{2j}{\tilde{\delta}})E'_F|_F \right\rceil \right| \end{aligned}$$

since

$$n\pi^*(K_X) - \frac{2}{\tilde{\delta}}E_{m_1} - \frac{1}{\mu}(1-\frac{2j}{\tilde{\delta}})E'_F \equiv (n - \frac{2m_1}{\tilde{\delta}} - \frac{1}{\mu}(1-\frac{2j}{\tilde{\delta}}))\pi^*(K_X) + \frac{2}{\tilde{\delta}}S_1$$

is simple normal crossing (by our assumption), nef and big. Then vanishing on F gives

$$\begin{aligned} &\left| K_F + \left\lceil n\pi^*(K_X)|_F - \frac{2}{\tilde{\delta}}E_{m_1}|_F - \frac{1}{\mu}(1-\frac{2j}{\tilde{\delta}})E'_F|_F \right\rceil \right| \\ &\succcurlyeq \left| K_F + \left\lceil n\pi^*(K_X)|_F - \frac{2}{\tilde{\delta}}E_{m_1}|_F - \frac{1}{\mu}(1-\frac{2j}{\tilde{\delta}})E'_F|_F - H_1 \right\rceil \right| \\ &= |K_C + \tilde{D}_n| \end{aligned}$$

where

$$\tilde{D}_n = \left\lceil n\pi^*(K_X)|_F - \frac{2}{\tilde{\delta}}E_{m_1}|_F - \frac{1}{\mu}(1-\frac{2j}{\tilde{\delta}})E'_F|_F - H_1 \right\rceil \Big|_C$$

with $\deg \tilde{D}_n > 2$. Hence $\varphi_{\lfloor \frac{2m_1}{\tilde{\delta}} + \frac{1}{\beta} + \frac{1}{\mu}\cdot(1-\frac{2j}{\tilde{\delta}}) \rfloor + 2, X}$ is birational. □

Theorem 3.2. *Let X be a minimal projective 3-fold of general type with $p_g(X) \geq 2$. Keep assumption (£). Let m_1 be a positive integer. Suppose that $M_{m_1} \geq j_1F + S_1$ for some moving divisor S_1 on X', $j_1 > 0$ is an integer and that $S_1|_F \geq j_2C + C'$ where C' is a moving irreducible curve on F with $C' \not\equiv C$, $j_2 > 0$ is an integer. Set $\delta_2 = (C' \cdot C)$. Then*

(1) if $j_1 \geq j_2$, then

(1.i) For any positive integer $n > \frac{m_1}{j_1} + \frac{j_1-j_2}{j_1} \cdot \frac{1}{\beta}$ *and*

$$(n - \frac{m_1}{j_1} - \frac{j_1 - j_2}{j_1} \cdot \frac{1}{\beta})\xi + \frac{\delta_2}{j_1} > 1, \text{ one has}$$

$$(n+1)\xi \geq \left\lceil (n - \frac{m_1}{j_1} - \frac{j_1 - j_2}{j_1} \cdot \frac{1}{\beta})\xi + \frac{\delta_2}{j_1} \right\rceil + 2.$$

(1.ii) $\varphi_{n+1,X}$ *is birational for*

$$n = \begin{cases} \left\lfloor \frac{1}{\xi(|G|)}(2 - \frac{\delta_2}{j_1}) + \frac{m_1}{j_1} + \frac{j_1-j_2}{j_1} \cdot \frac{1}{\beta} \right\rfloor + 1, & \delta_2 < 2j_1 \\ \left\lfloor \frac{2m_1}{\delta_2} + \frac{1}{\mu}(1 - \frac{2j_1}{\delta_2}) + \frac{1}{\beta}(1 - \frac{2j_2}{\delta_2}) \right\rfloor + 1, & \delta_2 \geq 2j_1. \end{cases}$$

(2) if $j_2 \geq j_1$*, then*

(2.i) For any positive integer $n > \frac{m_1}{j_2} + \frac{1}{\mu}(1 - \frac{j_1}{j_2})$ *and*

$$(n - \frac{m_1}{j_2} - \frac{1}{\mu}(1 - \frac{j_1}{j_2})) \cdot \xi + \frac{\delta_2}{j_2} > 1, \text{ one has}$$

$$(n+1)\xi \geq \left\lceil (n - \frac{m_1}{j_2} - \frac{1}{\mu}(1 - \frac{j_1}{j_2})) \cdot \xi + \frac{\delta_2}{j_2} \right\rceil + 2.$$

(2.ii) $\varphi_{n+1,X}$ *is birational for*

$$n = \begin{cases} \left\lfloor \frac{1}{\xi(|G|)}(2 - \frac{\delta_2}{j_2}) + \frac{m_1}{j_2} + \frac{1}{\mu} \cdot \frac{j_2-j_1}{j_2} \right\rfloor + 1, & \delta_2 < 2j_2 \\ \left\lfloor \frac{1}{\mu}(1 - \frac{2j_1}{\delta_2}) + \frac{1}{\beta}(1 - \frac{2j_2}{\delta_2}) \right\rfloor + 1, & \delta_2 \geq 2j_2. \end{cases}$$

Proof Modulo further birational modification, we may and do assume that $|S_1|$ is also base point free. Hence S_1 is nef. By assumption, we may find two effective $\mathbb{Q}$-divisors $\tilde{E}'_{m_1}$ and $\tilde{E}''_{m_1}$ such that

$$m_1\pi^*(K_X) \equiv j_1F + S_1 + \tilde{E}'_{m_1} \qquad \text{and} \qquad S_1|_F \equiv j_2C + C' + \tilde{E}''_{m_1}.$$

Set

$$n = \begin{cases} \left\lfloor \frac{1}{\xi(|G|)}(2 - \frac{\delta_2}{j_1}) + \frac{m_1}{j_1} + \frac{j_1-j_2}{j_1} \cdot \frac{1}{\beta} \right\rfloor + 1, & \delta_2 \leq 2j_1 \\ \left\lfloor \frac{2m_1}{\delta_2} + \frac{1}{\mu}(1 - \frac{2j_1}{\delta_2}) + \frac{1}{\beta}(1 - \frac{2j_2}{\delta_2}) \right\rfloor + 1, & \delta_2 > 2j_1. \end{cases}$$

By vanishing one has

$$
\begin{aligned}
|(n+1)K_{X'}||_F &\succcurlyeq \left|K_{X'} + \left\lceil n\pi^*(K_X) - \tfrac{1}{j_1}\tilde{E}'_{m_1}\right\rceil\right|\Big|_F \\
&\succcurlyeq \left|K_F + \left\lceil n\pi^*(K_X)|_F - \tfrac{1}{j_1}\tilde{E}'_{m_1}|_F\right\rceil\right| \\
&\succcurlyeq |K_F + \lceil Q_{1,m_1}\rceil|,
\end{aligned}
$$

$$
\begin{aligned}
\text{where} \qquad Q_{1,m_1} &= n\pi^*(K_X)|_F - \tfrac{1}{j_1}\tilde{E}'_{m_1}|_F - \tfrac{1}{j_1}\tilde{E}''_{m_1} \\
&\equiv \left(n - \tfrac{m_1}{j_1}\right)\pi^*(K_X)|_F + \tfrac{j_2}{j_1}C + \tfrac{1}{j_1}C'.
\end{aligned}
$$

Recall $\frac{1}{\beta}\pi^*(K_X)|_F \equiv C + H_1$, where H_1 is an effective $\mathbb{Q}$-divisor. Hence

$$
Q_{1,m_1} - \frac{j_1 - j_2}{j_1}H_1 \equiv \left(n - \frac{m_1}{j_1} - \frac{j_1 - j_2}{j_1}\cdot\frac{1}{\beta}\right)\pi^*(K_X)|_F + \frac{1}{j_1}C' + C.
$$

By vanishing once more, we have

$$
\begin{aligned}
\left|K_F + \left\lceil Q_{1,m_1} - \frac{j_1-j_2}{j_1}H_1\right\rceil\right|\Big|_C &= |K_C + D_{1,m_1}| \\
\text{where } D_{1,m_1} &= \left\lceil Q_{1,m_1} - \frac{j_1-j_2}{j_1}H_1 - C\right\rceil\Big|_C \\
\text{with } \deg(D_{1,m_1}) &\geq \left(n - \frac{m_1}{j_1} - \frac{j_1-j_2}{j_1}\cdot\frac{1}{\beta}\right)\xi + \frac{\delta_2}{j_1}.
\end{aligned}
$$

Using the same idea as in the previous theorem, we see that $\varphi_{n+1,X}$ is brational and *(1.ii)* ($\delta_2 \leq 2j_1$) follows; *(1.i)* follows routinely.

Now turn to *(1.ii)* where $\delta_2 > 2j_1$. By vanishing, we have

$$
\begin{aligned}
|(n+1)K_{X'}||_F &\succcurlyeq \left|K_{X'} + \left\lceil n\pi^*(K_X) - \frac{2}{\delta_2}\tilde{E}'_{m_1} - \frac{1}{\mu}\left(1 - \frac{2j_1}{\delta_2}\right)E'_F\right\rceil\right|\Big|_F \\
&\succcurlyeq \left|K_F + \left\lceil n\pi^*(K_X)|_F - \frac{2}{\delta_2}\tilde{E}'_{m_1}|_F - \frac{1}{\mu}\left(1 - \frac{2j_1}{\delta_2}\right)E'_F|_F\right\rceil\right|
\end{aligned}
$$

since $n\pi^*(K_X) - \frac{2}{\delta_2}\tilde{E}'_{m_1} - \frac{1}{\mu}(1 - \frac{2j_1}{\delta_2})E'_F$ is simple normal crossing (by definition), nef and big. By vanishing on F, we have

$$
\begin{aligned}
&\left|K_F + \left\lceil n\pi^*(K_X)|_F - \frac{2}{\delta_2}\tilde{E}'_{m_1}|_F - \frac{1}{\mu}\left(1 - \frac{2j_1}{\delta_2}\right)E'_F|_F\right\rceil\right|\Big|_C \\
&\quad\succcurlyeq \left|K_F + \left\lceil n\pi^*(K_X)|_F - \frac{2}{\delta_2}\tilde{E}'_{m_1}|_F - \frac{1}{\mu}\left(1 - \frac{2j_1}{\delta_2}\right)E'_F|_F - \right.\right. \\
&\qquad\qquad\qquad\left.\left. \frac{2}{\delta_2}\tilde{E}''_{m_1} - \left(1 - \frac{2j_2}{\delta_2}\right)H_1\right\rceil\right|\Big|_C \\
&\quad\succcurlyeq \left|K_C + \left\lceil \tilde{D}_{1,m_1}\right\rceil\right|
\end{aligned}
$$

where $\tilde{D}_{1,m_1} \equiv (n - \frac{2m_1}{\delta_2} - \frac{1}{\mu}(1 - \frac{2j_1}{\delta_2}) - \frac{1}{\beta}(1 - \frac{2j_2}{\delta_2}))\pi^*(K_X)|_C + \frac{2}{\delta_2}C'|_C$ with $\deg(\tilde{D}_{1,m_1}) > 2$. Thus $\varphi_{n+1,X}$ is birational.

One proves (*2*) in a similar way. $\square$

Remark 3.3. From the proof of Theorem 3.2, one clearly sees that the variant of Theorem 3.2 with $C' = 0$ is also true.

Theorem 3.4. *Let X be a minimal projective 3-fold of general type with $p_g(X) \geq 2$. Keep assumption (£). Let m_1 be a positive integer. Suppose that $M_{m_1} \geq j_1 F + S_1$ for some moving divisor S_1 on X', $j_1 > 0$ is an integer and that $S_1|_F \geq j_2 C$ where $j_2 > 0$ is an integer. The following statements hold:*

(1) if $j_1 \geq j_2$, then

(1.i) For any positive integer $n > \frac{m_1}{j_1} + \frac{j_1 - j_2}{j_1} \cdot \frac{1}{\beta}$ and

$$(n - \frac{m_1}{j_1} - \frac{j_1 - j_2}{j_1} \cdot \frac{1}{\beta})\xi > 1, \qquad \textit{one has}$$

$$(n+1)\xi \geq \left\lceil (n - \frac{m_1}{j_1} - \frac{j_1 - j_2}{j_1} \cdot \frac{1}{\beta})\xi \right\rceil + 2.$$

(1.ii) The map $\varphi_{n+1,X}$ is birational for

$$n = \left\lfloor \frac{2}{\xi(|G|)} + \frac{m_1}{j_1} + \frac{j_1 - j_2}{j_1} \cdot \frac{1}{\beta} \right\rfloor + 1.$$

(2) if $j_2 \geq j_1$, then

(2.i) For any positive integer $n > \frac{m_1}{j_2} + \frac{1}{\mu}(1 - \frac{j_1}{j_2})$ and

$$(n - \frac{m_1}{j_2} - \frac{1}{\mu}(1 - \frac{j_1}{j_2}))\xi > 1, \qquad \textit{one has}$$

$$(n+1)\xi \geq \left\lceil (n - \frac{m_1}{j_2} - \frac{1}{\mu}(1 - \frac{j_1}{j_2}))\xi \right\rceil + 2.$$

(2.ii) The map $\varphi_{n+1,X}$ is birational for

$$n = \left\lfloor \frac{2}{\xi(|G|)} + \frac{m_1}{j_2} + \frac{1}{\mu} \cdot \frac{j_2 - j_1}{j_2} \right\rfloor + 1.$$

Proof This follows directly from the proof of Theorem 3.2 by setting $C' = 0$. We omit the redundant details. $\square$

4 Threefolds of General Type with $p_g = 2$

This section is devoted to the classification of 3-folds of general type with $p_g(X) = 2$. In the notation of 2.1, we have an induced fibration $f: X' \to \Gamma$ of which the general fiber F is a nonsingular projective surface of general type. Denote by $\sigma: F \to F_0$ the contraction onto its minimal model. By [Che03, Theorem 3.3], we may assume that $\Gamma \cong \mathbb{P}^1$ (as $b = g(\Gamma) = 0$). Since $p_g(X) > 0$ and F is a general fiber, we have $p_g(F) > 0$. By the classification of surfaces, F belongs to one of:

(1) $(K_{F_0}^2, p_g(F_0)) = (1, 2)$;
(2) $(K_{F_0}^2, p_g(F_0)) = (2, 3)$;
(3) other surfaces with $p_g(F_0) > 0$.

Recall that we have

$$\pi^*(K_X) \sim_{\mathbb{Q}} F + E_1' \tag{4.1}$$

where E_1' is an effective $\mathbb{Q}$-divisor since $\mu \geq 1$. By Theorem 2.3, the natural restriction map

$$H^0(X', 3K_{X'} + 3F) \to H^0(F, 3K_F) \tag{4.2}$$

is surjective. Since $|6K_{X'}| \succcurlyeq |3K_{X'} + 3F|$, by (4.1) and (4.2), we have

$$\pi^*(K_X)|_F \equiv \tfrac{1}{2}\sigma^*(K_{F_0}) + Q' \equiv \tfrac{1}{2}C + \hat{E}_F \tag{4.3}$$

where Q' and $\hat{E}_F$ are effective $\mathbb{Q}$-divisors.

4.1 Non-$(1, 2)$-Surface Case

Lemma 4.1. *Let X be a minimal projective 3-fold of general type with $p_g(X) = 2$ and keep the setting in 2.1. Suppose $d_1 = 1$, $\Gamma \cong \mathbb{P}^1$ and F is neither a $(2, 3)$ surface nor a $(1, 2)$-surface. Then $\varphi_{6,X}$ is birational.*

Proof As $|6K_{X'}|$ distinguishes different general fibers of f, (4.2) implies that $\varphi_{6,X}$ is birational unless F is either a $(1, 2)$- or a $(2, 3)$-surface. □

Proposition 4.2. *Let X be a minimal projective 3-fold of general type with $p_g(X) = 2$, $d_1 = 1$, $\Gamma \cong \mathbb{P}^1$. Assume that F is a $(2, 3)$-surface. Then $\varphi_{6,X}$ is not birational if and only if $(\pi^*(K_X)|_F)^2 = \frac{1}{2}$.*

Proof We have $(\pi^*(K_X)|_F)^2 \geq \frac{1}{2}$ by (4.3) and by our assumption. By (4.1) and Kawamata–Viehweg vanishing, we have

$$|6K_{X'}||_F \succcurlyeq |K_{X'} + F + \lceil 4\pi^*(K_X) \rceil\, ||_F \succcurlyeq |K_F + \lceil 4\pi^*(K_X)|_F \rceil\, | \tag{4.4}$$

If $(\pi^*(K_X)|_F)^2 > \frac{1}{2}$, then $|K_F + \lceil 4\pi^*(K_X)|_F \rceil\,|$ is birational by [CHP21, Lemmas 2.3, 2.5] and (4.3). So $\varphi_{6,X}$ is birational.

If $(\pi^*(K_X)|_F)^2 = \frac{1}{2}$, we have $\pi^*(K_X)|_F \equiv \frac{1}{2}\sigma^*(K_{F_0})$ by the Hodge index theorem and (4.3). It is clear that $|3\sigma^*(K_{F_0})||_C \preccurlyeq |K_C + C|_C|$. Note that $|C| = |\sigma^*(K_{F_0})|$ is base point free, so

$$|6K_{X'}||_C \succcurlyeq (|\lfloor 6\pi^*(K_X)\rfloor\,||_F)|_C \succcurlyeq |3\sigma^*(K_{F_0})||_C.$$

Since $(\pi^*(K_X)\cdot C) = 1$, the vanishing theorem and (4.4) implies that $|M_6||_C = |K_C + D|$ with $\deg(D) \geq 2$. Hence $|M_6||_C = |K_C + C|_C|$. Since $|C|_C|$ gives a g_2^1, $|K_C + C|_C|$ is clearly non-birational.

Therefore $\varphi_{6,X}$ is non-birational if $(\pi^*(K_X)|_F)^2 = \frac{1}{2}$. □

4.2 The (1, 2)-Surface Case

From now on, we always assume that F is a (1, 2)-surface. We have $0 \leq \chi(\omega_X) \leq 1$ by our assumption and [Che04, Lemma 4.5]. Note that the proof of that lemma works for $p_g(X) = 2$, $b = g(\Gamma) = 0$ using the nefness of the vector bundle $f_*\omega_{X'/\Gamma}$. It is well known that $|K_{F_0}|$ has exactly one base point and that, after blowing up this point, F admits a canonical fibration of genus 2 with a unique section which we denote by H. Denote by C a general member in $|G| = \text{Mov}\,|\sigma^*(K_{F_0})|$.

Remark 4.3. By [Che14, Theorem 1.1], $\varphi_{7,X}$ is non-birational if and only if $\xi = \frac{2}{3}$. Since $p_g(X) > 0$, $\varphi_{6,X}$ is non-birational as well if $\xi = \frac{2}{3}$.

Lemma 4.4. *Let X be a minimal projective 3-fold of general type with $p_g(X) = 2$, $d_1 = 1$, $\Gamma \cong \mathbb{P}^1$. Assume that F is a (1, 2)-surface. Keep the setting in 2.1. Then $|6K_{X'}|$ distinguishes different generic irreducible elements of $|M|$ and $|6K_{X'}||_F$ distinguishes those of $|G|$.*

Proof Since $|M|$ is composed of a rational pencil, it is clear that $|6K_{X'}|$ distinguishes different generic irreducible elements of $|M|$. Notice that $|G|$ is composed of a rational pencil. The surjectivity of (4.2) implies that $|6K_{X'}||_F$ distinguishes generic irreducible elements of $|G|$. □

Lemma 4.5 ([Che07, 2.15]). *Let X be a minimal projective 3-fold of general type with $p_g(X) = 2$, $d_1 = 1$, $\Gamma \cong \mathbb{P}^1$. Assume that F is a (1, 2)-surface. Then $\beta \geq \frac{1}{2}$ and $\xi \geq \frac{2}{3}$.*

Considering the Zariski decomposition of K_F, we have $\pi^*K_X|_F \leq \sigma^*(K_{F_0})$. We deduce that $\xi \leq 1$.

Lemma 4.6. *Under the conditions of Lemma 4.5, assume that $\xi = 1$ and $(\pi^*(K_X)|_F)^2 > \frac{1}{2}$. Then $\varphi_{6,X}$ is birational.*

Proof Consider the Zariski decomposition of the following $\mathbb{Q}$-divisor $2\pi^*(K_X)|_F + 2\hat{E}_F \equiv (2\pi^*(K_X)|_F + N^+) + N^-$, where

(1) both N^+ and N^- are effective $\mathbb{Q}$-divisors with $N^+ + N^- = 2\hat{E}_F$
(2) the $\mathbb{Q}$-divisor $(2\pi^*(K_X)|_F + N^+)$ is nef
(3) $\big((2\pi^*(K_X)|_F + N^+)\cdot N^-\big) = 0$.

Step 1. $(\pi^*(K_X)|_F)^2 > \frac{1}{2}$ implies $(N^+ \cdot C) > 0$.

Since C is nef, we see $(N^+ \cdot C) \geq 0$. Assume that $(N^+ \cdot C) = 0$. Then $(N^+)^2 \leq 0$ as C is a fiber of the canonical fibration of F. Notice that

$$\tfrac{1}{2} < \big(\pi^*(K_X)|_F\big)^2 = \tfrac{1}{2}\big(\pi^*(K_X)\cdot C\big) + \big(\pi^*(K_X)|_F \cdot \hat{E}_F\big)$$

implies $\big(\pi^*(K_X)|_F \cdot \hat{E}_F\big) > 0$. We clearly have $\big(\pi^*(K_X)|_F \cdot N^+\big) > 0$ by the definition of Zariski decomposition. Hence the contradiction:

$$\begin{aligned}(N^+)^2 &= \big(N^+ \cdot (2\pi^*(K_X)|_F - C - N^-)\big)\\ &= 2\big(N^+ \cdot \pi^*(K_X)|_F\big) + \big(2\pi^*(K_X)|_F \cdot N^-\big) > 0.\end{aligned}$$

Step 2. $(N^+ \cdot C) > 0$ implies the birationality of $\varphi_{6,X}$.

By Kawamata–Viehweg vanishing, we have

$$|6K_{X'}||_F \succcurlyeq |K_{X'} + \lceil 5\pi^*(K_X) - E_1' \rceil\,||_F \succcurlyeq |K_F + \lceil (5\pi^*(K_X) - E_1')|_F \rceil\,|.$$

Noting that

$$\begin{aligned}(5\pi^*(K_X) - E_1')|_F &\equiv 4\pi^*(K_X)|_F \equiv 2\pi^*(K_X)|_F + C + 2\hat{E}_F\\ &\equiv (2\pi^*(K_X)|_F + N^+) + C + N^- \qquad (4.5)\end{aligned}$$

and that $2\pi^*(K_X)|_F + N^+$ is nef and big, the vanishing theorem gives

$$\big|K_F + \lceil (5\pi^*(K_X) - E_1')|_F - N^- \rceil\big|\big|_C = |K_C + D^+| \qquad (4.6)$$

where $\deg(D^+) \geq 2\xi + (N^+ \cdot C) > 2$. By Lemma 4.4, (4.5) and (4.6), $\varphi_{6,X}$ is birational. □

Lemma 4.7. *Under the conditions of Lemma 4.5, if $\beta > \frac{2}{3}$, then $\xi \geq \frac{5}{6}$ and $\varphi_{6,X}$ is birational.*

Proof Since $\alpha(5) \geq (5 - 1 - 1 - \frac{1}{\beta}) \cdot \xi > 1$, we have $\xi \geq \frac{4}{5}$ by (2.2). Now, as $\alpha(6) > 2$, we have $\xi \geq \frac{5}{6}$ by (2.2) and $\varphi_{6,X}$ is birational by Theorem 2.2. □

Lemma 4.8. *Under the conditions of Lemma 4.5, if $\mu > \frac{4}{3}$, then $\varphi_{6,X}$ is birational.*

Proof By assumption and (2.4), $\beta \geq \frac{4}{7}$. So $\alpha(5) = (5-1-\frac{1}{\mu}-\frac{1}{\beta})\cdot\xi > 1$, which implies $\xi \geq \frac{4}{5}$. Since $\alpha(6) > \frac{5}{2}\cdot\xi \geq 2$, $\varphi_{6,X}$ is birational by Theorem 2.2. □

Definition 4.9. For any integers $j \geq 0$, define the restriction maps:

$$H^0(X', M_{m_1} - jF) \overset{\theta_{m_1,-j}}{\longrightarrow} H^0(F, M_{m_1}|_F),$$
$$H^0(F, M_{m_1}|_F - jC) \overset{\psi_{m_1,-j}}{\longrightarrow} H^0(C, M_{m_1}|_C).$$

Set

$$U_{m_1,-j} = \mathrm{Im}(\theta_{m_1,-j}), \qquad u_{m_1,-j} = \dim U_{m_1,-j}$$
$$V_{m_1,-j} = \mathrm{Im}(\psi_{m_1,-j}), \qquad v_{m_1,-j} = \dim V_{m_1,-j}.$$

Lemma 4.10. *Let X be a minimal projective 3-fold of general type with $p_g(X) = 2$, $d_1 = 1$ and $\Gamma \cong \mathbb{P}^1$. Keep the notation in 2.1. Assume that F is a $(1,2)$-surface. Let $m_1 \geq 2$ be any integer. Then $\varphi_{6,X}$ is birational provided that one of the following holds:*

(1) $u_{m_1,0} = h^0(F, m_1K_F)$;
(2) $h^0(M_{m_1} - jF) \geq \lfloor \frac{4}{3}m_1 \rfloor - j + 2 \geq 2$ *and* $u_{m_1,-j} \leq 1$ *for some integer* $j \geq 0$.

Proof (*1*) Since $\theta_{m_1,0}$ is surjective, we have $m_1\pi^*(K_X)|_F \geq m_1C$, which means $\beta = 1$. Hence $\varphi_{6,X}$ is birational by Lemma 4.7.
(*2*) Since $|M_{m_1} - jF|$ and $|F|$ are composed of the same pencil,

$$M_{m_1} - jF \geq \big(\lfloor \tfrac{4}{3}m_1 \rfloor - j + 1\big)F.$$

So $\mu \geq \big(\lfloor \frac{4}{3}m_1 \rfloor + 1\big)/m_1 > \frac{4}{3}$ and $\varphi_{6,X}$ is birational by Lemma 4.8. □

Lemma 4.11. *Under the conditions of Lemma 4.5, suppose that $\xi = \frac{4}{5}$. Then $\varphi_{6,X}$ is not birational.*

Proof Write $|M_6| = \mathrm{Mov}\,|6K_{X'}|$. One may assume that $|M_6|$ is base point free. Denote by $G_{6,0}$ a general member of $|M_6|_F|$. Then

$$|G_{6,0}|_C| \succcurlyeq |G_{6,0}||_C \succcurlyeq |M_6||_C.$$

Since $\alpha(6) > 1$, we have $(6\pi^*(K_X)|_F \cdot C) \geq (G_{6,0} \cdot C) \geq 4$ by (2.2). Therefore $(G_{6,0}\cdot C) = 4$ by the assumption. On the other hand, vanishing gives

$$\begin{aligned}|6K_{X'}||_C &\succcurlyeq |K_F + \lceil 4\pi^*(K_X)|_F \rceil\,||_C\\ &\succcurlyeq \left|K_F + \left\lceil 4\pi^*(K_X)|_F - 2\hat{E}_F \right\rceil\right|\Big|_C \succcurlyeq |K_F + \hat{D}|\end{aligned}$$

where $\deg(\hat{D}) \geq 2$, which forces $|S_{6,0}||_F|_C = |G_{6,0}|_C|$. Take a general effective divisor $K \in |K_X|$. Then $\mathrm{supp}(\pi^*(K)|_F)|_C$ consists of just one point $P \in C$ since $\deg(\sigma^*(K_{F_0})|_C) = 1$. Here $2P \in |K_C|$. So $|4P| = |\lfloor(6\pi^*K|_F)|_C\rfloor| \succcurlyeq |G_{6,0}|_C|$. Therefore the restriction of the linear system $|S_{6,0}|$ on C is just $|2K_C|$, which implies that $\varphi_{6,X}$ is not birational. □

4.3 Effective Constraints on P_2, P_3, P_4, P_5 and P_6

This subsection is devoted to linking some numerical constrains on plurigenera $P_i(X)$ $(i = 1, \ldots, 6)$ to the birationality of φ_6. The following proposition is the prototype for Propositions 4.13–4.16.

Proposition 4.12. *Under the conditions of Lemma 4.5,*

(1) when $P_2(X) \geq 6$, $\varphi_{6,X}$ is birational;
(2) when $P_2(X) = 5$, $\varphi_{6,X}$ is not birational if and only if

$$\xi = 1 \qquad \text{and} \qquad (\pi^*(K_X)|_F)^2 = \tfrac{1}{2}.$$

Proof Set $m_1 = 2$. By Lemma 4.10, assume $u_{2,0} \leq h^0(2K_F) - 1 = 3$.
Case 1. $u_{2,0} \leq 3$ and $u_{2,-1} = 3$.

There is a moving divisor $S_{2,-1}$ on X' such that $M_2 \geq F + S_{2,-1}$ and $h^0(F, S_{2,-1}|_F) \geq 3$. Modulo further birational modification, we may assume that $|S_{2,-1}|$ is base point free. Denote by $C_{2,-1}$ a generic irreducible element of $|S_{2,-1}|_F|$. Then $|C_{2,-1}|$ is moving since $q(F) = 0$.

If $|S_{2,-1}|_F|$ and $|C|$ are composed of same pencil, then

$$M_2|_F \geq S_{2,-1}|_F \geq 2C$$

which means $\beta \geq 1$ and $\varphi_{6,X}$ is birational by Lemma 4.7. If $|S_{2,-1}|_F|$ and $|C|$ are not composed of the same pencil, then $\varphi_{6,X}$ is birational by [CHP21, Prop 3.6(1)].

Case 2. $u_{2,0} \leq 3$ and $u_{2,-1} = 2$.

If $|S_{2,-1}|_F|$ and $|C|$ are not composed of the same pencil, then $\varphi_{6,X}$ is birational by [CHP21, Prop 3.6(1)]. If $|S_{2,-1}|_F|$ and $|C|$ are composed of the same pencil, then we have $\xi \geq \frac{4}{5}$ by [CHP21, Prop 3.6(2.1)] (as $2\xi(|G|) \geq \frac{4}{3} > 1$). By [CHP21, Prop 3.6(2.2)] (taking $n = 3$), $\varphi_{6,X}$ is birational.

Case 3. $u_{2,0} \leq 3$, $u_{2,-1} \leq 1$ and $P_2(X) \geq 6$.

We have $h^0(M_2 - F) \geq 3$. Hence $\varphi_{6,X}$ is birational by Lemma 4.10 (*2*). This proves (*1*). For (*2*), suppose that $P_2(X) = 5$. Note that

$$P_2(X) = u_{2,0} + h^0(M_2 - F).$$

Suppose $\varphi_{6,X}$ is not birational. By the arguments above, $u_{2,0} \leq 3$ and $u_{2,-1} = 1$. If $u_{2,0} \leq 2$, one has $h^0(M_2 - F) \geq P_2(X) - u_{2,0} \geq 3$ since $P_2(X) = 5$. Thus $\mu \geq \frac{3}{2}$. Lemma 4.8 implies $\varphi_{6,X}$ is birational, a contradiction. So we have $u_{2,0} = 3$. If $|M_2|_F|$ and $|C|$ are composed of the same pencil, we get $\beta \geq 1$. Lemma 4.7 implies that $\varphi_{6,X}$ is birational. So $|M_2|_F|$ and $|C|$ are not composed of the same pencil. Thus

$$\xi = (\pi^*(K_X)|_F \cdot C) \geq \tfrac{1}{2}(M_2|_F \cdot C) \geq 1.$$

Lemma 4.6 implies that $(\pi^*(K_X)|_F)^2 = \frac{1}{2}$.

Conversely, assume that $\xi = 1$ and $(\pi^*(K_X)|_F)^2 = \frac{1}{2}$. If $\beta > \frac{1}{2}$ then $(\pi^*(K_X)|_F)^2 > \frac{1}{2}$, so $\beta = \frac{1}{2}$. Since $\xi = 1$, by the arguments above, $u_{2,0} \leq 3$ and $u_{2,-1} = 1$. If $u_{2,0} \leq 2$, we have $h^0(F, M_2 - F) \geq 3$. By [CHP21, Prop 3.5], one has $\beta > \frac{3}{5}$, which contradicts our assumption. So we have $u_{2,0} = 3$. Similarly $|M_2|_F|$ induces a generically finite morphism. Pick a generic irreducible element C_2 in $|M_2|_F|$. Since

$$(\pi^*(K_X)|_F)^2 \geq \tfrac{1}{2}(\pi^*(K_X)|_F \cdot C_2) \geq \tfrac{1}{4}C_2^2,$$

we have $C_2^2 = 2$ and $(\pi^*(K_X)|_F \cdot C_2) = 1$. Since $|C_2|_{C_2}|$ is generically finite, $|C_2|_{C_2}|$ is a g_2^1. By Kawamata–Viehweg vanishing, we have

$$\begin{aligned} |6K_{X'}||_F &\succcurlyeq |K_{X'} + F + 2M_2||_F \succcurlyeq |K_F + 2C_2| \text{ and} \\ |K_F + 2C_2||_{C_2} &= |K_{C_2} + C_2|. \end{aligned}$$

So $|M_6||_{C_2} \succcurlyeq |K_{C_2} + C_2|_{C_2}|$. As $6 = (6\pi^*(K_X)|_F \cdot C_2) \geq \deg(M_6|_{C_2})$,

$$|M_6|_{C_2}| = |M_6||_{C_2} = |K_{C_2} + C_2|_{C_2}|.$$

Since $|C_2|_{C_2}|$ is g_2^1, $\varphi_{6,X}$ is not birational. □

The proofs of the next four propositions are similar.

Proposition 4.13. *Under the conditions of Lemma 4.5, if $P_3(X) \geq 9$, then $\varphi_{6,X}$ is birational.*

Proof Set $m_1 = 3$. By Lemma 4.10, assume $u_{3,0} \leq h^0(3K_F) - 1 = 5$. Since $|3K_{X'}| \succcurlyeq |K_{X'} + F_1 + F|$ for two distinct general fibers of f, it follows by [Che04, Lemma 4.6] that $|M_3||_F \succcurlyeq |C|$, so $u_{3,0} \geq 2$.

Case 1. $u_{3,0} = 5$.

We have $h^0(M_3|_F - C) \geq 3$ since $v_{3,0} \leq 2$ by [CHP21, Prop 3.4]. If $v_{3,-1} \geq 2$, so $M_3|_F \geq C + C_{3,-1}$ where $C_{3,-1}$ is a moving curve on F with $(C_{3,-1} \cdot C) \geq 2$. By [CHP21, Prop 3.7(i)], $\varphi_{6,X}$ is birational. If $v_{3,-1} \leq 1$, then $|M_3|_F - C|$ and $|C|$ are composed of the same pencil. Then $M_3|_F \geq 3C$, so $\beta \geq 1$ and so $\varphi_{6,X}$ is birational by Lemma 4.7.

Case 2. $u_{3,0} = 4$

We have $h^0(M_3|_F - C) \geq 2$. If $v_{3,-1} \geq 2$, we conclude that $\varphi_{6,X}$ is birational by the same argument as in case 1. If $v_{3,-1} \leq 1$, we see that $|M_3|_F - C|$ and $|C|$ are composed of the same pencil. Then $M_3|_F \geq 2C$ which implies $\beta \geq \frac{2}{3}$.

Subcase 2.1. $u_{3,-1} \geq 3$.

If $|S_{3,-1}|_F|$ and $|C|$ are not composed of the same pencil, it follows that $(S_{3,-1}|_F \cdot C) \geq 2$ and so $\varphi_{6,X}$ is birational by [CHP21, Prop 3.6(1.2)] ($m_1 = 3, j = 1, \beta \geq \frac{2}{3}, \mu = 1, \delta = 2$). If $|S_{3,-1}|_F|$ and $|C|$ are composed of the same pencil, then $S_{3,-1}|_F \geq 2C$. We get $\beta \geq \frac{3}{4}$ by [CHP21, Prop 3.5] ($n_1 = 3$, $j_1 = 1$, $l_1 = 2$) and so $\varphi_{6,X}$ is birational by Lemma 4.7.

Subcase 2.2. $u_{3,-1} \leq 2$ and $u_{3,-2} = 2$.

If $|S_{3,-2}|_F|$ and $|C|$ are not composed of the same pencil, $\varphi_{6,X}$ is birational by Theorem 3.1(*1*). If $|S_{3,-2}|_F|$ and $|C|$ are composed of the same pencil, we hope to use Theorem 3.4 with $j_1 = 2$ and $j_2 = 1$. Recall that $\beta \geq \frac{2}{3}$, $\xi \geq \frac{2}{3}$ and $\mu = 1$. By taking $n = 7$ and applying (2.2), we get $\xi \geq \frac{5}{7}$. Similarly, one gets $\xi \geq \frac{4}{5}$ by (2.2) with $n = 6$. Finally Theorem 3.4(*1*) implies the birationality of $\varphi_{6,X}$.

Subcase 2.3. $u_{3,-1} \leq 2$, $u_{3,-2} = 1$ and $P_3(X) \geq 9$.

One has $h^0(M_3 - 2F) \geq 3$. As $u_{3,-2} = 1$, one has $M_3 \geq 4F$, which implies $\mu \geq \frac{4}{3}$. As $\alpha(5) > 1$, we get $\xi \geq \frac{4}{5}$ by (2.2). Since $\alpha(6) > 2$, $\varphi_{6,X}$ is birational.

Case 3. $u_{3,0} \leq 3$, $u_{3,-1} \leq 3$ and $u_{3,-2} \geq 2$.

We have $M_3 \geq 2F + S_{3,-2}$, where $S_{3,-2}$ is a moving divisor with $h^0(S_{3,-2}|_F) \geq 2$. Clearly $\beta \geq \frac{2}{3}$. If $|S_{3,-2}|_F|$ and $|C|$ are not composed of the same pencil, then $\xi \geq \frac{4}{5}$ by Theorem 3.1(*4*) ($n = 4$). Hence $\varphi_{6,X}$ is birational by Theorem 3.1(*1*). If $|S_{3,-2}|_F|$ and $|C|$ are composed of the same pencil, we get $\xi \geq \frac{4}{5}$ by Theorem 3.4(*1.i*) ($n = 4$) and $\varphi_{6,X}$ is birational by Theorem 3.4(*1.ii*).

Case 4. $u_{3,0} \leq 3$, $u_{3,-1} = 3$, $u_{3,-2} = 1$ and $P_3(X) \geq 9$.

We have $h^0(M_3 - 2F) \geq 3$. Since $u_{3,-2} = 1$, we have $M_3 \geq 4F$. Thus, by [CHP21, Prop 3.5], we have $\beta \geq \frac{4}{7}$. If $|S_{3,-1}|_F|$ and $|C|$ are not composed of the same pencil, $(S_{3,-1}|_F \cdot C) \geq 2$. [CHP21, Prop 3.6(1.2)] implies the birationality of $\varphi_{6,X}$ ($m_1 = 3$, $j = 1$, $\delta = 2$, $\mu = \frac{4}{3}$).

If $|S_{3,-1}|_F|$ and $|C|$ are composed of the same pencil, we have $\beta \geq \frac{3}{4}$ by [CHP21, Prop 3.5]. Thus $\varphi_{6,X}$ is birational by Lemma 4.7.

Case 5. $u_{3,0} \leq 3$, $u_{3,-1} \leq 2$, $u_{3,-2} = 1$ and $P_3(X) \geq 9$.

Clearly we have $h^0(M_3 - 2F) \geq 4$, which implies that $\mu \geq \frac{5}{3}$. Hence $\varphi_{6,X}$ is birational by Lemma 4.8. □

Proposition 4.14. *Under the conditions of Lemma 4.5,*

(1) when $P_4(X) \geq 15$, $\varphi_{6,X}$ *is birational;*
(2) when $P_4(X) = 14$, $\varphi_{6,X}$ *is non-birational if and only if one of the following holds:*
   (2.i) $\xi = \frac{4}{5}$;
   (2.ii) $\xi = 1$ *and* $(\pi^*(K_X)|_F)^2 = \frac{1}{2}$.

Proof Set $m_1 = 4$. By Lemma 4.7, assume that $u_{4,0} \leq h^0(4K_F) - 1 = 8$.

By [CHP21, Prop 3.4(2)], we know that $v_{4,0} \leq 3$. We claim that $v_{4,-1} = 3$ is impossible. Otherwise, we have $M_4|_F \geq C + C_{-1}$, where C_{-1} is a moving curve on F satisfying $h^0(C, C_{-1}|_C) \geq 3$. In particular, one has $(C_{-1} \cdot C) \geq 4$ by Riemann–Roch. Now we have

$$4 \geq (\sigma^*(K_{F_0}) \cdot M_4|_F) \geq (\sigma^*(K_{F_0}) \cdot (C + C_{-1})) \geq 5,$$

which is a contradiction. Hence we have $v_{4,-1} \leq 2$.

In the proof we choose an integer $j \geq 0$ such that $M_4 \geq jF + S_{4,-j}$ for a moving divisor $S_{4,-j}$ with $h^0(F, S_{4,-j}|_F) \geq 2$. Modulo further birational modifications, we may and do assume that $|S_{4,-j}|$ is base point free.

Case 1. $u_{4,0} \geq 7$.

Since $v_{4,0} \leq 3$ and $v_{4,-1} \leq 2$, we have $h^0(F, M_4|_F - 2C) \geq 2$. There is a moving curve C_{-2} such that $M_4|_F \geq 2C + C_{-2}$. When $|C|$ and $|C_{-2}|$ are composed of the same pencil, we get $M_4|_F \geq 3C$ and so $\beta \geq \frac{3}{4}$. Lemma 4.7 implies the birationality of $\varphi_{6,X}$. When $|C|$ and $|C_{-2}|$ are not composed of the same pencil, [CHP21, Prop 3.7(iii)] implies that $\xi \geq \frac{4}{5}$ ($n = 4$, $m_1 = 4$, $j = 2$, $\delta_1 = 2$, $\xi \geq \frac{2}{3}$, $\mu = 1$). So $\varphi_{6,X}$ is birational by [CHP21, Prop 3.7(i)] ($m_1 = 4$, $j = 2$, $\delta_1 = 2$, $\xi \geq \frac{4}{5}$, $\mu = 1$).

Case 2. $u_{4,0} = 6$.

The argument is organised according to the value of $v_{4,0}$.

Subcase 2.1. $u_{4,0} = 6$ and $v_{4,0} \leq 2$.

We have $M_4|_F \geq 2C + C_{-2}$, where $h^0(C_{-2}) = h^0(M_4|_F - 2C) \geq 2$. By the same argument as in case 1, $\varphi_{6,X}$ is birational.

Subcase 2.2. $u_{4,0} = 6$, $u_{4,-1} \geq 4$ and $v_{4,0} = 3$.

Since $v_{4,0} = 3$, we have $\xi = 1$. Clearly we have $h^0(F, S_{4,-1}|_F) \geq 4$ by assumption. If $|S_{4,-1}|_F|$ and $|C|$ are not composed of the same pencil and $(S_{4,-1}|_F \cdot C) \geq 4$. We have

$$(\pi^*(K_X)|_F)^2 \geq \tfrac{1}{5}(\xi + \tfrac{1}{2}(S_{4,-1}|_F \cdot C)) \geq \tfrac{3}{5} > \tfrac{1}{2}$$

by Theorem 3.1(*3*). So $\varphi_{6,X}$ is birational by Lemma 4.6. If $|S_{4,-1}|_F|$ and $|C|$ are not composed of the same pencil and $(S_{4,-1}|_F \cdot C) \leq 3$, then

$S_{4,-1}|_F \geq C + C_{-1}$, where C_{-1} is a moving curve on F. If $|C_{-1}|$ and $|C|$ are not composed of the same pencil, noting that

$$(\pi^*(K_X)|_F \cdot S_{4,-1}|_F) \geq \xi + \tfrac{1}{2}(C \cdot C_{-1}) \geq 2,$$

we still have $(\pi^*(K_X)|_F)^2 \geq \frac{3}{5}$ by Theorem 3.1(*3*). Hence $\varphi_{6,X}$ is birational by Lemma 4.6. If $|C_{-1}|$ and $|C|$ are composed of the same pencil, $\beta \geq \frac{3}{5}$ by [CHP21, Prop 3.5] ($n_1 = 4$, $j_1 = 1$, $l_1 = 2$). Since

$$(\pi^*(K_X)|_F)^2 \geq \beta \cdot \xi \geq \tfrac{3}{5},$$

$\varphi_{6,X}$ is birational by Lemma 4.6.

If $|S_{4,-1}|_F|$ and $|C|$ are composed of the same pencil, $S_{4,-1}|_F \geq 3C$. By [CHP21, Prop 3.5], we get $\beta \geq \frac{4}{5}$ and so $(\pi^*(K_X)|_F)^2 \geq \frac{4}{5}$. Hence $\varphi_{6,X}$ is birational by Lemma 4.7.

Subcase 2.3. $u_{4,0} = 6$, $u_{4,-1} \leq 3$, $u_{4,-2} \leq 3$, $u_{4,-3} \geq 2$ and $v_{4,0} = 3$.

$v_{4,0} = 3$ implies that $\xi = 1$. If $|S_{4,-3}|_F|$ and $|C|$ are not composed of the same pencil, we have $(\pi^*(K_X)|_F)^2 \geq \frac{4}{7} > \frac{1}{2}$ by Theorem 3.1(*3*) and $\varphi_{6,X}$ is birational by Lemma 4.6. If $|S_{4,-3}|_F|$ and $|C|$ are composed of the same pencil, we have $\beta \geq \frac{4}{7}$ by [CHP21, Prop 3.5]. Similarly, one has $(\pi^*(K_X)|_F)^2 \geq \frac{4}{7} > \frac{1}{2}$. Hence $\varphi_{6,X}$ is birational by Lemma 4.6.

Subcase 2.4. $u_{4,0} = 6$, $u_{4,-1} \leq 3$, $u_{4,-2} \leq 3$, $u_{4,-3} = 1$, $v_{4,0} = 3$ and $P_4(X) \geq 15$.

$v_{4,0} = 3$ implies that $\xi = 1$. As $|M_4 - 3F|$ and $|F|$ are composed of the same pencil and $h^0(M_4 - 3F) \geq 3$, we have $\mu \geq \frac{5}{4}$ and so $\beta \geq \frac{5}{9}$ by [CHP21, Prop 3.5]. Hence $\varphi_{6,X}$ is birational by Lemma 4.6 since $(\pi^*(K_X)|_F)^2 \geq \frac{5}{9} > \frac{1}{2}$.

Case 3. $u_{4,0} \leq 5$ and $u_{4,-1} \geq 4$.

If $|S_{4,-1}|_F|$ and $|C|$ are composed of the same pencil, then $\beta \geq \frac{4}{5}$ by [CHP21, Prop 3.5]. Hence $\varphi_{6,X}$ is birational by Lemma 4.7.

Assume that $|S_{4,-1}|_F|$ and $|C|$ are not composed of the same pencil. For the case $(S_{4,-1}|_F \cdot C) = 4$, $\varphi_{6,X}$ is birational by Theorem 3.1(*2*) ($m_1 = 4$, $j = 1$, $\tilde{\delta} = 4$, $\beta = \frac{1}{2}$, $\mu = 1$). For the case $(S_{4,-1}|_F \cdot C) \leq 3$, we have $S_{4,-1}|_F \geq C + C_{-1}$ where C_{-1} is a moving curve. When $|C_{-1}|$ and $|C|$ are not composed of the same pencil, then $\varphi_{6,X}$ is birational by Theorem 3.2(*1.ii*) ($m_1 = 4$, $j_1 = j_2 = 1$, $\mu = 1$, $\beta = \frac{1}{2}$, $\delta_2 = 2$).

When $|C_{-1}|$ and $|C|$ are composed of the same pencil, we have $\beta \geq \frac{3}{5}$ by [CHP21, Prop 3.5]. Since $\alpha(7) > 2$, we have $\xi \geq \frac{5}{7}$.

Subcase 3.1. $u_{4,0} \leq 5$, $u_{4,-1} = 4$ and $u_{4,-2} = 4$.

Since $|S_{4,-2}|_F| = |S_{4,-1}|_F|$, we have $S_{4,-2}|_F \geq 2C$. Hence $\varphi_{6,X}$ is birational by Theorem 3.4(*1.ii*) ($\xi = \frac{5}{7}$, $m_1 = 4$, $j_1 = j_2 = 2$, $\beta = \frac{3}{5}$).

Subcase 3.2. $u_{4,0} \leq 5$, $u_{4,-1} = 4$, $u_{4,-2} \leq 3$ and $u_{4,-3} \geq 2$.

If $|S_{4,-3}|_F|$ and $|C|$ are not composed of the same pencil, $\varphi_{6,X}$ is birational by Theorem 3.1(*1*) ($m_1 = 4$, $j = 3$, $\xi = \frac{5}{7}$, $\tilde{\delta} = 2$, $\beta = \frac{3}{5}$). If $|S_{4,-3}|_F|$ and $|C|$ are composed of the same pencil, $\xi \geq \frac{4}{5}$ by Theorem 3.4(*1.i*) ($n = 4$, $m_1 = 4$, $j_1 = 3$, $j_2 = 1$, $\beta = \frac{3}{5}$, $\xi = \frac{5}{7}$). Hence $\varphi_{6,X}$ is birational by Theorem 3.4(*1.ii*) ($m_1 = 4$, $j_1 = 3$, $j_2 = 1$, $\xi = \frac{4}{5}$, $\beta = \frac{3}{5}$).

Subcase 3.3. $u_{4,0} \leq 5$, $u_{4,-1} = 4$, $u_{4,-2} \leq 3$, $u_{4,-3} = 1$ and $P_4(X) \geq 15$.

Note that we have $h^0(M_4 - 3F) = P_4(X) - \sum_{i=0}^{2} u_{4,-i} \geq 3$ by our assumption. Since $u_{4,-3} = 1$, $|M_4 - 3F|$ and f are composed with the same pencil. Thus we have $M_4 \geq 5F$, which implies that $\mu \geq \frac{5}{4}$. Recall that we have $\beta \geq \frac{3}{5}$ in this case. We get $\xi \geq \frac{4}{5}$ since $\alpha(5) \geq \frac{46}{45} > 1$. Hence $\varphi_{6,X}$ is birational as $\alpha(6) \geq \frac{152}{75} > 2$.

Case 4. $u_{4,0} \leq 5$, $u_{4,-1} \leq 3$ and $u_{4,-3} \geq 3$.

If $|S_{4,-3}|_F|$ and $|C|$ are composed of the same pencil, $\beta \geq \frac{5}{7} > \frac{2}{3}$ by [CHP21, Prop 3.5]. So $\varphi_{6,X}$ is birational by Lemma 4.7.

Assume that $|S_{4,-3}|_F|$ and $|C|$ are not composed of the same pencil. When $(S_{4,-3}|_F \cdot C) \geq 3$, we have $\xi \geq \frac{6}{7}$ by Theorem 3.1(*3*). Theorem 3.1(*1*) ($\beta = \frac{1}{2}$) implies the birationality of $\varphi_{6,X}$. When $(S_{4,-3}|_F \cdot C) \leq 2$, we have $S_{4,-3}|_F \geq C$ and so $\beta \geq \frac{4}{7}$ by [CHP21, Prop 3.5]. Also we have $\xi \geq \frac{5}{7}$ by Theorem 3.1(*3*) ($j = 3$, $m_1 = 4$). Hence $\varphi_{6,X}$ is birational by Theorem 3.1(*1*) ($m_1 = 4$, $j = 3$, $\tilde{\delta} = 2$, $\xi = \frac{5}{7}$, $\beta = \frac{4}{7}$).

Case 5. $u_{4,0} \leq 5$, $u_{4,-1} \leq 3$, $u_{4,-3} \leq 2$ and $u_{4,-4} \geq 2$.

If $|S_{4,-4}|_F|$ and $|C|$ are composed of the same pencil, we get $\beta \geq \frac{5}{8}$ by [CHP21, Prop 3.5]. By Theorem 3.4(*1.i*), we have $\xi \geq \frac{4}{5}$ ($m_1 = 4$, $j_1 = 4$, $j_2 = 1$, $\beta = \frac{5}{8}$, $\xi = \frac{2}{3}$, $n = 4$). Then $\varphi_{6,X}$ is birational by Theorem 3.4(*1.ii*) ($m_1 = 4$, $j_1 = 4$, $j_2 = 1$, $\beta = \frac{5}{8}$, $\xi = \frac{4}{5}$).

If $|S_{4,-4}|_F|$ and $|C|$ are not composed of the same pencil, we have $\xi \geq \frac{3}{4}$ by Theorem 3.1(*3*). Furthermore one gets $\xi \geq \frac{7}{9}$ by Theorem 3.1(*4*) ($n = 8$). Finally $\varphi_{6,X}$ is birational by Theorem 3.1(*1*) ($m_1 = 4$, $j = 4$, $\beta = \frac{1}{2}$, $\xi = \frac{7}{9}$, $\tilde{\delta} = 2$).

Case 6. $u_{4,0} \leq 5$, $u_{4,-1} \leq 3$, $u_{4,-3} \leq 2$, $u_{4,-4} = 1$ and $P_4(X) \geq 15$.

As in the proof of subcase 3.3, the assumption implies that $\mu \geq \frac{5}{4}$. If $u_{4,0} \leq 4$, we get $h^0(M_4 - 4F) \geq 3$. One has $M_4 \geq 6F$ which means $\mu \geq \frac{3}{2} > \frac{4}{3}$. Hence $\varphi_{6,X}$ is birational by Lemma 4.8.

If $u_{4,0} = 5$, then either $(M_4|_F \cdot C) = 4$ or $M_4|_F \geq C + C_{-1}$, where C_{-1} is a moving curve with $h^0(C_{-1}) \geq 3$. We have $\xi = 1$ in first case and, since $\beta \geq \frac{5}{9}$ by (2.4), one has $(\pi^*(K_X)|_F)^2 \geq \frac{5}{9} > \frac{1}{2}$. Hence $\varphi_{6,X}$ is birational by Lemma 4.6. Now turn to the latter case. When $|C_{-1}|$ and $|C|$ are composed of the same pencil, we have $\beta \geq \frac{3}{4}$ and $\varphi_{6,X}$ is birational by Lemma 4.7. If $|C_{-1}|$ and $|C|$ are not composed of the same pencil, $\varphi_{6,X}$ is birational by [CHP21, Prop 3.7] ($j = 1$, $\delta_1 = 2$, $\mu = \frac{5}{4}$).

That concludes the proof of (*1*). For (*2*), now suppose $P_4(X) = 14$. Note $P_4(X) = u_{4,0} + h^0(M_4 - F)$. We first assume that $\varphi_{6,X}$ is not birational. By the arguments in cases 1–6, one of the following holds:

(a) $u_{4,0} = 6$, $u_{4,-1} \leq 3$, $u_{4,-2} \leq 3$, $u_{4,-3} = 1$, $v_{4,0} = 3$;
(b) $u_{4,0} \leq 5$, $u_{4,-1} \leq 3$, $u_{4,-2} \leq 3$ and $u_{4,-3} = 2$;
(c) $u_{4,0} \leq 5$, $u_{4,-1} \leq 3$, $u_{4,-2} \leq 3$ and $u_{4,-3} = 1$.

For (a), we have $\xi = 1$ since $v_{4,0} = 3$. Lemma 4.6 implies that we have $(\pi^*(K_X)|_F)^2 = \frac{1}{2}$. Thus (*2.ii*) holds.

For (b), if $|S_{4,-3}|_F|$ and $|C|$ are not composed of the same pencil, we have $(S_{4,-3}|_F \cdot C) \geq 2$. By Theorem 3.1(*4*), we have $\xi \geq \frac{4}{5}$ ($n = 4$). Theorem 3.1(*1*) implies that $\varphi_{6,X}$ is birational when $\xi > \frac{4}{5}$ ($\tilde{\delta} = 2$, $j = 3$, $\xi > \frac{4}{5}$, $\beta = \frac{1}{2}$). So $\xi = \frac{4}{5}$ holds.

If $|S_{4,-3}|_F|$ and $|C|$ are composed of the same pencil, then $\beta \geq \frac{4}{7}$ by [CHP21, Prop 3.5]. Since $\alpha(7) > \left(7 - 1 - 1 - \frac{1}{\beta}\right) \cdot \xi \geq \frac{13}{6} > 2$, we have $\xi \geq \frac{5}{7}$. By Theorem 3.4(*1.i*), we have $\xi \geq \frac{4}{5}$ ($m_1 = 4$, $j_1 = 3$, $j_2 = 1$, $\beta = \frac{4}{7}$, $\xi = \frac{5}{7}$). Theorem 3.4(*1.ii*) implies that $\varphi_{6,X}$ is birational when $\xi > \frac{4}{5}$ ($j_1 = 3$, $j_2 = 1$, $\xi > \frac{4}{5}$, $\beta = \frac{4}{7}$). Thus we have $\xi = \frac{4}{5}$.

We are left to treat (c). We claim that $u_{4,0} = 5$. Otherwise, one has $u_{4,0} \leq 4$. Thus we have $h^0(M_4 - 3F) \geq 4$. Then $\varphi_{6,X}$ is birational by Lemma 4.10(*2*) ($m_1 = 4$, $j = 3$), which contradicts our assumption. So we have $u_{4,0} = 5$. By our assumption, we have $h^0(M_4 - 3F) \geq 3$. Thus one has $\mu \geq \frac{5}{4}$ and $\beta \geq \frac{5}{9}$ by [CHP21, Prop 3.5].

If $v_{4,0} \leq 2$, one has $M_4|_F \geq C + C_{-1}$, where C_{-1} is a moving curve satisfying $h^0(F, C_{-1}) \geq 3$. If $|C_{-1}|$ and $|C|$ are composed of the same pencil, one gets $\beta \geq \frac{3}{4}$. Lemma 4.7 implies that $\varphi_{6,X}$ is birational, a contradiction. Thus $(C_{-1} \cdot C) \geq 2$. [CHP21, Prop 3.7(i) and (ii)] implies that $\varphi_{6,X}$ is birational ($m_1 = 4$, $j = 1$, $\mu = \frac{5}{4}$, $\beta = \frac{5}{9}$), which is a contradiction. So we have $v_{4,0} = 3$. In particular, one has $\xi = 1$. Lemma 4.6 implies that we have $(\pi^*(K_X)|_F)^2 = \frac{1}{2}$.

Now we consider the other direction. Lemma 4.11 implies that we only need to consider the case where $\xi = 1$ and $(\pi^*(K_X)|_F)^2 = \frac{1}{2}$. Observing that $\beta > \frac{1}{2}$ implies that $(\pi^*(K_X)|_F)^2 > \frac{1}{2}$. Thus we get $\mu = 1$ and $\beta = \frac{1}{2}$. By the argument in cases 1–6, one of the following holds:

(i) $u_{4,0} = 6$, $u_{4,-1} \leq 3$, $u_{4,-2} \leq 3$, $u_{4,-3} = 1$ and $v_{4,0} = 3$;
(ii) $u_{4,0} \leq 5$, $u_{4,-1} \geq 4$, $S_{4,-1}|_F \geq C + C_{-1}$, where C_{-1} is a moving curve. Moreover, $|C_{-1}|$, $|C|$ are not composed of the same pencil;

(iii) $u_{4,0} \leq 5$, $u_{4,-1} \leq 3$, $u_{4,-3} = 2$;
(iv) $u_{4,0} \leq 5$, $u_{4,-1} \leq 3$, $u_{4,-3} = 1$.

We first consider (i). Our assumption gives $M_4|_F \geq C + C_{-1}$, where C_{-1} is a moving curve on F satisfying $h^0(F, C_{-1}) \geq 3$. Since $q(F) = 0$, $\xi = 1$ and $(\pi^*(K_X)|_F)^2 = \frac{1}{2}$, $|C_{-1}|$ is not composed of pencil and we have $(\pi^*(K_X)|_F \cdot C_{-1}) = 1$. We may and do assume that $|C_{-1}|$ is base point free. Take a general member $C_{-1} \in |C_{-1}|$. One has $g(C_{-1}) \geq 3$ and $|C|_{C_{-1}}|$ is g_2^1. By Kawamata–Viehweg vanishing, we have

$$|6K_{X'}||_F \succcurlyeq |K_{X'} + M_4 + F||_F \succcurlyeq |K_F + C + C_{-1}|.$$

By the Ramanujam vanishing theorem, one has $h^1(F, K_F + C) = 0$. Thus

$$|K_F + C + C_{-1}||_{C_{-1}} = |K_{C_{-1}} + C|_{C_{-1}}|.$$

So $|M_6||_{C_{-1}} \succcurlyeq |K_{C_{-1}} + C|_{C_{-1}}|$. Since $\deg(K_{C_{-1}} + C|_{C_{-1}}) \geq 6$ and

$$(M_6 \cdot C_{-1}) \leq (6\pi^*(K_X)|_F \cdot C_{-1}) = 6,$$

we have $|M_6||_{C_{-1}} = |K_{C_{-1}} + C|_{C_{-1}}|$. Since $|C|_{C_{-1}}|$ is g_2^1, $\varphi_{6,X}$ is non-birational.

For (ii), by Theorem 3.1(*3*) ($j = 1$), we have

$$\pi^*(K_X)|_F \geq \tfrac{1}{5}C + \tfrac{1}{5}S_{4,-1}|_F \geq \tfrac{2}{5}C + \tfrac{1}{5}C_{-1}.$$

Since $\xi = 1$ and $\beta \geq \frac{1}{2}$, we have

$$(\pi^*(K_X)|_F)^2 \geq \tfrac{2}{5} + \tfrac{1}{5} \cdot \tfrac{1}{2}(C \cdot C_{-1}) \geq \tfrac{3}{5},$$

which contradicts our assumption.

For (iii), If $|S_{4,-3}|_F|$ and $|C|$ are composed of the same pencil, we have $\beta \geq \frac{4}{7}$ by [CHP21, Prop 3.5], which contradicts our assumption. Thus $|S_{4,-3}|_F|$ and $|C|$ are not composed of the same pencil. In particular, we have $(S_{4,-3}|_F \cdot C) \geq 2$. By Theorem 3.1(*3*) ($j = 3$), we have

$$\pi^*(K_X)|_F \geq \tfrac{3}{7}C + \tfrac{1}{7}S_{4,-3}|_F.$$

We get $(\pi^*(K_X)|_F)^2 \geq \frac{4}{7}$, which is a contradiction.

For (iv), $h^0(M_4 - 3F) = P_4(X) - \sum_{i=0}^{2} u_{4,-i} \geq 3$ by assumption. Since $u_{4,-3} = 1$, $|M_4 - 3F|$ and f are composed with the same pencil. Thus we have $M_4 \geq 5F$, which implies that $\mu \geq \frac{5}{4}$, which is a contradiction. □

Proposition 4.15. *Under the conditions of Lemma 4.5,*

(1) when $P_5(X) \geq 24$, then $\varphi_{6,X}$ is birational;
(2) when $22 \leq P_5(X) \leq 23$, $\varphi_{6,X}$ is non-birational if and only if $\xi = \frac{4}{5}$.

Proof Set $m_1 = 5$. By Lemma 4.10, assume $u_{5,0} \leq h^0(5K_F) - 1 = 12$.

By [CHP21, Prop 3.4], we know $v_{5,0} \leq 4$. When $\dim \psi_{5,0}(U_{5,0}) = 4$, we have $h^0(C, M_5|_C) \geq 4$, which implies $h^1(C, M_5|_C) = 0$. By Riemann–Roch and $(M_5 \cdot C) \leq 5\xi \leq 5$, we have $\deg(M_5|_C) = 5$, so $\xi = 1$. Moreover, we have $h^0(C, M_5|_C) = 4$. So $|M_5||_C$ is the complete linear system $|M_5|_C| = |K_C + D_1|$ with $\deg(D_1) = 3$. Thus $\varphi_{5,X}$ is birational, so $\varphi_{6,X}$ is birational. So we may assume that $\dim \psi_{5,0}(U_{5,0}) \leq 3$.

Suppose $v_{5,-1} \geq 3$. Then $M_5|_F \geq C + C_{-1}$ for some moving divisor C_{-1} satisfying $h^0(C, C_{-1}|_C) \geq 3$. By Riemann–Roch, $(C_{-1} \cdot C) \geq 4$. Then $\varphi_{6,X}$ is birational by [CHP21, Prop 3.7(ii)] ($\mu = 1$, $m_1 = 5$, $j = 1$, $\delta_1 = 4$, $\beta = \frac{1}{2}$). From now on, we may assume that $v_{5,-1} \leq 2$.

Case 1. $u_{5,0} \geq 8$.

Since $\dim \psi_{5,0}(U_{5,0}) \leq 3$, we have $h^0(M_5|_F - C) \geq 5$. As $v_{5,-1} \leq 2$, we have $M_5|_F \geq 2C + C_{-2}$, where C_{-2} is a moving curve on F satisfying $h^0(C_{-2}) \geq 3$. If $|C_{-2}|$ and $|C|$ are composed of the same pencil, we have $\beta \geq \frac{4}{5}$ and $\varphi_{6,X}$ is birational by Lemma 4.7. If $|C_{-2}|$ and $|C|$ are not composed of the same pencil, we have $(C_{-2} \cdot C) \geq 2$. We have

$$(C_{-2} \cdot C) \leq (C_{-2} \cdot \sigma^*(K_{F_0})) \leq ((M_5|_F - 2C) \cdot \sigma^*(K_{F_0})) \leq 3.$$

By [CHP21, Prop 3.7(iii)], we have $\xi \geq \frac{4}{5}$ ($n = 4$, $\mu = 1$, $m_1 = 5$, $j = 2$, $\xi(1, |C|) = \frac{2}{3}$). Then $\varphi_{6,X}$ is birational by [CHP21, Prop 3.7(i)] ($\mu = 1$, $m_1 = 5$, $j = 2$, $\delta_1 = 2$, $\xi = \frac{4}{5}$).

Case 2. $u_{5,0} = 7$.

If $\dim \psi_{5,0}(U_{5,0}) \leq 2$, we have $h^0(M_5|_F - C) \geq 5$. The same argument as in case 1 implies that $\varphi_{6,X}$ is birational. So we may assume that $\dim \psi_{5,0}(U_{5,0}) = 3$, which implies that $\xi \geq \frac{4}{5}$.

Since $v_{5,-1} \leq 2$, we have $M_5|_F \geq 2C + C_{-2}$, where C_{-2} is a moving curve satisfying $h^0(C_{-2}) \geq 2$. If $|C_{-2}|$ and $|C|$ are not composed of the same pencil, we have $2 \leq (C_{-2} \cdot C) \leq ((M_5|_F - 2C) \cdot \sigma^*(K_{F_0})) \leq 3$. [CHP21, Prop 3.7(i)] implies that $\varphi_{6,X}$ is birational ($\mu = 1$, $m_1 = 5$, $j = 2$, $\delta_1 = 2$, $\xi = \frac{4}{5}$).

Suppose $|C_{-2}|$ and $|C|$ are composed of the same pencil. We have $\xi \geq \frac{4}{5}$ and $\beta \geq \frac{3}{5}$ in this case.

Subcase 2.1. $u_{5,-1} \geq 6$.

If $\dim \psi_{5,0}(U_{5,-1}) = 3$, we have $(S_{5,-1}|_F \cdot C) \geq 4$. By Theorem 2.3, we have

$$|4K_{X'}||_F \succcurlyeq |2(K_{X'} + F)||_F \succcurlyeq |2\sigma^*(K_{F_0})|.$$

[CHP21, Lemma 3.1] implies that M_4 is a big divisor. So $S_{5,-1}$ is nef and big. Kawamata–Viehweg vanishing yields $|6K_{X'}||_F \succcurlyeq |K_F + S_{5,-1}|_F|$.

Thus $M_6|_F \geq C + S_{5,-1}|_F$. Then $\varphi_{6,X}$ is birational by [CHP21, Prop 3.7(ii)] ($\mu = 1$, $m_1 = 6$, $j = 1$, $\delta_1 = 4$, $\beta = \frac{3}{5}$, $\xi = \frac{4}{5}$).

If $\dim \psi_{5,0}(U_{5,-1}) \leq 2$, we have $S_{5,-1}|_F \geq 2C + C'$ where C' is a moving divisor on F. If $|C'|$ and $|C|$ are not composed of the same pencil, we have $2 \leq (C' \cdot C) \leq ((5\pi^*(K_X)|_F - 2C) \cdot \sigma^*(K_{F_0})) \leq 3$. Theorem 3.2(*2*) implies that $\varphi_{6,X}$ is birational ($\mu = 1$, $m_1 = 5$, $j_1 = 1$, $j_2 = 2$, $\delta_2 = 2$, $\xi = \frac{4}{5}$). If $|C'|$ and $|C|$ are composed of the same pencil, we have $S_{5,-1}|_F \geq 3C$. By Theorem 3.4(*2.ii*), $\varphi_{6,X}$ is birational ($\mu = 1$, $m_1 = 5$, $j_1 = 1$, $j_2 = 3$, $\xi = \frac{4}{5}$).

Subcase 2.2. $u_{5,-1} \leq 5$, $u_{5,-2} = 5$.

If $(S_{5,-2}|_F \cdot C) = 5$, we have $\xi = 1$. Since $\beta \geq \frac{3}{5}$, $\varphi_{6,X}$ is birational by Lemma 4.6, we may assume $(S_{5,-2}|_F \cdot C) \leq 4$. If $(S_{5,-2}|_F \cdot C) = 4$, $\varphi_{6,X}$ is birational by Theorem 3.1(*1*) ($m_1 = 5$, $j = 2$, $\tilde{\delta} = 4$, $\beta = \frac{3}{5}$, $\xi = \frac{4}{5}$).

We are left to treat the case when $(S_{5,-1}|_F \cdot C) \leq 3$. We have $S_{5,-2}|_F \geq C + C'$, where C' is a moving curve satisfying $h^0(F, C') \geq 3$. If $|C'|$ and $|C|$ are composed of the same pencil, we have $\beta \geq \frac{5}{7} > \frac{2}{3}$ by [CHP21, Prop 3.5]. By Lemma 4.7, $\varphi_{6,X}$ is birational. If $|C'|$ and $|C|$ are not composed of the same pencil, $\varphi_{6,X}$ is birational by Theorem 3.2(*1.ii*) ($m_1 = 5$, $j_1 = 2$, $j_2 = 1$, $\delta_2 = 2$, $\beta = \frac{3}{5}$, $\xi = \frac{4}{5}$).

Subcase 2.3. $u_{5,-1} \leq 5$, $u_{5,-2} \leq 4$ and $u_{5,-3} \geq 4$.

If $(S_{5,-3}|_F \cdot C) \geq 4$, $\varphi_{6,X}$ is birational by Theorem 3.1(*1*) ($m_1 = 5$, $j = 3$, $\tilde{\delta} = 4$, $\beta = \frac{3}{5}$, $\xi = \frac{4}{5}$).

If $(S_{5,-3}|_F \cdot C) \leq 3$, we have $S_{5,-3}|_F \geq C + C'$, where C' is a moving curve. If $|C'|$ and $|C|$ are composed of the same pencil, we have $\beta \geq \frac{5}{8}$ and $\varphi_{6,X}$ is birational by Theorem 3.4(*1.ii*) ($m_1 = 5$, $j_1 = 3$, $j_2 = 2$, $\beta = \frac{5}{8}$, $\xi = \frac{4}{5}$). If $|C'|$ and $|C|$ are not composed of the same pencil, we have $(C' \cdot C) \geq 2$ and $\varphi_{6,X}$ is birational by Theorem 3.2(*1.ii*) ($m_1 = 5$, $j_1 = 3$, $j_2 = 1$, $\delta_2 = 2$, $\beta = \frac{3}{5}$, $\xi = \frac{4}{5}$).

Subcase 2.4. $u_{5,-1} \leq 5$, $u_{5,-2} \leq 4$, $u_{5,-3} \leq 3$ and $u_{5,-4} \geq 3$.

If $|S_{5,-4}|_F|$ and $|C|$ are composed of the same pencil, we have $\beta \geq \frac{2}{3}$ by [CHP21, Prop 3.5]. Then $\varphi_{6,X}$ is birational by Theorem 3.4(*1.ii*) ($m_1 = 5$, $j_1 = 4$, $j_2 = 2$, $\beta = \frac{2}{3}$, $\xi = \frac{4}{5}$).

If $|S_{5,-4}|_F|$ and $|C|$ are not composed of the same pencil, we have $2 \leq (S_{5,-4}|_F \cdot C) \leq 5$. Then $\varphi_{6,X}$ is birational by Theorem 3.1(*1*) ($m_1 = 5$, $j = 4$, $\tilde{\delta} = 2$, $\beta = \frac{3}{5}$, $\xi = \frac{4}{5}$).

Subcase 2.5. $u_{5,-1} \leq 5$, $u_{5,-2} \leq 4$, $u_{5,-3} \leq 3$, $u_{5,-4} \leq 2$ and $u_{5,-5} \geq 2$.

If $|S_{5,-5}|_F|$ and $|C|$ are composed of the same pencil, $\varphi_{6,X}$ is birational by Theorem 3.4(*1.ii*) ($m_1 = 5$, $j_1 = 5$, $j_2 = 1$, $\beta = \frac{3}{5}$, $\xi = \frac{4}{5}$).

If $|S_{5,-5}|_F|$ and $|C|$ are not composed of the same pencil, $\varphi_{6,X}$ is birational by Theorem 3.1(*1*) ($m_1 = 5$, $j = 5$, $\tilde{\delta} = 2$, $\beta = \frac{3}{5}$, $\xi = \frac{4}{5}$).

Subcase 2.6. $u_{5,-1} \leq 5$, $u_{5,-2} \leq 4$, $u_{5,-3} \leq 3$, $u_{5,-4} \leq 2$, $u_{5,-5} = 1$ and $P_5(X) \geq 24$.

By our assumption, we have $h^0(M_5 - 5F) = P_5(X) - \sum_{i=0}^{4} u_{5,-i} \geq 3$. Since $u_{5,-5} = 1$, $|M_5 - 5F|$ and f are composed with the same pencil. We have $M_5 \geq 7F$, which implies that $\mu \geq \frac{7}{5} > \frac{4}{3}$. Thus $\varphi_{6,X}$ is birational by Lemma 4.8.

Case 3. $u_{5,0} \leq 6$, $u_{5,-1} \leq 6$, $u_{5,-2} \geq 5$.

If $(S_{5,-2}|_F \cdot C) \geq 5$, $\varphi_{6,X}$ is birational by Theorem 3.1(*2*) ($m_1 = 5$, $j = 2$, $\beta = \frac{1}{2}$, $\tilde{\delta} = 5$). If $(S_{5,-2}|_F \cdot C) = 4$, $\varphi_{6,X}$ is birational by Theorem 3.1(*1*) ($m_1 = 5$, $j = 2$, $\xi = \frac{4}{5}$, $\tilde{\delta} = 4$, $\beta = \frac{1}{2}$).

The remaining case is $(S_{5,-2}|_F \cdot C) \leq 3$. By Riemann–Roch, we have $\dim \psi_{5,0}(U_{5,-2}) \leq 2$. So we have $S_{5,-2}|_F \geq 2C$ and $S_{5,-2}|_F \geq C + C'$, where C' is a moving curve satisfying $h^0(F, C') \geq 3$. The former implies $\beta \geq \frac{4}{7}$ by [CHP21, Prop 3.5]. If $|C'|$ and $|C|$ are composed of the same pencil, we have $\beta \geq \frac{5}{7} > \frac{2}{3}$. Lemma 4.7 implies that $\varphi_{6,X}$ is birational. If $|C'|$ and $|C|$ are not composed of the same pencil, we have

$$2 \leq (C' \cdot C) \leq (C' \cdot \sigma^*(K_{F_0})) \leq ((S_{5,-2}|_F - C) \cdot \sigma^*(K_{F_0})) \leq 4.$$

By Theorem 3.2(*1.ii*), $\varphi_{6,X}$ is birational ($\xi = \frac{2}{3}$, $\delta_2 = 2$, $j_1 = 2$, $j_2 = 1$, $\beta = \frac{4}{7}$).

Case 4. $u_{5,0} \leq 6$, $u_{5,-1} \leq 6$, $u_{5,-2} \leq 4$, $u_{5,-3} \geq 4$.

If $(S_{5,-3}|_F \cdot C) \geq 4$, $\varphi_{6,X}$ is birational by Theorem 3.1(*1*) ($\xi = \frac{4}{5}$, $\tilde{\delta} = 4$, $j = 3$, $\beta = \frac{1}{2}$, $m_1 = 5$), so assume $(S_{5,-3}|_F \cdot C) \leq 3$. By Riemann–Roch, we have $S_{5,-3}|_F \geq C + C'$ where C' is a moving curve on F.

If $|C'|$ and $|C|$ are composed of the same pencil, $\beta \geq \frac{5}{8}$ by [CHP21, Prop 3.5]. We have $\alpha(7) \geq \frac{34}{15} > 2$, so $\xi \geq \frac{5}{7}$. Since $\alpha(8) \geq \frac{22}{7} > 3$, we have $\xi \geq \frac{3}{4}$. So $\alpha(5) \geq \frac{21}{20} > 1$ and $\xi \geq \frac{4}{5}$ follows. By Theorem 3.4(*1.ii*), $\varphi_{6,X}$ is birational ($\xi = \frac{4}{5}$, $m_1 = 5$, $j_1 = 3$, $j_2 = 2$, $\beta = \frac{5}{8}$).

If $|C'|$ and $|C|$ are not composed of the same pencil, we have that $2 \leq (C' \cdot C) \leq 4$. By Theorem 3.2(*1.i*), we have $\xi \geq \frac{5}{7}$ ($n = 6$, $\xi = \frac{2}{3}$, $\delta_2 = 2$, $j_1 = 3$, $j_2 = 1$, $m_1 = 5$, $\beta = \frac{1}{2}$). So $\varphi_{6,X}$ is birational by Theorem 3.2(*1.ii*) ($\xi = \frac{5}{7}$, $\delta_2 = 2$, $j_1 = 3$, $j_2 = 1$, $m_1 = 5$, $\beta = \frac{1}{2}$).

Case 5. $u_{5,0} \leq 6$, $u_{5,-1} \leq 6$, $u_{5,-2} \leq 4$, $u_{5,-3} \leq 3$, $u_{5,-4} \geq 3$.

If $(S_{5,-4}|_F \cdot C) \geq 4$, we have $\xi \geq \frac{4}{5}$. By Theorem 3.1(*1*), $\varphi_{6,X}$ is birational ($\tilde{\delta} = 4$, $\xi = \frac{4}{5}$, $j = 4$, $m_1 = 5$, $\beta = \frac{1}{2}$). We may assume that $(S_{5,-4}|_F \cdot C) \leq 3$. By Riemann–Roch, we have $S_{5,-4}|_F \geq C$. We get $\beta \geq \frac{5}{9}$ by [CHP21, Prop 3.5].

If $|S_{5,-4}|_F|$ and $|C|$ are composed of the same pencil, we have $\beta \geq \frac{2}{3}$ by [CHP21, Prop 3.5]. By Theorem 3.4(*1.i*), we have $\xi \geq \frac{4}{5}$ ($n = 4$, $\xi = \frac{2}{3}$,

$j_1 = 4$, $m_1 = 5$, $j_2 = 2$, $\beta = \frac{2}{3}$). Then $\varphi_{6,X}$ is birational by Theorem 3.4(*1.ii*) ($\xi = \frac{4}{5}$, $j_1 = 4$, $m_1 = 5$, $j_2 = 2$, $\beta = \frac{2}{3}$).

If $|S_{5,-4}|_F|$ and $|C|$ are not composed of the same pencil, we have $(S_{5,-4}|_F \cdot C) \geq 2$. By Theorem 3.1(*4*), we have $\xi \geq \frac{4}{5}$ ($n = 4$, $m_1 = 5$, $j = 4$, $\beta = \frac{5}{9}$, $\tilde{\delta} = 2$). Hence $\varphi_{6,X}$ is birational by Theorem 3.1(*1*) ($\xi = \frac{4}{5}$, $\tilde{\delta} = 2$, $j = 4$, $m_1 = 5$, $\beta = \frac{5}{9}$).

Case 6. $u_{5,0} \leq 6$, $u_{5,-1} \leq 6$, $u_{5,-2} \leq 4$, $u_{5,-3} \leq 3$, $u_{5,-4} \leq 2$, $u_{5,-5} \geq 2$.

If $|S_{5,-5}|_F|$ and $|C|$ are composed of the same pencil, we have $\beta \geq \frac{3}{5}$ by [CHP21, Prop 3.5]. By Theorem 3.4(*1.i*), we have $\xi \geq \frac{4}{5}$ ($n = 4$, $m_1 = 5$, $j_1 = 5$, $j_2 = 1$, $\beta = \frac{3}{5}$, $\xi = \frac{2}{3}$). By Theorem 3.4(*1.ii*), $\varphi_{6,X}$ is birational ($\xi = \frac{4}{5}$, $m_1 = 5$, $j_1 = 5$, $j_2 = 1$, $\beta = \frac{3}{5}$).

If $|S_{5,-5}|_F|$ and $|C|$ are not composed of the same pencil, we have $(S_{5,-5}|_F \cdot C) \geq 2$. By Theorem 3.1(*4*) , we have $\xi \geq \frac{4}{5}$. By Theorem 3.1(*1'*), $\varphi_{6,X}$ is birational ($m_1 = 5$, $j = 5$, $\tilde{\delta} = 2$, $\xi = \frac{4}{5}$).

Case 7. $u_{5,0} \leq 6$, $u_{5,-1} \leq 6$, $u_{5,-2} \leq 4$, $u_{5,-3} \leq 3$, $u_{5,-4} \leq 2$, $u_{5,-5} \leq 1$ and $P_5(X) \geq 24$.

We have $h^0(M_5 - 5F) = P_5(X) - \sum_{i=0}^{4} u_{5,-i} \geq 3$ by our assumption. Since $u_{5,-5} = 1$, $|M_5 - 5F|$ and f are composed with the same pencil. We have $M_5 \geq 7F$, which implies that $\mu \geq \frac{7}{5} > \frac{4}{3}$ by [CHP21, Prop 3.5]. Then $\varphi_{6,X}$ is birational by Lemma 4.8.

The proof of (*1*) is finished. Now we prove the second statement. Assume that $22 \leq P_5(X) \leq 23$. By Lemma 4.11, it suffices to consider the direction by assuming that $\varphi_{6,X}$ is not birational. By the arguments in cases 1–7, it suffices to consider one of the following situations:

(i) $u_{5,-2} = 4$;

(ii) $u_{5,0} = 7$, $u_{5,-1} \leq 5$, $u_{5,-2} \leq 3$, $u_{5,-3} \leq 3$, $u_{5,-4} \leq 2$, $\dim \psi_{5,0}(U_{5,0}) = 3$, $\xi \geq \frac{4}{5}$, $\beta \geq \frac{3}{5}$;

(iii) $u_{5,0} \leq 6$, $u_{5,-1} = 6$, $u_{5,-2} \leq 3$, $u_{5,-4} \leq 2$, $u_{5,-5} = 1$;

(iv) $u_{5,0} \leq 6$, $u_{5,-1} \leq 5$, $u_{5,-2} \leq 3$, $u_{5,-4} \leq 2$, $u_{5,-5} = 1$.

We first consider (i). If $(S_{5,-2}|_F \cdot C) \geq 4$, $\varphi_{6,X}$ is birational by Theorem 3.1(*1–2*) ($\tilde{\delta} \geq 4$, $m_1 = 5$, $j = 2$, $\beta = \frac{1}{2}$), which contradicts our assumption. So we have $(S_{5,-2}|_F \cdot C) \leq 3$. Then we have $S_{5,-2}|_F \geq C + C'$ for a moving curve C' on F. When $|C|$ and $|C'|$ are not composed of the same pencil, Theorem 3.2(*1.i*) implies $\xi \geq \frac{4}{5}$ ($n = 4$, $m_1 = 5$, $j_1 = 2$, $j_2 = 1$, $\delta_2 = 2$, $\beta = \frac{1}{2}$). Then $\varphi_{6,X}$ is birational by Theorem 3.2(*1.ii*), a contradiction. Otherwise, we have $S_{5,-2}|_F \geq 2C$. Thus $\beta \geq \frac{4}{7}$ by [CHP21, Prop 3.5]. Since $\alpha(7) \geq \frac{13}{6} > 2$, we have $\xi \geq \frac{5}{7}$. By Theorem 3.4(*1.i*), we have $\xi \geq \frac{4}{5}$ ($n = 4$, $m_1 = 5$, $j_1 = 2$, $j_2 = 2$, $\beta = \frac{4}{7}$, $\xi \geq \frac{5}{7}$). When $\xi > \frac{4}{5}$, by Theorem 3.4(*1.ii*), $\varphi_{6,X}$ is birational ($m_1 = 5$, $j_1 = 2$, $j_2 = 2$), which is a contradiction. So the only possibility is $\xi = \frac{4}{5}$.

For (ii), the condition $P_5(X) \geq 22$ and Lemma 4.10(*2*) imply that $u_{5,-4} = 2$. If $|S_{5,-4}|_F|$ and $|C|$ are not composed of the same pencil, Theorem 3.1(*1*) implies that $\varphi_{6,X}$ is birational ($m_1 = 5$, $j = 4$, $\tilde{\delta} = 2$, $\xi \geq \frac{4}{5}$, $\beta = \frac{3}{5}$), a contradiction. If $|S_{5,-4}|_F|$ and $|C|$ are composed of the same pencil. When $\xi > \frac{4}{5}$, Theorem 3.4(*1.ii*) implies that $\varphi_{6,X}$ is birational ($m_1 = 5$, $j_1 = 4$, $j_2 = 1$, $\beta = \frac{3}{5}$). The only possibility is $\xi = \frac{4}{5}$.

For (iii), we have $h^0(M_5 - 5F) = P_5(X) - \sum_{i=0}^{4} u_{5,-i} \geq 2$. Since $u_{5,-5} = 1$, $|M_5 - 5F|$ and f are composed with the same pencil. Thus we have $\mu \geq \frac{6}{5}$ and $\beta \geq \frac{6}{11}$ by (2.4). Since $\alpha(7) > 2$, we have $\xi \geq \frac{5}{7}$. If $(S_{5,-1} \cdot C) \geq 4$, $\varphi_{6,X}$ is birational by Theorem 3.1(*2*) ($m_1 = 5$, $\tilde{\delta} = 4$, $\mu \geq \frac{6}{5}$, $\beta \geq \frac{6}{11}$). So we may assume that $(S_{5,-1}|_F \cdot C) \leq 3$. Thus we have $S_{5,-1}|_F \geq 2C + C_{-2}$, where C_{-2} is a moving curve on F. If $|C_{-2}|$ and $|C|$ are not composed of the same pencil, $\varphi_{6,X}$ is birational by Theorem 3.2(*2.ii*) ($m_1 = 5$, $j_1 = 1$, $j_2 = 2$, $\delta_2 = 2$, $\mu = \frac{6}{5}$). If $|C_{-2}|$ and $|C|$ are composed of the same pencil, we have $\beta \geq \frac{2}{3}$ by [CHP21, Prop 3.5]. Since $\alpha(5) > 1$, we have $\xi \geq \frac{4}{5}$. Since $\alpha(6) > 2$, $\varphi_{6,X}$ is birational, which is a contradiction. Thus (iii) does not occur.

We are left to treat (iv). Similar as in the proof of (iii), we have $h^0(M_5 - 5F) \geq 3$. Lemma 4.10(*2*) implies that $\varphi_{6,X}$ is birational, which is a contradiction. Therefore we have $\xi = \frac{4}{5}$. □

Proposition 4.16. *Under the same assumption as that of Lemma 4.5,*

(1) when $P_6(X) \geq 35$, $\varphi_{6,X}$ is birational;
(2) when $32 \leq P_6(X) \leq 34$, $\varphi_{6,X}$ is non-birational if and only if $\xi = \frac{4}{5}$.

Proof Set $m_1 = 6$. By Lemma 4.10(*1*) assume $u_{6,0} \leq P_6(F) - 1 = 17$.

Reduction to $\dim \psi_{6,0}(U_{6,0}) \leq 4$, $v_{6,-1} \leq 3$, $v_{6,-2} \leq 2$ **and** $v_{6,-3} \leq 1$.

By [CHP21, Prop 3.4], we have $v_{6,0} \leq 5$. If $\dim \psi_{6,0}(U_{6,0}) = 5$, Riemann–Roch implies $\deg(M_6|_C) \geq 6$. Noting that $\deg(M_6|_C) \leq 6$, $|M_6||_C$ must be complete. So we can write $|M_6||_C = |K_C + D|$ where $\deg(D) = 4$. Thus $\varphi_{6,X}$ is birational. Hence assume $\dim \psi_{6,0}(U_{6,0}) \leq 4$.

Suppose $v_{6,-1} \geq 4$. Then $M_6|_F \geq C + C_{-1}$ for some moving curve C_{-1} on F satisfying $h^0(C, C_{-1}|_C) \geq 4$. In particular, we have $(C_{-1} \cdot C) \geq 5$. By [CHP21, Prop 3.7(ii)], $\varphi_{6,X}$ is birational ($\mu = 1$, $m_1 = 6$, $\delta_1 = 5$, $\beta = \frac{1}{2}$, $j = 1$). We may assume that $v_{6,-1} \leq 3$.

Suppose $v_{6,-2} \geq 3$. We have $M_6|_F \geq 2C + C_{-2}$, where C_{-2} is a moving curve satisfying $h^0(C, C_{-2}|_C) \geq 3$. By Riemann–Roch, one has

$(C_{-2} \cdot C) \geq 4$. We also have

$$(C_{-2} \cdot C) \leq (C_{-2} \cdot \sigma^*(K_{F_0})) \leq ((M_6|_F - 2C) \cdot \sigma^*(K_{F_0})) \leq 4.$$

So $(C_{-2} \cdot C) = 4$. By [CHP21, Prop 3.7(i)], $\varphi_{6,X}$ is birational ($\xi = \frac{2}{3}$, $\delta_1 = 4$, $\mu = 1$, $j = 2$, $m_1 = 6$). Thus we may assume that $v_{6,-2} \leq 2$.

Now assume that $v_{6,-3} \geq 2$. Then $M_6|_F \geq 3C + C_{-3}$ for some moving curve C_{-3} on F. In particular, we have $(C_{-3} \cdot C) \geq 2$. By [CHP21, Prop 3.7(iii)], we have $\xi \geq \frac{4}{5}$ ($n = 4$, $\mu = 1$, $m_1 = 6$, $j = 3$, $\xi = \frac{2}{3}$, $\delta_1 = 2$). Thus $\varphi_{6,X}$ is birational by [CHP21, Prop 3.7(i)] ($\xi = \frac{4}{5}$, $\delta_1 = 2$, $j = 3$, $\mu = 1$, $m_1 = 6$). So we may assume that $v_{6,-3} \leq 1$.

Case 1. $u_{6,0} \geq 11$.

If $\dim \psi_{6,0}(U_{6,0}) = 4$, one has $(M_6|_F \cdot C) \geq 5$ by Riemann–Roch. Hence $\xi \geq \frac{5}{6}$. By assumption, $M_6|_F \geq 4C$. Thus $\beta \geq \frac{2}{3}$. Since

$$\alpha(6) \geq \left(6 - 1 - 1 - \tfrac{1}{\beta}\right) \cdot \xi \geq \tfrac{25}{12} > 2,$$

$\varphi_{6,X}$ is birational by Theorem 2.2.

If $\dim \psi_{6,0}(U_{6,0}) \leq 3$, we get $M_6|_F \geq 5C$ by our assumption. In particular, we have $\beta \geq \frac{5}{6}$. By Lemma 4.7, $\varphi_{6,X}$ is birational.

Case 2. $u_{6,0} = 10$ and $P_6(X) \geq 31$.

If $\dim \psi_{6,0}(U_{6,0}) \leq 3$ and $v_{6,-1} = 3$, we have $h^0(F, M_6|_F - C) \geq 7$ and $M_6|_F \geq C + C_{-1}$, where $(C_{-1} \cdot C) \geq 4$. On the other hand, by our assumption ($u_{6,0} = 10$, $v_{6,-2} \leq 2$ and $v_{6,-3} \leq 1$), we have $M_6|_F \geq 4C$. In particular, we have $\beta \geq \frac{2}{3}$. By [CHP21, Prop 3.7(ii)], $\varphi_{6,X}$ is birational ($\mu = 1$, $m_1 = 6$, $\delta_1 = 4$, $j = 1$, $\beta = \frac{2}{3}$).

If $\dim \psi_{6,0}(U_{6,0}) \leq 3$ and $v_{6,-1} \leq 2$, we have $h^0(F, M_6|_F - C) \geq 8$. By assumption ($u_{6,0} = 10$, $v_{6,-2} \leq 2$ and $v_{6,-3} \leq 1$), we have $M_6|_F \geq 5C$. In particular, $\beta \geq \frac{5}{6}$. Lemma 4.7 implies that $\varphi_{6,X}$ is birational.

So we assume that $\dim \psi_{6,0}(U_{6,0}) = 4$ in this case. By Riemann–Roch, one has $\deg(M_6|_C) \geq 5$. When $\deg(M_6|_C) = 5$, then $|M_6||_C$ must be complete and clearly $\varphi_{6,X}$ is birational. Thus we can assume, from now on within this case, that $(M_6|_F \cdot C) = 6$. In particular, $\xi = 1$.

Subcase 2.1. $u_{6,-1} \geq 7$.

We first consider the case when $\dim \psi_{6,0}(U_{6,-1}) = 4$. By our assumption, we have $(S_{6,-1}|_F \cdot C) = (M_6|_F \cdot C) = 6$. [CHP21, Prop 3.6(1.2)] implies that $\varphi_{6,X}$ is birational ($\beta = \frac{1}{2}$, $m_1 = 6$, $\delta = 6$, $\mu = 1$).

So we assume that $\dim \psi_{6,0}(U_{6,-1}) \leq 3$. Thus $S_{6,-1}|_F \geq C + C_{-1}$, where C_{-1} is a moving curve on F which satisfies $h^0(F, C_{-1}) \geq 4$. If $\dim \psi_{6,-1}(H^0(F, C_{-1})) \geq 3$, we have $(C_{-1} \cdot C) \geq 4$ by Riemann–Roch. By Theorem 3.2(*1.ii*), $\varphi_{6,X}$ is birational ($\xi = 1$, $j_1 = j_2 = 1$, $\delta_2 = 4$,

$\mu = 1$, $\beta = \frac{1}{2}$). The case $\dim \psi_{6,-1}(H^0(F, C_{-1})) \leq 2$ remains. We have $S_{6,-1}|_F \geq 2C + C_{-2}$ where C_{-2} is a moving curve on F. If $|C_{-2}|$ and $|C|$ are composed of the same pencil, we get $\beta \geq \frac{4}{7}$ by [CHP21, Prop 3.5]. Since $\xi = 1$, $(\pi^*(K_X)|_F)^2 \geq \frac{4}{7}$. Lemma 4.6 implies $\varphi_{6,X}$ is birational. If $|C_{-2}|$ and $|C|$ are not composed of the same pencil, $\varphi_{6,X}$ is birational by Theorem 3.2(*2.ii*) ($m_1 = 6$, $j_1 = 1$, $j_2 = 2$, $\delta_2 = 2$, $\xi = 1$, $\mu = 1$).

Subcase 2.2. $u_{6,-1} \leq 6$, $u_{6,-3} \geq 4$.

If $\psi_{6,-3}(U_{6,-3}) \geq 3$, then $(S_{6,-3}|_F \cdot C) \geq 4$. So $(\pi^*(K_X)|_F)^2 \geq \frac{5}{9}$ by Theorem 3.1(*3*) ($j = 3$, $m_1 = 6$, $\tilde{\delta} = 4$). Then $\varphi_{6,X}$ is birational by Lemma 4.6.

If $\psi_{6,-3}(U_{6,-3}) \leq 2$, we have $S_{6,-3}|_F \geq C + C'$ where C' is a moving curve. Thus we still have $(\pi^*(K_X)|_F \cdot S_{6,-3}|_F) \geq 2$. By Theorem 3.1(*3*), we have $(\pi^*(K_X)|_F)^2 \geq \frac{5}{9}$. Hence $\varphi_{6,X}$ is birational by Lemma 4.6.

Subcase 2.3. $u_{6,-1} \leq 6$, $u_{6,-3} \leq 3$, $u_{6,-5} \geq 2$.

If $|S_{6,-5}|_F|$ and $|C|$ are composed of the same pencil, we have $\beta \geq \frac{6}{11}$ by [CHP21, Prop 3.5]. Thus $(\pi^*(K_X)|_F)^2 \geq \frac{6}{11}$. Lemma 4.6 implies that $\varphi_{6,X}$ is birational.

If $|S_{6,-5}|_F|$ and $|C|$ are not composed of the same pencil, we have $((\pi^*(K_X))|_F \cdot S_{6,-5}|_F) \geq 1$. By Theorem 3.1(*3*), $(\pi^*(K_X)|_F)^2 \geq \frac{6}{11}$ ($j = 5$). Lemma 4.6 implies that $\varphi_{6,X}$ is birational.

Subcase 2.4. $u_{6,-1} \leq 6$, $u_{6,-3} \leq 3$, $u_{6,-5} = 1$ and $P_6(X) \geq 31$.

We have $h^0(M_6 - 5F) = P_6(X) - \sum_{i=0}^{4} u_{6,-i} \geq 3$. Since $u_{6,-5} = 1$, $|M_6 - 5F|$ and f are composed with the same pencil. We have $\mu \geq \frac{7}{6}$ and $\beta \geq \frac{7}{13}$ by (2.4). So $(\pi^*(K_X)|_F)^2 \geq \frac{7}{13}$ and $\varphi_{6,X}$ is birational by Lemma 4.6.

Case 3. $u_{6,0} \leq 9$, $u_{6,-1} \geq 8$.

If $(S_{6,-1}|_F \cdot C) = 6$, we have $\xi = 1$. By [CHP21, Prop 3.6(1.2)] ($m_1 = 6$, $\beta = \frac{1}{2}$, $\mu = 1$, $\delta = 6$), $\varphi_{6,X}$ is birational. If $(S_{6,-1}|_F \cdot C) \leq 5$ and $\dim \psi_{6,0}(U_{6,-1}) \geq 4$, Riemann–Roch on C tells that $|S_{6,-1}||_C = |S_{6,-1}|_C| = |K_C + D|$, where $\deg(D) = 3$. Thus $\varphi_{6,X}$ is birational.

If $\dim(\psi_{6,0}(U_{6,-1})) \leq 3$, we have $S_{6,-1}|_F \geq C + C_{-1}$ where C_{-1} is a moving curve satisfying $h^0(F, C_{-1}) \geq 5$. By our reduction, we have $\dim \psi_{6,-1}(H^0(F, C_{-1})) \leq 3$. If $\dim \psi_{6,-1}(H^0(F, C_{-1})) = 3$, we have $(C_{-1} \cdot C) \geq 4$ by Riemann–Roch. By Theorem 3.2(*1.ii*), $\varphi_{6,X}$ is birational ($j_1 = 1$, $j_2 = 1$, $m_1 = 6$, $\mu = 1$, $\beta = \frac{1}{2}$, $\delta_2 = 4$). If $\dim \psi_{6,-1}(H^0(F, C_{-1})) \leq 2$, we have $S_{6,-1}|_F \geq 2C + C_{-2}$, where C_{-2} is a moving curve on F satisfying $h^0(F, C_{-2}) \geq 3$. If $|C_{-2}|$ and $|C|$ are not composed of the same pencil, we have $(C_{-2} \cdot C) \geq 2$. By Theorem 3.2(*2.i*), we have $\xi \geq \frac{4}{5}$ ($n = 4$, $m_1 = 6$, $j_1 = 1$, $j_2 = 2$, $\delta_2 = 2$,

$\xi = \frac{2}{3}$, $\mu = 1$). So $\varphi_{6,X}$ is birational by Theorem 3.2(*2.ii*) ($j_1 = 1$, $j_2 = 2$, $\delta_2 = 2$, $\mu = 1$, $\xi = \frac{4}{5}$). If $|C_{-2}|$ and $|C|$ are composed of the same pencil, we have $S_{6,-1}|_F \geq 4C$. By [CHP21, Prop 3.5], we have $\beta \geq \frac{5}{7}$. Lemma 4.7 implies that $\varphi_{6,X}$ is birational.

Case 4. $u_{6,0} \leq 9$, $u_{6,-1} \leq 7$ and $u_{6,-2} \geq 7$.

Note that $S_{6,-2}|_F \geq M_4|_F \geq 2\sigma^*(K_{F_0})$. So $S_{6,-2}|_F$ is a big divisor. If $(S_{6,-2}|_F \cdot C) \geq 5$, $\varphi_{6,X}$ is birational by Theorem 3.1(*2'*) ($m_1 = 6$, $j = 2$, $\beta = \frac{1}{2}$). If $(S_{6,-2}|_F \cdot C) \leq 4$, we have $S_{6,-2}|_F \geq C + C_{-1}$, where C_{-1} is a moving curve satisfying $h^0(F, C_{-1}) \geq 4$. If $h^0(C, C_{-1}|_C) \geq 3$, we have $(C_{-1} \cdot C) = 4$ by Riemann–Roch and our assumption $(S_{6,-2}|_F \cdot C) \leq 4$. By Theorem 3.2(*1.ii*), $\varphi_{6,X}$ is birational ($j_1 = 2$, $j_2 = 1$, $\delta_2 = 4$, $m_1 = 6$, $\xi = \frac{2}{3}$, $\beta = \frac{1}{2}$, $\mu = 1$) . We may assume that $h^0(C, C_{-1}|_C) \leq 2$. Thus we have $S_{6,-2}|_F \geq 2C + C_{-2}$, where C_{-2} is a moving curve on F. If $|C_{-2}|$ and $|C|$ are not composed of the same pencil, we have $(C_{-2} \cdot C) \geq 2$. By Theorem 3.2(*1.ii*), $\varphi_{6,X}$ is birational ($j_1 = 2$, $j_2 = 2$, $\xi = \frac{2}{3}$, $\delta_2 = 2$, $m_1 = 6$, $\beta = \frac{1}{2}$). If $|C_{-2}|$ and $|C|$ are composed of the same pencil, we have $S_{6,-2}|_F \geq 3C$. By Theorem 3.4(*2.i*), we have $\xi \geq \frac{4}{5}$ ($n = 4$, $m_1 = 6$, $j_1 = 2$, $j_2 = 3$, $\mu = 1$, $\xi = \frac{2}{3}$). Thus $\varphi_{6,X}$ is birational by Theorem 3.4(*2.ii*) ($j_1 = 2$, $j_2 = 3$, $m_1 = 6$, $\mu = 1$, $\xi = \frac{4}{5}$).

Case 5. $u_{6,0} \leq 9$, $u_{6,-1} \leq 7$, $u_{6,-2} \leq 6$ and $u_{6,-3} \geq 5$.

If $(S_{6,-3}|_F \cdot C) \geq 4$, we have $\xi \geq \frac{7}{9}$ by Theorem 3.1(*3*) ($j = 3$, $m_1 = 6$, $\tilde{\delta} = 4$). By Theorem 3.1(*1–2*), $\varphi_{6,X}$ is birational ($j = 3$, $\tilde{\delta} \geq 4$, $\xi = \frac{7}{9}$, $m_1 = 6$, $\beta = \frac{1}{2}$).

If $(S_{6,-3}|_F \cdot C) \leq 3$, we have $S_{6,-3}|_F \geq C + C_{-1}$, where C_{-1} is a moving curve satisfying $h^0(F, C_{-1}) \geq 3$. If $|C_{-1}|$ and $|C|$ are composed of the same pencil, we have $S_{6,-3}|_F \geq 3C$. By [CHP21, Prop 3.5], we have $\beta \geq \frac{2}{3}$. By Theorem 3.4(*1.i*), we have $\xi \geq \frac{4}{5}$ ($n = 4$, $m_1 = 6$, $j_1 = j_2 = 3$, $\xi = \frac{2}{3}$, $\beta = \frac{2}{3}$). Then $\varphi_{6,X}$ is birational by Theorem 3.4(*1.ii*) ($j_1 = 3$, $j_2 = 3$, $\xi = \frac{4}{5}$, $m_1 = 6$, $\beta = \frac{2}{3}$). If $|C_{-1}|$ and $|C|$ are not composed of the same pencil, we have $(C_{-1} \cdot C) \geq 2$. By Theorem 3.2(*1.ii*), $\varphi_{6,X}$ is birational ($j_1 = 3$, $j_2 = 2$, $\xi = \frac{2}{3}$, $m_1 = 6$, $\delta_2 = 2$, $\beta = \frac{1}{2}$).

Case 6. $u_{6,0} \leq 9$, $u_{6,-1} \leq 7$, $u_{6,-2} \leq 6$, $u_{6,-3} \leq 4$ and $u_{6,-4} \geq 4$.

If $(S_{6,-4}|_F \cdot C) \geq 4$, we get $\xi \geq \frac{4}{5}$ by Theorem 3.1(*3*) ($j = 4$, $\tilde{\delta} = 4$, $m_1 = 6$). Hence $\varphi_{6,X}$ is birational by Theorem 3.1(*1*) ($\tilde{\delta} = 4$, $j = 4$, $\xi = \frac{4}{5}$, $\beta = \frac{1}{2}$).

If $(S_{6,-4}|_F \cdot C) \leq 3$, we get $S_{6,-4}|_F \geq C + C_{-1}$, where C_{-1} is a moving curve. If $|C_{-1}|$ and $|C|$ are composed of the same pencil, we have $S_{6,-4}|_F \geq 2C$. By [CHP21, Prop 3.5], we have $\beta \geq \frac{3}{5}$. By Theorem 3.4(*1.i*), we have $\xi \geq \frac{4}{5}$ ($n = 4$, $m_1 = 6$, $j_1 = 4$, $j_2 = 2$, $\xi = \frac{2}{3}$,

$\beta = \frac{3}{5}$). Then $\varphi_{6,X}$ is birational by Theorem 3.4(*1.ii*) ($j_1 = 4$, $j_2 = 2$, $\xi = \frac{4}{5}$, $\beta = \frac{3}{5}$). If $|C_{-1}|$ and $|C|$ are not composed of the same pencil, we have $(C_{-1} \cdot C) \geq 2$. By Theorem 3.2(*1.i*), we have $\xi \geq \frac{4}{5}$ ($n = 4$, $m_1 = 6$, $j_1 = 4$, $j_2 = 1$, $\beta = \frac{1}{2}$, $\xi = \frac{2}{3}$, $\delta_2 = 2$). Theorem 3.2(*1.ii*) implies that $\varphi_{6,X}$ is birational ($j_1 = 4$, $j_2 = 1$, $m_1 = 6$, $\xi = \frac{4}{5}$, $\beta = \frac{1}{2}$).

Case 7. $u_{6,0} \leq 9$, $u_{6,-1} \leq 7$, $u_{6,-2} \leq 6$, $u_{6,-3} \leq 4$, $u_{6,-4} \leq 3$ and $u_{6,-5} \geq 3$.

If $(S_{6,-5}|_F \cdot C) \geq 4$. By the same argument as in case 6, $\varphi_{6,X}$ is birational (note that we have $S_{6,-4}|_F \geq S_{6,-5}|_F$). If $2 \leq (S_{6,-5}|_F \cdot C) \leq 3$, we have $S_{6,-5}|_F \geq C$ by Riemann–Roch. By [CHP21, Prop 3.5], we have $\beta \geq \frac{6}{11}$. By Theorem 3.1(*4*), we have $\xi \geq \frac{4}{5}$ ($n = 4$, $\xi = \frac{2}{3}$, $m_1 = 6$, $j = 5$, $\beta = \frac{6}{11}$, $\tilde{\delta} = 2$). Then $\varphi_{6,X}$ is birational by Theorem 3.1(*1*) ($j = 5$, $m_1 = 6$, $\tilde{\delta} = 2$, $\xi = \frac{4}{5}$, $\beta = \frac{6}{11}$).

If $|S_{6,-5}|_F|$ and $|C|$ are composed of the same pencil, $S_{6,-5}|_F \geq 2C$. By [CHP21, Prop 3.5], we have $\beta \geq \frac{7}{11}$. By Theorem 3.4(*1.i*), we have $\xi \geq \frac{4}{5}$ ($n = 4$, $m_1 = 6$, $j_1 = 5$, $j_2 = 2$, $\beta = \frac{7}{11}$, $\xi = \frac{2}{3}$). Then $\varphi_{6,X}$ is birational by Theorem 3.4(*1.ii*) ($j_1 = 5$, $j_2 = 2$, $m_1 = 6$, $\xi = \frac{4}{5}$, $\beta = \frac{7}{11}$).

Case 8. $u_{6,0} \leq 9$, $u_{6,-1} \leq 7$, $u_{6,-2} \leq 6$, $u_{6,-3} \leq 4$, $u_{6,-4} \leq 3$, $u_{6,-5} \leq 2$, $u_{6,-6} = 2$.

If $|S_{6,-6}|_F|$ and $|C|$ are composed of the same pencil, we get $\beta \geq \frac{7}{12}$ by [CHP21, Prop 3.5]. By Theorem 3.4(*1.i*), we have $\xi \geq \frac{4}{5}$ ($n = 4$, $m_1 = 6$, $j_1 = 6$, $j_2 = 1$, $\beta = \frac{7}{12}$, $\xi = \frac{2}{3}$). Then $\varphi_{6,X}$ is birational by Theorem 3.4(*1.ii*) ($j_1 = 6$, $j_2 = 1$, $m_1 = 6$, $\xi = \frac{4}{5}$, $\beta = \frac{7}{12}$).

If $|S_{6,-6}|_F|$ and $|C|$ are not composed of the same pencil, we have $(S_{6,-6}|_F \cdot C) \geq 2$. By Theorem 3.1(*4*), we have $\xi \geq \frac{5}{7}$ ($n = 6$, $m_1 = 6$, $j = 6$, $\beta = \frac{1}{2}$, $\xi = \frac{2}{3}$). One has $\xi \geq \frac{4}{5}$ by Theorem 3.1(*4*) ($n = 4$, $m_1 = 6$, $j = 6$, $\beta = \frac{1}{2}$, $\xi = \frac{5}{7}$). Thus $\varphi_{6,X}$ is birational by Theorem 3.1(*1'*) ($\tilde{\delta} = 2$, $j = 6$, $\xi = \frac{4}{5}$, $m_1 = 6$, $\beta = \frac{1}{2}$) .

Case 9. $u_{6,0} \leq 9$, $u_{6,-1} \leq 7$, $u_{6,-2} \leq 6$, $u_{6,-3} \leq 4$, $u_{6,-4} \leq 3$, $u_{6,-5} \leq 2$, $u_{6,-6} = 1$ and $P_6(X) \geq 35$.

By assumption, $h^0(M_6 - 6F) = P_6(X) - \sum_{i=0}^{5} u_{6,-i} \geq 4$. As $u_{6,-6} = 1$, $|M_6 - 6F|$ and f are composed with the same pencil. Thus $M_6 \geq 9F$, which implies that $\mu \geq \frac{3}{2} > \frac{4}{3}$. So $\varphi_{6,X}$ is birational by Lemma 4.8.

Now suppose $32 \leq P_6(X) \leq 34$. By Lemma 4.11, we only consider the direction assuming the non-birationality of $\varphi_{6,X}$. By the arguments in cases 1–9, it suffices to consider one of the following cases: cases i–iii.

Case i. $u_{6,0} = 9$, $\dim \psi_{6,0}(U_{6,0}) \leq 4$.

We first treat $\dim \psi_{6,0}(U_{6,0}) = 4$. We have $(M_6|_F \cdot C) \geq 5$ by Riemann–Roch. If $(M_6|_F \cdot C) = 5$, Riemann–Roch implies that $|M_6||_C$ is a complete linear system $|K_C + D|$, where $\deg D = 3$. So $\varphi_{6,X}$ is birational, which is a contradiction. So we have $(M_6|_F \cdot C) = 6$, which implies $\xi = 1$. We will prove that this can not happen at all. By Lemma 4.6, we have $\beta = \frac{1}{2}$.

Subcase i.a. $u_{6,-1} = 7$, $\dim \psi_{6,0}(U_{6,0}) = 4$.

If $\dim \psi_{6,0}(U_{6,-1}) \geq 4$, we have $(S_{6,-1}|_F \cdot C) \geq 5$ by Riemann–Roch. When $(S_{6,-1}|_F \cdot C) \geq 6$, $\varphi_{6,X}$ is birational by [CHP21, Prop 3.6(1.2)] ($m_1 = 6$, $\delta = 6$, $\beta = \frac{1}{2}$, $\mu = 1$). When $(S_{6,-1}|_F \cdot C) = 5$, the linear system $|S_{6,-1}||_C$ must be complete, that is $|S_{6,-1}|_C|$, by Riemann–Roch. In fact, $|S_{6,-1}|_C| = |K_C + D|$ with $\deg(D) = 3$. Clearly, $\varphi_{6,X}$ is birational.

If $\dim \psi_{6,0}(U_{6,-1}) \leq 3$, we have $S_{6,-1}|_F \geq C + C'_{-1}$, where C'_{-1} is a moving curve on F satisfying $h^0(F, C'_{-1}) \geq 4$. When $(C'_{-1} \cdot C) \geq 4$, then $\varphi_{6,X}$ is birational by Theorem 3.2(*1.ii*) ($m_1 = 6$, $j_1 = j_2 = 1$, $\delta_2 = 4$, $\beta = \frac{1}{2}$, $\mu = 1$), which is a contradiction. So we have $(C'_{-1} \cdot C) \leq 3$, which implies that $S_{6,-1}|_F \geq 2C + C'_{-2}$, where C'_{-2} is a moving curve on F. When $|C'_{-2}|$ and $|C|$ are composed of the same pencil, we have $\beta \geq \frac{4}{7}$ by [CHP21, Prop 3.5]. Then we have $(\pi^*(K_X)|_F)^2 \geq \frac{4}{7} > \frac{1}{2}$, which means $\varphi_{6,X}$ is birational by Lemma 4.6, a contradiction.

When $|C'_{-2}|$ and $|C|$ are not composed of the same pencil, we have $(C'_{-2} \cdot C) \geq 2$. Theorem 3.2(*2.ii*) implies that $\varphi_{6,X}$ is birational ($m_1 = 6$, $j_1 = 1$, $j_2 = 2$, $\delta_2 = 2$, $\xi = 1$, $\mu = 1$), which contradicts our assumption.

In a word, subcase i.a does not occur.

Subcase i.b. $u_{6,-1} \leq 6$, $u_{6,-3} \leq 4$, $u_{6,-4} \leq 3$, $u_{6,-5} \leq 2$, $u_{6,-6} = 1$ and $\dim \psi_{6,0}(U_{6,0}) = 4$.

By our assumption, we have $h^0(M_6 - 6F) \geq P_6(X) - \sum_{i=0}^{5} u_{6,-i} \geq 2$. Since $u_{6,-6} = 1$, $|M_6 - 6F|$ and f are composed with the same pencil. Thus we have $M_6 \geq 7F$. By (2.4), $\beta \geq \frac{7}{13}$ and $(\pi^*(K_X)|_F)^2 > \frac{1}{2}$, a contradiction by Lemma 4.6. Hence this subcase does not occur either.

Next we treat the case when $\dim \psi_{6,0}(U_{6,0}) \leq 3$. The map $\varphi_{6,X}$ is generically finite, since $\alpha(6) \geq \left(6-1-1-\frac{1}{\beta}\right)\cdot\xi > 1$, so $\dim \psi_{6,0}(U_{6,0}) \geq 3$. We assume that $\dim \psi_{6,0}(U_{6,0}) = 3$ from here on. We have $M_6|_F \geq C + C_{-1}$, where C_{-1} is a moving curve on F satisfying $h^0(F, C_{-1}) \geq 6$.

If $(C_{-1} \cdot C) \leq 3$, we have $C_{-1} \geq 2C + C_{-2}$, where C_{-2} is a moving curve on F. If $|C_{-2}|$ and $|C|$ are not composed of the same pencil, we have $(C_{-2} \cdot C) \geq 2$. By [CHP21, Prop 3.7(iii)], we have $\xi \geq \frac{4}{5}$ ($m_1 = 6$, $j = 3$, $\delta_1 = 2$, $\mu = 1$, $\xi \geq \frac{2}{3}$, $n = 4$). [CHP21, Prop 3.7(i)] implies that $\varphi_{6,X}$ is birational ($m_1 = 6$, $j = 3$, $\delta_1 = 2$, $\mu = 1$, $\xi \geq \frac{4}{5}$), which contradicts our assumption. Thus $|C_{-2}|$ and $|C|$ are composed of the same pencil. We get $M_6|_F \geq 4C$. In particular, we have $\beta \geq \frac{2}{3}$. Since $\alpha(7) > 2$, we have $\xi \geq \frac{5}{7}$. We have $\alpha(5) \geq \frac{15}{14} > 1$. So $\xi \geq \frac{4}{5}$. We can get $\alpha(6) > 2$ when $\xi > \frac{4}{5}$. Thus we have $\xi = \frac{4}{5}$.

If $(C_{-1} \cdot C) \geq 4$, [CHP21, Prop 3.7(ii)] implies that $\varphi_{6,X}$ is birational whenever $\beta > \frac{1}{2}$. Thus we need to study the situation with $\beta = \frac{1}{2}$. Taking $n = 9$, 10, 11, 12, 13, respectively, and run [CHP21, Prop 3.7(iii)], one finally gets $\xi \geq \frac{6}{7}$. So we will work under the constraints: $\xi \geq \frac{6}{7}$ and $\beta = \frac{1}{2}$, throughout the rest of this case.

Subcase i.c. $u_{6,-1} = 7$, $\dim \psi_{6,0}(U_{6,0}) \leq 3$.

Clearly, one has $\dim \psi_{6,0}(U_{6,-1}) \leq 3$, which is parallel to the second part of subcase i.a. We have $S_{6,-1}|_F \geq C + C'_{-1}$, where C'_{-1} is a moving curve on F satisfying $h^0(F, C'_{-1}) \geq 4$. When $(C'_{-1} \cdot C) \geq 4$, then $\varphi_{6,X}$ is birational by Theorem 3.2(*1.ii*) ($m_1 = 6$, $j_1 = j_2 = 1$, $\delta_2 = 4$, $\beta = \frac{1}{2}$, $\mu = 1$), which is a contradiction. So we have $(C'_{-1} \cdot C) \leq 3$, which implies that $S_{6,-1}|_F \geq 2C + C'_{-2}$, where C'_{-2} is a moving curve on F. When $|C'_{-2}|$ and $|C|$ are composed of the same pencil, we have $\beta \geq \frac{4}{7}$ by [CHP21, Prop 3.5], a contradiction to our assumption $\beta = \frac{1}{2}$. When $|C'_{-2}|$ and $|C|$ are not composed of the same pencil, we have $(C'_{-2} \cdot C) \geq 2$. Theorem 3.2(*2.ii*) implies that $\varphi_{6,X}$ is birational ($m_1 = 6$, $j_1 = 1$, $j_2 = 2$, $\delta_2 = 2$, $\xi = \frac{6}{7}$, $\mu = 1$), which contradicts our assumption.

Subcase i.d. $u_{6,-1} \leq 6$, $u_{6,-3} = 4$.

If $(S_{6,-3}|_F \cdot C) \geq 4$, Theorem 3.1(*1*) implies that $\varphi_{6,X}$ is birational ($m_1 = 6$, $j = 3$, $\beta = \frac{1}{2}$, $\tilde{\delta} = 4$, $\xi = \frac{6}{7}$), which contradicts our assumption.

If $(S_{6,-3}|_F \cdot C) \leq 3$, we have $S_{6,-3}|_F \geq C + C'$, where C' is a moving curve on F. When $|C'|$ and $|C|$ are composed of the same pencil, we have $\beta \geq \frac{5}{9} > \frac{1}{2}$ by [CHP21, Prop 3.5], which contradicts assumption. Then $|C'|$ and $|C|$ are not composed of the same pencil, we get $(C' \cdot C) \geq 2$. Then $\varphi_{6,X}$ is birational by Theorem 3.2(*1.ii*) ($m_1 = 6$, $j_1 = 3$, $j_2 = 1$, $\delta_2 = 2$, $\beta = \frac{1}{2}$, $\xi \geq \frac{6}{7}$), which contradicts our assumption.

Subcase i.e. $u_{6,-1} \leq 6$, $u_{6,-3} \leq 3$, $u_{6,-5} = 2$.

The assumption $\beta = \frac{1}{2}$ implies that $|S_{6,-5}|_F|$ and $|C|$ are not composed of the same pencil. Then $(S_{6,-5}|_F \cdot C) \geq 2$. Theorem 3.1(*1'*) implies that $\varphi_{6,X}$ is birational ($m_1 = 6$, $j = 5$, $\tilde{\delta} = 2$, $\xi = \frac{6}{7}$), which contradicts our assumption. Thus this subcase does not occur.

Subcase i.f. $u_{6,-1} \leq 6$, $u_{6,-3} \leq 3$, $u_{6,-5} = 1$, $P_6(X) \geq 30$.

We have $h^0(M_6 - 5F) \geq P_6(X) - \sum_{i=0}^{4} u_{6,-i} \geq 3$. Since $u_{6,-5} = 1$, $|M_6 - 5F|$ and f are composed with the same pencil. Thus we have $\mu \geq \frac{7}{6}$ and $\beta \geq \frac{7}{13} > \frac{1}{2}$ by (2.4), which contradicts our assumption.

Case ii. $u_{6,0} \leq 8$, $u_{6,-1} \leq 7$, $u_{6,-2} = 6$, $u_{6,-3} \leq 4$, $u_{6,-4} \leq 3$, $u_{6,-5} \leq 2$, $u_{6,-6} = 1$.

If $(S_{6,-2}|_F \cdot C) \leq 3$, then $S_{6,-2}|_F \geq 2C + C_{-2}$, where C_{-2} is a moving curve on F. When $|C_{-2}|$ and $|C|$ are composed of the same pencil, $\xi \geq \frac{4}{5}$

by Theorem 3.4(*2.i*) ($n = 4$, $m_1 = 6$, $j_1 = 2$, $j_2 = 3$, $\xi = \frac{2}{3}$, $\mu = 1$). Thus $\varphi_{6,X}$ is birational by Theorem 3.4(*2.ii*) ($m_1 = 6$, $j_1 = 2$, $j_2 = 3$, $\xi = \frac{4}{5}$, $\mu = 1$), a contradiction. When $|C_{-2}|$ and $|C|$ are not composed of the same pencil, $(C_{-2} \cdot C) \geq 2$. By Theorem 3.2(*1.ii*), $\varphi_{6,X}$ is birational ($m_1 = 6$, $j_1 = 2$, $j_2 = 2$, $\xi = \frac{2}{3}$, $\delta_2 = 2$, $\beta = \frac{1}{2}$), a contradiction.

Thus $(S_{6,-2}|_F \cdot C) \geq 4$. Theorem 3.1(*1–2*) imply that $\varphi_{6,X}$ is birational if $\beta > \frac{1}{2}$ ($m_1 = 6$, $j = 2$, $\tilde{\delta} \geq 4$, $\xi = \frac{2}{3}$, $\mu = 1$). So we have $\beta = \frac{1}{2}$ by our assumption. But our assumption in this case gives $h^0(M_6 - 6F) \geq 2$. Since $u_{6,-6} = 1$, we have $\beta \geq \frac{7}{13}$, which is a contradiction.

Case iii. $u_{6,0} \leq 8$, $u_{6,-1} \leq 7$, $u_{6,-2} \leq 5$, $u_{6,-3} \leq 4$, $u_{6,-4} \leq 3$, $u_{6,-5} \leq 2$, $u_{6,-6} = 1$.

We have $h^0(M_6 - 6F) \geq P_6(X) - \sum_{i=0}^{5} u_{6,-i} \geq 3$ since $P_6(X) \geq 32$. As $u_{6,-6} = 1$, $|M_6 - 6F|$ and f are composed with the same pencil. Thus $\mu \geq \frac{4}{3}$. Then one gets $\beta \geq \frac{4}{7}$ by (2.4). We have $\alpha(7) \geq \frac{7}{3} > 2$, so $\xi \geq \frac{5}{7}$. Since $\alpha(5) > 1$, we have $\xi \geq \frac{4}{5}$. When $\xi > \frac{4}{5}$, we have $\alpha(6) > 2$, which implies that $\varphi_{6,X}$ is birational. So the only possibility is $\xi = \frac{4}{5}$. □

4.4 Estimation of the Canonical Volume

We go on working under the same assumption as that of Lemma 4.5.

Lemma 4.17. *Let $\pi\colon X' \to X$ be any birational morphism where X' is nonsingular and projective. Assume that $|M|$ is any base point free linear system on X'. Denote by S a general member of $|M|$. Then*

$$\left((\pi^*(K_X)|_S)^2\right)^2 \geq K_X^3 \cdot (\pi^* K_X|_S \cdot S|_S).$$

Proof Take a sufficiently large and divisible integer m such that the linear system $|\pi^*(mK_X)|$ is base point free. Denote by S_m a general member of $|\pi^*(mK_X)|$. By Bertini's theorem, S_m is a smooth projective surface of general type. On the surface S_m, by the Hodge index theorem,

$$(\pi^*(K_X)|_{S_m} \cdot S|_{S_m})^2 \geq (\pi^*(K_X)|_{S_m})^2 \cdot (S|_{S_m})^2. \qquad \square$$

Let $0 \leq l \leq m$ be integers. Assume that $h^0(M_m - lF) \geq 2$. Denote the moving part of $|M_m - lF|$ by $|S_{m,-l}|$. Modulo further blowups, we may assume that $|S_{m,-l}|$ is base point free. Multiplying the inequality $M_m \geq jF + S_{m,-l}$ with $\pi^*(K_X)^2$ and applying Lemma 4.17 gives

$$\begin{aligned} K_X^3 &\geq \tfrac{1}{m}\left(l(\pi^*(K_X)|_F)^2 + \sqrt{K_X^3 \cdot (\pi^*(K_X)|_{S_{m,-l}} \cdot S_{m,-l}|_{S_{m,-l}})}\right) \\ &\geq \tfrac{1}{m}\left(l(\pi^*(K_X)|_F)^2 + \sqrt{(m-l)K_X^3 \cdot (\pi^*(K_X)|_F \cdot S_{m,-l}|_F)}\right). \end{aligned} \tag{4.7}$$

For the last inequality, note that $M_m \geq mF$, implying $S_{m,-l} \geq (m-l)F$.

Proposition 4.18. *Keep the same assumption as that of Lemma 4.5. Suppose that $\varphi_{6,X}$ is not birational, $\xi \neq \frac{2}{3}$ and $\xi \neq \frac{4}{5}$. Then:*

(1) $K_X^3 \geq \frac{5}{14}$;
(2) when $P_6(X) \geq 26$, $K_X^3 \geq \frac{11}{28}$;
(3) when $P_6(X) \geq 27$, $K_X^3 > 0.4328$;
(4) when $P_6(X) \geq 28$, $K_X^3 > 0.4714$;
(5) when $P_6(X) \geq 31$, $K_X^3 \geq \frac{8}{15}$.

Proof Since $\alpha(7) \geq (7-1-1-\frac{1}{\beta}) \cdot \xi > 2$, we have $\xi \geq \frac{5}{7}$. (*1*) follows from (2.3) with $\beta \geq \frac{1}{2}$ and $\xi \geq \frac{5}{7}$. By the arguments in cases 1–9 in the proof of Proposition 4.16, one of the following holds:

Case I. $u_{6,0} = 10$, $\dim \psi_{6,0}(U_{6,0}) = 4$ and $P_6(X) \leq 30$ ($\Rightarrow K_X^3 \geq \frac{1}{2}$).

We have $(M_6|_F \cdot C) \geq 5$ by Riemann–Roch. If $(M_6|_F \cdot C) = 5$, $|M_6||_C$ is a complete linear system whose general member has degree 5, which implies that $\varphi_{6,X}$ is birational, which contradicts our assumption. Thus we have $(M_6|_F \cdot C) = 6$. We get $\xi = 1$. Therefore we have $K_X^3 \geq \frac{1}{2}$.

Case IIa. $u_{6,0} = 9$, $\dim \psi_{6,0}(U_{6,0}) = 4$ ($\Rightarrow K_X^3 \geq \frac{1}{2}$).

As we have seen, $(M_6|_F \cdot C) \geq 5$ and equality implies the birationality of $\varphi_{6,X}$. Hence $(M_6|_F \cdot C) = 6$. In particular, $\xi = 1$. By Lemma 4.6 and the assumption, we have $\beta = \frac{1}{2}$ and $(\pi^*(K_X)|_F)^2 = \frac{1}{2}$. Thus $K_X^3 \geq \frac{1}{2}$.

Claim. We have $u_{6,-1} \leq 6$, $u_{6,-3} \leq 3$, $u_{6,-5} = 1$, $P_6(X) \leq 29$.

$u_{6,-1} \leq 6$ follows from the proof of Proposition 4.16 (see subcase i.a). From the proof of Proposition 4.16, we have $u_{6,-3} \leq 4$ and $u_{6,-5} \leq 2$.

Suppose $u_{6,-3} = 4$. If $(S_{6,-3}|_F \cdot C) \geq 4$, by Theorem 3.1(*3*) ($m_1 = 6$, $j = 3$), we have $\pi^*(K_X)|_F \geq \frac{1}{3}C + \frac{1}{9}S_{6,-3}|_F$. Since $\xi = 1$, we have $(\pi^*(K_X)|_F)^2 \geq \frac{5}{9} > \frac{1}{2}$, which contradicts our assumption. So $(S_{6,-3}|_F \cdot C) \leq 3$, which gives $S_{6,-3}|_F \geq C + C_{-1}$, where C_{-1} is a moving curve on F. Using the same argument as above, we can get $(\pi^*(K_X)|_F)^2 \geq \frac{5}{9} > \frac{1}{2}$, which is a contradiction. So $u_{6,-3} \leq 3$. By a similar argument, we also see that $u_{6,-5} = 1$.

If $P_6(X) \geq 30$, we have $h^0(M_6 - 5F) \geq P_6(X) - \sum_{i=0}^{4} u_{6,-i} \geq 3$. Since $u_{6,-5} = 1$, $|M_6 - 5F|$ and f are composed with the same pencil. Thus $M_6 \geq 7F$, and so $\mu \geq \frac{7}{6}$. By (2.4), we have $\beta \geq \frac{7}{13} > \frac{1}{2}$, a contradiction.

Case IIb. $u_{6,0} = 9$, $\dim \psi_{6,0}(U_{6,0}) \leq 3$ ($\Rightarrow K_X^3 \geq \frac{10}{21}$).

One has $M_6|_F \geq C + C_{-1}$, where C_{-1} is a moving curve on F satisfying $h^0(F, C_{-1}) \geq 6$. By the argument in case i of Proposition 4.16 and the assumption $\xi \neq \frac{4}{5}$, we know $(C_{-1} \cdot C) \geq 4$, $\xi \geq \frac{6}{7}$, $\beta = \frac{1}{2}$, $u_{6,-1} \leq 6$, $u_{6,-3} \leq 3$, $u_{6,-5} = 1$ and $P_6(X) \leq 29$. We have

$$(\pi^*(K_X)|_F)^2 \geq \frac{(\pi^*(K_X)|_F \cdot C) + (\pi^*(K_X)|_F \cdot C_{-1})}{6} \geq \frac{10}{21}.$$

In particular, we have $K_X^3 \geq \frac{10}{21}$.

Case IIIa. $u_{6,0} = 8$, $\dim \psi_{6,0}(U_{6,0}) = 4$ ($\Rightarrow K_X^3 \geq \frac{1}{2}$).

We have $(M_6|_F \cdot C) = 6$ and so $\xi = 1$. By the same argument as in case IIa, we have $u_{6,-1} \leq 6$, $u_{6,-3} \leq 3$, $u_{6,-5} = 1$, $P_6(X) \leq 28$. In particular, we have $K_X^3 \geq \frac{1}{2}$.

Case IIIb. $u_{6,0} = 8$, $\dim \psi_{6,0}(U_{6,0}) \leq 3$ and $M_6|_F \geq C + C_{-1}$ with $(C_{-1} \cdot C) = 4$ ($\Rightarrow K_X^3 \geq \frac{10}{21}$).

By the same argument as in case i of Proposition 4.16 (subcases i.c–f), we know $\xi \geq \frac{6}{7}$, $\beta = \frac{1}{2}$, $u_{6,-1} \leq 6$, $u_{6,-3} \leq 3$, $u_{6,-5} = 1$, $P_6(X) \leq 28$. By the same argument as in case IIb, we have $K_X^3 \geq \frac{10}{21}$.

Case IIIc. $u_{6,0} = 8$, $\dim \psi_{6,0}(U_{6,0}) \leq 3$ and $M_6|_F \geq C + C_{-1}$ with $(C_{-1} \cdot C) \leq 3$ ($\Rightarrow K_X^3 \geq \frac{5}{12}$).

Since $h^0(F, C_{-1}) \geq 5$, we have $C_{-1} \geq C + C'$, where C' is a moving curve on F satisfying $h^0(F, C') \geq 3$. If $|C'|$ and $|C|$ are composed of the same pencil, we have $\beta \geq \frac{2}{3}$. Since $\alpha(7) > 2$, we have $\xi \geq \frac{5}{7}$. We get $\alpha(5) \geq (5 - 1 - 1 - \frac{1}{\beta}) \cdot \xi \geq \frac{15}{14} > 1$. Thus $\xi \geq \frac{4}{5}$. By assumption, we have $\xi > \frac{4}{5}$, which gives $\alpha(6) > 2$. Then $\varphi_{6,X}$ is birational, which contradicts our assumption. So $|C'|$ and $|C|$ are not composed of the same pencil.

Therefore we have $M_6|_F \geq 2C + C'$. By [CHP21, Prop 3.7(iii)], we have $\xi \geq \frac{3}{4}$ ($m_1 = 6$, $j = 2$, $\delta_1 = 2$) by successively taking $n = 6$, 7. Thus

$$(\pi^*(K_X)|_F)^2 \geq \frac{2\xi + 1}{6} \geq \frac{5}{12}.$$

In particular, we have $K_X^3 \geq \frac{5}{12}$.

Case IVa. $u_{6,-1} = 7$, $\dim \psi_{6,0}(U_{6,-1}) \leq 3$ ($\Rightarrow K_X^3 > 0.4714$).

When $(S_{6,-1}|_F \cdot C) \leq 3$, we have $S_{6,-1}|_F \geq 2C + C_{-2}$, where C_{-2} is a moving curve on F satisfying $h^0(F, C_{-2}) \geq 3$. By the same argument as the end of case 3 of Proposition 4.16, $\varphi_{6,X}$ is birational, a contradiction. Thus we only consider the case $(S_{6,-1}|_F \cdot C) \geq 4$. We have $S_{6,-1}|_F \geq C + C_{-1}$, where C_{-1} is a moving curve on F satisfying $h^0(F, C_{-1}) \geq 4$.

If $(C_{-1} \cdot C) \geq 4$, $\varphi_{6,X}$ is birational by Theorem 3.2(*1.ii*) ($m_1 = 6$, $j_1 = j_2 = 1$, $\delta_2 = 4$, $\mu = 1$, $\beta = \frac{1}{2}$), which is a contradiction.

Thus $(C_{-1} \cdot C) \leq 3$. So $C_{-1} \geq C + C'$, where C' is a moving curve on F. When $|C'|$ and $|C|$ are not composed of the same pencil, we have

$(C' \cdot C) \geq 2$. By Theorem 3.2(*2.i*), $\xi \geq \frac{5}{7}$ ($n = 6$, $m_1 = 6$, $j_1 = 1$, $j_2 = 2$, $\mu = 1$, $\xi = \frac{2}{3}$, $\delta_2 = 2$). Theorem 3.2(*2.ii*) implies that $\varphi_{6,X}$ is birational ($m_1 = 6$, $j_1 = 1$, $j_2 = 2$, $\delta_2 = 2$, $\mu = 1$, $\xi = \frac{5}{7}$), a contradiction.

So $C' \sim C$ and so $S_{6,-1}|_F \geq 3C$. By [CHP21, Prop 3.5], $\beta \geq \frac{4}{7}$. Since $\alpha(7) > 2$, we have $\xi \geq \frac{5}{7}$. Since $\alpha(8) \geq (8 - 1 - 1 - \frac{1}{\beta}) \cdot \xi > 3$, we have $\xi \geq \frac{3}{4}$. Denote by $\xi_{6,-1}$ the intersection number $(\pi^*(K_X)|_F \cdot S_{6,-1}|_F)$. We have $\xi_{6,-1} \geq \beta(C \cdot S_{6,-1}|_F) \geq \frac{16}{7}$. By Kawamata–Viehweg vanishing

$$|K_{X'} + M_6||_F \succcurlyeq |K_F + S_{6,-1}|_F| \succcurlyeq |C + S_{6,-1}|_F|,$$

which directly implies $7\pi^*(K_X)|_F \geq C + S_{6,-1}|_F$ since $|C + S_{6,-1}|_F|$ is base point free. Noting that a general $S_{6,-1}|_F$ can be smooth, nef and big, we may use the the similar method to that of [CHP21, Prop 3.6(2.1)] to obtain the following inequality, for any $n \geq 8$,

$$(n+1)\xi_{6,-1} \geq \lceil (n-6)\xi_{6,-1} \rceil + 16,$$

where one notes that $((K_F + S_{6,-1}|_F) \cdot S_{6,-1}|_F) \geq 16$. Take $n = 8$, we get $\xi_{6,-1} \geq \frac{7}{3}$. Take $n = 10$, we get $\xi_{6,-1} \geq \frac{26}{11}$. Take $n = 9$, we get $\xi_{6,-1} \geq \frac{12}{5}$. Since $7\pi^*(K_X)|_F \geq C + S_{6,-1}|_F$ and $\xi \geq \frac{3}{4}$, we have $(\pi^*(K_X)|_F)^2 \geq \frac{63}{140}$. By (4.7), we have $K_X^3 > 0.4714$.

Case IVb. $u_{6,-1} = 6$ and $\dim \psi_{6,0}(U_{6,-1}) \leq 3$ ($\Rightarrow K_X^3 \geq \frac{11}{28}$).

If $(S_{6,-1}|_F \cdot C) \geq 4$, we have $\xi \geq \frac{3}{4}$ by [CHP21, Prop 3.6(1.1)] ($n = 11$, $\delta = 4$, $m_1 = 6$, $\beta = \frac{1}{2}$, $\xi = \frac{2}{3}$). As in case IVa, $7\pi^*(K_X)|_F \geq C + S_{6,-1}|_F$. Thus we get $K_X^3 \geq (\pi^*(K_X)|_F)^2 \geq \frac{11}{28}$.

If $(S_{6,-1}|_F \cdot C) \leq 3$, we have $S_{6,-1}|_F \geq 2C + C_{-2}$, where C_{-2} is a moving curve on F. When $|C_{-2}|$ and $|C|$ are composed of the same pencil, we have $\beta \geq \frac{4}{7}$ by [CHP21, Prop 3.5]. Besides, $\alpha(7) > 2$ implies $\xi \geq \frac{5}{7}$. Thus we have $(\pi^*(K_X)|_F)^2 \geq \frac{20}{49}$. When $|C_{-2}|$ and $|C|$ are not composed of the same pencil, we have $(C_{-2} \cdot C) \geq 2$. By Theorem 3.2(*2.i*), we have $\xi \geq \frac{5}{7}$ ($n = 6$, $m_1 = 6$, $j_1 = 1$, $j_2 = 2$, $\mu = 1$, $\xi = \frac{2}{3}$, $\delta_2 = 2$). Theorem 3.2(*2.ii*) implies that $\varphi_{6,X}$ is birational ($n = 6$, $m_1 = 6$, $j_1 = 1$, $j_2 = 2$, $\mu = 1$, $\xi = \frac{5}{7}$, $\delta_2 = 2$), which contradicts our assumption.

Case V. $u_{6,-2} = 6$ ($\Rightarrow K_X^3 > 0.4771$).

If $(S_{6,-2}|_F \cdot C) \leq 3$, we have $S_{6,-2}|_F \geq 2C + C_{-2}$, where C_{-2} is a moving curve on F. By the same argument as in the last part of case 4 of Proposition 4.16, we conclude that $\varphi_{6,X}$ is birational, a contradiction.

So we have $(S_{6,-2} \cdot C) \geq 4$ in this case. In fact, the case $(S_{6,-2} \cdot C) \geq 5$ has been treated in case 4 of Proposition 4.16, which shows that $\varphi_{6,X}$ is birational (a contradiction). Hence $(S_{6,-2}|_F \cdot C) = 4$. Theorem 3.1(*1*) implies that $\varphi_{6,X}$ is birational if $\beta > \frac{1}{2}$. Thus we have $\beta = \frac{1}{2}$. By Theorem

3.1(*3–4*), we have $\xi \geq \frac{3}{4}$ and, for any $n \geq 6$,

$$(n+1)\xi \geq \lceil (n-5)\xi \rceil + 4.$$

Take $n=8$, we get $\xi \geq \frac{7}{9}$. Take $n=9$, we get $\xi \geq \frac{4}{5}$. By our assumption, we have $\xi > \frac{4}{5}$. Take $n=10$ in the above inequality, $\xi \geq \frac{9}{11}$. By Theorem 3.1(*3*) ($m_1=6$, $j=2$, $(S_{6,-2}|_F \cdot C)=4$), we have $(\pi^*(K_X)|_F)^2 \geq \frac{5}{11}$. By (4.7), we have $K_X^3 > 0.4771$.

Case VI. $u_{6,-3}=4$ ($\Rightarrow K_X^3 > 0.4734$).

If $(S_{6,-3}|_F \cdot C) \geq 4$, $\varphi_{6,X}$ is birational by Theorem 3.1(*1*) ($m_1=6$, $j=3$, $\tilde{\delta}=4$, $\beta=\frac{1}{2}$, $\xi=\frac{5}{7}$), a contradiction. Thus $(S_{6,-3}|_F \cdot C) \leq 3$. So $S_{6,-3}|_F \geq C + C_{-1}$, where C_{-1} is a moving curve on F.

If $|C_{-1}|$ and $|C|$ are not composed of the same pencil, $(C_{-1} \cdot C) \geq 2$. By Theorem 3.2(*1.i*), we have $\xi \geq \frac{4}{5}$ ($n=4$, $m_1=6$, $j_1=3$, $j_2=1$, $\beta=\frac{1}{2}$, $\delta_2=2$). By assumption, $\xi > \frac{4}{5}$. Theorem 3.2(*1.ii*) implies that $\varphi_{6,X}$ is birational ($m_1=6$, $j_1=3$, $j_2=1$, $\xi > \frac{4}{5}$, $\beta=\frac{1}{2}$), a contradiction.

Thus $|C_{-1}|$ and $|C|$ are composed of the same pencil. So $S_{6,-3}|_F \geq 2C$. By [CHP21, Prop 3.5], we have $\beta \geq \frac{5}{9}$. By Theorem 3.4(*1.i*), we have $\xi \geq \frac{4}{5}$ ($n=4$, $m_1=6$, $j_1=3$, $j_2=2$, $\beta=\frac{5}{9}$). So $(\pi^*(K_X)|_F)^2 \geq \frac{4}{9}$. By (4.7), we have $K_X^3 > 0.4734$.

Case VII. $u_{6,-5}=2$ ($\Rightarrow K_X^3 > 0.4362$).

If $|S_{6,-5}|_F|$ and $|C|$ are not composed of the same pencil, we have $(S_{6,-5}|_F \cdot C) \geq 2$. By Theorem 3.1(*4*), we get $\xi \geq \frac{5}{7}$ by taking $n=6$ and $\xi \geq \frac{3}{4}$ by taking $n=7$. By Theorem 3.1(*3*), we have $(\pi^*(K_X)|_F)^2 \geq \frac{19}{44}$. By (4.7), we have $K_X^3 > 0.4746$.

If $|S_{6,-5}|_F|$ and $|C|$ are composed of the same pencil, we have $\beta \geq \frac{6}{11}$. By Theorem 3.4(*1.i*), we get $\xi \geq \frac{5}{7}$ by taking $n=6$ and $\xi \geq \frac{3}{4}$ by taking $n=7$. We have $(\pi^*(K_X)|_F)^2 \geq \frac{9}{22}$. By (4.7), we obtain $K_X^3 > 0.4362$.

Now we prove (*2*). By the results of cases I–VII, we only need to consider the case where $u_{6,0} \leq 7$, $u_{6,-1} \leq 5$, $u_{6,-3} \leq 3$, $u_{6,-5}=1$. We have $h^0(M_6-5F) \geq P_6(X) - \sum_{i=0}^{4} u_{6,-i} \geq 3$. Since $u_{6,-5}=1$, $|M_6-5F|$ and f are composed with the same pencil. We have $\mu \geq \frac{7}{6}$ and $\beta \geq \frac{7}{13}$. As $\alpha(7) > 2$, we have $\xi \geq \frac{5}{7}$. So we get $K_X^3 \geq \frac{35}{78} > \frac{11}{28}$.

For (*3*), by cases I–VII, we are left to treat the following:

(3.i) $u_{6,0}=8$, $\xi \geq \frac{3}{4}$, $(\pi^*(K_X)|_F)^2 \geq \frac{5}{12}$, $u_{6,-4}=3$;

(3.ii) $u_{6,0}=8$, $\xi \geq \frac{3}{4}$, $(\pi^*(K_X)|_F)^2 \geq \frac{5}{12}$, $u_{6,-1} \leq 6$, $u_{6,-2} \leq 5$, $u_{6,-3} \leq 3$, $u_{6,-4} \leq 2$, $u_{6,-5}=1$;

(3.iii) $u_{6,0} \leq 7$, $u_{6,-1} \leq 6$, $u_{6,-2} \leq 5$, $u_{6,-3} \leq 3$, $u_{6,-4} \leq 2$, $u_{6,-5}=1$;

(3.iv) $u_{6,0} \leq 7$, $u_{6,-1} \leq 6$, $u_{6,-2} \leq 5$, $u_{6,-3} \leq 3$, $u_{6,-4}=3$, $u_{6,-5}=1$.

We first treat (*3.i*). If $|S_{6,-4}|_F|$ and $|C|$ are composed of the same pencil, we have $\beta \geq \frac{3}{5}$. We get $K_X^3 \geq \beta\xi \geq \frac{9}{20} > 0.4328$. If $|S_{6,-4}|_F|$ and $|C|$ are not composed of the same pencil, we have $(\pi^*(K_X)|_F \cdot S_{6,-4}|_F) \geq 1$. We have $K_X^3 > 0.4328$ by (4.7). Next we treat (*3.ii*). Similar as in the proof of (2), we have $\mu \geq \frac{7}{6}$. By (2.4), we have $\beta \geq \frac{7}{13}$. Then we get $K_X^3 \geq \frac{49}{104} > 0.4328$. For (*3.iii*), we have $\mu \geq \frac{4}{3}$ and $\beta \geq \frac{4}{7}$. Since $\alpha(5) > 1$, we get $\xi \geq \frac{4}{5}$. Our assumption implies that we have $\xi > \frac{4}{5}$. Thus we have $\alpha(6) > 2$, which implies that $\varphi_{6,X}$ is birational, which is a contradiction. So (*3.iii*) does not occur. Finally, for (*3.iv*), we have $\mu \geq \frac{7}{6}$ and $\beta \geq \frac{7}{13}$. Since we have $\xi \geq \frac{5}{7}$, so we get $K_X^3 \geq \frac{35}{78} > 0.4328$.

For (*4*), by cases I–VII, we only need to treat the following:

(4.i) $u_{6,0} = 8,\ u_{6,-1} \leq 6,\ u_{6,-2} = 5,\ u_{6,-3} \leq 3,\ u_{6,-5} = 2;$
(4.ii) $u_{6,0} \leq 8,\ u_{6,-1} \leq 6,\ u_{6,-2} \leq 4,\ u_{6,-3} \leq 3,\ u_{6,-5} = 2;$
(4.iii) $u_{6,0} \leq 8,\ u_{6,-1} \leq 6,\ u_{6,-2} = 5,\ u_{6,-3} \leq 3,\ u_{6,-5} = 1;$
(4.iv) $u_{6,0} \leq 8,\ u_{6,-1} \leq 6,\ u_{6,-2} \leq 4,\ u_{6,-3} \leq 3,\ u_{6,-5} = 1;$
(4.v) $u_{6,0} \leq 7,\ u_{6,-1} \leq 6,\ u_{6,-2} = 5,\ u_{6,-3} \leq 3,\ u_{6,-5} = 2.$

For (*4.i*), by the argument in case VII we need to treat the case when $|S_{6,-5}|_F|$ and $|C|$ are composed of the same pencil. In particular, $\beta \geq \frac{6}{11}$.

We claim that $(S_{6,-2}|_F \cdot C) \leq 3$. Otherwise, Theorem 3.1(*1–2*) imply that $\varphi_{6,X}$ is birational ($m_1 = 6$, $j = 2$, $\tilde{\delta} \geq 4$, $\xi \geq \frac{2}{3}$, $\beta \geq \frac{6}{11}$), which contradicts our assumption. Thus we have $S_{6,-2}|_F \geq C + C_{-1}$, where C_{-1} is a moving curve on F satisfying $h^0(F, C_{-1}) \geq 3$.

If $|C_{-1}|$ and $|C|$ are composed of the same pencil, $S_{6,-2}|_F \geq 3C$. By Theorem 3.4(*2.i*), we have $\xi \geq \frac{4}{5}$ ($n = 4$). Theorem 3.4(*2.ii*) implies that $\varphi_{6,X}$ is birational ($m_1 = 6$, $j_1 = 2$, $j_2 = 3$, $\xi = \frac{4}{5}$, $\mu = 1$), which contradicts our assumption. Thus $|C_{-1}|$ and $|C|$ are not composed of the same pencil. In particular, we have $(C_{-1} \cdot C) \geq 2$. By Theorem 3.2(*1.i*), we have $\xi \geq \frac{4}{5}$ ($n = 4$, $m_1 = 6$, $j_1 = 2$, $j_2 = 1$, $\beta = \frac{6}{11}$, $\xi = \frac{5}{7}$, $\delta_2 = 2$) .

By the arguments in cases IIIa–IIIc, we have $M_6|_F \geq 2C + C'$, where C' is a moving curve on F satisfying $h^0(F, C') \geq 3$. Moreover, $|C'|$ and $|C|$ are not composed of the same pencil. Thus

$$(\pi^*(K_X)|_F)^2 \geq \frac{2\xi + \beta(C \cdot C')}{6} \geq \frac{74}{165}.$$

By (4.7) ($m = 6$, $l = 5$), we have $K_X^3 > 0.4766$.

For (*4.ii*), by the argument in case VII we only need to consider the case when $|S_{6,-5}|_F|$ and $|C|$ are composed of the same pencil. By [CHP21, Prop 3.5], we have $\beta \geq \frac{6}{11}$. Since $P_6(X) \geq 28$, we have $\mu \geq \frac{7}{6}$ by our assumption. Note that we have $\xi \geq \frac{5}{7}$. Since $\alpha(8) > 3$, we get $\xi \geq \frac{3}{4}$. Thus we have $K_X^3 \geq \mu\beta\xi > 0.4772$.

For (*4.iii*), we have $h^0(M_6 - 5F) \geq P_6(X) - \sum_{i=0}^{4} u_{6,-i} \geq 3$. Since $u_{6,-5} = 1$, $|M_6 - 5F|$ and f are composed with the same pencil. We deduce that $M_6 \geq 7F$, which implies that $\mu \geq \frac{7}{6}$. We have $\mu \geq \frac{7}{6}$ by our assumption. By (2.4), we have $\beta \geq \frac{7}{13}$. Since $u_{6,-2} = 5$, by the same argument as in (*4.i*), we get $\xi \geq \frac{4}{5}$. Hence $K_X^3 \geq \mu\beta\xi \geq \frac{98}{195} > \frac{1}{2}$.

For (*4.v*), we have $h^0(M_6 - 6F) \geq P_6(X) - \sum_{i=0}^{5} u_{6,-i} \geq 2$ by our assumption. If $u_{6,-6} = 1$, by the same argument as in the proof of (*4.iii*), again (*4*) holds. The case when $u_{6,-6} = 2$ remains. If $|S_{6,-6}|_F|$ and C are composed with the same pencil, we have $\beta \geq \frac{7}{12}$ by [CHP21, Prop 3.5]. Thus we have $(\pi^*K_X|_F)^2 \geq \beta \cdot \xi \geq \frac{7}{18}$ by Lemma 4.5. Thus

$$K_X^3 \geq (\pi^*K_X|_F)^2 + \tfrac{1}{6}(\pi^*K_X|_F \cdot S_{6,-6}|_F) \geq \tfrac{7}{18} + \tfrac{1}{6}\xi \geq \tfrac{1}{2},$$

which again implies (*4*). If $|S_{6,-6}|_F|$ and C are not composed with the same pencil, we have $(\pi^*K_X|_F \cdot S_{6,-6}|_F) \geq \frac{1}{2}(C \cdot S_{6,-6}|_F) \geq 1$. By Theorem 3.1(*3*), $(\pi^*K_X|_F)^2 \geq \frac{1}{2}\xi + \frac{1}{12}(\pi^*K_X|_F \cdot S_{6,-6}|_F) \geq \frac{5}{12}$. So

$$K_X^3 \geq (\pi^*K_X|_F)^2 + \tfrac{1}{6}(\pi^*K_X|_F \cdot S_{6,-6}|_F) \geq \tfrac{5}{12} + \tfrac{1}{6} \geq \tfrac{7}{12}.$$

For (*4.iv*), we have $\mu \geq \frac{4}{3}$ by assumption. By (2.4), $\beta \geq \frac{4}{7}$. Since $\alpha(5) > 1$, we have $\xi \geq \frac{4}{5}$. By assumption, $\xi > \frac{4}{5}$. Thus $\alpha(6) > 2$, which implies that $\varphi_{6,X}$ is birational, a contradiction. So (*4.iv*) does not occur.

For (*5*), by case V ($\Rightarrow \beta = \frac{1}{2}$) and case VI ($\Rightarrow \beta \geq \frac{5}{9}$), we see that $u_{6,-2} = 6$ and $u_{6,-3} = 4$ cannot hold simultaneously. Combining all the arguments in cases I–VII, we only need to consider the following:

(5.i) $u_{6,0} \leq 8$, $u_{6,-1} = 7$, $\beta \geq \frac{4}{7}$, $u_{6,-2} \leq 5$, $u_{6,-3} = 4$, $u_{6,-4} \leq 3$, $u_{6,-5} \leq 2$, $u_{6,-6} = 1$;

(5.ii) $u_{6,0} \leq 8$, $u_{6,-1} = 7$, $\beta \geq \frac{4}{7}$, $u_{6,-2} \leq 5$, $u_{6,-3} \leq 3$, $u_{6,-4} \leq 3$, $u_{6,-5} \leq 2$, $u_{6,-6} = 1$;

(5.iii) $u_{6,0} \leq 8$, $u_{6,-1} \leq 6$, $u_{6,-2} = 6$, $u_{6,-3} \leq 3$, $u_{6,-4} \leq 3$, $u_{6,-5} \leq 2$, $u_{6,-6} = 1$;

(5.iv) $u_{6,0} \leq 8$, $u_{6,-1} \leq 6$, $u_{6,-2} \leq 5$, $u_{6,-3} = 4$, $u_{6,-4} \leq 3$, $u_{6,-5} \leq 2$, $u_{6,-6} = 1$.

For (*5.i*), by the argument in case VI, we have $\xi \geq \frac{4}{5}$. By our assumption, we have $h^0(M_6 - 6F) \geq P_6(X) - \sum_{i=0}^{5} u_{6,-i} \geq 2$. Since $u_{6,-6} = 1$, we have $M_6 \geq 7F$, which implies that $\mu \geq \frac{7}{6}$. Thus we have

$$K_X^3 \geq \mu\beta\xi \geq \tfrac{8}{15}.$$

For (*5.ii*), as in the proof of (*5.i*), we have $M_6 \geq 8F$, which implies that $\mu \geq \frac{4}{3}$. As in the case (*4.iv*), we see that $\alpha(5) > 1$ and $\alpha(6) > 2$, which implies the birationality of $\varphi_{6,X}$, a contradiction.

Cases (*5.iii*) and (*5.iv*) are similar to (*5.ii*). One has $\mu \geq \frac{4}{3}$, which gives a contradiction. □

4.5 The Classification of $\mathbb{B}^{(5)}(X)$

Lemma 4.19. *Let X be a minimal projective 3-fold of general type with $p_g(X) = 2$, $d_1 = 1$, $\Gamma \cong \mathbb{P}^1$. Assume that F is a $(1,2)$-surface. Then*

(1) $P_3(X) \geq P_2(X) + 2$;
(2) $P_4(X) \geq P_3(X) + 4$;
(3) $P_5(X) \geq P_4(X) + 4$;
(4) $P_6(X) \geq P_5(X) + 7$.

Proof As $|3K_{X'}||_F \succcurlyeq |K_{X'} + F_1 + F|$ for two distinct general fibers of f, it follows from [Che04, Lemma 4.6] that $|M_3||_F \succcurlyeq |C|$, which implies (*1*). Since $|4K_{X'}| \succcurlyeq |2(K_{X'} + F)|$ for a general fiber of f, by Theorem 2.3, $|M_4||_F \succcurlyeq |2\sigma^*(K_{F_0})|$, which implies (*2*). Note that we have $|M_5||_F \succcurlyeq |M_4||_F$. Thus $P_5(X) \geq P_4(X) + 4$, hence (*3*). As in (*2*), $|M_6||_F \succcurlyeq \mathrm{Mov}\,|\sigma^*(3K_{F_0})|$. $\alpha(6) > 1$ implies that $(M_6 \cdot C) \geq 4$. Thus $|M_6||_F \neq \mathrm{Mov}\,|\sigma^*(3K_{F_0})|$, which implies (*4*). □

Collecting all the results above, we must classify those baskets $\mathbb{B}(X)$ which satisfy the following properties:

(1) $3 \leq P_2(X) \leq 4$;
(2) $P_2(X) + 2 \leq P_3(X) \leq 8$;
(3) $P_3(X) + 4 \leq P_4(X) \leq 13$;
(4) $P_4(X) + 4 \leq P_5(X) \leq 21$;
(5) $P_5(X) + 7 \leq P_6(X) \leq 31$;
(6) $\chi(\mathcal{O}_X) = 0, -1$;
(7) $K_X^3 \geq \frac{5}{14}$;
(8) If $P_6(X) \geq 26$, then $K_X^3 \geq \frac{11}{28}$;
(9) If $P_6(X) \geq 27$, we have $K_X^3 > 0.4328$;
(10) If $P_6(X) \geq 28$, we have $K_X^3 > 0.4714$;
(11) If $P_6(X) \geq 31$, we have $K_X^3 \geq \frac{8}{15}$.

Remark 4.20. If we only use (*7*), there will be a lot of baskets. So it is necessary to use (*8–11*) to rule out many baskets.

This situation naturally fits into the hypothesis of [CC10b, (3.8)] from which we can list all the possibilities for $B^{(5)}(X)$. To be precise,

$$B^{(5)} = \{n^5_{1,2}\times(1,2), n^5_{2,5}\times(2,5), n^5_{1,3}\times(1,3), n^5_{1,4}\times(1,4), n^5_{1,5}\times(1,5), \ldots\}$$

where

$$\begin{aligned}
n^5_{1,2} &= 3\chi(\mathcal{O}_X) + 6P_2 - 3P_3 + P_4 - 2P_5 + P_6 + \sigma_5,\\
n^5_{2,5} &= 2\chi(\mathcal{O}_X) - P_3 + 2P_5 - P_6 - \sigma_5,\\
n^5_{1,3} &= 2\chi(\mathcal{O}_X) + 2P_2 + 3P_3 - 3P_4 - P_5 + P_6 + \sigma_5,\\
n^5_{1,4} &= \chi(\mathcal{O}_X) - 3P_2 + P_3 + 2P_4 - P_5 - \sigma_5,\\
n^5_{1,r} &= n^0_{1,r}, r \geq 5,
\end{aligned}$$

and $\sigma_5 = \sum_{r\geq 5} n^0_{1,r} \geq 0$, $\sigma_5 \leq 2\chi(\mathcal{O}_X) - P_3 + 2P_5 - P_6$. Note also that, by our definition, each of the above coefficients satisfies $n^0_{\cdot,\cdot} \geq 0$.

With these constraints, a computer program produced a raw list for $\{B^{(5)}_X, P_2(X), \chi(\mathcal{O}_X)\}$. Taking into account those possible packings, we finally get the list $\mathbb{S}_2$ of 263 elements, which can be found at

http://www.dima.unige.it/~penegini/publ.html

Proof of Theorem 1.2 This follows from Lemma 4.1, Proposition 4.2, Remark 4.3, Lemma 4.11, and Propositions 4.12–4.16 and 4.18. □

Acknowledgements

The first author was supported by National Natural Science Foundation of China (#12071078, #11731004) , Program of Shanghai Subject Chief Scientist (#16XD1400400) and National Key Research and Development Program of China (#2020YFA0713200). The second author was supported by National Researcher Program of National Research Foundation of Korea (Grant No. 2010-0020413). The third author was supported by PRIN 2015 "Geometry of Algebraic Varieties" and by GNSAGA of INdAM.

References

[CC10a] Jungkai A. Chen and Meng Chen. Explicit birational geometry of 3-folds of general type, II. *J. Differential Geom.*, 86(2):237–271, 2010.

[CC10b] Jungkai A. Chen and Meng Chen. Explicit birational geometry of threefolds of general type, I. *Ann. Sci. Éc. Norm. Supér. (4)*, 43(3):365–394, 2010.

[Che03] Meng Chen. Canonical stability of 3-folds of general type with $p_g \geq 3$. *Internat. J. Math.*, 14(5):515–528, 2003.

[Che04] Meng Chen. Inequalities of Noether type for 3-folds of general type. *J. Math. Soc. Japan*, 56(4):1131–1155, 2004.

[Che07] Meng Chen. A sharp lower bound for the canonical volume of 3-folds of general type. *Math. Ann.*, 337(4):887–908, 2007.

[Che14] Meng Chen. Some birationality criteria on 3-folds with $p_g > 1$. *Sci. China Math.*, 57(11):2215–2234, 2014.

[CHP21] Meng Chen, Yong Hu, and Matteo Penegini. On projective threefolds of general type with small positive geometric genus. *Electronic Research Archive*, 29(3):2293–2323, 2021.

[CZ16] Meng Chen and Qi Zhang. Characterization of the 4-canonical birationality of algebraic threefolds, II. *Math. Z.*, 283(3-4):659–677, 2016.

[IF00] A. R. Iano-Fletcher. Working with weighted complete intersections. In *Explicit birational geometry of 3-folds*, volume 281 of *London Math. Soc. Lecture Note Ser.*, pages 101–173. Cambridge Univ. Press, Cambridge, 2000.

[Kaw82] Yujiro Kawamata. A generalization of Kodaira-Ramanujam's vanishing theorem. *Math. Ann.*, 261(1):43–46, 1982.

[Kaw99] Yujiro Kawamata. On the extension problem of pluricanonical forms. *Algebraic geometry: Hirzebruch 70 (Warsaw, 1998)*, volume 241 of *Contemp. Math.*, pages 193–207. Amer. Math. Soc., Providence, RI, 1999.

[Rei87] Miles Reid. Young person's guide to canonical singularities. In *Algebraic geometry, Bowdoin, 1985 (Brunswick, Maine, 1985)*, volume 46 of *Proc. Sympos. Pure Math.*, pages 345–414. Amer. Math. Soc., Providence, RI, 1987.

[Vie82] Eckart Viehweg. Vanishing theorems. *J. Reine Angew. Math.*, 335:1–8, 1982.

15-Nodal Quartic Surfaces. Part I: Quintic del Pezzo Surfaces and Congruences of Lines in $\mathbb{P}^3$

Igor V. Dolgachev

To Miles Reid

Abstract

We explain a classical construction of a del Pezzo surface of degree $d = 4$ or 5 as a smooth order 2 congruence of lines in $\mathbb{P}^3$ whose focal surface is a quartic surface X_{20-d} with $20 - d$ ordinary double points. We also show that X_{15} can be realized as a hyperplane section of the Castelnuovo–Richmond–Igusa quartic hypersurface in $\mathbb{P}^4$. This leads to the proof of rationality of the moduli space of 15-nodal quartic surfaces. We discuss some other birational models of X_{15}: quartic symmetroids, 5-nodal quartic surfaces, 10-nodal sextic surfaces in $\mathbb{P}^4$ and nonsingular surfaces of degree 10 in $\mathbb{P}^6$. Finally we study some birational involutions of a 15-nodal quartic surface which, as it is shown in Part II of the paper jointly with I. Shimada [DS20], belong to a finite set of generators of the group of birational automorphisms of a general 15 nodal quartic surface.

1 Introduction

The geometry of 16-nodal Kummer quartic surfaces is widely discussed in classical as well as in modern literature. This paper addresses less known but nevertheless very rich geometry of 15-nodal quartic surfaces. What unites the two classes of nodal quartic surfaces is the fact that surfaces from both classes are realized in 6 different ways as focal surfaces of smooth congruences of lines in $\mathbb{P}^3$, congruences of bidegree $(2, 2)$ in the former case and bidegree $(2, 3)$ in the latter case. The line geometry realization of the quartic surfaces brings an unexpected dividend: it defines a certain involution on a smooth K3 model of such surface with quotient a del Pezzo surface of degree 4 in the Kummer case and degree 5 in the other case. Other interesting involutions come from a realization of the surface as a quartic symmetroid, the discriminant surface of a web of quadrics in $\mathbb{P}^3$. The quotient by these involutions is a Coble surface with the bi-anticanonical divisor consisting of six disjoint smooth rational curves in the Kummer case and five such curves in the other case.

Of course, a Kummer surface comes as a specialization of a 15-nodal quartic surface by smoothing one of its 16 ordinary nodes. Explicitly, it can be seen by realizing any 15-nodal quartic surface as a hyperplane section of the notorious Castelnuovo–Richmond–Igusa quartic hypersurface in $\mathbb{P}^4$ with 15 double lines, the tangent hyperplanes give a Kummer surface. The fact that such realization is always possible (but not generically possible as it is well known) which we prove here (also proved independently by A. Avilov in [Avi19]) is used to prove the rationality of the moduli space of 15-nodal quartic surfaces. The relationship between 15-nodal quartic surfaces X_{15} and the beautiful geometry of the Castelnuovo–Richmond–Igusa quartic hypersurface and its dual Segre cubic hypersurface goes back to H. F. Baker [Bak10, Chapter V]. We employ this relationship to describe some other interesting birational models of X_{15} such as a 5-nodal quartic surface with 20 lines, a sextic model with 10 nodes, a smooth model as a quadric section of the Fano threefold of genus 5 and index 2.

The first half of the paper is devoted to summarizing some classical results about congruences of lines in $\mathbb{P}^3$, and especially congruences of order 2 from [Jes69, Stu92]. Following the classical approach, we do not use the assumption of smoothness of the congruence assuming only isolated non-normal singularities leaving a possibility to extend some of our results to quartic surfaces with less than 15 number of nodes.

The group of birational automorphisms of a general 16-nodal Kummer quartic surface was described by S. Kondo [Kon98]. In Part II [DS20], we find a finite set of generators that contains some of the involutions discussed in the paper.

We work over $\mathbb{C}$, however most of the results which we discuss here are valid when the ground field is an algebraically closed field of characteristic $p \neq 2, 3$.

The paper originates from an attempt to answer some questions of S. Mukai related to 15-nodal quartic surfaces. I am thankful to him for many discussions, questions and insights on this and on other closely related questions about geometry of Enriques and Coble surfaces. I am also grateful to A. Verra and I. Shimada for helpful consultations.

2 Generalities on Congruences of Lines

2.1. Following classical terminology, a congruence of lines is a complete irreducible 2-dimensional family of lines in $\mathbb{P}^3$. It is parameterized by a surface S in the Grassmannian variety $\mathbb{G} = \mathrm{Gr}_1(\mathbb{P}^3)$ of lines in $\mathbb{P}^3$. The lines ℓ_s corresponding to points $s \in S$ are called *rays* to distinguish them from any line in $\mathbb{P}^3$.

The cohomology class $[S]$ in $H^4(\mathbb{G}, \mathbb{Z}) \cong \mathbb{Z}^2$ is equal to $m[\sigma_x] + n[\sigma_\pi]$, where σ_x is a plane in $\mathbb{G}$ of rays containing a point $x \in \mathbb{P}^3$ and σ_π is a plane in $\mathbb{G}$ parametrizing lines in a plane $\pi \subset \mathbb{P}^3$. The number m (resp. n) is called the *order* (resp. *class*) of S and the pair (m, n) is referred to as the bidegree of S. The sum $m + n$ is the degree of S in the Plücker space $\mathbb{P}^5$. The natural duality between lines in $\mathbb{P}^3$ and lines in the dual $\mathbb{P}^3$ defines an isomorphism between the Grassmannians of lines, the image S^* of S under this isomorphism is called the *dual congruence*. Its bidegree is (n, m). For brevity, we will assume that $m, n \neq 1$ (a modern treatment of other cases can be found in [Ran86]).

The universal family of lines $Z_S = \{(x, s) \in \mathbb{P}^3 \times S \colon x \in \ell_s\}$ comes with two projections:

$$\mathbb{P}^3 \xleftarrow{p} Z_S \xrightarrow{q} S.$$

The restriction of p to a fiber $q^{-1}(s)$ is an isomorphism onto the ray ℓ_s. This allows us to identify the fibers of q with the rays of the congruence. A confusing classical terminology defines a *singular point* of S as a point $x \in \mathbb{P}^3$ such that the fiber $p^{-1}(x)$ consists of infinitely many rays. A one-dimensional irreducible component of the set of singular points is called a *fundamental curve* of S.

We will assume that S has no fundamental curves and it is smooth or has only isolated singular points such that the normalization S' of S is smooth. The rays corresponding to singular points of S are called *multi-rays*. This kind of singularities arise from projections of smooth surfaces S' in $\mathbb{P}^6$ from a point that is contained in a finitely many secant lines to S'. If S is not smooth and $\phi \colon S' \to S$ is the normalization map, we replace Z_S with the base change $Z_{S'}$, and get the following diagram

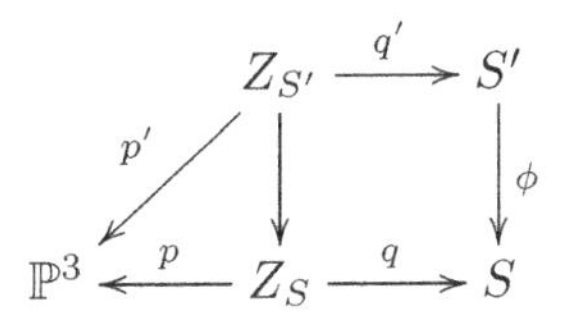

Under these assumptions, $Z_{S'}$ is smooth and the map p' is a generically finite map of degree m of smooth 3-dimensional varieties. The ramification divisor of p'

$$R_{S'} := \mathrm{Ram}(p') \subset Z_{S'}$$

is a closed subscheme of $Z_{S'}$ equal to the support of $\Omega^1_{Z_{S'}/\mathbb{P}^3}$. The usual Hurwitz type formula gives

$$\omega_{Z_{S'}} \cong p'^*\omega_{\mathbb{P}^3} \otimes \mathcal{O}_{Z_{S'}}(R_{S'}) \cong p'^*\mathcal{O}_{\mathbb{P}^3}(-4) \otimes \mathcal{O}_{Z_{S'}}(R_{S'}).$$

The projection $q'\colon Z_{S'} \to S'$ is the projectivization of the pre-image of the universal rank 2 tautological vector bundle over $\mathbb{G}$ under the map $\phi'\colon S' \xrightarrow{\phi} S \hookrightarrow \mathbb{G}$. Its first Chern class is equal to $h := c_1(\mathcal{O}_{S'}(1))$, where $\mathcal{O}_{S'}(1)$ is the pre-image of $\mathcal{O}_{\mathbb{G}}(1)$ under the map ϕ'. The usual formula for the canonical class of a projective bundle gives another formula for the canonical sheaf of $Z_{S'}$

$$\omega_{Z_{S'}} \cong q'^*(\omega_{S'} \otimes \mathcal{O}_{S'}(-1)) \otimes p'^*(\mathcal{O}_{\mathbb{P}^3}(-2)). \tag{2.1}$$

Comparing the two formulas, we find

$$[R_{S'}] = p'^*(2H) + q'^*(h + K_{S'}) \tag{2.2}$$

where $H = c_1(\mathcal{O}_{\mathbb{P}^3}(1))$. The image of $R_{S'}$ under the projection to $\mathbb{P}^3$ is the branch divisor of p'. It is called the *focal surface* of S and will be denoted by $\mathrm{Foc}(S)$. Intersecting $R_{S'}$ with a general fiber of q', we deduce from (2.2) that

$$\deg q'|_{R_{S'}} = 2. \tag{2.3}$$

If the fiber $q'^{-1}(s')$ is contained in $R_{S'}$, its image ℓ_s in $\mathbb{P}^3$ lies in $\mathrm{Foc}(S)$. Otherwise $q'^{-1}(s')$ contains two critical points in $R_{S'}$ and the ray ℓ_s intersects $\mathrm{Foc}(S)$ at the images of these points with multiplicity ≥ 2. It follows from (2.3) that the focal surface $\mathrm{Foc}(S)$ consists of at most two irreducible components. We will assume that it is irreducible. We will also assume that $\mathrm{Foc}(S)$ contains only finitely many rays of S, so that the surface $R_{S'}$ is irreducible too. Hence:

(1) *S is an irreducible component of the surface of bitangents of* $\mathrm{Foc}(S)$.

For any curve C on S, the image of $q^{-1}(C)$ under the map $p\colon Z_S \to \mathbb{P}^3$ is a *ruled surface.* Its degree is equal to the degree of C in the Plücker embedding. For any line l in $\mathbb{P}^3$, the set of rays intersecting l is a (special) hyperplane section $R(l) = \mathbb{T}_l(\mathbb{G}) \cap S$ of S. Following the classical notation we denote by (l) the corresponding ruled surface in $\mathbb{P}^3$.

Let $\mathcal{P}_l \cong \mathbb{P}^1$ be the pencil of planes containing a general line l in $\mathbb{P}^3$. Consider a map

$$f\colon R(l) \longrightarrow \mathcal{P}_l \times l \cong \mathbb{P}^1 \times \mathbb{P}^1$$

that assigns to a point $s \in R(l)$, the pair $(\langle l, \ell_s\rangle, \ell_s \cap l)$ that consists of the plane $\langle l, \ell_s\rangle$ generated by the line l and the ray ℓ_s and the intersection point $\ell_s \cap l$. The composition of f with the first projection (resp. the second projection) is a map of degree n (resp. m). For a general line l, the curve $R(l)$ is smooth and its genus is called the *sectional genus* of S. The image of the map f is a curve of bidegree (m, n) on $\mathbb{P}^1 \times \mathbb{P}^1$ and arithmetic genus $(m-1)(n-1)$. The map f is not an isomorphism in general. In fact, two rays ℓ_1, ℓ_2 in $R(l)$ go to the same point $(\pi, x) \in \mathcal{P}_l \times l$ if and only if $\pi = \langle \ell_1, l\rangle = \langle \ell_2, l\rangle$ and $\ell_1 \cap \ell_2 = \{x\}$. Hence the Schubert line $\sigma_{x,\pi} := \sigma_x \cap \sigma_\pi$ is a secant line of S passing through the point $x \in \mathbb{G}$.

Let r be the number of secant lines of S in $\mathbb{G}$ passing through a general point $x \in \mathbb{G}$. It is called the *rank* of S. Note that, since $\mathbb{G}$ is a quadric hypersurface in $\mathbb{P}^5$, a secant line of S through a general point in $\mathbb{G}$ is contained in $\mathbb{G}$. Obviously $r = 0$ if S is degenerate as a subvariety of $\mathbb{P}^5$. It is shown in [Ver88] that, if S is nondegenerate, then $r = 0$ if and only if it is a Veronese surface in $\mathbb{P}^5$ realized as the congruence of lines of bidegree $(1, 3)$, the congruence of secant lines of a twisted cubic.

It follows from above that $R(l)$ is the normalization of a curve of arithmetic genus $(m-1)(n-1)$ with r double points. Therefore its genus g is given by the formula

$$g = (m-1)(n-1) - r. \tag{2.4}$$

The composition of f and the second projection is a degree m cover of $\mathbb{P}^1$ of a curve of genus g with $\deg(\mathrm{Foc}(S))$ ordinary branch points. Applying Hurwitz formula and (2.4), we obtain

(2) *The degree of the focal surface is given by the formula*

$$\deg \mathrm{Foc}(S) = 2m + 2g - 2 = 2n(m-1) - 2r, \tag{2.5}$$

where g is the sectional genus of S and r is the rank of S.

2.2. Let $\sigma_{x,\pi} \subset \mathbb{G}$ be a secant line of S. The plane π contains two rays of S intersecting at the point x. It is called the *null-plane* of the congruence and the point x is its *center*. Let $\mathrm{Sec}(S) \subset \mathrm{Gr}_1(\mathbb{P}^5)$ be the variety of secant lines of S. The projection $\alpha\colon \mathrm{Sec}(S) \to \mathbb{P}^3, \sigma_{x,\pi} \to x$, is a map of degree $m(m-1)/2$, the projection $\beta\colon \mathrm{Sec}(S) \to \check{\mathbb{P}}^3, \sigma_{x,\pi} \to \pi$, is a map of degree $n(n-1)/2$:

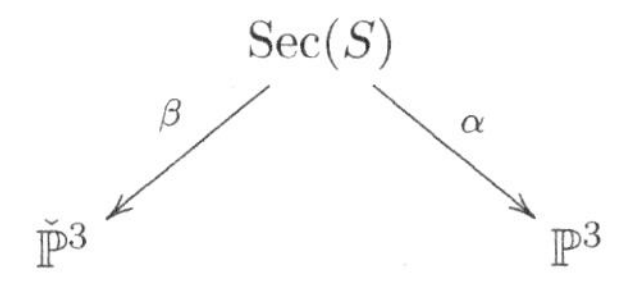

Note that, if $m = 2$, the map α (resp. β) is an isomorphism over the set of nonsingular points of $\mathrm{Foc}(S)$.

A general plane π in the pencil $\mathcal{P}_l$ contains $n(n-1)/2$ intersection points of n rays in this plane. The locus of such points when π moves in $\mathcal{P}_l$ is a curve, denoted in classical literature by $|l|$. A plane π containing l contains $n(n-1)/2$ points equal to the pre-image of the corresponding point $\pi^* \in l^*$ under β and also $r = \operatorname{rank} S$ points on l that are the centers of null-planes containing l. In other words, the planes through l define a pencil in $|l|$ with $n(n-1)/2$ moving points and r fixed point. Thus

$$\deg(|l|) = \tfrac{1}{2}n(n-1) + r. \tag{2.6}$$

Let $x \in \mathrm{Foc}(S)$ be a singular point. We denote by $K(x)$ the cone of rays through x. Its degree is equal to $h(x)$.

Obviously, $K(x)$ is an irreducible component of the ruled surface (l) when $x \in l$. The residual part is of degree $m + n - h(x)$. Let $C(x)$ be the curve in S parameterizing the lines in the ruling of $K(x)$. Its degree in the Plücker embedding is equal to $h(x)$. Since $R(l)$ is a hyperplane section of S, it intersects $C(x)$ with multiplicity $h(x)$. Thus (l) and $K(x)$ intersect in S with multiplicity $h(x)$. If $h(x) > 1$, then each pair of common rays lies in the same plane $\langle l, x\rangle$, hence $x \in |l|$. This shows that x is a $\frac{1}{2}h(x)(h(x)-1)$-multiple point of $|l|$. It is also a point of multiplicity $h(x)$ on (l).

As each point of $|l|$ lies on a ray intersecting l, the curve $|l|$ is contained in the ruled surface (l). Under the normalization map $q'^{-1}(C(l)) \to (l)$ the pre-image of a general point on $|l|$ consists of $\frac{1}{2}n(n-1)$ points.

(3) *The curve $|l|$ passes through singular points x with multiplicity $\frac{1}{2}h(x)(h(x)-1)$. It is contained in the locus of singular points of multiplicity $\frac{1}{2}n(n-1)$ on (l).*

Next we take a plane P^* dual to a point $P \in \mathbb{P}^3$ and define the surface $(P) = \alpha(\beta^{-1}(P^*))$ in $\mathbb{P}^3$. It follows from the definition, that the surface (P) is the locus of points on some line through P which are centers of null-planes containing the line. The duality argument of the proof of (2.6) shows that

$$\deg(P) = \tfrac{1}{2}m(m-1) + r. \tag{2.7}$$

For any singular point x of $\mathrm{Foc}(S)$ with $h(x) = 1$, the plane-cone $K(x)$ does not contain a general point P. Thus[1] the surface (P) does not contain x. On the other hand, if $h(x) > 1$, the plane π spanned by P and a ray ℓ passing through x contains $h(x) - 1$ additional rays ℓ' through x, hence x lies in (P). The null planes spanned by ℓ and the remaining $n - h(x)$ rays in π have centers belonging to (P). Since a general ℓ intersects (P) at $\deg(P)$ points, we obtain the following.

(4) *Each singular point $x \in \mathrm{Foc}(S)$ is a singular point of (P) of multiplicity $\frac{1}{2}m(m-1) + r - n + h(x)$.*

2.3. Let B be the branch curve of the degree 2 map $q' \colon R_{S'} \to S'$. Let $\deg B = h \cdot [B]$ be the degree of the image of B on S in the Plücker embedding. Using (2.1) and (2.2) and applying the adjunction formula, we get

$$K_{R_{S'}} = (K_{Z_{S'}} + R_{S'}) \cdot R_{S'} = q'^*(2K_{S'} + 2h). \tag{2.8}$$

The Hurwitz type formula now gives $2K_{R_{S'}} = q'^*(2K_{S'} + B)$, hence

$$[B] = 2K_{S'} + 4h. \tag{2.9}$$

A ray ℓ_s may be tangent to $\mathrm{Foc}(S)$ with multiplicity 4 at a nonsingular point. It is clear that in this case $q'^{-1}(\phi^{-1}(s))$ intersects $R_{S'}$ at one point, hence $s' = \phi^{-1}(s)$ belongs to B. Conversely, if s' belongs to B, then the ray $\ell_{\phi(s')}$ intersects $\mathrm{Foc}(S)$ at one point with multiplicity 4. Let $\mathrm{Foc}(S)_0$ be the set of such points in $\mathrm{Foc}(S)$. The pre-image of $\mathrm{Foc}(S)_0$ under the projection $p' \colon R_{S'} \to \mathrm{Foc}(S)$ is equal to the ramification locus of $q' \colon R_{S'} \to S'$. Since $q'_* p'^* \mathcal{O}_{\mathbb{P}^3}(1) = \mathcal{O}_{S'}(1)$, the projection formula shows that the degree of $\mathrm{Foc}(S)_0$ is equal to $\deg B$. Now (2.9) gives

$$\deg B = 2K_{S'} \cdot h + 4h^2 = 2(K_{S'} + h) \cdot h + 2h^2 = 2(2g-2) + 2(m+n).$$

Taking into account (2.4), we obtain

[1] Jessop on p. 262 of [Jes69] apparently mistakenly says that (P) passes through all singular points.

(5) *The degree of the locus of points* $\mathrm{Foc}(S)_0$ *on* $\mathrm{Foc}(S)$ *where a ray touches* S *with multiplicity* 4 *is equal to* $\deg B$.

For any singular point x of $\mathrm{Foc}(S)$, the fiber $q^{-1}(x)$ is a connected curve equal to the intersection $C(x) = \sigma_x \cap S$. Thus $K(x)$ is a cone over $C(x)$ of degree $h(x)$. Its pre-image under $p\colon Z_S \to \mathbb{P}^3$ is equal to the ruled surface $q^{-1}(C(x))$ with base curve $C(x) \subset S$. Its pre-image in $Z_{S'}$ is the ruled surface $q'^{-1}(\tilde{C}(l))$, where $\tilde{C}(l) \subset S'$ is the pre-image of $C(l)$ under the normalization map $\phi\colon S' \to S$. Assuming that a singular point of $\mathrm{Foc}(S)$ is an ordinary node, then the exceptional curve $E_x = p'^{-1}(x) \subset R_{S'}$ must be a smooth rational curve and, under the projection $q'\colon Z_{S'} \to S'$, it is mapped isomorphically onto $\tilde{C}(x)$ and defines the normalization map $E_x \cong \tilde{C}(x) \to C(x)$. Since $C(x)$ is a rational plane curve of degree $h(x)$, it is singular if $h(x) > 2$, and hence $\phi\colon S' \to S$ is not an isomorphism over singular points of $C(x)$.

(6) *Suppose* S *is smooth and* x *is an ordinary node of* $\mathrm{Foc}(S)$. *Then* $h(x) \le 2$.

In fact, under the generality assumption that all singular points of the curve $C(x) = \sigma_x \cap S$ are ordinary nodes, we obtain that the number of singular points of S is equal to

$$s = \frac{1}{2} \sum_{x \in \mathrm{Foc}(S)} (h(x) - 1)(h(x) - 2). \tag{2.10}$$

3 Congruences of Degree 2 and Class n

3.1. We specialize assuming that $m = 2, n \ge 2$. We also assume that the congruence does not have fundamental curves. This excludes only one case when $(m, n) = (2, 2)$ and S is a quartic scroll (see [AG93]). We have in this case $r = 1, g = 0$ and $\deg \mathrm{Foc}(S) = 2$. So, we may assume that $\deg \mathrm{Foc}(S) \ge 4$.

(7) $\deg \mathrm{Foc}(S) = 4$ *and* $g = 1, r = n - 2$.

Let us see that $\deg \mathrm{Foc}(S) = 4$, the rest follows from (2.5). Suppose the degree is larger. Then a general ray ℓ is tangent to the focal surface at two points, hence ℓ intersects it at some other (nonsingular) point x. Since $m = 2$ and x belongs to $\mathrm{Foc}(S)$, there will be another ray ℓ' tangent to $\mathrm{Foc}(S)$ at the point x. Thus there are two different rays through x contradicting the definition of the focal surface.

Since $g = 1$, a general hyperplane section H of S is an elliptic curve. By the adjunction formula, the anti-canonical sheaf $\omega_{S'}^{-1} \cong \mathcal{O}_{S'}(1)$ is ample. By definition, S' is a del Pezzo surface. The anti-canonical linear system $|-K_{S'}|$ embeds S' into $\mathbb{P}^d$, where $d = K_{S'}^2$ is equal to the degree of the image. It follows from (5) that $B \in |-2K_{S'}|$ and $\deg B = 2K_{S'}^2 = 2(2+n)$, hence we obtain the following.

(8) *The normalization S' of S is a del Pezzo surface of degree $d = 2+n$. The normalization map $\phi\colon S' \to S$ is an isomorphism if $n = 2, 3$ and is the projection of the surface S' anti-canonically embedded in $\mathbb{P}^{2+n}$ to a surface of degree $2+n$ in $\mathbb{G} \subset \mathbb{P}^5$. The branch curve B of $q'\colon R_{S'} \to S'$ belongs to $|-2K_{S'}|$.*

Collecting the previous formulas, we obtain the following.

(9) *The degree of the curve $|l|$ is equal to $\frac{1}{2}(n^2+n-4)$. The degree of the surface (P) is equal to $n-1$.*

(10) *The surface (P) is the locus of centers of null-planes that contain the point P. It passes through singular points x of* $\mathrm{Foc}(S)$ *with multiplicity $h(x)-1$.*

Since the degree d of a del Pezzo surface in $\mathbb{P}^d$ satisfies $d \leq 9$, we see that $n \leq 7$. Moreover, there are two different types of congruences of class 6. They correspond to two types of del Pezzo surfaces of degree 8: re-embedded quadrics and blow-ups of one point in $\mathbb{P}^2$.

We know from (2.8) that $\omega_{R_{S'}} \cong \mathcal{O}_{R_{S'}}$. Since $\mathrm{Foc}(S)$ is a quartic surface in $\mathbb{P}^3$, we have $\omega_{\mathrm{Foc}(S)} \cong \mathcal{O}_{\mathrm{Foc}(S)}$, hence $p'^*\omega_{\mathrm{Foc}(S)} = \omega_{R_{S'}}$, hence all singular points of the quartic surface $\mathrm{Foc}(S)$ are rational double points. Since $Z_{S'}$ is smooth, all singular points of $\mathrm{Foc}(S)$ are singular points of the congruence S. Also, no ray ℓ is contained in $\mathrm{Foc}(S)$ (otherwise, each $x \in \ell$ is contained in two rays, ℓ and another ray from $C(\ell)$ passing through x). This implies that the map $q'\colon R_{S'} \to S'$ is a finite morphism of degree 2, and hence the branch curve B is smooth.

(11) *The surface $R_{S'}$ is a K3 surface. All singular points of* $\mathrm{Foc}(S)$ *are double rational points and the map $p'\colon R_{S'} \to$* $\mathrm{Foc}(S)$ *is a minimal resolution of singularities. The double cover $q'\colon R_{S'} \to S'$ is a finite morphism of degree 2 with smooth branch curve $B \in |-2K_{S'}|$.*

Since a ray ℓ_s is tangent to $\mathrm{Foc}(S)$ at two points, $p^{-1}(\ell)$ splits into the union of two smooth rational curves in Z_S, one of them is the fiber $q^{-1}(s)$.

Table 4.1 *Singular points of congruences of bidegree* $(2, n)$

	$(2,2)$	$(2,3)$	$(2,4)$	$(2,5)$	$(2,6)_I$	$(2,6)_{II}$	$(2,7)$
α_1	16	10	6	3	1	0	0
α_2		5	6	6	4	8	0
α_3			2	3	6	0	10
α_4				1	0	4	0
α_5					1		
α_6							1
$\sum \alpha_i$	16	15	14	13	12	12	11

The other component parametrizes rays that intersect ℓ_s. It is projected to a hyperplane section of S with a singular point at x. This curve is cut out by a hyperplane tangent to $\mathbb{G}$ at the point $s \in S$.

From now on we assume that $\mathrm{Foc}(S)$ has only ordinary nodes as singularities.

Two ruled surfaces (l) and (l') intersect at $n + m = n + 2$ common rays. Thus the two surfaces residually intersect along a curve of degree $(n + 2)^2 - (n + 2) = n^2 + 3n + 2$. If x is a point on the intersection curve not lying on the common rays, then it is the intersection point of two different rays, hence no other ray passes through it. This shows that three ruled surfaces can intersect only at singular points of $\mathrm{Foc}(S)$ and at $3(n+2)^2$ points on the common rays (a ray common to two must meet the third). Since each singular point $x \in \mathrm{Foc}(S)$ has multiplicity $h(x)$ on (l), the intersection curve has multiplicity $h(x)^2$ at x. Intersecting with the third ruled surface (l''), we get

$$(n + 2)^3 = 3(n + 2)^2 + \sum i^3 \alpha_i,$$

where α_i is the number of singular points x with $h(x) = i$. The Table 4.1 of possible solutions can be found in [Jes69, p. 280].

Applying (2.10), we see that all congruences S are singular if $n > 3$, and the number of singular points is equal to $\frac{1}{2}(n - 2)(n - 3)$.

3.2. Let x be a singular point of $\mathrm{Foc}(S)$. Then a general ray of the cone $K(x)$ is tangent to $\mathrm{Foc}(S)$ at some point outside x. The closure of the locus of the tangency points is a *trope-curve* $T(x)$ on $\mathrm{Foc}(S)$ of degree $2h(x)$. It is cut out in $\mathrm{Foc}(S)$ with multiplicity 2 by the cone $K(x)$. If $h(x) = 1$ (resp. $h(x) = 2$), it is a conic (resp. quartic curve) called a *trope-conic* (resp. *trope-quartic*) of $\mathrm{Foc}(S)$. Since by *(11)* the projection

$R_{S'} \to S'$ is a finite morphism, no ray lies in $\mathrm{Foc}(S)$. This implies the the trope-curves are irreducible.

The orbits of the birational involution of $\mathrm{Foc}(S)$ corresponding to the deck transformation σ of the cover $q' \colon R_{S'} \to S'$ are pairs of tangency points of a ray of the congruence. If $x \in \mathrm{Sing}(\mathrm{Foc}(S))$, then the exceptional curve $E_x \subset R_{S'}$ is mapped under the involution to some other curve $\sigma(E_x)$. The curve $E_x + \sigma(E_x)$ is equal to the pre-image of the curve $\tilde{C}(x) = \phi^{-1}(C(x)) \subset S'$ under the covering map q'. This shows that the curve $\tilde{C}(x)$ splits under the cover, and hence is everywhere tangent to the branch curve B. The degree of $\tilde{C}(x)$ in the anti-canonical embedding of S' is equal to $h(x)$.

Two singular points of $\mathrm{Foc}(S)$ are called *conjugate* if the line joining these points is a ray from the congruence. Suppose x, x' are two conjugate singular points. Then the ray $\ell = \langle x, x' \rangle$ is contained in the intersection of the cones $K(x)$ and $K(x')$ and joins their vertices. A general plane π containing ℓ intersects $K(x)$ along $h(x)$ rays and intersects $K(x')$ along $h(x')$ rays. Since π contains n rays, we see that the points x and x' are conjugate if $h(x) + h(x') > n$. In this case, the ray $\langle x, x' \rangle$ must have multiplicity $h(x) + h(x') - n$. Note that the inequality $h(x) + h(x') > n$ is only a sufficient condition for the conjugacy.

(12) *Suppose $h(x) + h(x') > n$, then the points x, x' are conjugate and the ray joining the two points has multiplicity $h(x) + h(x') - n$.*

The *conjugacy graph* of S is a graph whose vertices are singular points of $\mathrm{Foc}(S)$ and two conjugate vertices x, x' are joined by an edge. We also mark a vertex x with the integer equal to $h(x)$ (if its larger than 1) and mark the edge with $(h(x) + h(x') - n$ if $h(x) + h(x') - n > 1$.

(13) *Let $\tilde{C}(x)$ be the pre-image in S' of the curve $C(x)$ on S, where x is a singular point of* $\mathrm{Foc}(S)$. *Then $\tilde{C}(x)$ is everywhere tangent to the branch curve B and has degree $h(x)$ in the anti-canonical embedding of S'. The conjugacy graph of S is the dual intersection graph of the curves $\tilde{C}(x)$ on S'.*

Example 3.1. Assume $n = 4$. The surface $\mathrm{Foc}(S)$ has two singular points x_1, x_2 with $h(x_i) = 3$. Thus $C(x_1)$ and $C(x_2)$ are irreducible singular cubic curves and S is singular with one singular point corresponding to the multi-ray that passes through the points x_1 and x_2. It is a common line of the cones $K(x_1)$ and $K(x_2)$. The normalization S' of S is a del Pezzo surface of degree 6 anti-canonically embedded in $\mathbb{P}^6$. The normalization map $\phi \colon S' \to S$ is the projection from a general point on the secant hypersurface of S'. The projection from a general point of $\mathbb{P}^6$ is a smooth del Pezzo surface of degree 6 that is contained in a smooth quadric. If we realize this quadric as the Klein quadric $\mathbb{G}$, we obtain a realization of a smooth del Pezzo surface of degree 6 as a smooth congruence of lines if bidegree $(3,3)$.

4 Congruences of Bidegree $(2,3)$

4.1. Let us now summarize what we have found in the case $n = 3$. We assume that S is general enough so that all singularities of the focal surface are ordinary nodes.

(14) *The surface S is a smooth quintic del Pezzo surface anti-canonically embedded in $\mathbb{G} \subset \mathbb{P}^5$.*

(15) *The focal surface* $\mathrm{Foc}(S)$ *is a quartic surface with 15 nodes, ten of them with $h(x) = 1$ and remaining ones with $h(x) = 2$.*

(16) *The surfaces (l) are quintic elliptic ruled surfaces. A curve $|l|$ is a singular curve of multiplicity* 3 *on (l), it coincides with the base curve of the pencil of quadrics generated by (P) and (P').*

(17) *The surface $Y = R_{S'}$ is a K3 surface. The morphism $p' = p\colon Y \to$* $\mathrm{Foc}(S)$ *is a minimal resolution of singularities and the morphism $q' = q\colon Y \to S' = S$ is a finite morphism of degree* 2 *with a smooth branch curve $B \in |-2K_S|$.*

Let us denote the set of nodes x of $\mathrm{Foc}(S)$ with $h(x) = 1$ (resp. $h(x) = 2$) by $\mathcal{L}$ (resp. $\mathcal{C}$). The curves $C(x), x \in \mathcal{L}$, are lines on S tangent to B at one point. The curves $C(x), x \in \mathcal{C}$, are conics in S tangent to B at two points. The curves $T(x) = (K(x) \cap \mathrm{Foc}(S))_{\mathrm{red}}$ are trope-conics if $x \in \mathcal{L}$ and trope-quartics if $x \in \mathcal{C}$.

Since $\sigma(E_x) \cdot E_x = h(x)$, we see that x is a simple point of $T(x)$ if $x \in \mathcal{L}$ and a double point if $x \in \mathcal{C}$. We know that two singular points x, x' with $h(x) = h(x') = 2$ are conjugate, i.e. the cones $K(x)$ and $K(x')$ have a common ray. This implies that $T(x)$ and $T(x')$ intersect and the corresponding conics $C(x)$ and $C(x')$ belong to different pencils of conics on S.

Let us recall that a quintic del Pezzo surface S is isomorphic to the blow-up of four points p_1, p_2, p_3, p_4 in the plane, in general linear position. Let e_0 be the class of the pre-image of a line and $e_1, \ldots, e_4$ be the classes of the exceptional curves over the points p_i. The anti-canonical

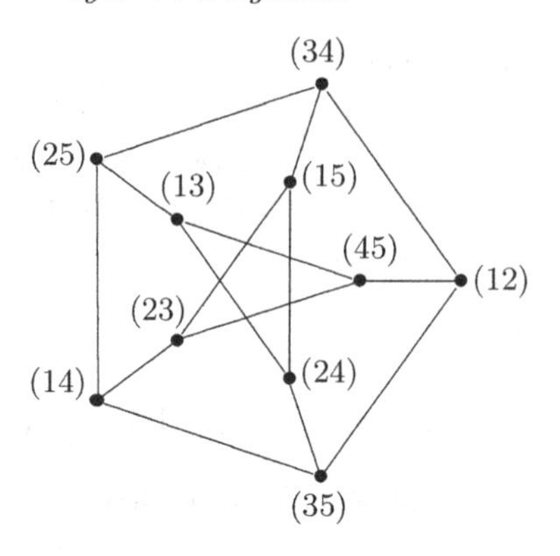

Figure 4.1 The Petersen graph

embedding is given by the linear system $|3e_0 - e_1 - e_2 - e_3 - e_4|$ represented in the plane by the linear system of cubic curves passing through the points $p_1, \ldots, p_4$. In the anti-canonical embedding, the surface has 10 lines with the classes $e_1, \ldots, e_4$ and $e_0 - e_i - e_j$. It has 5 pencils of conics with classes $e_0 - e_i$ and $2e_0 - e_1 - e_2 - e_3 - e_4$.

We denote by $\phi_{\mathcal{L}\mathcal{L}} \colon \mathcal{L} \to 2^{\mathcal{L}}$ the map that assigns to a point $x \in \mathcal{L}$ the set of points from $\mathcal{L}$ contained in $\sigma(E_x) \setminus \{x\}$. We have similar maps $\phi_{\mathcal{C}\mathcal{C}} \colon \mathcal{C} \to 2^{\mathcal{C}}, \phi_{\mathcal{L}\mathcal{C}} \colon \mathcal{L} \to 2^{\mathcal{C}}$ and $\phi_{\mathcal{C}\mathcal{L}} \colon \mathcal{C} \to 2^{\mathcal{L}}$. It follows that

$$\#\phi_{\mathcal{L}\mathcal{L}}(x) = 3, \quad \#\phi_{\mathcal{C}\mathcal{L}}(x) = 4, \quad \#\phi_{\mathcal{L}\mathcal{C}}(x) = 2, \quad \#\phi_{\mathcal{C}\mathcal{L}}(x) = 4.$$

In the sequel we will be indexing the set $\mathcal{C}$ by the set $\{16, 26, 36, 46, 56\}$ and the set $\mathcal{L}$ by subsets of $\{1, 2, 3, 4, 5\}$ of cardinality 2 as in the Petersen graph in Figure 4.1. We find that

$$\begin{aligned}
\phi_{\mathcal{L}\mathcal{L}}(x_{ab}) &= \{x_{cd} \colon (cd) \cap (ab) = \emptyset\},\\
\phi_{\mathcal{C}\mathcal{C}}(x_{a6}) &= \{x_{b6} \colon a \neq b\},\\
\phi_{\mathcal{L}\mathcal{C}}(x_{ab}) &= \{x_{a6}, x_{b6}\},\\
\phi_{\mathcal{C}\mathcal{L}}(x_{a6}) &= \{x_{ab} \colon b \neq a, 6\}.
\end{aligned}$$

(18) *The conjugacy graph of the congruence S is equal to the dual intersection graph of 10 lines l_{ab} on S and 5 conics C_i, each taken from one of the five pencils of conics. It is equal to the union of the Petersen graph and the complete graph $K(5)$ on the set $\{1, 2, 3, 4, 5\}$. Each vertex (ab) of the Petersen graph is connected to two vertices a, b of $K(5)$ and each vertex a of $K(5)$ is connected to 4 vertices (ab), $b \neq a$, of the Petersen graph.*

4.2. The plane $K(x), x \in \mathcal{L}$, that cuts out a trope-conic passes through x and the nodes from $\phi_{\mathcal{L}}(x)$ and $\phi_{\mathcal{LC}}(x)$. This shows that, for any $x \in \mathcal{L}$,

$$2\sigma(E_x) \in \left|\eta_H - E_x - \sum_{x' \in \phi_l(x)} E_{x'} - \sum_{y \in \phi_{lc}(x)} E_y\right| \tag{4.1}$$

where $\eta_H = c_1(p^*\mathcal{O}_{\mathbb{P}^3}(1))$.

Similarly, we find, that for any $E_y, y \in \mathcal{C}$,

$$2\sigma(E_y) \in \left|2\eta_H - 2E_y - \sum_{y' \in \mathcal{C}, y' \neq y} E_{y'} - \sum_{x \in \phi_{cl}(y)} E_x\right|. \tag{4.2}$$

Applying (4.1) and (4.2), we find that

$$\sigma(\eta_H) \in \left|4\eta_H - \sum_{x \in \mathcal{L}} E_x - 2\sum_{y \in \mathcal{C}} E_y\right|.$$

The linear system $|\sigma(\eta_H)|$ is the restriction to $\mathrm{Foc}(S)$ of the 4-dimensional linear system of quartic surfaces with nodes on $\mathcal{C}$ and simple points at $\mathcal{L}$. By choosing an appropriate basis of the linear system, it defines a birational self-map of $\mathrm{Foc}(S)$ that blows down $\sigma(E_x), x \in \mathcal{L} \cup \mathcal{C}$, to the nodes.

Let $\tilde{B} \cong B$ be the ramification curve of $q \colon Y \to S$. We know that its image $\mathrm{Foc}(S)_0$ in $\mathrm{Foc}(S)$ is equal to the locus of points where a ray is touching $\mathrm{Foc}(S)$ with multiplicity 4. It follows from (5) that it is a curve in $\mathbb{P}^3$ of degree $4(mn - r) - 2(m + n) = 10$. Since $\tilde{B}$ intersects each $E_x, x \in \mathcal{C}$, at two points and intersects the curves $E_y, y \in \mathcal{L}$, at one point, we see that it passes through all singular points of $\mathrm{Foc}(S)$. If we assume that the congruence is *moduli general* in the sense that the rational Picard group $\mathrm{Pic}(S)_{\mathbb{Q}}$ is generated by the curves $E_x, x \in \mathcal{L} \cup \mathcal{C}$, and η_H (see the explanation of this assumption in section 8), we obtain

$$\begin{aligned}\tilde{B} \sim \frac{1}{2}\left(5\eta_H - 2\sum_{x \in \mathcal{C}} E_x - \sum_{x \in \mathcal{L}} E_x\right) &\sim \frac{1}{2}\left(\sum_{x \in \mathcal{L}} (E_x + \sigma(E_x))\right) \\ &\sim \frac{1}{2}\left(\sum_{x \in \mathcal{C}} (E_x + \sigma(E_x))\right).\end{aligned} \tag{4.3}$$

Applying (4.1), we see that $\tilde{B}$ is a curve of degree 10 cut out with multiplicity 2 by a quintic surface with nodes at $y, y \in \mathcal{C}$, that passes through the remaining nodes with multiplicity 1.

The linear system $|\tilde{B}|$ defines a σ-equivariant map from Y to a surface of degree 10 in $\mathbb{P}^6$. We will study this birational model in Section 7.

4.3. Let (P) be the quadric surface corresponding to a nonsingular point P on $\mathrm{Foc}(S)$ and ℓ be the unique ray passing through P. Since a general point x on ℓ intersects another ray, ℓ is contained in a null-plane. Since $P \in \ell$, we see that ℓ is contained in (P). Conversely, if l is a line on (P), a general point on ℓ is the intersection point of two rays, hence l lies on an infinite set of secants of S. This shows that l is a ray. So, we obtain that the quadric surface (P) contains only one line passing through P, hence it must be a quadric cone. Since the degree of the discriminant surface of a web of quadrics is equal to 4, we obtain that $\mathrm{Foc}(S)$ coincides with this surface.

Also observe that, if P is a singular point P with $h(P) = 1$, then any ray through P is contained in (P), hence the plane $\langle T(x) \rangle = K(x)$ is contained in (P). Thus the quadric (P) is reducible. The singular line of the quadric (P) is the tangent line to the trope-conic at the point P.

(19) *The web W of quadrics (P), $P \in \mathbb{P}^3$, has base locus equal to the set $\mathcal{C}$ of singular points of* $\mathrm{Foc}(S)$ *with $h(x) = 2$. The surface* $\mathrm{Foc}(S)$ *coincides with the* discriminant surface D_W *of W, i.e. the locus of singular quadrics in W. The 10 quadrics (x), $x \in \mathcal{L}$, are reducible quadrics, one of the irreducible components is the plane $K(x)$.*

We will return to the discussion of the geometry related to the web of quadrics W in Section 6.

5 The Segre Cubic and Castelnuovo–Richmond–Igusa Quartic

5.1. Let us recall some known facts about the Segre cubic primal and the Castelnuovo–Richmond–Igusa quartic hypersurface that can be found, for example, in [Dol12, 9.4]. Let $|L|$ be the linear system of quadrics through the singular points $p_1, \ldots, p_5 \in \mathbb{P}^3$ in general linear position. The linear system $|L|$ defines a rational map

$$\Phi \colon \mathbb{P}^3 \dashrightarrow \mathbb{P}^4 \tag{5.1}$$

with the image a cubic threefold in $\mathbb{P}^4$ projectively isomorphic to the *Segre cubic primal* S_3 given by equations

$$\sum_{i=1}^{6} z_i^3 = \sum_{i=1}^{6} z_i = 0,$$

where we identify $\mathbb{P}^4$ with the hyperplane $V(\sum_{i=1}^6 z_i)$ in $\mathbb{P}^5$. The permutation group $\mathfrak{S}_6$ acts naturally on S_3 via permuting the coordinates. We denote this action by ${}^I\mathfrak{S}_6$. There is a natural isomorphism between S_3 and the GIT-quotient of $(\mathbb{P}^1)^6$ by PGL(2). It follows from an explicit map from $(\mathbb{P}^1)^6$ to S_3 given by the Joubert functions [Dol12, Theorem 9.4.10] that the action of $\mathfrak{S}_6$ on S_3 via permuting the factors of $(\mathbb{P}^1)^6$ differs from the previous action by an outer automorphism of $\mathfrak{S}_6$. We denote this action by ${}^{II}\mathfrak{S}_6$. The linear action of $\mathfrak{S}_5$ in $\mathbb{P}^3$ leaving invariant the set of points $p_1, \ldots, p_5$ embeds $\mathfrak{S}_5$ onto a transitive subgroup of ${}^I\mathfrak{S}_6$ and onto a non-transitive subgroup of ${}^{II}\mathfrak{S}_6$ (recall that there are two conjugacy classes of subgroups of $\mathfrak{S}_6$ isomorphic to $\mathfrak{S}_5$.)

Recall that a *duad* is a subset of the set $[1,6] = \{1, \ldots, 6\}$ of cardinality 2. A *syntheme* is a set of three complementary duads. A *total* is a set of 5 synthemes that contains all duads. A subgroup of $\mathfrak{S}_6$ fixing a total is isomorphic to $\mathfrak{S}_5$ and acts transitively on the set $[1,6]$.

The images P_{ij} of 10 lines $\langle p_i, p_j \rangle$ are singular points of S_3, all ordinary nodes. The images Π_i of the exceptional divisors of the blow-up of $\mathbb{P}^3$ at the points p_i and the planes $\langle p_i, p_j, p_k \rangle$ are planes contained in S_3. The 15 planes and 10 nodes form an abstract configuration $(15_4, 10_6)$.

It is also known that the projectively dual hypersurface of S_3 is projectively isomorphic to the *Castelnuovo–Richmond–Igusa quartic hypersurface* CR_4. The equations of CR_4 are

$$4\sum_{i=1}^{6} y_i^4 - \sum_{i=1}^{6} y_i^2 = \sum_{i=1}^{6} y_i = 0.$$

The projective duality map

$$\Psi \colon \mathsf{S}_3 \dashrightarrow \mathrm{CR}_4$$

is given by the choice of a basis in the linear system of polar quadrics of S_3. The map Ψ is equivariant with respect to the action ${}^{II}\mathfrak{S}_6$ on S_3 and the action of $\mathfrak{S}_6$ on CR_4 via permuting the coordinates y_i.

The images of planes in S_3 under Ψ are 15 double lines of CR_4. Each line intersects three other lines, and three lines pass through each of 15 intersection points forming a symmetric abstract configuration (15_3), the *Cremona—Richmond configuration.* The images of the tangent cones at singular points of S_3 are quadric surfaces in CR_4 cut out by hyperplanes everywhere tangent to CR_4 (*cardinal hyperplanes*).

The Segre cubic S_3 is characterized by the property that it contains 10 singular points, the maximal number for a normal cubic hypersurface in $\mathbb{P}^4$. The singular points are ordinary nodes forming a ${}^I\mathfrak{S}_6$-orbit of the point $[1,1,1,-1,-1,-1]$. They are indexed by subsets (abc) of cardinality 3 of the set $[1,6]$, up to a complementary sets. The set of planes is the $\mathfrak{S}_6$-orbit of the plane $\pi_{12,34,56}\colon z_1+z_2=z_3+z_4=0$. They are indexed by synthemes (ab,cd,ef). Two planes intersect along a line if and only if the synthemes share a duad. Otherwise they intersect at a singular point of S_3.

Dually, we have 15 double lines in CR_4 forming a $\mathfrak{S}_6$-orbit of the line given by equations $y_1-y_2=y_3-y_4=y_5-y_6=0$. They are also indexed by synthemes. The intersection points of the double lines form a $\mathfrak{S}_6$-orbit of the point $[-2,2,1,1,1,1]$. They are indexed by duads. Two double lines intersect if they share a duad.

A cardinal hyperplane is dual to a singular point of S_3. The set of cardinal planes form one $\mathfrak{S}_6$-orbit of the hyperplane $y_1+y_2+y_3-y_4-y_5-y_6=0$. They are indexed by subsets of $[1,6]$ of cardinality 3 up to complementary set.

The cardinal hyperplane $H(123)$ intersects CR_4 along the quadric given by equation $y_1y_2+y_1y_3+y_2y_3-y_4y_5-y_4y_6-y_5y_6=0$. A cardinal hyperplane $H(abc)$ contains 6 double lines corresponding to the synthemes $(ai)(bj)(ck)$, where $\{i,j,k\}$ is the complementary set to the set $\{a,b,c\}$.

We will denote by $\mathbb{P}^4_{\mathrm{Seg}}$ the projective space $|L|^*$, where S_3 lies, and denote by $\mathbb{P}^4_{\mathrm{CR}}$ the dual space $|L|$, where CR_4 lies,

A general hyperplane H in $\mathbb{P}^4_{\mathrm{CR}}$ cuts out CR_4 along a 15-nodal quartic surface X^H. Its pre-image under Ψ is cut out in S_3 by a quadric Q^H. It is the polar quadric of S_3 with the pole h corresponding to H via the projective duality.

The pre-image of $X^H_{\mathrm{Seg}}=\mathsf{S}_3\cap Q^H$ under the rational map Φ from (5.1) is a quartic surface K^H with nodes at the points $p_1,\ldots,p_5$. The linear system of such surfaces defines the composition map

$$\Psi\circ\Phi\colon \mathbb{P}^3 \dashrightarrow \mathbb{P}^4_{\mathrm{CR}}.$$

The surface K^H contains two sets of 10 lines, ten of them are the lines $\langle p_i,p_j\rangle$ and the other ten are the residual lines in the sections of K^H by the planes $\langle p_i,p_j,p_k\rangle$. In a minimal resolution of K^H we get two sets of disjoint (-2)-curves, each one from one set intersects three members of the second set.

(20) *A 15-nodal quartic surface X^H is birationally isomorphic to a quartic surface K^H in $\mathbb{P}^3$ with 5 nodes and two sets of disjoint lines forming a symmetric configuration* (10_3).

5.2. It is known that the Fano variety $F(\mathsf{S}_3)$ of lines in S_3 consists of 21 irreducible components. Fifteen of them parameterize lines in 15 planes. The remaining six components are isomorphic to the quintic del Pezzo surface dP_5 [Dol16]. We denote them by $F(\mathsf{S}_3)_1, \ldots, F(\mathsf{S}_3)_6$. The six components are transitively permuted by ${}^{II}\mathfrak{S}_6$. The stabilizer subgroup of $F(\mathsf{S}_3)_i$ in ${}^{II}\mathfrak{S}_6$ is isomorphic to $\mathfrak{S}_5 = \mathrm{Aut}(\mathrm{dP}_5)$. The five del Pezzo components $F(\mathsf{S}_3)_i, i = 1, \ldots, 5$, parameterize lines in S_3 that meet the planes Π_i. Under the map $\Phi\colon \mathbb{P}^3 \dashrightarrow \mathsf{S}_3$ from (5.1), they are realized as the blow-ups of 4 points on the plane Π_i equal to the images of lines $\langle p_i, p_j\rangle$, $j \neq i$. The component $F(\mathsf{S}_3)_6$ parameterizes lines that meet five planes $\Pi_1, \ldots, \Pi_5$. By the *Theorem on the fifth associated plane* (see [SR85, Chapter X, Theorem XXIV], or [Dol12, Theorem 8.5.3]) this is a quintic del Pezzo surface.

Thus the choice of a non-transitive subgroup $\mathfrak{S}_5$ of ${}^{II}\mathfrak{S}_6$ or a transitive subgroup of ${}^{I}\mathfrak{S}_6$ defines the choice of 5 planes among the 15 planes contained in S_3. As we noted earlier, this choice is equivalent to the choice of a total.

(21) *The choice of a total in the set* $[1, 6]$ *defines the choice of* 5 *planes in* S_3, *or, equivalently,* 5 *double lines in* CR_4, *or equivalently,* 5 *nodes of a* 15-*nodal quartic* X^H. *There are six such choices.*

The intersection of the quadric Q^H with a general line on S_3 corresponding to a point in $F(\mathsf{S}_3)_i$ consists of two points. This defines a birational involution on X^H_{Seg} (or X^H) and hence a biregular involution on the K3-surface Y^H birationally isomorphic to X^H_{Seg} and X^H. Its orbits correspond to points in $F(\mathsf{S}_3)_i$.

(22) *The choice of a total defines a birational involution on* X^H *and a biregular involution on* Y^H *with quotient isomorphic to the quintic del Pezzo surface* dP_5.

5.3. Let us recall some known facts about a geometric interpretation of irreducible representations of $\mathfrak{S}_5$ (see, for example, [CKS20, DFL18]). Let D be the union of 10 lines on a quintic del Pezzo surface dP_5 and $\Omega_{\mathrm{dP}_5}(\log D)$ be the rank 2 vector bundle of logarithmic 1-forms. It is invariant with respect to the group of automorphisms of dP_5 isomorphic to $\mathfrak{S}_5$. The linear space $H^0(\mathrm{dP}_5, \Omega_{\mathrm{dP}_5}(\log D))$ of global sections is an irreducible 5-dimensional linear representation W_5 of $\mathfrak{S}_5$. It generates the vector bundle and defines a $\mathfrak{S}_5$-equivariant anti-canonical embedding of dP_5 into the Grassmannian $G(3, W_5) \cong G(2, W_5^\vee)$ by taking the kernels (or cokernels) of the evaluation maps. The composition of this map with the Plücker embedding defines an embedding $\mathrm{dP}_5 \hookrightarrow \mathbb{P}(\bigwedge^2 W_5^\vee)$ which is contained in the projectivization of a 6-dimensional irreducible summand of $\bigwedge^2 W_5^\vee$.

The linear system $|\mathcal{O}_{\mathbb{P}(W_5^\vee)}(2)| = \mathbb{P}(S^2(\bigwedge^2 W_5))$ of quadrics in $\mathbb{P}(W_5^\vee)$ contains the subspace $|I_{\mathrm{dP}_5}(2)|$ of quadrics with base locus dP_5. It corresponds to an irreducible summand of dimension 5 of the linear representation $S^2(\bigwedge^2 W_5)$ isomorphic to W_5. We will use that the representation W_5 of $\mathfrak{S}_5$ is self-dual so that we may identify $G(2, W_5^\vee)$ with $G(2, W_5)$. The embedding $\mathrm{dP}_5 \hookrightarrow G(2, W_5)$ is now defined by assigning to a point $x \in \mathrm{dP}_5$ the pencil of quadrics from $|I_{\mathrm{dP}_5}(2)|$ with singular point at x. Under the projection map of dP_5 to $\mathbb{P}^4$ with center at x, this pencil is the pre-image of the pencil of quadrics containing the quartic del Pezzo surface equal to the image of the projection map (see [KV19]).

The universal family Z_{dP_5} of lines in $\mathbb{P}(W_5)$ parameterized by $\mathrm{dP}_5 \subset G(2, W_5)$ can be identified now with the incidence variety

$$Z_{\mathrm{dP}_5} = \{(Q, x) \in |I_{\mathrm{dP}_5}(2)| \times \mathrm{dP}_5 \colon x \in \mathrm{Sing}(Q)\}$$

that comes with two projection maps

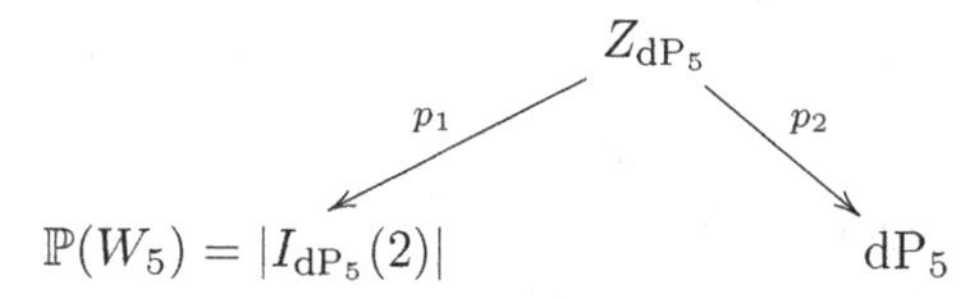

The image of the first projection p_1 consists of quadrics that have a singular point at some point in dP_5. In fact, any singular quadric from $|I_{\mathrm{dP}_5}(2)|$ has this property. Otherwise, if we project from its singular point not in dP_5, the image of a quintic surface dP_5 will be contained in a quadric surface, obviously impossible (see [KV19, Lemma 5.2]). Thus the image of p_1 is the locus Δ of singular quadrics in $|I_{\mathrm{dP}_5}(2)|$. The scheme-theoretical locus $\tilde{\Delta}$ of a linear system quadrics in $\mathbb{P}^5$ is given by vanishing of the discriminant polynomial of degree 6. It is known that the projective tangent space to a quadric Q of corank 1 considered as a point on the discriminant hypersurface of a linear system of quadrics is equal to the linear subspace of quadrics that pass through the singular point of Q [Dol12, (1.4.5)]. Since all quadrics in $|I_{\mathrm{dP}_5}(2)|$ pass through dP_5, we see that a general quadric in Δ is a singular point of the discriminant scheme. This shows that $\tilde{\Delta} = 2\Delta$ and

$$\deg \Delta = 3.$$

The restriction of the logarithmic bundle $\Omega_{\mathrm{dP}_5}(\log D)$ to a line l on dP_5 is a rank two vector bundle of determinant $-K_{\mathrm{dP}_5}\cdot l = 1$. It follows that the pre-image $p^{-1}(l)$ is the minimal ruled surface $\mathbb{F}_1$ and the projection p_1 blows down its exceptional section to a singular point of the image Δ of p_1. Thus we see that Δ is an irreducible cubic hypersurface in $|I_{\mathrm{dP}_5}(2)|$ with 10 isolated singular points. It must be isomorphic to the Segre cubic S_3.

(23) *There is a $\mathfrak{S}_5$-equivariant isomorphism between the locus Δ of singular quadrics in $|I_{\mathrm{dP}_5}(2)|$ and the Segre cubic S_3. The universal family of lines in S_3 parameterized by a del Pezzo component $F(\mathsf{S}_3)_i$ is $\mathfrak{S}_5$-equivariantly isomorphic to the $\mathbb{P}^1$-bundle $Z_{\mathrm{dP}_5}\to \mathrm{dP}_5$.*

(24) *The projection $p_1\colon Z_{\mathrm{dP}_5}\to \mathsf{S}_3$ is a small projective resolution of the Segre cubic.*

Our proof is a slight modification of the proofs from [CKS20, Lemma 2.31] and [KV19, Theorem 5.6].

Remark 5.1. In [Fin87] Finkelnberg proves that the Segre cubic admits six non-isomorphic small projective resolutions. Ours represents one isomorphism class. It is one of the two isomorphism classes of small resolutions that have the property that its automorphism group is isomorphic to $\mathfrak{S}_5$.

We have already used the fact that a quintic del Pezzo surface has 5 pencils $\mathcal{Q}_i$ of conics. Each one defines a conic fibration $\pi_i\colon S\to\mathbb{P}^1$. In the anti-canonical embedding the planes spanned by the members of the pencil $\mathcal{Q}_i$ sweep a scroll Z_i. The scroll Z_i is the image of the projective rank 2-bundle $\mathbb{P}(\mathcal{E}_i)$, where $\mathcal{E}_i = \pi_{i*}(\mathcal{O}_S(-K_S))$. Applying the relative Riemann-Roch, we easily find that $c_1(\mathcal{E}_i)\cong\mathcal{O}_{\mathbb{P}^1}(1)^{\oplus 3}$, i.e. the projective bundle is isomorphic to $\mathbb{P}^2\times\mathbb{P}^1\to\mathbb{P}^1$. The map from this bundle to $\mathbb{P}^5$ is given by $|\mathcal{O}_{\mathbb{P}(\mathcal{E}_i)}(1)|$ and coincides with the Segre map $\mathbb{P}^2\times\mathbb{P}^1\to\mathbb{P}^5$.

Let $|\mathcal{O}_{\mathbb{P}^5}(2)|\to|\mathcal{O}_{Z_i}(2)|$ be the restriction map. Since $h^0(\mathcal{O}_{Z_i}(2)) = h^0(S^2(\mathcal{E}_i)) = h^0(\mathbb{P}^1,\mathcal{O}_{\mathbb{P}^1}(2)^{\oplus 6}) = 18$, each scroll Z_i is contained in a net of quadrics. After we identify the Segre variety $s(\mathbb{P}^2\times\mathbb{P}^1)$ with the projective space of (2×3)-matrices of rank 1, we easily see that the space of quadrics containing it consists of quadrics of rank 4.

(25) [KV19] *The space $|I_S(2)|$ of quadrics containing an anti-canonical quintic del Pezzo surface S contains 5 planes Π_i of quadrics of corank 2.*

Using the projective duality, we can also embed $\mathrm{dP}_5\subset\mathrm{Gr}(2,W_5)$ in the dual Grassmannian $\mathrm{Gr}(3,W_5^\vee)$ of planes in the dual space $\mathbb{P}^4_{\mathrm{CR}}$.

The universal family of planes parameterized by dP_5 now becomes a $\mathbb{P}^2$-bundle over dP_5. Its projection to $\mathbb{P}^4_{\mathrm{CR}} = \mathbb{P}(W_5^\vee)$ is a double cover branched along the Castelnuovo–Richmond–Igusa quartic CR_4 [CKS20, Proposition 2.4.5].

6 15-Nodal Quartic Surfaces

6.1. It is a natural guess from reading the previous section that the choice of a total defines an isomorphism between the 15-nodal quartic surface X^H and the focal surface $\mathrm{Foc}(S)$ of a congruence of lines S of bidegree $(2,3)$. Note that this is not quite obvious since the construction gives a birational model of X^H inside the universal family of lines in $\mathbb{P}^4$ (not in $\mathbb{P}^3$) parameterized by S considered as a surface of lines in $\mathrm{Gr}_1(4)$ (not in $\mathrm{Gr}_1(3)$).

Fix a component $F(\mathsf{S}_3)_i$ of $F(\mathsf{S}_3)$ and let l be a line in it. We consider the plane $\langle l,h\rangle$ spanned by this line and the point $h \in \mathbb{P}^4_{\mathrm{Seg}}$ (dual to the hyperplane H). The dual of this plane in $\mathbb{P}^4_{\mathrm{CR}}$ is a line l^* contained in $H \subset \mathbb{P}^4_{\mathrm{CR}}$. The intersection $\mathsf{S}_3 \cap \langle l,h\rangle$ consists of l and a conic C that intersect l at two points. The intersection of the polar quadric Q^H with the plane $\langle l,h\rangle$ is equal to the intersection of the polar conic of $\mathsf{S}_3 \cap \langle l,h\rangle$ with respect to the point h. It passes through its singular points, i.e. the intersection points of l and C. Since the image of Q^H under the duality map Ψ is H, we obtain that the line l^* intersects the quartic X^H at two points, the images of $l \cap C$. In other words l^* is a bitangent line of X^H. Thus we realize $F(\mathsf{S}_3)_i$ as an irreducible congruence of bitangent lines of the quartic surface X^H.

Take a general point $x \in H$. The set of lines $l^* \subset H$ containing x corresponds to the set of planes $\langle \ell, h\rangle$ in the dual space that are contained in the hyperplane $x^\perp$ dual to the point x. This shows that the order m of the congruence $F(\mathsf{S}_3)_i$ is equal to the number of lines in S_3 that lie in the hyperplane H. If we consider the map (5.1) and take $i = 6$ then this number is equal to the number of twisted cubic curves parameterized by $F(\mathsf{S}_3)_6$ that lie in a quadric through the five points. This number is equal to 2 and consists of the unique curves of bidegree $(1,2)$ and $(2,1)$ that pass through general 5 points on the quadric. Similar count gives $m = 2$ for other components $F(\mathsf{S}_3)_i$ (in fact, they all permuted by $\mathfrak{S}_6$).

Now (2.4) and (2.5) give that the class n of $F(\mathsf{S}_3)_i$ is equal to 3. Alternatively, we can compute n as follows. If we take a general plane π contained in H, then its dual $\pi^\perp$ will be a line passing through h.

It intersects S_3 at three points. Through each of these points passes a unique line on S_3 from the family $F(\mathsf{S}_3)_i$. These define three planes $\langle l, h\rangle$ and give three lines l^* contained in π.

(26) *The choice of a total realizes X^H as the focal surface of a congruence of bidegree* $(2,3)$ *in* $\mathrm{Gr}_1(H)$.

The surface X^H also contains 10 tropes, the intersections with 10 cardinal hyperplanes. This gives ten more components of bi-degree $(0,1)$. The bidegree of the bitangent surface of a normal quartic surface is equal to $(12,28)$ [Sal58, p. 281]. Since $(12,28) = 10(0,1)+6(2,3)$ we have found all irreducible components.

6.2. Let X_{15} be any 15-nodal quartic surface in $\mathbb{P}^3$. Let us see how it arises as the focal surface of a congruence of bidegree $(2,3)$ and hence it is isomorphic to a surface X^H (another proof of this fact can be found in [Avi19]). Projecting X_{15} from a node, we obtain a double cover of $\mathbb{P}^2$ branched along a curve of degree 6 with 14 nodes. The only way to realize this curve is to take the union of four lines l_1, l_2, l_3, l_4 in general linear position and a conic C intersecting each line transversally outside the intersection points. The fifteenth node of X_{15} arises, in the usual way, from a conic K which is everywhere tangent to the sextic curve. The linear system on the double cover $\pi\colon X_{15} \dashrightarrow \mathbb{P}^2$ that maps it to a quartic surface is equal to $|\pi^*(l)+K'|$, where $\pi^{-1}(K) = K'+K''$ and l is a line on $\mathbb{P}^2$.

The reduced pre-images of the line components of the branch curves are conics passing through the fifteenth node P. These are our trope-conics $\sigma(E_x), x \in \mathcal{L}$, passing through the node. In this way we see the configuration of 10 trope-conics and 15 nodes forming an abstract configuration of type $(15_4, 10_6)$. The pre-image of the conic component C of the branch locus is a curve of degree 4 passing doubly through P and simply through 7 other nodes. Thus X_{15} contains 15 nodes, 10 trope-conics passing through 6 nodes, and 5 quartic curves with a double point at one of the nodes.

Consider the linear system of polar cubics of the 15-nodal quartic X_{15}. On a minimal resolution Y of X_{15} it is defined by the linear system $|D| = |3\eta - \sum E_i|$, where E_i are the exceptional curves of the resolution and η is the pre-image of the class of a hyperplane section of X_{15}. We check that its dimension is equal to 4 and its degree is equal to 6.

We have $(D-E_i)^2 = 0$, so that $|D-E_i|$ is an elliptic pencil and E_i is mapped under the map $\phi_D\colon Y \to \mathbb{P}^4$ given by the linear system $|D|$ to a plane conic. The members of $|D - E_i|$ are curves of arithmetic genus one and the map $\phi_{|D|}$ maps these curves one-to-one to curves of arithmetic genus one spanning a $\mathbb{P}^3$ that contains the plane spanned by the conic image of E_i. This shows that the map $\phi_{|D|}$ is of degree 1 and its image Z is a surface of degree 6, the intersection of a quadric Q and a cubic (not uniquely defined). The quadric Q is tangent to a cubic at the points $q_1, \dots, q_{10}$, the images of the 10 trope-conics on X_{15}. Let us consider the 4-dimensional linear system of quadrics through $q_1, \dots, q_{10}$. It defines a rational map $f\colon \mathbb{P}^4 \dashrightarrow \mathbb{P}^4$. Its restriction to Z is given by the linear system $|A| = |2\eta_Z - R_1 - \cdots - R_{10}|$ on the minimal resolution $\tilde{Z}$ of Z, where η_Z is the pre-image of the class of a hyperplane section of Z and R_i are the exceptional curves over the points q_i. Since $A^2 = 4$, it maps Z (and $\tilde{Z}$) to a quartic surface in $\mathbb{P}^4$ contained in the hyperplane H whose pre-image under the map is the quadric Q. Thus the image Z' of Z is the intersection of a quartic hypersurface V_4 with the hyperplane H'. The preimage of V_4 under f is a cubic hypersurface that contains 15 planes spanned by the 15 conics on Z. The intersection points of the 15 planes give 10 singular points of this hypersurface. By the projective uniqueness of the Segre cubic, the cubic hypersurface must be projectively equivalent to S_3. Since CR_4 is the image of S_3 under the map f given by quadrics through its nodes, we see that $f(\mathsf{S}_3)$ is projectively equivalent to CR_4 and Z' is projectively equivalent to $X^H = \mathrm{CR}_4 \cap H$ for some hyperplane H projectively equivalent to H'.

Let η_H be the class of a hyperplane section of X^H. We have

$$\eta_Z = 3\eta - \sum E_i \quad \text{and} \quad \eta_H = 2\eta_Z - \sum R_i = 6\eta - 2\sum E_i - \sum R_i$$

in $\mathrm{Pic}(Y)$. By expressing the classes R_i corresponding to 10 trope-conics on X_{15} in terms of η and the classes of E_i, and substituting into η_H we obtain $\eta_H = \eta$. Thus we have constructed a projective isomorphism from X_{15} to X^H.

(27) *A 15-nodal quartic surface X_{15} is projectively isomorphic to a hyperplane section X^H of the Castelnuovo–Richmond–Igusa quartic.*

(28) *The congruence* $\mathrm{Bit}(X_{15})$ *of bitangent lines of X_{15} consists of 16 irreducible components, six of them are congruences of bidegree* $(2,3)$ *and 10 are planes of bidegree* $(0,1)$.

Remark 6.1. The last assertion should be compared with the similar assertion that the congruence of bitangent lines of a quartic Kummer surface with 16 nodes consists of 16 plane components and 6 components of bidegree $(2,2)$. The minimal resolution R_S of the Kummer surface

$\mathrm{Foc}(S)$ is a double cover of a quartic del Pezzo surface with branch curve $B \in |-2K_S|$. In the isomorphic model of S as the blow-up of 5 points $p_1, \ldots, p_5$ in the plane, the curve B is the proper transform of a plane curve of degree 6 with cusps at the five points and tangent to all lines $\langle p_i, p_j \rangle$ and the conic through the five points. The analog of the linear system $|\tilde{B}|$ from (4.3) defines a smooth octic model of the focal surface $\mathrm{Foc}(S)$. It is equal to the intersection of three diagonal quadrics:

$$\sum_{i=0}^{5} z_i^2 = \sum_{i=0}^{5} a_i z_i^2 = \sum_{i=0}^{5} a_i^2 z_i^2 = 0.$$

The sections by the hyperplanes $z_i = 0$ are the canonical models of the curve B. The six choices of the hyperplane correspond to the six realizations of the Kummer surface as the focal surface of a congruence of bidegeree $(2,2)$. The curve B is a *Humbert curve* of genus 5 (see [Edg91]). It has a special property that it contains a maximal number ($= 10$) of vanishing even theta characteristics for a curve of genus 5 [Var86]. The number of moduli of Humbert curves is equal to 2.

6.3. Let $\mathcal{M}_{4,k}$ be the moduli space of projective equivalence classes of k-nodal quartic surfaces in $\mathbb{P}^3$. It is a locally closed subset of the GIT-quotient of $|\mathcal{O}_{\mathbb{P}^3}(4)|//\mathrm{SL}(4)$. It is an interesting problem to find out whether the irreducible components of $\mathcal{M}_{4,k}$ are rational varieties. It follows from the above that any 15-nodal quartic surface is birationally isomorphic to $X^H = \mathrm{CR}_4 \cap H$. It is known that the group of projective automorphisms of S_3 and CR_4 is isomorphic to $\mathfrak{S}_6$ (e.g. [Fin87, §3]). Thus the stabilizer subgroup of X^H in $\mathrm{Aut}(\mathbb{P}^4_{\mathrm{CR}}) \cong \mathrm{PGL}(5)$ is isomorphic to $\mathfrak{S}_6$. This gives

$$\mathbb{P}^4_{\mathrm{Seg}}/\mathfrak{S}_6 = \check{\mathbb{P}}^4_{\mathrm{CR}}/\mathfrak{S}_6 \dashrightarrow \mathcal{M}_{4,15}, \ H \mapsto H \cap \mathrm{CR}_4,$$

admits the inverse. It is known [Dol12, 9.4.5] that

$$\mathbb{P}^4_{\mathrm{Seg}}/\mathfrak{S}_6 \cong \mathbb{P}(2,3,4,5,6).$$

Therefore we obtain the following.

(29) *The moduli space $\mathcal{M}_{4,15}$ of 15-nodal quartic surfaces is a rational 4-fold.*

Note that the moduli space of Kummer surfaces $\mathcal{M}_{4,16}$ is realized as the hypersurface in $\mathcal{M}_{4,15}$ birationally isomorphic to the quotient

$$\mathrm{CR}_4/\mathfrak{S}_6 \cong \mathbb{P}(2,3,5,6).$$

Of course, the rationality of $\mathcal{M}_{4,16}$ is a well-known fact.

6.4. Now we know that X_{15} is projectively isomorphic to a quartic surface in $\mathbb{P}^4_{\mathrm{CR}}$ equal to a hyperplane section X^H of CR_4. So, we can identify the ambient space $\mathbb{P}^3$ with a hyperplane H in $\mathbb{P}^4_{\mathrm{CR}}$. We can also identify X_{15} with the focal surface $\mathrm{Foc}(S)$ of a congruence S of lines in H of bidegree $(2,3)$ that defines a partition $\mathcal{L}+\mathcal{C}$ of its 15 nodes and defines the involution σ such that $\sigma(E_x)$, $x\in\mathcal{L}$, are its trope-conics and $\sigma(E_x)$, $x\in\mathcal{C}$, are its trope-quartics. There are six ways to do it and we fix one of them. We will use the expressions of different linear systems on X_{15} in terms of the divisor classes of $E_x,\sigma(E_x)$, on the minimal resolution Y of X_{15} given in subsection 4.2.

Recall from (*19*) that there is a web W of quadrics (P) in H through the nodes $x\in\mathcal{C}$ parameterized by points P in H. So, we can identify W with H and consider the map $f\colon X_{15}\dashrightarrow H^*$ given by the linear system W. It is a hyperplane in the linear system $|2\eta_H-\sum_{x\in\mathcal{C}}E_x|$, where η_H is a hyperplane section of X_{15} (lifted to Y, as usual). We know from (*17*) that X_{15} coincides with the discriminant surface D_W of W. Recall that the *Steinerian surface* (or the *Jacobian surface*) of W is the locus of singular points of quadrics from W (see [Dol12, 1.1.7]). The rational map

$$\mathrm{st}\colon \mathsf{D}_W \dashrightarrow \mathrm{St}(W)$$

that assigns to a singular quadric from W its singular point is called the *Steinerian map*. It is given by the linear system $|\eta_{\mathrm{st}}|$ such that

$$2\eta_{\mathrm{st}} = 3\eta_H - \sum_{x\in\mathcal{L}} E_x.$$

Since $\eta_{\mathrm{st}}^2=4$, the Steinerian surface is a quartic surface in H. We see that it has 5 nodes, the images of E_x, $x\in\mathcal{C}$, and 20 lines, the images of E_x, $\sigma(E_x)$, $x\in\mathcal{L}$. Thus it coincides with the 5-nodal quartic model K^H of X_{15} and we can identify the source space $\mathbb{P}^3$ in the map Φ from (5.1) with the space $|\eta_{\mathrm{st}}|^*$. We have just described a *Cremona transformation* of H that transforms a 15-nodal quartic to a 5-nodal quartic.

The map Φ is given now by the linear system

$$\Big|2\eta_{\mathrm{st}} - \sum_{x\in\mathcal{C}} E_x\Big|, \tag{6.1}$$

so W is identified with a hyperplane in this linear system. It maps X_{15} to the projection of the sextic surface $X^H_{\mathrm{Seg}}=\mathsf{S}_3\cap Q^H$ from the point

$h \in \mathbb{P}^4_{\text{Seg}}$ corresponding to H via the projective duality. It is a sextic surface in the dual space H^*.

Next we identify this sextic surface with the dual surface X_{15}^* of X_{15}. The pre-image of a plane in H^* through a point x^* under the map f is a quadric in H that contains the fiber $f^{-1}(x^*)$. Thus this fiber is the base scheme of the net of quadrics in H defined by the point x^* considered as a plane in H. A general net of quadrics in $\mathbb{P}^3$ has 8 distinct base points. In our case all quadrics pass through 5 points from $\mathcal{C}$. So, our map f is of degree 3. A net of quadrics has less than 8 base points if one of the quadrics in the net has a singular point at a base point. This shows that the ramification divisor $\text{Ram}(f)$ coincides with the set of singular points of quadrics from W. This is the Steinerian surface $\text{St}(W)$. On the other hand, a quadric is singular if it does not intersect transversally the branch divisor $f(\text{Ram}(f))$. Considered as a point in H, the quadric belongs to the discriminant surface D_W of W. We know from fact (*19*) that D_W coincides with our quartic surface X_{15}. Thus the branch divisor is the dual surface $X_{15}{}^*$ of X_{15}. The image of the Steinerian surface is the branch divisor. This gives an identification of the sextic surface in H^* with the dual surface of X_{15}.

(30) *If we identify a* 15-*nodal quartic surface* X_{15} *with* X^H, *then its projectively dual surface* X_{15}^* *is a surface of degree* 6 *in* H^* *obtained by projection of its sextic model* X^H_{Seg} *from the point* h *dual to* H.

(31) *The* 5-*nodal quartic model* K^H *is identified with the Steinerian quartic surface in* H *of the web of quadrics* W *whose discriminant surface is equal to* X_{15}.

The degree of the dual surface agrees with the Plücker formula [Dol12, Theorem 1.2.7], according to which the degree of the dual of a k-nodal quartic surface is equal to $36 - 2k$. We refer to [Bak10, page 160] for another more explicit map from an X_{15} to its 5-nodal quartic model.

Remark 6.2. Note that $f^{-1}(f(\text{Ram}(f))) = 2\text{Ram}(f) + F$, and comparing the degrees, we obtain that the residual surface F is of degree 10. The map f defines a birational isomorphism $F \cong X_{15}^*$, so it realizes a birational model of X_{15} of degree 10 in $\mathbb{P}^3$. In the next section we will study a nonsingular degree 10 model of X_{15} in $\mathbb{P}^6$. The surface F is its projection to $\mathbb{P}^3$. The surfaces F and $\text{St}(W) = \text{Ram}(f)$ intersect along a curve of degree 40. We believe (but no proof yet) that this curve is birationally isomorphic to the curve $\text{Foc}(S)_0$ when we identify X_{15} with the focal surface of a congruence S of bidegree $(2, 3)$.

6.5. Recall that a web W of quadrics in $\mathbb{P}^3$ defines a *Reye congruence* $\mathrm{Rey}(W)$ of lines in $\mathbb{P}^3$ (reducible if W has base points). Its rays are lines (called *Reye lines*) which are contained in a subpencil of the web (see [Cos83], [Dol12, 1.1.7] and [DK, Chapter 2]). If W has no base points, then $\mathrm{Rey}(W)$ is irreducible and its bidegree is equal to $(7,3)$. The lines through the base points are Reye lines, so, in our case, the Reye congruence has 5 components of bidegree $(1,0)$. The remaining irreducible component $\mathrm{Rey}(W)_0$ is of bidegree $(2,3)$.

Since $m = 2$, each quadric surface $(P) \in W$ is equal to the locus of centers of null-planes containing P. Take a ray l_s from S that does not pass through the base points of W. Any general point $x \in \ell_s$ lies on another ray $\ell_{s'}$ such that $x = \ell_s \cap \ell_{s'}$ becomes the center of the null-plane $\langle \ell_s, \ell_{s'} \rangle$. This shows that the quadric (x) contains the line ℓ_s, and hence ℓ_s is a Reye line. This implies that $\mathrm{Rey}(W)_0$ coincides with S.

A pencil $\mathcal{P}_\ell$ of quadrics from W containing a Reye line ℓ not through a base point has the base locus equal to the union of the line and a twisted cubic curve intersecting it at two points and passing through the base points. We again see the congruence as a del Pezzo surface parameterizing twisted cubics through the set $\mathcal{C}$.

The pencil of quadrics containing a general Reye line ℓ has exactly two singular quadrics (P) and (P') with singular points at the singular points of the base curve of the pencil. This shows that ℓ is tangent to the discriminant surface X_{15} of W at two points P and P'. But it also intersects the Steinerian surface at two points. The first pair of points is an orbit of the involution σ on X_{15} defined by the congruence. The second pair of points is an orbit of an involution on $\mathrm{St}(W)$ which we denote by τ_{Rey} and call the *Reye involution.* Note that they are different involutions on a nonsingular model Y of both surfaces. As we know the former involution has the fixed locus an irreducible curve $\mathrm{Foc}(S)_0$ of degree 10. The latter involution has fixed locus consisting of 5 nodes of $\mathrm{St}(S)$ which become the disjoint union of 5 smooth rational curves on Y. The base locus of the pencil in W through a Reye line intersecting $\mathrm{Foc}(S)$ at a point in $\mathrm{Foc}(S)_0$ consists of this line and a twisted cubic tangent to the line. The base locus of the quadrics containing a Reye line passing through a base point p_i consists of this line and a twisted cubic passing through all base points and tangent to the line at the point p_i.

Finally, observe that the proper transforms $\sigma(E_x), x \in \mathcal{L}$, of ten trope-conics on Y are mapped under the Steinerian map to lines joining two base points of W. They are Reye lines invariant under the Reye involution.

Let $Z = Y/(\tau_{\mathrm{Rey}})$ be the quotient surface by the involution τ_{Rey}. Since its set of fixed points is the union of 5 smooth rational curves E_x, $x \in \mathcal{C}$, the usual formulas tell us that Z is a smooth rational surface with Euler–Poincaré characteristic equal to 18. The branch curve of the quotient map $q\colon Y \to Z$ is the disjoint union of smooth rational curves $\bar{E}_x$ with self-intersection -4. The Hurwitz type formula $K_Y = q^*(K_Z) + \sum_{x\in\mathcal{C}} E_x$ implies that $\sum_{x\in\mathcal{C}} \bar{E}_x \in |-2K_Y|$. Since $h^0(-2K_Z) = 1$, we see that $h^0(-K_Z) = 0$. Thus Z is a *Coble surface*, a smooth projective rational surface with empty anti-canonical linear system and nonempty anti-bicanonical linear system. As we observed in the previous paragraph, the exceptional curves $\sigma(E_x)$, $x \in \mathcal{L}$, are invariant under τ_{Rey}. They intersect the ramification locus at two points from different irreducible components. Their images on Z are 10 disjoint (-1)-curves. After we blow down these curves we obtain a quintic del Pezzo surface $\bar{Z}$. The image of the branch locus is the union of 5 conics on it (in an anti-canonical model). Since we identified Reye lines with rays of the congruence S, the surface $\bar{Z}$ can be identified with S and the conics with the conics $C(y)$, $y \in \mathcal{C}$, corresponding to trope-conics on $\mathrm{Foc}(S) = X_{15}$. Fix a realization of X_{15} as the focal surface $\mathrm{Foc}(S)$ of a quintic del Pezzo congruence of lines in $\mathbb{P}^3$.

(32) *The blow up of* 10 *intersection points of* 5 *conics* $C(y)$, $h(y) = 2$, *on* S *is a Coble surface* Z *with* $K_Z^2 = -5$. *The double cover of* Z *branched at the proper transforms of the conics is isomorphic to a K3 model* Y *of* X_{15}. *The covering involution coincides with the Reye involution* τ_{Rey} *associated to the web* W *of quadrics* (P), $P \in \mathbb{P}^3$, *with discriminant surface* X_{15}.

Recall that any quintic del Pezzo surface S contains 5 pencils of conics. If we choose one conic C_i from each pencil and blow up the ten intersection points we obtain a Coble surface Z with $|-2K_Z|$ represented by the union of proper transforms $\bar{C}_i$ of the conics. The double cover branched along this divisor is a K3 surface with 5 isolated (-2)-curves, the reduced pre-images of the curves $\bar{C}_i$. There are also 10 disjoint (-2)-curves, the pre-images of the exceptional curves over the intersection points. The construction depends on 5 parameters defined by a choice of a conic from each pencil. The isomorphism classes of our K3 surfaces depend on 4 parameters, so they define a hypersurface in $(\mathbb{P}^1)^5/\mathfrak{S}_5 \cong \mathbb{P}^5$, we do not know its degree.

Note that degenerating the five conics to the union of 5 reducible conics, we obtain a 'most algebraic' K3 surface with Picard number 20 studied by E. Vinberg [Vin83].

Remark 6.3. It is known that in the case when a web W of quadrics has no base points and its Steinerian surface is nonsingular, the Reye congruence $\mathrm{Rey}(W)$ is an Enriques surface realized as a congruence of lines of bidegree $(7,3)$ and the discriminant surface D_W is a Cayley quartic symmetroid with 10 nodes. The map

$$\nu\colon \mathrm{Rey}(W) \longrightarrow \mathrm{Bit}(\mathsf{D}_W),$$

from $\mathrm{Rey}(W)$ to the surface $\mathrm{Bit}(\mathsf{D}_W)$ of bitangents of D_W that assigns to a Reye line ℓ the pencil of quadrics $\mathcal{P}_\ell$ in W containing ℓ is the normalization map [DK, 7.4]. In our case, $\mathrm{Rey}(W)$ and $\mathrm{Bit}(\mathsf{D}_W)$ are reducible surfaces. The former has 5 components isomorphic to a plane and one component isomorphic to a quintic del Pezzo surface. They are mapped isomorphically to irreducible components of $\mathrm{Bit}(\mathsf{D}_W)$.

Remark 6.4. The fact that a 15-nodal quartic surface is a quartic *symmetroid* (i.e. equal to the determinant of a symmetric matrix with linear forms as its entries) should be compared with the well known fact that the Kummer surface X_{15} is also a quartic symmetroid with the Steinerian surface equal to the 6-nodal *Weddle quartic surface.* However, in the Kummer case the web of quadrics is not formed by surfaces (P) because it follows from (2.7) that, in this case they are planes but not quadrics.

Let us explain the relationship between the Weddle and the Kummer surface. Recall that we have a map $\Phi\colon \mathbb{P}^3 \dashrightarrow \mathsf{S}_3$ given by the linear system of quadrics through 5 points $p_1,\ldots,p_5$. If we add one more point p_6, then the web W of quadrics through the six points defines a map $\phi_W\colon \mathbb{P}^3 \dashrightarrow W^* \cong \mathbb{P}^3$ and its image is equal to the projection of the image S_3 of the map $\Phi\colon \mathbb{P}^3 \dashrightarrow \mathsf{S}_3$ from the point $\Psi(p_6) \in \mathsf{S}_3$. As we explained in subsection 6.4, the Steinerian surface $\mathrm{St}(W)$ is the ramification divisor of the map ϕ_W. It is a quartic surface with 6 nodes at $p_1,\ldots,p_6$, classically called a *Weddle surface.* The branch divisor of the map is the dual surface of the discriminant surface D_W of W.

On the other hand, we know that, the image of the ramification divisor under the map Ψ is the intersection $Q \cap \mathsf{S}_3$, where Q is the polar quadric with the pole at $x = \Phi(p_6)$ that corresponds to the hyperplane $H(y)$ that is tangent to CR_4 at the point $y = \Psi(x)$ and cuts out our Kummer surface X_{16}. This identifies W^* with $H(y)$ and the branch divisor with X_{16}, hence the discriminant surface D_W with X_{16}^*. Since X_{16} and X_{16}^* are projectively isomorphic, we get a symmetric determinantal realization of any Kummer surface. There are many classical sources that explain this construction (see, for example, [Cob82, page 124]).

We have described the birational map from the Weddle surface $\mathrm{St}(W)$ to the Kummer surface X_{16} as the map from the ramification divisor of ϕ_W to the branch divisor of the degree two map given by the web W quadrics through the nodes of $\mathrm{St}(W)$. The inverse map from X_{16} to $\mathrm{St}(W)$ is given by cubic surfaces that pass through 10 nodes of X_{16} complementary to a set of nodes on one of the trope-conic that contains the node y of X_{16} (a *odd trope-conic*). It maps the six nodes to the points $p_1, \ldots, p_6$ which are the six nodes of the Weddle surface $\mathrm{St}(W)$. It maps the trope-conic to the twisted cubic through the points $p_1, \ldots, p_6$ and it maps all other 15 trope-conics to lines $\langle p_i, p_j \rangle$ [Hud90, Chapter XI, §97]. The Reye involution is now the composition

$$\tau_{\mathrm{Rey}} = \sigma \circ p_y \circ \sigma,$$

where σ is the switch involution corresponding to one of the six realizations of X_{16} as the focal surface of a congruence of lines of bidegree $(2, 2)$ and p_y is the projection involution of X_{16} from the point y. The locus of fixed points of the Reye involution is the union of 6 tropes passing through the node y. The quotient by the Reye involution is a Coble surface obtained by blowing up 15 intersection points of six lines in general position in the plane. The surface is birationally isomorphic to the double cover of the plane branched along the six lines. This is of course a familiar birational model of a Kummer surface [Dol12, Theorem 10.3.16].

Note that the Kummer surface admits another symmetric determinant realizations as the Hessian surface of a nonsingular cubic surface (see [DK02, Theorem 4.1]) which has no analog for a 15-nodal quartic surface.

6.6. Let us introduce some elliptic pencils on Y invariant with respect to the Reye involution (recall that we have six of them corresponding to a choice of an isomorphism $X_{15} \cong \mathrm{Foc}(S)$).

Let

$$\mathfrak{r} = 2\eta_H - \sum_{x \in \mathcal{L}} E_x = 2\eta_{\mathrm{st}} - \eta_H \in \mathrm{Pic}(Y),$$

where we use the definition of η_{st} from (6.1). It defines a reflection involution of $\mathrm{Pic}(Y)$

$$s_{\mathfrak{r}} \colon v \mapsto v + \frac{x \cdot \mathfrak{r}}{2} \mathfrak{r}.$$

One checks that $\tau_{\mathrm{Rey}}^* = s_{\mathfrak{r}}$ [DK, 8.5]. In particular, we have

$$\begin{aligned}
\tau_{\mathrm{Rey}}^*(E_x) &\sim 2\eta_H - \sum_{x' \in \mathcal{L}, x' \neq x} E_x && \text{if } x \in \mathcal{L},\\
\tau_{\mathrm{Rey}}^*(E_y) &= E_y && \text{if } y \in \mathcal{C},\\
\tau_{\mathrm{Rey}}^*(\sigma(E_x)) &= \sigma(E_x) && \text{if } x \in \mathcal{L},\\
\tau_{\mathrm{Rey}}(\sigma(E_y)) &\sim \sigma(E_y) + 4\eta_H - 2\sum_{x \in \mathcal{L}} E_x && \text{if } y \in \mathcal{C},\\
\tau_{\mathrm{Rey}}^*(\eta_H) &= 8\eta_{\mathrm{st}} - 3\eta_H,\\
\tau_{\mathrm{Rey}}^*(\eta_{\mathrm{st}}) &= 3\eta_{\mathrm{st}} - \eta_H.
\end{aligned}$$

For any line l on $\mathrm{St}(W)$, the pencil of plane sections of $\mathrm{St}(W)$ containing l defines the residual pencil of cubic curves. It gives rise to an elliptic pencil $|\eta_{\mathrm{st}} - L|$, where L is the proper transform of l on Y. Thus we have 10 elliptic pencils $|F_x| = |\eta_{\mathrm{st}} - E_x|, x \in \mathcal{L}$ and 10 pencils $|F_x'| = |\eta_{\mathrm{st}} - \sigma(E_x)|$. Using the formulas from the above, we verify that

$$\tau_{\mathrm{Rey}}^*(F_x) = \tau_{\mathrm{Rey}}^*(\eta_{\mathrm{st}} - E_x) = \eta_{\mathrm{st}} - E_x.$$

This shows that the pencils $|F_x|$ are invariant with respect to the Reye involution. We have

$$\sum_{x \in \mathcal{L}} F_x = 10\eta_{\mathrm{st}} - \sum_{x \in \mathcal{L}} E_x = 10\eta_{\mathrm{st}} + 2\eta_{\mathrm{st}} - 3\eta_H = 3(4\eta_{\mathrm{st}} - \eta_H).$$

The linear system

$$\left|\frac{1}{3}\sum_{x \in \mathcal{L}} F_x\right| = |4\eta_{\mathrm{st}} - \eta_H| \tag{6.2}$$

maps Y to a surface of degree 20 in $\mathbb{P}^{11}$. The ramification divisor of the projection $Y \to Z$ is equal to $R = \sum_{y \in \mathcal{C}} E_y$. Since $F_x \cdot E_y = 0$, each pencil $|F_x|$ descends to an elliptic pencil $\bar{F}_x$ on the Coble surface Z. The linear system (6.2) descends to a linear system

$$|\Delta| = \left|\frac{1}{3}\sum_{x \in \mathcal{L}} \bar{F}_x\right|$$

that maps Z to a rational surface of degree 10 in $\mathbb{P}^5$, a *Fano model* of the Coble surface Z (see [DK, Chapter 9]). It has 5 ordinary nodes, the images of the curves $E_y, y \in \mathcal{C}$.

Note that the involution σ defined by the congruence of lines S transforms $|F_x|$ to the pencil $|F_x'|$. It is invariant with respect to the conjugate involution $\sigma \circ \tau_{\mathrm{Rey}} \circ \sigma$.

Let us now consider other elliptic pencils invariant with respect to the Reye involution. They are the pre-images $|F_y|, y \in \mathcal{C}$, of the pencils of conics on S under the composition map $f\colon Y \to Z = Y/(\tau_{\mathrm{Rey}}) \to S$. Recall that each pencil contains the curve $C(y)$, $y \in \mathcal{C}$, defined by the trope-quartic on $X_{15} = \mathrm{Foc}(S)$.

Take one conic $C(y), y \in \mathcal{C}$, from each of the five pencils. Since the map ramifies over $C(y)$, its pre-image on Y is the curve $E_y, y \in \mathcal{C}$, that enters in the fiber with multiplicity 2. The other components are the pre-images of four curves $\sigma(E_x), x \in \mathcal{L}$, that intersect $\sigma(E_y)$ with multiplicity one. This shows that the fiber is of type $\tilde{D}_4$ (or type I_0^* in Kodaira's notation). In our duad notations of the curves $E_x, x \in \mathcal{L} \cup \mathcal{C}$, we re-denote $|F_y|$ by $|F_a|$, where $y = (a6)$. Using formula from subsection 4.2, we find that

$$F_a \sim \frac{1}{2}\Big(4\eta_H - \sum_{b\neq 6} E_y - \sum_{b\neq a} E_{b6} - 2\sum_{c,d\neq a,6} E_{cd}\Big)$$

and $F_a \cdot F_b = 2$. Adding up, we get a linear system

$$\begin{aligned} \Big|\frac{1}{2}(F_1 + \cdots + F_5)\Big| &= \Big|5\eta_H - \sum_{y\in\mathcal{C}} E_y - 2\sum_{x\in\mathcal{L}} E_x\Big| \\ &= \Big|\sum_{x\in\mathcal{L}} \sigma(E_x) + \sum_{y\in\mathcal{C}} E_y\Big| \end{aligned} \tag{6.3}$$

which defines a τ_{Rey}-equivariant birational morphism to a degree 10 surface in $\mathbb{P}^6$. It blows down the curves $\sigma(E_x)$, $x \in \mathcal{L}$, to 10 singular points, maps the curves E_x, $x \in \mathcal{L}$, to rational quartic curves, maps the curves E_y, $y \in \mathcal{C}$, to conics and maps the curves $\sigma(E_y)$, $y \in \mathcal{C}$, to rational curves of degree 6.

(33) *The surface X_{15} admits 15 elliptic pencils $|F_x|$, $x \in \mathcal{L} \cup \mathcal{C}$, invariant with respect to the Reye involution τ_{Rey}. The linear system $|\frac{1}{3}\sum_{x\in\mathcal{L}} F_x|$ descends to a linear system on the Coble surface $Z = Y/(\tau_{\mathrm{Rey}})$ that defines a Fano model of Z as a rational* 5*-nodal surface of degree* 10 *in* $\mathbb{P}^5$. *The linear system $|\frac{1}{2}\sum_{y\in\mathcal{C}} F_y|$ defines a* 10*-nodal degree* 10 *birational model of Y in* $\mathbb{P}^6$.

7 Degree 10 Model of a 15-Nodal Quartic Surface

7.1. We continue to identify a 15-nodal quartic surface X_{15} with the focal surface $\mathrm{Foc}(S)$ of a quintic del Pezzo congruence of lines S. We have two linear systems $|\tilde{B}|$ given by equation (4.3) and given by equation (6.3). The former linear system is σ-invariant and ample. It maps the minimal resolution Y of X_{15} onto a nonsingular K3 surface of degree 10 in $\mathbb{P}^6$. The latter is τ_{Rey}-invariant and maps Y onto a 10-nodal degree 10 surface in $\mathbb{P}^6$. In this section we will only consider the nonsingular degree 10 model Y_{10} of Y obtained in this way.

In the anti-canonical embedding of S the curve $B \in |-2K_S|$ is cut out in S by a quadric. Since S is a linear intersection of the Grassmannian $\mathrm{Gr}_1(4)$, the curve B is a canonical curve in $\mathbb{P}^5$ equal to the intersection of quadrics. By Petri Theorem, it is neither hyperelliptic, nor trigonal, no contains a g_2^5. It also embeds in S by global sections of the logarithmic bundle on S. It follows from [Muk93] that it has only finitely many g_4^1, i.e. it is not a bielliptic curve. By another result of Mukai [Muk88], the image of Y under the map defined by $|B|$ is the intersection of the Fano threefold V_5 of genus 6 (degree 5) and index 2 in $\mathbb{P}^6$ with a quadric Q. The images of all curves E_x, $\sigma(E_x)$, $x \in \mathcal{L}$, are lines and the images of E_x, $\sigma(E_x)$, $x \in \mathcal{L}$, are conics.

We know that, for any $y \in \mathcal{C}$, we have $E_y \cdot \sigma(E_y) = 2$. This means that $G_y = E_y + \sigma(E_y)$ varies in an elliptic pencil $|G_x|$ on Y. The curves E_x, $E_{x'}$, $\sigma(E_x)$, $\sigma(E_{x'})$, x, $x' \in \mathcal{L}$, such that $\sigma(E_x) \cdot E_y = \sigma(E_{x'}) \cdot E_y = 0$ form a reducible fiber of $|G_y|$ of type $\tilde{A}_3$. There are 3 such fibers. The curves $E_x, x \in \mathcal{L}$, such that $E_x \cdot E_y = 1$ define 4 sections of the elliptic fibration.

We have $G_y \cdot G_{y'} = 2$ and it follows from (4.3) that

$$2\tilde{B} \sim \sum_{y \in \mathcal{C}} G_y.$$

The elliptic pencils descend to pencils of conics on the quotient S by the involution σ. It follows that the curve $\tilde{B}$ passes through the intersection points $E_i \cap E_i'$ and two opposite singular points in each reducible fiber of the pencil. Since $\tilde{B} \cdot G_y = 4$, the image of each member of $|G_y|$ in the degree 10 model $\mathsf{V}_5 \cap Q \subset \mathbb{P}^6$ of X_{15} is a quartic curve of arithmetic genus one, and as such it spans a $\mathbb{P}^3$ in $\mathbb{P}^6$. Since $G_y \cdot G_{y'} = 2$, we see that any pair of the $\mathbb{P}^3$'s intersects along a subspace of dimension ≥ 1.

7.2. Consider the scroll $\mathcal{S}_i$ swept by the 3-planes spanned by the images of the members of the elliptic pencil $|G_y|$. It is the image of the projective bundle $\mathbb{P}(\mathcal{E}_i)$, where $\mathcal{E}_i = (f_i)_*(\mathcal{O}_Y(\tilde{B}))$ is a rank 4 vector bundle on $\mathbb{P}^1$. We have already observed in section 5.4 that a quintic del Pezzo surface S equipped with a conic bundle structure $\pi_i \colon S \to \mathbb{P}^1$ embeds in the trivial projective bundle $\mathbb{P}^1 \times \mathbb{P}^2 = \mathbb{P}(\mathcal{N}_i) \to \mathbb{P}^1$, where $\mathcal{N}_i = \pi_{i*}(\omega_S^{-1}) \cong \mathcal{O}_{\mathbb{P}^1}(1)^{\oplus 3}$. The quotient projections $q\colon Y \to S$ show that

$$\mathcal{E}_i \cong \pi_{i*}(q_*(\mathcal{O}_Y(\tilde{B}))) = \pi_{i*}(\omega_S^{-1} \oplus \mathcal{O}_S) = \mathcal{O}_{\mathbb{P}^1}(1)^{\oplus 3} \oplus \mathcal{O}_{\mathbb{P}^1}.$$

The scroll $\mathcal{S}_i$ is equal to the image of $\mathbb{P}(\mathcal{E}_i^\vee)$ under the map given by the linear system $|\mathcal{O}_{\mathbb{P}(\mathcal{E})}(1)|$ (see also [JK04] for the confirmation of this fact[2]).

The image of the section $\mathbb{P}^1 \to \mathbb{P}(\mathcal{E}_i)$ given by the projection $\mathcal{E}_i \to \mathcal{O}_{\mathbb{P}^1}$ is blown down to a singular point s_i of $\mathcal{S}_i$ which is contained in all rulings of the scroll. The projections $\mathbb{P}^6 \dashrightarrow \mathbb{P}^5$ with center at s_i maps the scroll to the scroll $\mathbb{P}(\mathcal{N})$. Its restriction to Y_{10} is the projection $q\colon Y_{10} \to S$.

The canonical sheaf of $\mathbb{P} = \mathbb{P}(\mathcal{E}^\vee)$ is equal to $\mathcal{O}_{\mathbb{P}}(-4) \otimes \pi_i^*\omega_{\mathbb{P}^1}$, where π_i is the projection to the base. Using the adjunction formula, we see that $Y_{10} \subset \mathcal{S}_i$ is equal to a complete intersection of the images of two divisors $D_1, D_2 \in |\mathcal{O}_{\mathbb{P}}(2) \otimes \pi_i^*\mathcal{O}_{\mathbb{P}^1}(-1)|$ under the projection to $\mathbb{P}^6$ [JK04, 9.2]. The linear system $|\mathcal{O}_{\mathbb{P}}(2) \otimes \pi_i^*\mathcal{O}_{\mathbb{P}^1}(-1)|$ is cut out by quadrics in $\mathbb{P}^6$ that contain a ruling 3-plane of $\mathcal{S}_i$. Since a smooth quadric in $\mathbb{P}^6$ does not contain a 3-plane, all quadrics that cut out Y_{10} in $\mathcal{S}_i$ must be singular. They all pass through the singular points s_i of $\mathcal{S}_i$.

(34) *The nonsingular σ-invariant degree 10 model Y_{10} of a 15-nodal quartic Y is contained in five scrolls $\mathcal{S}_i$ swept by 3-planes spanning a fiber of the elliptic fibrations $f_i\colon Y \to \mathbb{P}^1$ given by the pencil $|G_y|$, $y \in \mathcal{C}$. Each scroll is the image of the projective bundle $\mathbb{P}(\mathcal{E}) = \mathbb{P}(\mathcal{O}_{\mathbb{P}^1}(1)^{\oplus 3} \oplus \mathcal{O}_{\mathbb{P}^1})$ in $\mathbb{P}^6$ under the linear system $|\mathcal{O}_{\mathbb{P}(\mathcal{E})}(1)|$. The section defined by the surjection $\mathcal{E} \to \mathcal{O}_{\mathbb{P}^1}$ is blown down to a singular points s_i of $\mathcal{S}_i$. The surface Y_{10} is cut out in $\mathcal{S}_i$ by two quadrics singular at the point s_i.*

7.3. The Fano variety V_5 is isomorphic to a transversal intersection of the Grassmannian $G = \mathrm{Gr}_1(\mathbb{P}(V_5))$ of lines in $\mathbb{P}^4 = |V_5|$ with a linear subspace $|W_7|$ of dimension 6 in the Plücker space $|\bigwedge^2 V_5|$. The linear system $|I_G(2)|$ of quadrics containing G is of dimension 4 and consists of quadrics of rank 6. It can be identified with the projective space $|V_5|$ via assigning to a vector $v \in V_5$ the quadratic form $q(x) = v \wedge x \wedge x$. After we choose a volume form on V_5, this corresponds to an isomorphism $I_G(2) \cong \bigwedge^4 V_5^\vee \cong V_5$.

[2] Note that there is a misprint in the inequality for $c = \mathrm{Cliff}(C) < [\frac{g-1}{2}|$ for the Clifford index for a curve C of genus g in [JK04, Theorem 3.2]. According to [GL87], it should be $\mathrm{Cliff}(C) \le [\frac{g-1}{2}]$. In our case $C = \tilde{B}$ is a curve of genus 6, and we get $c = 2$.

Let Y be a transversal intersection of V_5 with a quadric $Q = V(q)$. It is a smooth K3 surface of degree 10 and genus 6 in $\mathbb{P}^6 = |W_7|$. We have

$$I_Y(2) = I_{\mathsf{V}_5}(2) \oplus \langle q \rangle. \tag{7.1}$$

Since a general quadric in $I_{\mathsf{V}_5}(2)$ is of corank 1, we obtain that the discriminant hypersurface of degree 7 of singular quadrics in $|I_Y(2)|$ is equal to the union of a hyperplane $\mathcal{H} = |I_{\mathsf{V}_5}(2)|$ and a hypersurface $\mathcal{D}_Y$ of degree 6.

Now let $Y = Y_{10}$. Consider the restriction of the linear system of quadrics in $\mathbb{P}^6$ to a scroll $\mathcal{S}_i$. We have

$$S^2(\mathcal{E}) = S^2(\mathcal{O}_{\mathbb{P}^1}(1)^{\oplus 3}) \oplus S^2(\mathcal{O}_{\mathbb{P}^1}) \oplus \mathcal{O}_{\mathbb{P}^1}(1)^{\oplus 3}.$$

This implies that

$$\dim H^0(\mathbb{P}(\mathcal{E}), \mathcal{O}_{\mathbb{P}(\mathcal{E})}(2)) = 6h^0(\mathcal{O}_{\mathbb{P}^1}(2)) + 3h^0(\mathcal{O}_{\mathbb{P}^1}(1) + 1 = 25.$$

This shows that $\dim |I_{\mathcal{S}_i}(2)| = 28 - 25 - 1 = 2$, hence each scroll $\mathcal{S}_i$ is contained in a net of quadrics. Since each quadric containing $\mathcal{S}_i$ contains a pencil of 3-planes and also contains Y_{10}, we obtain that the discriminant hypersurface of the 5-dimensional hyperweb of quadrics $|I_{Y_{10}}(2)|$ contains 5 planes of quadrics of codimension 2. In fact, projecting a quadric $\mathcal{Q}$ containing $\mathcal{S}_i$ from the singular point s_i of $\mathcal{S}_i$, we find that the fibers contain $\mathbb{P}^3$ as a codimension one subvariety, hence the corank of $\mathcal{Q}$ must be greater than or equal to 2.

(35) *The discriminant hypersurface of the linear system $|I_{Y_{10}}(2)|$ of quadrics containing Y_{10} contains 5 planes in its singular locus parameterizing quadrics of corank ≥ 2.*

This should be compared with an analogous fact for a quintic del Pezzo surface S discussed in subsection 7.2.

7.4. Let Y be again any transversal intersection $\mathsf{V}_5 \cap Q$. Using (7.1), we obtain a decomposition

$$\bigwedge^3 I_Y(2) = \bigwedge^3 I_{\mathsf{V}_5}(2) \oplus \Big(\bigwedge^2 I_{\mathsf{V}_5}(2) \otimes \langle q \rangle\Big) = \bigwedge^3 V_5 \oplus \Big(\bigwedge^2 V_5 \otimes \langle q \rangle\Big).$$

If we view $G = \mathrm{Gr}_1(\mathbb{P}^4)$ as $\mathrm{Gr}(2, V_5)$ for some linear space V_5 of dimension 5, then we have a natural (means $\mathrm{SL}(V_5)$-equivariant) isomorphism

$$V_5 \to I_G(2), \quad v \mapsto v \wedge x \wedge x.$$

We view the quadric $Q = V(q)$ as a restriction of a quadric on the Plücker space $\bigwedge^2 V_5$, so, via polarization with respect to a vector $v \in \bigwedge^2 V_5$, it defines a linear form $L_v \in \bigwedge^2 V_5^\vee \cong \bigwedge^3 V_5$. We associate to Q the linear subspace

$$A_Q := \{(L_v, v \otimes q), v \in \bigwedge^2 V_5\} \subset \bigwedge^3 I_Y(2).$$

One checks that A_Q is a 10-dimensional Lagrangian subspace of $\bigwedge^3 I_Y(2)$ with respect to the wedge-product pairing, and as such defines an *EPW-sextic hypersurface* in $|I_Y(2)|$ isomorphic to the sextic hypersurface $\mathcal{D}_Y$ [IM11, Proposition 2.1 and its proof].

It is known that the irreducible family of EPW-sextics coming from a general degree 10 K3 surfaces via the construction from above form a codimension 1 subvariety in the moduli space of EPW-sextics. The dual EPW-sextics of those coming from K3's have a singular point of multiplicity 3 [O'G08, Proposition 3.4].

Let $Y^{[2]}$ be the *Hilbert scheme* of 0-cycles ξ on Y of length 2. For any such cycle ξ, its span $\langle \xi \rangle$ in $\mathbb{P}^6$ is a line. The linear subspace of quadrics in $|I_Y(2)|$ containing this line is a hyperplane, defining a rational map

$$\alpha \colon Y^{[2]} \dashrightarrow |I_Y(2)|^*. \tag{7.2}$$

It is clear that it is not defined on the set of cycles ξ that are contained in a line in Y. This set is the union of planes $P_l = l^{[2]}$, where l is a line in Y. Obviously any line on Y is contained in V_5. The Hilbert scheme $\mathrm{Fano}_1(\mathsf{V}_5)$ of lines in V_5 is isomorphic to $\mathbb{P}^2$ [Isk77, Corollary (6.6)]. Each line $l \in \mathrm{Fano}_1(\mathsf{V}_5)$ either intersects Y at a 0-cycle $\xi \in Y^{[2]}$ or is contained in Y. Thus $\mathrm{Fano}_1(\mathsf{V}_5)^0 = \mathrm{Fano}_1(\mathsf{V}_5) \setminus \{\text{lines in } Y\}$ embeds in $Y^{[2]}$. Let $\widetilde{Y^{[2]}}$ be the blow-up of the planes P_l and $\tilde{P}$ be the closure in $\widetilde{Y^{[2]}}$ of $\mathrm{Fano}_1(\mathsf{V}_5)^0$.

It is known that the Hilbert scheme of conics $\mathrm{Fano}_2(\mathsf{V}_5)$ is isomorphic to $\mathbb{P}^4$ [Ili94]. By [O'G05, Lemma 4.20], for any $\xi \in Y^{[2]}$ not contained in a line in V_5, there is a unique conic C_ξ on V_5 that contains ξ. The intersection of $C_\xi \cap Y$ is cut out by the quadric Q, thus one can define the residual set 0-cycle ξ' of length 2 (O'Grady shows that the residual 0-cycle is well defined for a non-reduced Z). This defines a birational involution ι of $Y^{[2]}$.

Since C_ξ is contained in V_5, the set of quadrics in $I_Y(2)$ containing the plane $\langle C_\xi \rangle$ is a hyperplane. Thus $\alpha(\xi) = \alpha(\iota(\xi))$ is the same point in W. This shows that the degree of α is at least 2.

The following results can be found in [O'G08].

(36) *For a general* $Y = \mathsf{V}_5 \cap Q$, *the involution* ι *extends to a biregular involution of the blow-up of* $Y^{[2]}$ *at the image of* $\mathrm{Fano}_1(\mathsf{V}_5)$ *under*

the map $l \mapsto l \cap Y$. The quotient by this involution is equal to the image W of α. It is a EPW-sextic with a unique triple point equal to the image of the exceptional divisor of the blow-up. The EPW-sextic is dual to the EPW-sextic $\mathcal{D}_Y$.

In our case, the situation is more complicated because Y contains lines and conics. Note that if ℓ_s contains a singular point x of X_{15}, then a line passing through this point intersects X_{15} at a point on $T(x)$, hence its image is a cycle of two points $(y, \sigma(y)) \in E_x \cup \sigma(E_x)$. The surface $\alpha(Y^{[2]})$ must have other singularities.

It is known that the Picard group of $Y^{[2]}$ is generated by $\mathrm{Pic}(Y)$ and $\frac{1}{2}[\Delta]$, where Δ is the exceptional divisor of $\pi\colon Y^{[2]} \to Y^{(2)}$, the pre-image of the diagonal. Here $\mathrm{Pic}(Y)$ is embedded in $\mathrm{Pic}(Y^{[2]})$ by $D \mapsto \pi^*(p_1^*(D) + p_2^*(D))$, where $p_i\colon Y \times Y \to Y$ are the projection maps. We denote the image of D in $\mathrm{Pic}(Y^{[2]})$ by D_2. Thus we see that $\alpha^*((E_x)_2)$ is the class of a line on S if $x \in \mathcal{L}$ and the class of a conic if $x \in \mathcal{C}$. Moreover, $\alpha^*(\frac{1}{2}\Delta) = \tilde{B} = -K_S$.

The involution σ of $Y_{10}^{[2]}$ extends to an involution $\tilde{\sigma}$ of $Y_{10}^{[2]}$ with fixed locus equal to the image of S in $Y_{10}^{[2]}$ defined by the isomorphism $Y_{10}/(\sigma) = S$. Composing the inclusion S in $Y_{10}^{[2]}$, we get a map

$$f\colon S \to Y_{10}^{[2]}.$$

(37) *The map $\alpha \circ f\colon S \to |I_{Y_{10}}(2)|^* = \mathbb{P}^5$ coincides with the anti-canonical map of the del Pezzo surface S.*

Note that there is also a rational map $\mathrm{Foc}(S) \dashrightarrow \Delta = \mathbb{P}(\Omega^1_{Y_{10}}) \subset Y_{10}^{[2]}$ that assigns to a nonsingular point of $\mathrm{Foc}(S)$ the cycle ξ that consists of the point and the infinitely near point corresponding to the tangent direction defined by the ray passing through this point. We extend this map to Y_{10} by assigning to a point $y \in E_x$ the point (y, t_y), where t_y is the tangent line to the line (conic) E_x at y. Composing this map with the map (7.2), we obtain a rational map

$$Y_{10} \to \mathbb{P}(\Omega^1_{Y_{10}}) \dashrightarrow |I_2(Y_{10})|^* \cong \mathbb{P}^5.$$

8 The Picard Group

8.1. In this section, we compute the Picard group of the minimal resolution Y of a *moduli general* 15*-nodal quartic surface* X_{15}. Since X_{15}

varies in a 4-dimensional family, we expect that the rank of $\mathrm{Pic}(Y)$ is equal to 16. Obvious generators over $\mathbb{Q}$ are the classes of exceptional curves E_i and the class of a hyperplane section η_H.

Recall that a subset N of the set $\mathcal{N}$ of nodes of X_{15} is called *even* (resp. *weakly even*) if the sum e_N of the divisor classes of the exceptional curves E_x, $x \in N$, is divisible by 2 (resp. is divisible by 2 after adding the class η_H). Let N be the set of 15 nodes of X_{15} and $\mathcal{N}$ be the sublattice of $\mathrm{Pic}(Y)$ generated by the classes of the exceptional curves E_x, $x \in N$, and η_H. We have $V = \mathcal{N}/2\mathcal{N} \cong \mathbb{F}_2^{16}$ and the kernel of the natural homomorphism $V \to \mathrm{Pic}(Y)/2\,\mathrm{Pic}(Y)$ is a linear code C generated by the projections of the classes e_N, η_H to V. It is known that for a 15-nodal quartic surface the code C is of dimension 5 and it contains ten words of weight 6, six words of weight 10 and 15 words of weight 8 (see [End99, Theorem 3.3] and [Roh87, p.48]). After we fix a splitting $N = \mathcal{L}+\mathcal{C}$, (4.1) gives an explicit description of 10 weakly even sets defining words of weight 6 and (4.2) gives an expression of weakly even sets defining words of weight 8. One chooses a basis of C consisting of 5 words of weight 6 from (4.1) by taking $B_C = (\sigma(E_{12}), \sigma(E_{23}), \sigma(E_{34}), \sigma(E_{45}), \sigma(E_{15}))$ (modulo $2\,\mathrm{Pic}(Y)$).

Let $\tilde{\mathcal{N}}$ be the even overlattice of $\mathcal{N}$ that corresponds to the isotropic subgroup H of the discriminant group $D_{\mathcal{N}} = \mathcal{N}^\vee/\mathcal{N}$ generated by the basis B_C of the code C. It contain $\mathcal{N}$ as a sublattice of finite index and its discriminant group is isomorphic to $A = H^\perp/H$ [Nik79, Proposition 1.4.1]. It is generated by

$$\begin{aligned}
\tfrac{1}{4}\eta_H - \tfrac{1}{2}(E_{14} + E_{25} + E_{35} + E_{56}), &\quad \tfrac{1}{2}(E_{13} + E_{16} + E_{26} + E_{36}),\\
\tfrac{1}{2}(E_{13} + E_{25} + E_{34} + E_{56}), &\quad \tfrac{1}{2}(E_{13} + E_{24} + E_{12} + E_{46}),\\
\tfrac{1}{2}(E_{13} + E_{35} + E_{16} + E_{56}), &\quad \tfrac{1}{2}(E_{14} + E_{24} + E_{16} + E_{26}).
\end{aligned}$$

Comparing this discriminant group and the quadratic form defined on it with that of the lattice $T = U(2) \oplus U(2) \oplus A_1(2) \oplus A_1$, we find that they are isomorphic. It follows from [Nik79, Theorem 1.4.2] that the genus of this lattice consists of one isomorphism class. It also follows from [Nik79, Theorem 1.4.4] that the lattice T admits a primitive embedding into the K3 lattice $L_{K3} = U^{\oplus 3} \oplus E_8^{\oplus 2}$ and all such embeddings are equivalent with respect to the group of isometries of L_{K3}. Let M be the orthogonal complement of T in L_{K3}. Then it is an even lattice of signature $(1, 15)$. Its discriminant group is isomorphic to the discriminant group of T (with the quadratic form multiplied by -1). Again applying Nikulin's result, we find that $\tilde{\mathcal{N}} \cong M$ and M admits a unique (up to isometry of L_{K3}) primitive embedding into L_{K3}. The theory of periods of lattice polarized

K3 surfaces shows that there exists an irreducible 4 dimensional moduli space of K3 surfaces containing the lattice M primitively embedded into their Picard lattice. Moreover, the Picard lattice of a moduli general surface isomorphic to M. Since the isomorphism class of Y belongs to this moduli space, a general 15-nodal quartic surface has the Picard lattice isomorphic to M.

(38) *The transcendental lattice of a K3 surface Y birationally isomorphic to a general 15-nodal quartic surface X_{15} is isomorphic to the lattice $U(2) \oplus U(2) \oplus A_1(2) \oplus A_1$. The Picard lattice is isomorphic to the orthogonal complement of this lattice in the K3-lattice $U^{\oplus 3} \oplus E_8^{\oplus 2}$.*

(39) *The Picard lattice* $\mathrm{Pic}(Y)$ *is generated by* 15 *classes of the curves E_x, $x \in \mathcal{L} \cup \mathcal{C}$, and 5 classes of the curves $\sigma(E_{12})$, $\sigma(E_{23})$, $\sigma(E_{34})$, $\sigma(E_{45})$, $\sigma(E_{15})$.*

It is natural to embed $\mathrm{Pic}(Y)$ into the Picard lattice of a minimal resolution Y of the Kummer surface $\mathrm{Kum}(J(C))$ of the Jacobian of a general curve C of genus 2. It is known that the latter is generated by the classes of a hyperplane section η_H, 16 exceptional curves N_α and 16 classes of trope-conics T_β indexed by subsets of cardinality 2 of the set $[1,6]$ with two additional curves N_0 and T_0. We have $N_\alpha \cdot T_\beta = 1$ if and only if $\alpha + \beta \in \{(0), (16), (26), (36), (46), (56)\}$, where the addition of subsets is the symmetric sum modulo the complementary set.

We continue to index our curves $E_x, x \in \mathcal{L}$, by 2-element subsets of $[1,5] = \{1, \ldots, 5\}$ and $E_x, x \in \mathcal{C}$, by elements of $[1,5]$. Then the embedding is as follows:

$$\eta_H \mapsto \eta_H, \qquad [E_{ab}] \mapsto [N_{ab}], \qquad [E_a] \mapsto [N_{a6}],$$
$$[\sigma(E_{ab})] \mapsto [T_{ab}], \qquad [\sigma(E_a)] \mapsto [T_{a6} + T_0 + N_0].$$

It is easy to see that under this embedding $\mathrm{Pic}(Y)$ is equal to the orthogonal complement of $r = [N_0]$ with $r^2 = -2$. This embedding corresponds to the fact that, under the specialization, the trope-quartic $\sigma(E_a)$ splits into the union of two trope-conics $T_{a6} + T_0$ passing through the new node with the exceptional curve N_0.

Recall that the group of birational automorphisms of a general 16-nodal Kummer quartic surface contains 16 *switch involutions* which interchange nodes with tropes. One of these involutions defines a biregular involution of Y that interchanges N_α with T_α [Kon98]. It follows that the involution σ is equal to the composition of this switch involution and the reflection with respect to the (-2)-vector $\alpha_0 = [T_0 + E_0]$.

8.2. We know that X_{15} contains 10 trope-conics and 5 trope-quartics. They are realized as the images of the exceptional curves E_x under the involution σ defined by the choice of a syntheme. Let us see that a general 15-nodal quartic surface X_{15} has no more conics or quartic rational curves lying on it.

Suppose we have an irreducible conic C on X_{15}. Let $\bar{C}$ be its proper transform on Y. Then $(\eta_H - \bar{C})^2 = -2$, $(\eta_H - \bar{C}) \cdot \eta_H = 2$, hence $\eta_H = \bar{C} + \bar{C}'$ for some other conic C'. Suppose $C = C'$, then the plane $\langle C \rangle$ is tangent to X_{15} along C and hence $2[\bar{C}] = \eta_H - \sum_{x\in A} E_x$, where A is a set of 6 exceptional curves (to make the self-intersection equal to -8). It is easy to see that we can find a trope-conic that passes through 3 nodes in A. Computing the intersection number, we conclude that $\bar{C}$ coincides with this conic. Suppose $C \neq C'$. Then $[\bar{C}] + [\bar{C}'] = \eta_H - \sum_{x\in B} E_x$. Since $\mathrm{Pic}(Y)_{\mathbb{Q}}$ is generated by η_H, E_x, we see that

$$2[\bar{C}] = H - \sum_{x\in B'} E_x, 2[\bar{C}'] = H - \sum_{x\in B''} E_x.$$

We find a contradiction by computing the intersection with some trope-conic.

Suppose we have an irreducible rational quartic curve C with a node at some singular point x_0 on X_{15}. Then $\bar{C} \sim \eta_H - E_{x_0} - \frac{1}{2}\sum_{i\in I} E_{x_i}$, where $\#I = 8$, as above. Without loss of generality we may assume that $E_{x0} = E_{16}$. Consider 4 trope-conics $\sigma(E_{1a})$, $a = 2,3,4,5$ passing through x_0. Let $Z(a)$ be the set of nodes contained in the conic except the node x_0. Let us identify nodes of X_{15} and the exceptional curves on Y with their indices (ab). We have $Z(a) \cap Z(b) = \{(cd)\}$, where $\{a,b,c,d\} = [1,4]$. It is easy to see that the set I of 8 nodes has at least 3 nodes in common with one of the subsets $Z(a)$ unless $I = \{(26),(36),(46),(56),(12),(13),(14),(15)\}$. In the latter case $\bar{Q}$ coincides with the quartic curve $\sigma(E_{16})$. So, we have proved the following.

(40) *A general* 15*-nodal quartic surface contains* 10 *conics which coincide with the trope-conics* $\sigma(E_x)$, $x \in \mathcal{L}$. *It also contains* 15 *rational quartic curves with a double point at one of the nodes. They coincide with the curves* ${}^s\sigma(E_x)$, $x \in \mathcal{C}$, *where* ${}^s\sigma$ *is one of the six involutions obtained from* σ *by conjugation by a permutation* $s \in \mathfrak{S}_6$.

(41) *The surface does not contain curves of odd degree.*

The last property follows from the fact that η_H intersects evenly any divisor class on Y since this is true for generators of $\mathrm{Pic}(Y)$.

9 Admissible Pentads

9.1. So far, we have located the following birational involutions of X_{15}:

- 6 involutions σ_i with quotient a del Pezzo surface of degree 5 with the branch curve a nonsingular curve of genus 6;
- 6 Reye involutions τ_{Rey}^i corresponding to the six structures of a quartic symmetroid with quotient a Coble surface Z with $K_Z^2 = -5$ and the branch curve equal to the union of five smooth rational curves.

Also, we have obvious involutions defined by the projections from nodes. In this section, we will study a new set of involutions defined by pentads of nodes, no four of the them are coplanar. Such a pentad will be called *admissible.* For a general 15-nodal quartic surface this condition is equivalent to the property that no four of the points lie on a trope-conic.

Let $\mathcal{P} = \{x_1, \ldots, x_5\}$ be an admissible pentad. For a general point $x \in X_{15}$, there exists a unique twisted cubic γ (maybe degenerate) that passes through $\mathcal{P} \cup \{x\}$. It intersects the surface at one more point (may be equal to x). This defines a birational involution of X_{15} that extends to a biregular involution $\tau_{\mathcal{P}}$ of Y.

Note that if we drop the condition of admissibility of a pentad, then a pentad with four nodes on a trope-conic will still define an involution but it will coincide with the involution defined by the projection from the remaining fifth node. If we put all 5 nodes on a trope-conic, the involution is not defined.

Let C be a quartic curve on X_{15} cut out by a plane Π containing three nodes, say p_1, p_2, p_3, from the pentad. Any conic γ' in H passing through these nodes intersects C at two additional points x, x'. The union of γ' and the line $\langle p_4, p_5\rangle$ is a degenerate twisted cubic γ intersecting X_{15} at $p_1, \ldots, p_5$ and x, x'. Thus the pair (x, x') is an orbit of the involution $\tau_{\mathcal{P}}$ and hence C is invariant curve with respect to this involution.

In the special case when the plane H cuts out a trope-conic, we see that any point on it must be fixed under the involution. Also, if there is a trope-quartic C with a double point at x_1 and passing through the remaining points of $\mathcal{P}$, then a twisted cubic γ passing through $\mathcal{P}$ and a general point x on a trope-quartic C will be contained in the cone $K(x_1)$ and hence it will intersect X_{15} at x with multiplicity two. This shows that C is fixed pointwise too.

(42) *The involution $\tau_{\mathcal{P}}$ associated with an admissible pentad $\mathcal{P}$ leaves any plane section H of X_{15} containing three nodes from the pentad*

invariant. A trope-conic through 3 points in $\mathcal{P}$ or a trope-quartic passing through $\mathcal{P}$ with a double point at one of the points in $\mathcal{P}$ is fixed pointwise.

One checks that the sublattice $M(\mathcal{P})$ spanned by the ten divisor classes $\eta_H - E_{x_i} - E_{x_j} - E_{x_k}$ is of rank 5. Since $\tau_{\mathcal{P}}$ obviously leaves invariant the exceptional curves over the remaining 10 nodes, we see that it acts as a reflection on the Picard lattice of Y. In fact, we check that the vector

$$r_{\mathcal{P}} := 3\eta_H - 2\sum_{x\in\mathcal{P}} E_x \in \mathrm{Pic}(Y)$$

has $r_{\mathcal{P}}^2 = -4$ and it is orthogonal to the sublattice $M(\mathcal{P})$, hence $\tau_{\mathcal{P}}^*$ coincides with the reflection with respect to $r_{\mathcal{P}}$ and acts as

$$\tau_{\mathcal{P}}^*(v) = v + \tfrac{1}{2}(r_{\mathcal{P}} \cdot v) \cdot r_{\mathcal{P}}.$$

In particular, $\tau_{\mathcal{P}}^*$ leaves invariant the divisor class

$$F_i^{\mathcal{P}} = 2\eta_H - 2E_{x_i} - \sum_{j\neq i} E_{x_j} \tag{9.1}$$

of self-intersection 0. Since we can write

$$F_i^{\mathcal{P}} = (\eta_H - E_{x_1} - E_{x_2} - E_{x_3}) + (\eta_H - E_{x_1} - E_{x_4} - E_{x_5}),$$

and each of the summand represents a nef class, we see that the divisor class $F_i^{\mathcal{P}}$ is nef and represents the class of a general fiber of an elliptic pencil $|F_i^{\mathcal{P}}|$ on Y.

We have $F_i^{\mathcal{P}} \cdot F_j^{\mathcal{P}} = 2$ and

$$\left|\tfrac{1}{2}(F_1^{\mathcal{P}} + \cdots + F_5^{\mathcal{P}})\right| = \left|5\eta_H - 3\sum_{x\in\mathcal{P}} E_x\right|. \tag{9.2}$$

It is clear that ten curves $E_x, x \notin \mathcal{P}$, are irreducible components of fibers of $|F_i^{\mathcal{P}}|$.

Suppose x_1, x_2, x_3 and x_1, x_3, x_4 do not lie on a trope-conic. Then the plane section of X_{15} containing the points x_1, x_2, x_3 or x_1, x_4, x_5 cut out two smooth rational curve whose proper transform on Y are linearly equivalent to $\eta_H - E_{x_1} - E_{x_2} - E_{x_3}$ and $\eta_H - E_{x_1} - E_{x_4} - E_{x_5}$. Their sum is linearly equivalent to $F_1^{\mathcal{P}}$. This shows that the elliptic pencil $|F_1^{\mathcal{P}}|$ acquires a reducible fiber of type $\tilde{A}_1$ (or I_2 in Kodaira's notation) that consists of two irreducible components (they may be tangent so the fiber becomes of type III in Kodaira's notation).

On the other hand suppose that x_1, x_2, x_3 lie on a trope-conic passing through 3 more points $y_1, y_2, y_3 \notin \mathcal{P}$, and x_1, x_4, x_5 do not lie on a trope-conic. Then $F_1^{\mathcal{P}}$ is linearly equivalent to the sum

$$2\Big(\tfrac{1}{2}(\eta_H - E_{x_1} - E_{x_2} - E_{x_3} - E_{y_1} - E_{y_2} - E_{y_3}) \\ + E_{y_1} + E_{y_2} + E_{y_3} + (\eta_H - E_{x_1} - E_{x_4} - E_{x_5})\Big),$$

where each of the summands represents the divisor class of a (-2)-curve. This shows that the elliptic pencil $|F_1^{\mathcal{P}}|$ contains a reducible fiber of type $\tilde{D}_4$ (or I_0^* in Kodaira's notation).

Similar argument shows that if x_1, x_2, x_3 and x_1, x_4, x_5 lie on trope-conics, then $|F_1^{\mathcal{P}}|$ acquires a fiber of type $\tilde{D}_6$. Also, if

$$\frac{1}{2}\Big(2\eta_H - 2E_{x_1} - \sum_{i=2}^{5} E_{x_i} - \sum_{i=1}^{4} E_{y_i}\Big)$$

represents a trope-quartic C, then $2C + \sum_{i=1}^{4} E_{y_i} \in |F_1^{\mathcal{P}}|$ is a fiber of type $\tilde{D}_4$. One can check that no other types of reducible fibers are possible because $\mathcal{P}$ is admissible.

Let $\Gamma(\mathcal{P})$ be the graph with the set of vertices $[1, 6]$ and the set of 5 edges corresponding to the indices (ab) of points from $\mathcal{P}$. A trope-conic passes through 6 points corresponding to a duad (ab), three duads in $[1, 6] \setminus \{a, b, c\}$ and two duads $(ac), (bc)$. It can be represented by the following graph

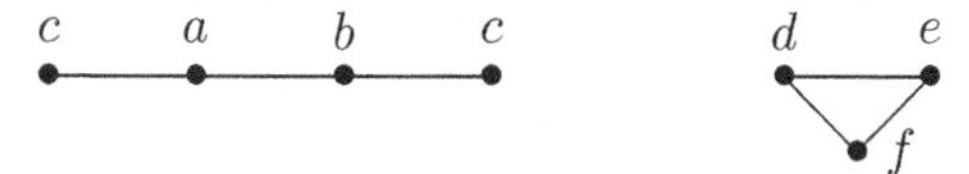

By definition, a pentad is admissible if and only if $\Gamma(\mathcal{P})$ does not contain a subgraph of this graph with four edges. Also three nodes in $\mathcal{P}$ are on a trope-conic if and only if $\Gamma(\mathcal{P})$ contain a subgraph of this graph with three edges.

Table 4.2 shows all possible graphs $\Gamma(\mathcal{P})$, the number c of subsets of three nodes on a trope-conic, possible types of reducible fibers of the elliptic pencils $|F_i^{\mathcal{P}}|$, possible singularities of the branch curve of the double plane realization of X_{15} defined by $\tau_{\mathcal{P}}$ (see next subsection) and the number of orbits of $\mathfrak{S}_6$ on the set of pentads (the last two columns were kindly provided by I. Shimada).

Note that an elliptic pencil $|F_i^{\mathcal{P}}|$ has a section if and only if there exists a trope-conic passing through the node x_i and one other node of $\mathcal{P}$. Not all elliptic pencils have a section. In fact, one of the fibrations

Table 4.2 *Admissible pentads*

Type	Graph	c	Pencils	Double planes	#orbits
1		0	$(9\tilde{A}_1 + \tilde{D}_4) \times 5$	$(14A_1) \times 5$	6
2		4	$(5\tilde{A}_1 + 2\tilde{D}_4) \times 4$ $(\tilde{A}_1 + 2\tilde{D}_6) \times 1$	$(6A_1 + 2D_4) \times 4$ $(14A_1) \times 1$	45
3		5	$(3\tilde{A}_1 + \tilde{D}_4 + \tilde{D}_6) \times 5$	$(6A_1 + 2D_4) \times 5$	72
4		2	$(9\tilde{A}_1 + \tilde{D}_4) \times 4$ $(3\tilde{A}_1 + \tilde{D}_4 + \tilde{D}_6) \times 1$	$(10A_1 + D_4) \times 4$ $(14A_1) \times 1$	90
5		3	$(5\tilde{A}_1 + 2\tilde{D}_4) \times 3$ $(\tilde{A}_1 + 3\tilde{D}_4) \times 1$ $(9\tilde{A}_1 + \tilde{D}_4) \times 1$	$(10A_1 + D_4) \times 3$ $(14A_1) \times 1$ $(2A_1 + 3D_4) \times 1$	120
6		2	$(5\tilde{A}_1 + 2\tilde{D}_4) \times 2$ $(9\tilde{A}_1 + \tilde{D}_4) \times 2$ $(7\tilde{A}_1 + \tilde{D}_6) \times 1$	$(10A_1 + D_4) \times 4$ $(14A_1) \times 1$	180
7		3	$(5\tilde{A}_1 + 2\tilde{D}_4) \times 2$ $(9\tilde{A}_1 + \tilde{D}_4) \times 2$ $(3\tilde{A}_1 + \tilde{D}_4 + \tilde{D}_6) \times 1$	$(6A_1 + 2D_4) \times 2$ $(10A_1 + D_4) \times 2$ $(14A_1) \times 1$	360
8		3	$(5\tilde{A}_1 + 2\tilde{D}_4) \times 3$ $(7\tilde{A}_1 + \tilde{D}_6) \times 2$	$(10A_1 + D_4) \times 4$ $(6A_1 + 2D_4) \times 1$	360
9		4	$(5\tilde{A}_1 + 2\tilde{D}_4) \times 2$ $(7\tilde{A}_1 + \tilde{D}_6) \times 2$ $(3\tilde{A}_1 + \tilde{D}_6 + \tilde{D}_4) \times 1$	$(10A_1 + D_4) \times 3$ $(6A_1 + 2D_4) \times 2$	360

of types 2, 4 and 7 has no sections. Table 4.2 shows that the rank of the sublattice of $\mathrm{Pic}(Y)$ generated by irreducible components of each pencil is equal to 14. Applying the Shioda–Tate formula gives

(43) *The Mordell-Weyl group of sections of the Jacobian fibration of the elliptic fibration defined by $|F_i^{\mathcal{P}}|$ is of rank 1.*

9.2. Consider the quotient Z of Y by the involution $\tau_{\mathcal{P}}$. Since any admissible pentad $\mathcal{P} = \{x_1, \ldots, x_5\}$ consists of 5 points in general linear position in $\mathbb{P}^3$, we can use them to define the map (5.1) and identify a twisted cubic through $\mathcal{P}$ with a point on the irreducible component $F(\mathsf{S}_3)_6$ of the surface of lines on the Segre cubic. This defines a $\tau_{\mathcal{P}}$-equivariant degree

2 regular map $f\colon X_{15} \to \mathrm{dP}_5$ to a quintic del Pezzo surface dP_5 that factors through a birational morphism

$$\tilde{f}\colon Y/(\tau_{\mathcal{P}}) \longrightarrow \mathrm{dP}_5\,.$$

Let $|C_i|$, $i = 1, \ldots, 5$, be the pencils of conics on dP_5. It is easy to see that in our identification of dP_5 with $F(\mathsf{S}_3)_6$, each pencil consists of twisted cubics that are tangent to a ruling of a quadratic cone with vertex at one of the points x_i and passing through three remaining points x_j. If we compare this with our definition of the pencils $|F_i^{\mathcal{P}}|$ in (9.1), we find that the pre-image of the pencil of conics $|C_i|$ on dP_5 is the pencil $|F_i^{\mathcal{P}}|$. Since $C_1 + \cdots + C_5 \sim -2K_{\mathrm{dP}_5}$, we see that the pre-image of a hyperplane section of dP_5 in the anti-canonical embedding is a member of the linear system $|D|$ defined in (9.2). Comparing the dimensions of the linear systems we find that the linear subsystem $\mathbb{P}(H^0(Y, \mathcal{O}_Y(D))^{\tau_{\mathcal{P}}})$ of $\tau_{\mathcal{P}}$-invariant sections is a hyperplane and defines the $\tau_{\mathcal{P}}$-equivariant map $\tilde{f}$.

(44) *The linear system* (9.2) *contains a hyperplane defining a* $\tau_{\mathcal{P}}$*-equivariant map* $Y \to \mathrm{dP}_5$*. The pre-image of a pencil of conics on* dP_5 *is an elliptic pencil* $|F_i^{\mathcal{P}}|$ *on* Y*.*

Suppose three of the nodes in $\mathcal{P}$ lie on a trope-conic. Then intersecting the divisor class of its proper transfer on Y with the divisor class defining the linear system (9.2), we obtain that its image on the quotient del Pezzo surface is a line in the anti-canonical embedding. Since the trope-conic is contained in the fixed locus of $\tau_{\mathcal{P}}$, we see that the involutions defined by pentads of types 2–9 in Table 4.2 are different from the Reye involutions (whose components of the fixed locus are mapped to conics on the del Pezzo surface). Similar computation shows that the image of a trope-quartic component of the fixed locus is a conic on the del Pezzo surface.

(45) *The image on* dP_5 *of a component of the fixed locus of* $\tau_{\mathcal{P}}$ *formed by a trope-conic (resp. trope-quartic) is a line (resp. conic) on* D_5*.*

The quotient surface $Z = Y/(\tau_{\mathcal{P}})$ is a smooth rational surface and the map $\tilde{f}\colon Z \to \mathrm{dP}_5$ is a birational morphism. Let R be an exceptional curve that is blown down to a point in dP_5. Then R intersects each $F_i^{\mathcal{P}}$ with multiplicity zero, hence is contained in a reducible fiber of each elliptic pencil $|F_i^{\mathcal{P}}|$. Of course, the converse is also true. We have 10 such curves $E_x, x \notin \mathcal{P}$. We can deduce from our description of reducible fibers of the elliptic pencils $|F_i^{\mathcal{P}}|$ that there are no more such curves, hence Z is obtained from dP_5 by blowing up 10 points and $K_Z^2 = -5$.

Since the fixed locus $Y^{\tau_{\mathcal{P}}}$ contains a curve, for example a multiple component of a fiber of type $\tilde{D}_4$, the involution acts as the minus identity on the transcendental part of $H^2(Y, \mathbb{C})$. This shows that the invariant part of the involution $\tau_{\mathcal{P}}^*$ on $H^2(Y, \mathbb{C})$ coincides with the invariant part on $\mathrm{Pic}(Y)$. Since $\tau_{\mathcal{P}}$ acts as a reflection on $\mathrm{Pic}(Y)$, the dimension of the invariant part $\mathrm{Pic}(Y)_{\mathbb{C}}^{\tau_{\mathcal{P}}}$ is equal to 15, and therefore the Lefschetz number $\mathrm{Lef}(\tau_{\mathcal{P}})$ is equal to $15 - 7 + 2 = 10$. Applying the Lefschetz fixed-point-formula, we obtain that the Euler–Poincaré characteristic of the fixed locus $Y^{\tau_{\mathcal{P}}}$ of $\tau_{\mathcal{P}}$ is equal to 10. It follows that the fixed locus contains at least 5 disjoint smooth rational curves $R_1, \ldots, R_5$. Using the Hurwitz formula, we also see that the Euler–Poincaré characteristic of $Z = Y/(\tau_{\mathcal{P}})$ is equal to 17. This confirms our previous assertion that Z is obtained from dP_5 by blowing up 10 points.

(46) *The quotient* $Z = Y/(\tau_{\mathcal{P}})$ *is a rational surface with* $K_Z^2 = -5$. *The branch curve* B *of the cover* $Y \to Z$ *belongs to* $|-2K_Z|$ *and contains* 5 *disjoint smooth rational curves with self-intersection* -4. *The surface* Z *is isomorphic to the blow-up of 10 points on a quintic del Pezzo surface, the points are the images of nodes on* X_{15} *which do not belong to* $\mathcal{P}$. *The image of* B *on* Z *belongs to* $|-2K_{\mathrm{dP}_5}|$.

We will see in Example 9.4 that the fixed locus may contain other non-rational component and the quotient surface Z may not be a Coble surface as in the case of a Reye involution.

Remark 9.1. Recall that a quintic del Pezzo surface dP_5 admits 5 non-projectively equivalent blow-down morphism $\pi\colon \mathrm{dP}_5 \to \mathbb{P}^2$. They are defined by the linear system $|-K_{\mathrm{dP}_5} - C_i|$, where $|C_i|$ is pencil of conics. Since we know that the pre-image of $|-K_{\mathrm{dP}_5}|$ on Y is linear system (9.2) and the pre-image of $|C_i|$ is the elliptic pencil $|F_i^{\mathcal{P}}|$, we obtain that

$$\Big|3\eta_H - E_{x_i} - 2\sum_{j\neq i} E_{x_j}\Big| \tag{9.3}$$

defines a morphism $Y \to \mathrm{dP}_5 \to \mathbb{P}^2$. Thus we have five double plane realizations of Y corresponding to a choice of an admissible pentad. The branch curve of these double planes are reducible plane sextics and the possible types of their singularities are given in Table 4.2.

We have to warn that, in general, the double plane with the branch curve described in Table 4.2 is not isomorphic to a 15-nodal quartic surface. In fact, (9.3) shows that the set of the images of the curves E_{x_i}, $x_i \in \mathcal{P}$, on the del Pezzo quotient surface are smooth rational curves R_i of degree 6 not passing through the singular points of the branch curve. We have $R \cdot K_{\mathrm{dP}_5} = -6$, hence $R_i^2 = 4$. In order that R_i be the image of a (-2)-curve on Y, it must split under the cover. So we need to find 6 such curves which are everywhere tangent to the branch curve B. It is easy to see that a smooth rational sextic of dP_5 is equal to the proper transform under the blow-up $\mathrm{dP}_5 \to \mathbb{P}^2$ of either a smooth conic not passing through the fundamental points or a quartic curve with nodes at three of the fundamental points. In both cases the class of R_i is divisible by 2. As soon as we have found such a set of curves $R_1, \ldots, R_5$ everywhere tangent to the branch curve, we use (9.1) to define the linear system $|\eta_H|$ that will map the double cover to X_{15}.

Example 9.2. Let $\mathcal{P}$ be an admissible pentad of type 1. This is the only type such that all subsets of 3 nodes in $\mathcal{P}$ do not lie on a trope-conic. We call such pentad a *Göpel pentad* (by analogy with a Göpel tetrad on a Kummer surface [Hud90, Chapter VII, §53]). We may assume that it coincides with the set $\mathcal{C}$ which we index by $\{(16), (26), (36), (46), (56)\}$. The fixed locus of $\tau_{\mathcal{P}}$ is equal to the union of the trope-quartics $\sigma(E_x)$, $x \in \mathcal{C}$. The surface Y is the double cover of the blow-up of 10 intersection points of the union of 5 conics on a quintic del Pezzo surface. The branch curve is equal to the proper transform of these conics on the blow-up. We recognize a similar construction of X_{15} from a Reye involution τ_{Rey}. In that case, the fixed locus was the union of the curves E_y, $y \in \mathcal{C}$. The six involutions defined by Göpel pentads are, in fact, conjugate to the six Reye involutions

$$\tau_{\mathcal{P}_i} = \sigma_i \circ \tau_{\mathrm{Rey}_i} \circ \sigma_i, \qquad i = 1, \ldots, 6.$$

Example 9.3. Assume $\mathcal{P}$ is of type 2. We can index its points by (12), (13), and (24), (34), (56). There are 4 trope-conics passing through three nodes from $\mathcal{P}$. They are $\sigma(E_{15})$, $\sigma(E_{25})$, $\sigma(E_{35})$, $\sigma(E_{45})$. The node (56) is contained in all trope-conics. Each pair of them passing through two complementary pairs of the remaining nodes form a fiber of type $\tilde{D}_6$ of F_{56}. The remaining fibers have two reducible fibers of type $\tilde{D}_4$.

Let us consider the images of the four trope-conics on dP_5. We observe that E_{14} and E_{23} are mapped to the intersection points of the two disjoint pairs of lines $\{\ell_1, \ell_2\}$ and $\{\ell_3, \ell_4\}$, the images of $\sigma(E_{25})$, $\sigma(E_{35})$ and the images of $\sigma(E_{15})$, $\sigma(E_{45})$. The remaining 8 exceptional curves are blown down to 8 points, two on each of the lines. Since the image of the branch curve B on Z belongs to $|-2K_{\mathrm{dP}_5}|$, we see that B has another component C of degree 6 in the anti-canonical embedding. Let us consider a dP_5 as the blow-up of points p_1, p_2, p_3, p_4 in $\mathbb{P}^2$. It is easy to see that the images of the four lines $\ell_1, \ldots, \ell_4$ are either two pairs of lines joining two complementary pairs of points or two lines $\langle p_i, p_j\rangle$ and $\langle p_i, p_k\rangle$ intersecting at one of the points p_i. There will be 4 non-projectively equivalent ways to blow down to obtain the first possibility and only one to obtain the second possibility. In the first case, the remaining irreducible component C of B is the proper transform of a nonsingular conic not passing through $p_1, \ldots, p_4$. In the second case, the curve C is the proper transform of a quartic curve with three nodes at p_j, p_k and p_l, where $\{i, j, k, l\} = \{1, 2, 3, 4\}$. Thus we see that $|-2K_Z|$ consists of 5 disjoint smooth rational curves and Z is a Coble surface.

Example 9.4. Let us consider a pentad of type 3. By (*46*), the image of the fixed locus of τ_{Rey} on dP_5 contains five lines among its irreducible components. They are represented by a pentagon subgraph of the Petersen graph. There are five projective equivalence classes of blowing down morphisms $Z \to \mathrm{dP}_5$. The image of the pentagon of lines is formed by three lines $\langle p_i, p_j\rangle, \langle p_i, p_k\rangle, \langle p_i, p_l\rangle$. The remaining component of the image of the fixed locus is the proper transform of a cubic curve through the four points. It is a quintic elliptic curve on dP_5. Thus we see that the fixed locus of $\tau_{\mathcal{P}}$ may contain an elliptic curve besides the five (-2)-curves. Since the pentagon of lines belongs to $|-K_{\mathrm{dP}_5}|$ and passes through all ten points which we blow-up to obtain Z, we see that $|-K_Z| \neq \emptyset$ and Z *is not a Coble surface.*

References

[AG93] Enrique Arrondo and Mark Gross. On smooth surfaces in $\mathrm{Gr}(1, \mathbf{P}^3)$ with a fundamental curve. *Manuscripta Math.*, 79(3-4):283–298, 1993.

[Avi19] A. Avilov. Biregular and birational geometry of double coverings of a projective space with ramification in a quartic with 15 ordinary double points. *Izv. Ross. Akad. Nauk Ser. Mat.*, 83(3):5–14, 2019.

[Bak10] H. F. Baker. *Principles of geometry. Volume 4. Higher geometry.* Cambridge Library Collection. Cambridge University Press, Cambridge, 2010. Reprint of the 1925 original.

[CKS20] Ivan Cheltsov, Alexander Kuznetsov, and Konstantin Shramov. Coble fourfold, $\mathfrak{S}_6$-invariant quartic threefolds, and Wiman-Edge sextics. *Algebra Number Theory*, 14(1):213–274, 2020.

[Cob82] Arthur B. Coble. *Algebraic geometry and theta functions*, volume 10 of *American Mathematical Society Colloquium Publications*. American Mathematical Society, Providence, R.I., 1982. Reprint of the 1929 edition.

[Cos83] François R. Cossec. Reye congruences. *Trans. Amer. Math. Soc.*, 280(2):737–751, 1983.

[DFL18] Igor Dolgachev, Benson Farb, and Eduard Looijenga. Geometry of the Wiman-Edge pencil, I: algebro-geometric aspects. *Eur. J. Math.*, 4(3):879–930, 2018.

[DK] Igor Dolgachev, and Shigeyuki Kondō. Enriques surfaces, vol. 2. Available online: http://www.math.lsa.umich.edu/~idolga/lecture notes.html

[DK02] Igor Dolgachev and Jonghae Keum. Birational automorphisms of quartic Hessian surfaces. *Trans. Amer. Math. Soc.*, 354(8):3031–3057, 2002.

[Dol12] Igor V. Dolgachev. *Classical algebraic geometry: A modern view.* Cambridge University Press, Cambridge, 2012.

[Dol16] Igor Dolgachev. Corrado Segre and nodal cubic threefolds. In *From classical to modern algebraic geometry*, Trends Hist. Sci., pages 429–450. Birkhäuser/Springer, Cham, 2016.

[DS20] Igor Dolgachev and Ichiro Shimada. 15-nodal quartic surfaces. Part II: the automorphism group. *Rend. Circ. Mat. Palermo (2)*, 69(3):1165–1191, 2020.

[Edg91] W. L. Edge. A plane sextic and its five cusps. *Proc. Roy. Soc. Edinburgh Sect. A*, 118(3-4):209–223, 1991.

[End99] Stephan Endraß. On the divisor class group of double solids. *Manuscripta Math.*, 99(3):341–358, 1999.

[Fin87] Hans Finkelnberg. Small resolutions of the Segre cubic. *Nederl. Akad. Wetensch. Indag. Math.*, 49(3):261–277, 1987.

[GL87] Mark Green, and Robert Lazarsfeld. Special divisors on curves on a $K3$ surface. *Invent. Math.*, 89(2):357–370, 1987.

[Hud90] R. W. H. T. Hudson. *Kummer's quartic surface.* Cambridge Mathematical Library. Cambridge University Press, Cambridge, 1990. With a foreword by W. Barth, Revised reprint of the 1905 original.

[Ili94] Atanas Iliev. The Fano surface of the Gushel threefold. *Compositio Math.*, 94(1):81–107, 1994.

[IM11] Atanas Iliev, and Laurent Manivel. Fano manifolds of degree ten and EPW sextics. *Ann. Sci. Éc. Norm. Supér. (4)*, 44(3):393–426, 2011.

[Isk77] V. A. Iskovskih. Fano threefolds. I. *Izv. Akad. Nauk SSSR Ser. Mat.*, 41(3):516–562, 717, 1977.

[Jes69] C. M. Jessop. *A treatise on the line complex.* Chelsea Publishing Co., New York, 1969.

[JK04] Trygve Johnsen and Andreas Leopold Knutsen. *K3 projective models in scrolls*, volume 1842 of *Lecture Notes in Mathematics.* Springer-Verlag, Berlin, 2004.

[Kon98] Shigeyuki Kondō. The automorphism group of a generic Jacobian Kummer surface. *J. Algebraic Geom.*, 7(3):589–609, 1998.

[KV19] Grzegorz Kapustka, and Alessandro Verra. On Morin configurations of higher length. *Int. Math. Res. Not., no. 1:* 727–772, 2022.

[Muk88] Shigeru Mukai. Curves, $K3$ surfaces and Fano 3-folds of genus ≤ 10. In *Algebraic geometry and commutative algebra, Vol. I*, pages 357–377. Kinokuniya, Tokyo, 1988.

[Muk93] Shigeru Mukai. Curves and Grassmannians. In *Algebraic geometry and related topics (Inchon, 1992)*, Conf. Proc. Lecture Notes Algebraic Geom., I, pages 19–40. Int. Press, Cambridge, MA, 1993.

[Nik79] V. V. Nikulin. Integer symmetric bilinear forms and some of their geometric applications. *Izv. Akad. Nauk SSSR Ser. Mat.*, 43(1):111–177, 238, 1979.

[O'G05] K. G. O'Grady. Involutions and linear systems on holomorphic symplectic manifolds. *Geom. Funct. Anal.*, 15(6):1223–1274, 2005.

[O'G08] Kieran G. O'Grady. Dual double EPW-sextics and their periods. *Pure Appl. Math. Q.*, (Special Issue: In honor of Fedor Bogomolov. Part 1) 4(2):427–468, 2008.

[Ran86] Ziv Ran. Surfaces of order 1 in Grassmannians. *J. Reine Angew. Math.*, 368:119–126, 1986.

[Roh87] Karl Rohn. Die Flächen vierter Ordnung hinsichtlich ihrer Knotenpunkte und ihrer Gestaltung. *Math. Ann.*, 29(1):81–96, 1887.

[Sal58] George Salmon. *A treatise on the analytic geometry of three dimensions.* Revised by R. A. P. Rogers. 7th ed. Vol. 1. Edited by C. H. Rowe. Chelsea Publishing Company, New York, 1958.

[SR85] J. G. Semple and L. Roth. *Introduction to algebraic geometry.* Oxford Science Publications. The Clarendon Press, Oxford University Press, New York, 1985. Reprint of the 1949 original.

[Stu92] R. Sturm. *Die Gebilde ersten und zweiten Grades der Liniengeometrie in synthetischer Behandlung.* vol. 1-3. Leipzig, 1892.

[Var86] Robert Varley. Weddle's surfaces, Humbert's curves, and a certain 4-dimensional abelian variety. *Amer. J. Math.*, 108(4):931–951, 1986.

[Ver88] Alessandro Verra. Smooth surfaces of degree 9 in $G(1,3)$. *Manuscripta Math.*, 62(4):417–435, 1988.

[Vin83] È. B. Vinberg. The two most algebraic $K3$ surfaces. *Math. Ann.*, 265(1):1–21, 1983.

Mori Flips, Cluster Algebras and Diptych Varieties Without Unprojection

Tom Ducat

To Miles Reid

Abstract

We give a survey on the connections between terminal 3-fold flips and cluster algebras. In particular we observe that Mori's algorithm for generating the relations defining a type $k2A$ flipping neighbourhood is a form of generalised cluster algebra mutation. We then use the Laurent phenomenon for this cluster algebra structure to give an alternative proof of the existence of Brown and Reid's diptych varieties. Finally we briefly explain how Hacking, Tevelev and Urzúa's universal family of antiflips can be obtained from the diptych variety.

1 Introduction

Mori's algorithm. A *Mori flip* is a type of elementary birational map appearing in the Minimal Model Program (MMP) for terminal 3-folds. Mori [Mor02] introduced a *continued division algorithm* that can be used to generate global sections on a particular type of Mori flip, known as a $k2A$ neighbourhood. Hacking, Tevelev and Urzúa [HTU17] observed that a special case of Mori's algorithm (the case $d_1 = d_2 = 1$ in the notation of §3) is given by mutation in a rank 2 cluster algebra. In this paper, we extend this observation to show that, for a $k2A$ neighbourhood $f\colon X \to Y$, the general case of Mori's algorithm is given by mutation in a *generalised* cluster algebra $\mathcal{A}_f$. In particular, we show that the graded ring R_f defining the affine cover of the flip is given by the upper cluster algebra of $\mathcal{A}_f$.

Diptych varieties. Inspired by Mori's algorithm, Brown and Reid [BR13] introduced *diptych varieties* in order to understand the entire structure of the graded ring defining a Mori flip. In particular, for a special class of $k2A$ neighbourhoods, they provide a generating set for R_f by an inductive process called *serial unprojection*. We recover their explicit generating set for the diptych variety using the Laurent phenomenon for the cluster algebra $\mathcal{A}_f$, rather than unprojection.

Universal families of flips. Hacking, Tevelev and Urzúa used Mori's algorithm to construct a family of surfaces over a toric surface $\mathcal{M}_d$ containing an infinite chain of rational curves of self-intersection $-d$. This family can be used to construct families of $k1A$ and $k2A$ neighbourhoods. In the final section, we give a brief explanation as to the existence of this universal family from the point of view of diptych varieties.

Main result. Our main result is as follows.

Theorem 1.1. *Fix some integers $d_1, d_2 \geq 2$, $k \geq 4$, let $d_i := d_{(i \bmod 2)}$ for all $i \in \mathbb{Z}$ and define $l_1, l_2, m_1, m_2 \in \mathbb{Z}_{\geq 0}$ by the following matrix product.*

$$\begin{pmatrix} l_1 & m_1 \\ l_2 & m_2 \end{pmatrix} = \begin{pmatrix} d_1 & -1 \\ 1 & 0 \end{pmatrix} \begin{pmatrix} d_2 & -1 \\ 1 & 0 \end{pmatrix} \cdots \begin{pmatrix} d_{k-3} & -1 \\ 1 & 0 \end{pmatrix} \begin{pmatrix} 0 & 1 \\ -1 & 0 \end{pmatrix}$$

Let $\mathcal{A}$ be the cluster algebra of rank two defined over the coefficient ring $R = \mathbb{C}[A, B, L, M]$ and determined by the initial exchange relations

$$x_0 x_2 = A x_1^{d_1} + L^{l_1} M^{m_1}, \qquad x_1 x_3 = B x_2^{d_2} + L^{l_2} M^{m_2}.$$

Then the diptych variety V (described in §5) is given by $V = \operatorname{Spec}\mathcal{U}$ where $\mathcal{U}$ is the upper cluster algebra of $\mathcal{A}$. In particular, $\mathcal{U}$ is finitely generated and an explicit (but not necessarily minimal) generating set for $\mathcal{U}$ is given in Definition 6.4.

Conventions. We let $\mu_r \subset \mathbb{C}^*$ be the group of rth roots of unity. We use the standard notation $\frac{1}{r}(a_1, \ldots, a_n)$ for cyclic quotient singularities. Inequalities between vectors $u, v \in \mathbb{R}^k$ are assumed to hold componentwise, although we interpret strict inequality $v > u$ to mean $v_i \geq u_i$ for all $i = 1, \ldots, k$ and $v_i > u_i$ for at least one i.

2 Mori Flips

2.1 Extremal Neighbourhoods

Suppose that X belongs to the *Mori category* of normal quasiprojective 3-folds over $\mathbb{C}$ with at worst terminal $\mathbb{Q}$-factorial singularities. Running the MMP produces a curve $C \subset X$ with $K_X \cdot C < 0$, and an *extremal contraction*, i.e. a proper morphism $f\colon X \to Y$ with $f_*\mathcal{O}_X = \mathcal{O}_Y$ and such that, for any irreducible curve $C' \subset X$, $f(C') = \mathrm{pt}$ if and only if $C \sim_{\mathrm{num}} C'$, up to taking multiples.

We are interested in the local behaviour of f around the point $P = f(C) \in Y$, where C is contracted. Therefore we take the localisation at $P \in Y$ and the completion of X along $Z = f^{-1}(P)_{\mathrm{red}}$ to reduce to the case of an *extremal neighbourhood* $f\colon (Z \subset X) \to (P \in Y)$. In practical cases, we can usually regard Y to be affine.

If the exceptional locus of f is purely 1-dimensional then f is called a *small contraction*. In this case, by a theorem of Kawamata, we can always factor f analytically into a sequence of contractions with an irreducible exceptional curve. Therefore we assume that $Z = C$, which is the usual point of view taken in these situations. Unfortunately, a small contraction takes us out of the Mori category, since $P \in Y$ will always be a non-$\mathbb{Q}$-factorial singularity. We look for the *flip of* f, that is, $f^+\colon (C^+ \subset X^+) \to (P \in Y)$ a proper birational morphism where X^+ is in the Mori category, $C^+ = \mathrm{exc}(f^+)$ is purely 1-dimensional, $K_{X^+} \cdot C^+ > 0$ and the induced birational map $X \dashrightarrow X^+$ gives an isomorphism $X^+ \setminus C^+ \cong X \setminus C$.

2.2 The Affine Cover of an Extremal Neighbourhood

Suppose that $f\colon X \to Y$ is a small extremal neighbourhood with relative Weil divisor class group $\mathrm{Cl} = \mathrm{Cl}(X/Y) \cong \mathbb{Z}[A] \oplus T$, generated by a relatively ample divisor A and torsion subgroup T. We consider the following Cl-graded R_0-algebra over the ring $R_0 = \mathcal{O}_Y$

$$R_f = \bigoplus_{D \in \mathrm{Cl}} R_D := \bigoplus_{D \in \mathrm{Cl}} \mathcal{O}_Y(f_*D)$$

and let $R_f^+ = \bigoplus_{D \cdot C \geq 0} R_D$ and $R_f^- = \bigoplus_{D \cdot C \leq 0} R_D$. Then $Y = \mathrm{Spec}\, R_0$, $X = \mathrm{Proj}_Y\, R_f^-$ and the flip is given by $X^+ = \mathrm{Proj}_Y\, R_f^+$. It follows that X^+ is unique if it exists, and the question of existence boils down to showing the finite generation of R_f. We call $W = \mathrm{Spec}_Y\, R_f$ the *affine cover of* f. This is a 4-fold, affine over Y, with a cyclic group action $G = \mathrm{Hom}(\mathrm{Cl}, \mathbb{C}^*) \curvearrowright W$ coming from the Cl-grading on R_f. We can reconstruct X, Y and X^+ as a variation of GIT-quotients for three different characters of the action $G \curvearrowright W$.

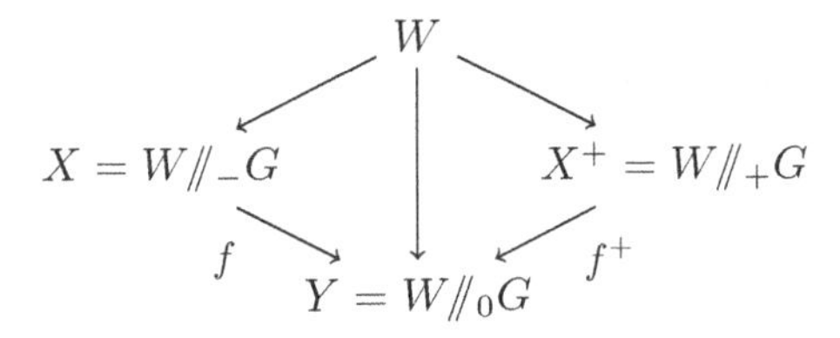

Following the philosophy of Reid [Rei92], we want to study the geometry of extremal neighbourhoods by constructing W with this G-action. Then by varying the equations of W or by varying the G-action, we can study flips (or other types of extremal contraction) as they degenerate in families and eventually hope to classify them.

2.3 Type A Flips

The general elephant. The first key theorem towards the classification of 3-fold flips is known as Reid's general elephant conjecture (GEC) and proven in our setting by Kollár and Mori [KM92, Theorem 1.7]. The GEC states that if $f\colon (C \subset X) \to (P \in Y)$ is a small (or divisorial) contraction then a general element $E_X \in |-K_X|$ has at worst Du Val singularities along C. Similarly $E_Y = f(E_X) \in |-K_Y|$ has at worst a Du Val singularity at $P \in Y$ and, by adjunction, the map $f|_{E_X} : E_X \to E_Y$ is a partial crepant resolution.

Division into cases. As a consequence of the GEC, Kollár and Mori divide extremal neighbourhoods into cases according to the singularities along $C \subset X$, the Du Val type of $P \in E_Y$ and the curve (if any) extracted by the partial crepant resolution $f|_{E_X} : E_X \to E_Y$. In particular, if $P \in E_Y$ is a Du Val singularity of type A there are precisely two cases.

Theorem 2.1 (See [KM92, Theorem 2.2]). *If $P \in E_Y$ is Du Val of type A then either*

(1) $f|_{E_X} : E_X \to E_Y$ is an isomorphism, $P \in E_Y$ is of type A_{kr-1} and X has one singularity of type cA/r along C (and possibly one more cA singularity of index 1); or

(2) $f|_{E_X} : E_X \to E_Y$ is a partial crepant resolution which extracts C from E_Y, $P \in E_Y$ is of type $A_{k_1r_1+k_2r_2-1}$ and X has two singularities of type cA/r_1 and cA/r_2 along C (and no other singular points).

Extremal neighbourhoods of the first kind are called $k1A$ *neighbourhoods* and those of the second kind are called $k2A$ *neighbourhoods*. The exceptional cases, for which $P \in E_Y$ is of type D or E, are completely classified by Kollár and Mori.

3 Cluster Algebras

We make a brief detour to review the definition of (generalised) cluster algebra mutation. For convenience, we restrict to the rank 2 case.

3.1 The Cluster Algebra $\mathcal{A}$

We fix a $\mathbb{C}$-algebra R which is a unique factorisation domain to be used as a coefficient ring, and the following data, called an *initial seed*:

(1) two *cluster variables* x_1, x_2 (viewed as elements of $\operatorname{Frac} R[x_1, x_2]$),
(2) two *exchange polynomials* $F_1(T), F_2(T) \in R[T]$ of degree $\deg F_i = d_i \geq 0$, for $i = 1, 2$, with relatively coprime coefficients.

Remark 3.1. In the original set up (as introduced by Fomin and Zelevinsky), the exchange polynomials were required to be of the form $F_i(T) = T^{d_i} + 1$. A *generalised* cluster algebra is allowed to have arbitrary exchange polynomials. From now on, we use the term *'cluster algebra'* to mean *'generalised cluster algebra'*.

Mutation. We can produce new seeds from the initial seed by *mutation*. We let $d_i := d_{(i \bmod 2)}$ for all $i \in \mathbb{Z}$. Given $S_{i-1,i} = \{(x_{i-1}, x_i), (F_{i-1}, F_i)\}$ we can produce $S_{i,i+1} = \{(x_i, x_{i+1}), (F_i, F_{i+1})\}$ by the formulae

$$x_{i+1} = \frac{F_i(x_i)}{x_{i-1}} \qquad \text{and} \qquad F_{i+1}(T) = \frac{T^{d_{i-1}} F_{i-1}\big(T^{-1} F_i(0)\big)}{c_{i-1,i+1}} \tag{3.1}$$

where $c_{i-1,i+1} \in R$ is a greatest common divisor of the coefficients of the polynomial appearing in the numerator. By switching the roles of i and $i+1$ we can also mutate backwards $S_{i,i+1} \to S_{i-1,i}$. Each mutation involves making a choice of $c_{i-1,i+1}$ up to multiplying by a unit in R. We make any choice consistent with the system of relations $c_{i-1,i+1} c_{i+1,i-1} = F_i(0)^{d_{i-1}}$ for all $i \in \mathbb{Z}$, in which case mutation becomes an involution on seeds.

The cluster algebra. Elements of the sequence $(x_i \in \operatorname{Frac} R[x_1, x_2] : i \in \mathbb{Z})$ are called *cluster variables* and the equations that they satisfy $x_{i-1}x_{i+1} = F_i(x_i)$ *exchange relations.* A *cluster* is a pair $\{x_i, x_{i+1}\}$ and the *cluster algebra* is the ring $\mathcal{A} = R[x_i : i \in \mathbb{Z}]$. The behaviour of $\mathcal{A}$ is strongly related to whether the discriminant $\Delta := d_1 d_2 - 4$ is positive, negative or zero.

Periodicity. It is fun exercise to check that mutation is periodic if and only if $\Delta < 0$, and that the period is $4, 5, 6, 8$ for $\Delta = -4, -3, -2, -1$ respectively.

The Laurent phenomenon. The most important property enjoyed by $\mathcal{A}$ is the *Laurent phenomenon*, which states that $x_j \in R[x_i^{\pm 1}, x_{i+1}^{\pm 1}]$ for any cluster $\{x_i, x_{i+1}\}$ and any cluster variable x_j.

3.2 The Upper Cluster Algebra $\mathcal{U}$

Cluster tori. We let $U_i := R[x_i^{\pm 1}, x_{i+1}^{\pm 1}]$, which we think of as the coordinate ring of a *cluster torus* $T_i := \operatorname{Spec} U_i = \operatorname{Spec} R \times (\mathbb{C}^*)^2$. We can glue all such cluster tori together with the birational maps obtained by mutating, to obtain a variety $V_0 = \bigcup_{i \in \mathbb{Z}} T_i$. We call the ideal $I_{\mathcal{A}} = (x_i x_{i+1} : i \in \mathbb{Z}) \subset \mathcal{A}$, generated by all pairs of consecutive cluster variables, the *deep ideal of* $\mathcal{A}$. Then $V_0 = (\operatorname{Spec} \mathcal{A} \setminus V(I_{\mathcal{A}})) \subseteq \operatorname{Spec} \mathcal{A}$.

Definition 3.2. The intersection $\mathcal{U} = \bigcap_{i \in \mathbb{Z}} U_i$ is called the *upper cluster algebra* of $\mathcal{A}$.

By the Laurent phenomenon, we have $\mathcal{A} \subseteq \mathcal{U}$ but, as we will see, the inclusion can be strict. Geometrically, $\mathcal{U}$ is the ring of regular functions on V_0. It is not clear that $\mathcal{U}$ is finitely generated, but when it is we call $V = \operatorname{Spec} \mathcal{U}$ a *cluster variety.* In this case $V_0 \subset V$ is the 'affinisation' of V_0. Note that $\mathcal{U}$ is integrally closed since it is an intersection of integrally closed rings, so V is normal.

Criterion for $\mathcal{U}$. Following Matherne and Muller [MM15, Lemma 4.3.1], we have the following useful criteria to test for equality with $\mathcal{U}$.

Proposition 3.3. *Let S be a Noetherian ring such that $\mathcal{A} \subseteq S \subseteq \mathcal{U}$. Then $S = \mathcal{U}$ if and only if S is normal and* $\operatorname{codim}(SI_{\mathcal{A}}) \leq 2$.

Moreover, although $\mathcal{U}$ is defined to be the intersection of infinitely many rings, it can be rewritten as a finite intersection.

Lemma 3.4. *For any i, we have $\mathcal{U} = U_{i-1} \cap U_i \cap U_{i+1}$. Therefore to test membership in $\mathcal{U}$ it is enough to test membership in U_i for any three consecutive clusters.*

A key assumption of our set-up is that $\mathcal{A}$ is defined over a coefficient *ring* R. Lemma 3.4 appears as [BFZ05, Corollary 1.7]. However, the authors of that paper use a different convention concerning coefficients and it was not clear whether or not their proof assumes that the coefficients are invertible. Therefore we provide a stand-alone proof.

Proof The statement follows after showing that $U_{i-1} \cap U_i \cap U_{i+1}$ is independent of i, and this follows from showing that $U_{i-1} \cap U_i \cap U_{i+1} = U_i \cap U_{i+1} \cap U_{i+2}$ for any i. Suppose that $z \in U_{i-1} \cap U_i \cap U_{i+1}$. It is now enough to show that $z \in U_{i+2}$ since this implies $U_i \cap U_{i+1} \cap U_{i+2} \subseteq U_{i-1} \cap U_i \cap U_{i+1}$ and, by symmetry, we have equality.

Since $z \in U_{i+1}$, for some $a, b \in \mathbb{Z}_{\geq 0}$ and $G_{jk} \in R$, we can write z as

$$z = \sum_{j,k \geq 0} G_{jk} x_{i+1}^{-a+j} x_{i+2}^{-b+k} = \sum_{j,k \geq 0} G_{jk} \left(x_{i+3}^{-1} F_{i+2}(x_{i+2}) \right)^{-a+j} x_{i+2}^{-b+k}.$$

To show that $z \in U_{i+2}$, it is sufficient to show that $F_{i+2}(x_{i+2})^{a-j}$ divides $\sum_{k \geq 0} G_{jk} x_{i+2}^k$ for $0 \leq j \leq a$. Now, we also have

$$z = \sum_{j,k \geq 0} G_{jk} \left(x_{i-1}^{-1} F_i(x_i) \right)^{-a+j} x_{i+2}^{-b+k}$$

where $x_{i+2} = x_i^{-1} F_{i+1} \left(x_{i-1}^{-1} F_i(x_i) \right)$. Since $z \in U_{i-1}$, we conclude that $F_i(x_i)^{a-j}$ divides $\sum_{k \geq 0} G_{jk} x_{i+2}^k$ for $0 \leq j \leq a$. By the mutation rule (3.1), we have $F_i \left(T^{-1} F_{i+1}(0) \right) = c T^{-d_i} F_{i+2}(T)$ for some $c \in R$. Since

$$F_i(x_i) = F_i \left(x_{i+2}^{-1} F_{i+1} \left(x_{i+3}^{-1} F_{i+2} \left(x_{i+2} \right) \right) \right),$$

this gives

$$F_i(x_i) \equiv F_i \left(x_{i+2}^{-1} F_{i+1}(0) \right) \equiv c x_{i+2}^{-d_i} F_{i+2}(x_{i+2}) \equiv 0 \mod F_{i+2}(x_{i+2}),$$

and therefore

$$F_i(x_i)^{a-j} \,\Big|\, \sum_{k \geq 0} G_{jk} x_{i+2}^k \implies F_{i+2}(x_{i+2})^{a-j} \,\Big|\, \sum_{k \geq 0} G_{jk} x_{i+2}^k. \qquad \square$$

Combining Proposition 3.3 and Lemma 3.4, we obtain the following.

Lemma 3.5. *For any $i \in \mathbb{Z}$, let $W_{i,i+1} \subset \operatorname{Spec} R \times \mathbb{A}^4$ be the affine complete intersection of codimension 2 defined by two consecutive exchange relations*

$$W_{i,i+1} = V(x_{i-1} x_{i+1} - F_i(x_i),\ x_i x_{i+2} - F_{i+1}(x_{i+1})).$$

Then $\mathcal{U}$ is equal to the ring of regular functions on $W^0_{i,i+1} := W_{i,i+1} \setminus V(x_i, x_{i+1})$.

Proof Let S be the ring of regular functions on $W^0_{i,i+1}$, which is a normal algebraic variety. Clearly $\mathcal{A} \subseteq S$ by the Laurent phenomenon and $S \subseteq \mathcal{U}$ since, by Lemma 3.4, $\mathcal{U}$ is the ring of regular functions on $W_{i,i+1} \setminus V(x_{i-1}x_i, x_i x_{i+1}, x_{i+1}x_{i+2})$ and

$$\begin{aligned}(x_i, x_{i+1}) &\supseteq (x_{i-1}x_i, x_i x_{i+1}, x_{i+1}x_{i+2}) \\ &= (x_{i-1}, x_{i+1}) \cap (x_i, x_{i+1}) \cap (x_i, x_{i+2}) \supseteq I_{\mathcal{A}}.\end{aligned}$$

By Proposition 3.3, it is now sufficient to show that $V(x_{i-1}, x_{i+1})$ and $V(x_i, x_{i+2})$ define sets of codimension ≥ 2 in $W^0_{i,i+1}$. But $V(x_{i-1}, x_{i+1}) \cap W_{i,i+1} = V(x_{i-1}, x_{i+1}, F_i(x_i), x_i x_{i+2} - F_{i+1}(0))$ is a complete intersection of codimension 4 in $\operatorname{Spec} R \times \mathbb{A}^4$, and hence has codimension 2 in $W^0_{i,i+1}$. Similarly for (x_i, x_{i+2}). □

In particular, a component $\Pi \subseteq V(I_{\mathcal{A}}) \cap W_{i,i+1}$ of codimension one in $W_{i,i+1}$ must lie in the locus $V(x_i, x_{i+1})$, and therefore such a component looks like $\Pi = V(x_i, x_{i+1}, r)$ where $r \in R$ divides both $F_i(0)$ and $F_{i+1}(0)$.

3.3 Expanding in a Cluster

We fix a choice of cluster to be used as a frame of reference in which to expand all elements of $\mathcal{U}$ as Laurent polynomials. Without loss of generality we take $\{x_1, x_2\}$.

Definition 3.6. Given an element $z \in \mathcal{U}$, we will write z as a Laurent polynomial $z = x_1^{-a} x_2^{-b} G_z(x_1, x_2)$ in terms of the cluster variables x_1, x_2, where $G_z \in R[x_1, x_2]$ is not divisible by x_1 or x_2. We write $x_1^c x_2^d \in z$ to mean that the monomial $x_1^c x_2^d$ appears in z with nonzero coefficient. Then

(1) the *numerator polynomial of* z is $G_z(x_1, x_2)$,
(2) the *denominator vector of* z is $v(z) := (a, b) \in \mathbb{Z}^2$,
(3) the *Newton polytope* $\operatorname{Newt}(z)$ of z is the convex hull of

$$\{(-c, -d) \in \mathbb{Z}^2 \colon x_1^c x_2^d \in z\},$$

(4) the *support polytope* of z is $\operatorname{supp}(z) := \operatorname{Newt}(z) + \mathbb{Z}^2_{\leq 0}$, the Minkowski sum.

Note that $v(x_1) = (-1, 0)$ and $v(x_2) = (0, -1)$. We use a slightly unusual convention with our choice of signs in the definition of $\operatorname{Newt}(z)$ so that $v(z) \in \operatorname{Newt}(z)$ if $G_z(0, 0) \neq 0$.

4 Mori's Algorithm

We now return to examine Type A flipping contractions in more detail, concentrating on $k2A$ neighbourhoods in particular. With the exception of Theorem 4.2 and §4.4, this section is essentially just a summary of Mori's paper [Mor02], albeit rephrased in the language of cluster algebras. We follow [Mor02] very closely and the reader is referred to the original paper for proofs and further details. However, be warned — no attempt has been made to reconcile differences in notation (e.g. what we call ε Mori calls Δ, and what Mori calls ε we call Δ).

4.1 Local Model of a $k2A$ Neighbourhood

Suppose that $f\colon (C \subset X) \to (P \in Y)$ is a $k2A$ neighbourhood which, as we recall from Theorem 2.1, has precisely two singularities $Q_i \in C \subset X$ of type cA/r_i for $i = 1, 2$. Up to replacement by formal neighbourhoods, Mori shows that we can obtain a model for $C \subset X$ by simply gluing together affine neighbourhoods for Q_1 and Q_2.

Theorem 4.1 ([Mor02] Theorems 2.2 and 2.9). *Any $k2A$ neighbourhood can be obtained by the following construction, up to taking the completion along $C \subset X$. For $i = 1, 2$, let U_i be the affine cA/r_i coordinate chart*

$$(Q_i \in U_i) = \Big(0 \in V\big(\xi_i\eta_i - f_i(\zeta_i^{r_i}, u)\big)\Big) \subset \mathbb{A}^4_{\xi_i,\eta_i,\zeta_i,u} \Big/ \tfrac{1}{r_i}(1, -1, a_i, 0) \tag{4.1}$$

where $\gcd(r_i, a_i) = 1$ and $f_i \in \mathbb{C}[\![u]\!][\zeta^{r_i}]$ is a squarefree, monic polynomial in ζ^{r_i} of degree d_i such that $f_i(\zeta_i^{r_i}, 0) = \zeta_i^{r_i d_i}$. Let $X = U_1 \cup U_2$ be the variety obtained by glueing these charts by identifying $\xi_1^{r_1} = \xi_2^{-r_2}$ and $\zeta_1\xi_1^{-a_1} = \zeta_2\xi_2^{r_2-a_2}$. In each chart $E_X \in |-K_X|$ is given by $E_X|_{U_i} = V(\zeta_i)/\frac{1}{r_i}$ and $C|_{U_i}$ is given by the ξ_i-axis.

The general hypersurface section H_X. In addition to the general elephant $E_X \subset X$, it is also useful to consider a general hypersurface section $P \in H_Y \subset Y$ through P, and the preimage $H_X = f^{-1}(H_Y)$. In the setting of Theorem 4.1, we may take $H_X = V(u)$ and the proof of the theorem proceeds by showing that the statements hold modulo u, i.e. for the map of surfaces $f|_{H_X} : (C \subset H_X) \to (P \in H_Y)$, and then lifting the equations to X. In fact, X is a 1-parameter $\mathbb{Q}$-Gorenstein smoothing of H_X, where H_X is a toric surface with two cyclic quotient singularities of the form $\frac{1}{r_i^2 d_i}(1, r_i d_i a_i - 1)$ (known as *T-singularities*). Moreover H_X and E_X intersect along their toric boundary strata. We

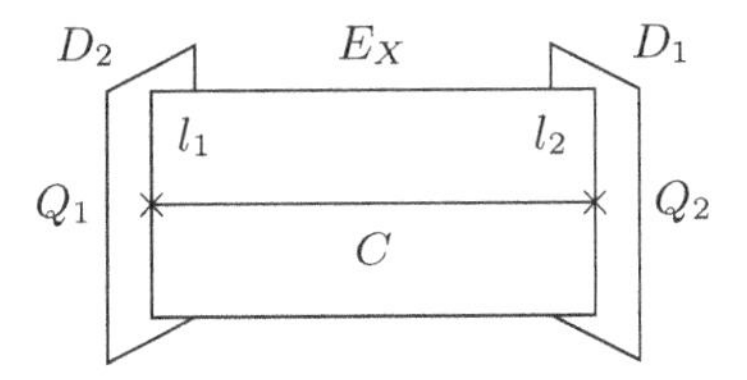

Figure 5.1 The divisors D_{3-i} generating $\mathrm{Cl}(U_i, Q_i)$, for $i = 1, 2$.

write $H_X \cap E_X = l_1 \cup C \cup l_2$ where $l_i = \mathbb{A}^1_{\eta_i} \subset U_i$ is a (non-complete) curve contained in the chart U_i (see Figure 5.1).

Two key invariants. Since X is a $\mathbb{Q}$-Gorenstein smoothing of H_X, it follows that $K_X \cdot C = K_{H_X} \cdot C$. Standard calculations in toric geometry for $C \subset H_X$ show that $K_X \cdot C = -\frac{\delta}{r_1 r_2}$, where $\delta = r_1 a_2 + r_2 a_1 - r_1 r_2$. This number δ is an important integer invariant associated to f, known as the *Fano index.* Note that $\delta > 0$ since $K_X \cdot C < 0$.

Since $C \subset H_X$ is a contractible curve, there is a second integer invariant $\varepsilon = d_1 r_1^2 - \delta d_1 d_2 r_1 r_2 + d_2 r_2^2 > 0$ such that $(C|_{H_X})^2 = -\frac{\varepsilon}{r_1^2 r_2^2 d_1 d_2} < 0$. This is the index of the quotient singularity $P \in H_Y$.

4.2 Mori's Continued Division Algorithm

The two equations (4.1) already look similar to the exchange relations in a cluster algebra. However, at the moment these equations can not be compared because they are written in terms of two different open sets of X. Instead, we start writing things in terms of global variables.

Generators for $\mathrm{Cl}(X, Q_i)$**.** Let $D_{3-i} = (\xi_i = 0)/\frac{1}{r_i}$ be a Weil divisor generating the local Weil divisor class group $\mathrm{Cl}(X, Q_i)$ for $i = 1, 2$ (where the swap in indices is deliberate). The glueing relations for X give that $r_2 D_1 \sim r_1 D_2$ and $E_X \sim a_1 D_2 - (r_2 - a_2) D_1$, and therefore we have $r_1 E_X \sim \delta D_1$ and $r_2 E_X \sim \delta D_2$.

Now pick sections $x_i \in H^0(X, D_i)$ for $i = 1, 2$ and $z \in H^0(X, E_X)$. Since $r_1 E_X \sim \delta D_1$, we are able to evaluate the expression

$$F_1(x_1^\delta, z^{r_1}) := x_1^{\delta d_1} f_1(z^{r_1} x_1^{-\delta}, u)$$

which is divisible in U_1 by ξ_1, and hence divisible on X by x_2. We get a new global section $x_0 \in H^0(X, \delta d_1 D_1 - D_2)$ given by the equation $x_0 x_2 = F_1(x_1^\delta, z^{r_1})$. In a similar vein, we obtain $x_3 \in H^0(X, \delta d_2 D_2 - D_1)$ given by the equation $x_1 x_3 = F_2(x_2^\delta, z^{r_2})$, where $F_2(x_2^\delta, z^{r_2}) = x_1^{\delta d_1} f_2(z^{r_2} x_2^{-\delta}, u)$.

The division algorithm. Mori's continued division algorithm [Mor02, §3] is now equivalent to cluster algebra mutation in a rank 2 cluster algebra $\mathcal{A}_f$, defined over the coefficient ring $\mathbb{C}[\![u]\!][z]$ and determined by the initial seed $\{x_1, x_2\}$ and exchange polynomials $F_1(T^\delta, z_1^{r_1})$ and $F_2(T^\delta, z_2^{r_2})$ in $\mathbb{C}[\![u]\!][z][T]$.

Two integer sequences. We introduce the following useful sequences. Let $(r_i : i \in \mathbb{Z})$ be the sequence extending the initial values (r_1, r_2) by the recurrence relation $r_{i-1} + r_{i+1} = \delta d_i r_i$, and let $(b_i : i \in \mathbb{Z})$ be the sequence extending $(b_1, b_2) = (a_1, r_2 - a_2)$ analogously.

The relative class group. Let $m = \gcd(r_1, r_2)$. Since $\gcd(r_i, a_i) = 1$ for $i = 1, 2$, we can write $\mathrm{Cl}(X/Y) = \mathbb{Z}[A] \oplus \mathbb{Z}/m\mathbb{Z}[B]$ for an ample class A and m-torsion class B where

$$E_X \sim \delta A, \quad D_1 \sim r_1 A + a_1 B \quad \text{and} \quad D_2 \sim r_2 A + (r_2 - a_2)B.$$

We see that $x_i \in H^0(X, r_i A + b_i B)$ for $i = 0, 1, 2, 3$.

The affine cover of f. The affine cover $W = \operatorname{Spec} R_f$ is determined by the graded ring

$$R_f = \bigoplus_{(r,b) \in \mathbb{Z} \oplus \mathbb{Z}/m\mathbb{Z}} \mathcal{O}_Y(f_*(rA + bB))$$

and for $(g, \epsilon) \in G = \mathbb{C}^* \times \mu_m$ the group action $G \curvearrowright W$ is given by $(g, \epsilon) \cdot y = g^r \epsilon^b y$ for $y \in H^0(X, rA + bB)$. In particular, for the first few cluster variables and the coefficients this action is given by the following table of weights.

	x_0	x_1	x_2	x_3	z	u
r	r_0	r_1	r_2	r_3	δ	0
b	b_0	b_1	b_2	b_3	0	0

We have an inclusion $\mathcal{A}_f \subseteq R_f$, however the cluster variables are not sufficient to generate the whole of R_f in general.

Theorem 4.2. *The ring R_f is equal to $\mathcal{U}_f$, the upper cluster algebra of $\mathcal{A}_f$.*

Proof In Theorem 4.1, X is constructed by glueing together cyclic quotients of two affine patches $X = U_1 \cup U_2$. In terms of the global sections, the open sets are given by $U_i = \{x_i \neq 0\}$ for $i = 1, 2$ and therefore

$X = W_{12}^0/G$ is a quotient of $W_{12}^0 = W_{12} \setminus V(x_1, x_2)$ where W_{12} is the following complete intersection of codimension 2,

$$W_{12} = V\left(x_0x_2 - F_1(x_1^\delta, z^{r_1}), x_1x_3 - F_2(x_2^\delta, z^{r_2})\right) \subset \mathbb{A}^6_{x_0,x_1,x_2,x_3,z,u}.$$

By Lemma 3.5, R_f is the ring of regular functions $\mathcal{U}_f$ on W_{12}^0. □

Whilst this is a nice theoretical result, it does not immediately give a very explicit description of R_f. Instead of trying to determine the whole graded ring R_f directly, Mori shows that the sequence of cluster variables contains enough information to obtain a description of the flip.

4.3 Describing the Flip

We now analyse the sequence of cluster variables in more detail.

A game of ping-pong. Note that (r_1, r_2) is an integral point on the real conic

$$\Gamma = V(d_1u^2 - \delta d_1d_2uv + d_2v^2 - \varepsilon) \subset \mathbb{R}^2_{u,v}.$$

Given any point $(u_0, v_0) \in \Gamma$, we find $(\delta d_2v_0 - u_0, v_0) \in \Gamma$ by fixing v_0 and solving a quadratic in u_0 (known as 'Vieta jumping'). Similarly $(u_0, \delta d_1u_0 - v_0) \in \Gamma$ and therefore we have $(r_{2i\pm1}, r_{2i}) \in \Gamma$ for all $i \in \mathbb{Z}$. Since $\varepsilon > 0$ our conic Γ resembles one of the two lefthand cases of Figure 5.2, depending on the value of Δ (or two parallel lines if $\Delta = 0$). By Vieta jumping we can ping-pong between points of Γ until we reach $r_i \leq 0$.

Now, up to switching $i \leftrightarrow 3 - i$ for all i, we may now assume that there exists $k > 2$, the first integer for which $r_k \leq 0$.

The exchange relations. We are now in a position to write out the system of exchange relations for the cluster algebra $\mathcal{A}_f$ determined by the initial seed above. We start with the initial exchange relations $x_0x_2 = F_1(x_1^\delta, z^{r_1})$ and $x_1x_3 = F_2(x_2^\delta, z^{r_2})$. Under the assumption that we have defined $F_i(T^\delta, z^{r_i})$, we introduce the notation $\alpha_i, \beta_i \in \mathbb{C}[\![u]\!]$ for $F_i(x_i^\delta, 0) = \alpha_i x_i^{\delta d_i}$ and $F_i(0, z^{r_i}) = \beta_i z^{d_ir_i}$, where we necessarily must have $\gcd(\alpha_i, \beta_i) = 1$. By the assumptions of Theorem 4.1 we have $\beta_1 = \beta_2 = 1$. Working inductively for $i \geq 3$, by the rule for cluster algebra mutation (3.1) we introduce

$$\begin{aligned} F_{i+1}(T^\delta, z^{r_{i+1}}) &:= c_{i-1,i+1}^{-1} T^{\delta d_{i-1}} F_{i-1}(T^{-\delta} F_i(0, z^{r_i})^\delta, z^{r_{i-1}}) \\ &= c_{i-1,i+1}^{-1} F_{i-1}(\beta_i^\delta z^{r_{i+1}+r_{i-1}}, z^{r_{i-1}} T^\delta) \end{aligned}$$

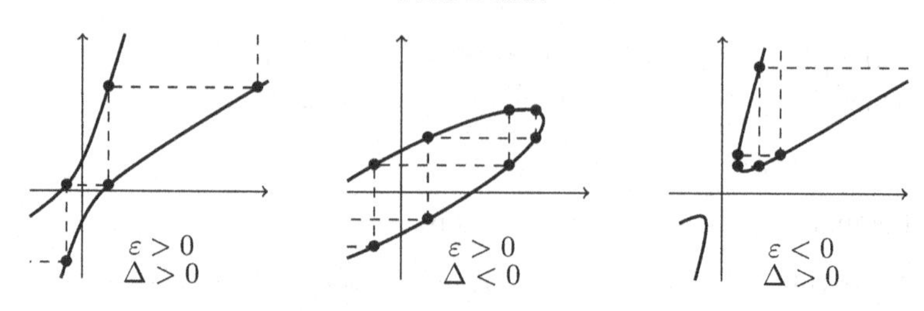

Figure 5.2 The sequence $(r_{2i\pm1}, r_{2i})$ on the conic $\Gamma \subset \mathbb{R}^2$. (If both $\varepsilon, \Delta < 0$ then $\Gamma = \emptyset$.)

where $c_{i-1,i+1} \in R$ is yet to be chosen. This follows from the fact that $F_{i-1}(T^\delta, z^{r_{i-1}})$ is homogeneous of degree d_i as a polynomial in T^δ and $z^{r_{i-1}}$. Now, after cancelling the highest common power of z, for some $c \in \mathbb{C}[\![u]\!]$ we obtain

$$F_{i+1}(T^\delta, z^{r_{i+1}}) = \begin{cases} c^{-1}F_{i-1}(\beta_i^\delta z^{r_{i+1}}, T^\delta) & \text{if } r_{i+1} \geq 0, \\ c^{-1}F_{i-1}(\beta_i^\delta, z^{-r_{i+1}}T^\delta) & \text{if } r_{i+1} < 0. \end{cases}$$

The coefficients $\alpha_{i+1}, \beta_{i+1} \in \mathbb{C}[\![u]\!]$ are relatively coprime and satisfy $\alpha_{i+1}c = \beta_{i-1}$ and $\beta_{i+1}c = \alpha_{i-1}\beta_i^{\delta d_{i-1}}$. Hence $c = \gcd(\beta_{i-1}, \alpha_{i-1}\beta_i^{\delta d_{i-1}})$ $= \gcd(\beta_{i-1}, \beta_i^{\delta d_{i-1}})$. Now, working inductively we find that (α_i, β_i) are given by

i	1	2	3	4	$\cdots$	i
α_i	α_1	α_2	1	1	$\cdots$	1
β_i	1	1	α_1	$\alpha_1^{\delta d_0}\alpha_2$	$\cdots$	$\alpha_1^{p_i}\alpha_2^{q_i}$

where $p_{i-1} + p_{i+1} = \delta d_{i-1} p_i$ and similarly for q_i. In particular $\alpha_i = 1$ for $i \geq 3$.

It is convenient to define a modified sequence $(s_i \colon i \in \mathbb{Z})$ by setting $s_i = r_i$ if $i \leq k$ and $s_i = -r_{i-2}$ if $i \geq k+1$. Note that $s_k, s_{k+1} \leq 0$ and $s_i \geq 0$ otherwise. We find that we get the system of exchange relations

$$x_{i-1}x_{i+1} = \begin{cases} F_i(x_i^\delta, z^{s_i}) & i < k \text{ or } i > k+1 \\ F_i(x_i^\delta z^{-s_i}, 1) & i = k, k+1, \end{cases}$$

and from these equations, $x_i \in H^0(X, s_iA_i + b_iB)$ for all $i \in \mathbb{Z}$.

Cutting out C as a complete intersection. The key result that lets Mori construct a model for the flip X^+ is that he is able to show that the two sections x_k, x_{k+1} cut out C as a set-theoretic complete intersection, i.e. $C = V(x_k, x_{k+1})$. To see this, consider the three curve classes $[l_1]$, $[l_2]$, $[C] \in \mathrm{Cl}(H_X)$ and let $\mathrm{div}_H(\phi)$ denote the divisor of ϕ restricted to H_X. By setting $u = 0$ in the exchange relations we obtain the following equations

$$x_{i-1}x_{i+1} = F_i|_{u=0} = \begin{cases} z^{s_i d_i} & i = 1,2 \\ x_i^{\delta d_i} & 2 < i < k \\ x_i^{\delta d_i} z^{-s_i d_i} & i = k, k+1 \end{cases}$$

using the fact that $\beta_1 = \beta_2 = 1$, $\alpha_i = 1$ for $i \geq 3$ and all other coefficients in the exchange relations vanish modulo u. Since $H_X \cap E_X = l_1 \cup C \cup l_2$, we have the following relations

$$\mathrm{div}_H(z) = [l_1] + [C] + [l_2], \quad \mathrm{div}_H(x_1) = s_2 d_2 [l_2], \quad \mathrm{div}_H(x_2) = s_1 d_1 [l_1].$$

By using the equations above, it follows from a simple induction that

$$\begin{aligned} \mathrm{div}_H(x_i) &= \mathrm{div}_H(F_{i-1}|_{u=0}) - \mathrm{div}_H(x_{i-2}) \\ &\equiv \begin{cases} s_{i-1} d_i [l_1] & 2 \leq i \leq k \\ s_{i+1} d_i [l_2] & i \geq k+1 \end{cases} \quad \mathrm{mod}\ [C] \end{aligned}$$

where we don't worry about the coefficient of $[C]$ (other than that it is positive) and, by the definition of s_i, we note that $\mathrm{div}_H(x_i) \geq 0$ for all i. Now consider the set-theoretic intersection $V(x_k, x_{k+1})$ inside X. If $x_k = x_{k+1} = 0$ then

$$x_{k-1}x_{k+1} = F_k(z^{-r_k} x_k^\delta, 1) \implies F_k(0,1) = 0 \implies u = 0$$

and thus $V(x_k, x_{k+1}) = V(x_k, x_{k+1}, u) = C$, since x_k is nonvanishing along l_2 and x_{k+1} is nonvanishing along l_1. We have found two sections x_k, x_{k+1} that cut out C set-theoretically.

Constructing the flip. Using the existence of these sections x_k, x_{k+1}, [Mor02, Corollary 4.1] shows that the curve $C \subset X$ is contractible (in the 3-fold) and constructs a model for the flip $f^+ \colon X^+ \to Y$, albeit only over some open set of Y that contains P. We consider the following complete intersection of codimension 2

$$W_{k,k+1} = V\Big(x_{k-1}x_{k+1} - F_k(x_k^\delta z^{-s_k}, 1),\ x_k x_{k+2} - F_{k+1}(x_{k+1}^\delta z^{-s_{k+1}}, 1)\Big)$$

and the open set $W^0_{k,k+1} = W_{k,k+1} \setminus V(x_k, x_{k+1})$. Now [Mor02, Theorems 4.5 and 4.7] give the following construction of the flip.

Theorem 4.3. *The k2A neighbourhood $f \colon (C \subset X) \to (P \in Y)$ either has a flip or is a divisorial contraction to a curve. A model for the flip* *or divisorial contraction is given by* $X^+ \cong W^0_{k,k+1}/G$, *and we can decide which case we are in as follows.*

(1) If $s_k < 0$ *then* f *is a flipping contraction with* $C^+ = \mathbb{P}^1_{(x_k\,:\,x_{k+1})}$ *and* $K_{X^+}\cdot C^+ = \frac{\delta}{s_k s_{k+1}} > 0$. *Moreover* X^+ *has two terminal* cA/r *singularities along* C^+ *of index* $-s_i$ *for* $i = k, k+1$, *given in local coordinates by* $(x_{i-1}x_{i+1} - F_i(z^{-s_i},1))\,/\frac{1}{-s_i}(1,-1,b_i,0)$.

(2) If $s_k = 0$ *then* f *is a divisorial contraction, we have* $-s_{k+1} = \delta = \gcd(r_1, r_2)$ *and* f *contracts the exceptional divisor* $F = V(x_{k+1}) \subset X$ *onto the curve* $\Gamma = V(z, F_k(x_k^\delta,1))/\frac{1}{\delta}$ *contained in the* cA/δ *singularity* $\left(x_k x_{k+2} - F_{k+1}(z^\delta,1)\right)/\frac{1}{\delta}(1,-1,b_{k+1},0)$.

4.4 Key Varieties for Mori Flips

Mori's algorithm gives an attractive and explicit method for calculating $k2A$ flips, as we will now demonstrate with an example. Moreover, by generalising some of the features of Mori's algorithm, we can obtain a larger cluster variety V which we can use to construct $W = \operatorname{Spec} R_f$ by '*regular pullback*'. This variety V is called a '*key variety*' for f.

Example 4.4. Suppose that $f\colon (C \subset X) \to (P \in Y)$ is the $k2A$ neighbourhood determined by taking the following equations in Theorem 4.1

$$\left(\xi_1\eta_1 = \alpha(u) + \beta(u)\zeta_1^{33} + \zeta_1^{66}\right)/\tfrac{1}{33}(1,-1,14,0)$$
$$\left(\xi_2\eta_2 = \gamma(u) + \zeta_2^{19}\right)/\tfrac{1}{19}(1,-1,11,0)$$

for some functions $\alpha(u), \beta(u), \gamma(u) \in \mathbb{C}[u]$. In particular, X is a smoothing of the surface $H_X = V(u)$ with T-singularities $\frac{1}{33^2\cdot 2}(1, 33\cdot 14\cdot 2 - 1)$ and $\frac{1}{19^2}(1, 19\cdot 11 - 1)$. Fixing a minimal resolution $\mu\colon \widetilde{H}_X \to H_X$, we have the following dual intersection diagram of rational curves in $\widetilde{H}_X$

2 4 6 3 2 2 3 2 3 1 2 4 5 3 2 3

●—●—●—●—●—●—●—●—●—○—●—●—●—●—●—●

where the black nodes are contracted to the T-singularities, the white node corresponds to $\widetilde{C}$, the strict transform of C, and the labels are the negative self-intersection numbers of the corresponding curve. The restriction $f|_{H_X}\colon (C \subset H_X) \to (P \in H_Y)$ is a contraction of C onto a $\frac{1}{31}(1,18)$ quotient singularity, where $\frac{31}{18} = [2,4,3,2]$ is found by blowing down this configuration of curves. The first key invariant is $\delta = 2$ and we write down the following table.

i	0	1	2	3	4	5	6
δd_i	2	4	2	4	2	4	2
r_i	113	33	19	5	1	−3	−13
b_i	46	14	8	2	0	−2	−8

The sign of r_i changes at $k = 5$, so we write out the sequence of exchange relations for the cluster variables up to and including $k + 1 = 6$.

$$x_0x_2 = \alpha x_1^4 + \beta x_1^2 z^{33} + z^{66} \qquad x_3x_5 = x_4^2 + \alpha^2\gamma z$$
$$x_1x_3 = \gamma x_2^2 + z^{19} \qquad x_4x_6 = z^6x_5^2 + \alpha^3\beta\gamma^2z^3 + \alpha^7\gamma^4$$
$$x_2x_4 = x_3^4 + \beta x_3^2z^5 + \alpha z^{10} \qquad x_5x_7 = zx_6^2 + \alpha^{12}\gamma^7$$

Since $\gcd(33, 19) = 1$, the group $G \cong \mathbb{C}^*$ is simply a torus, which acts by the following weights.

x_0	x_1	x_2	x_3	x_4	x_5	x_6	x_7	z	u
113	33	19	5	1	−3	−1	3	2	0

Since $r_k = -3 < 0$, the neighbourhood is a flipping contraction and, as in Theorem 4.3, we can obtain a model for the flip

$$\phi\colon \left(W_{12} \subset \mathbb{A}^6_{x_0,x_1,x_2,x_3,u,z}\right)/\mathbb{C}^* \dashrightarrow \left(W_{56} \subset \mathbb{A}^6_{x_4,x_5,x_6,x_7,u,z}\right)/\mathbb{C}^*.$$

The map ϕ could be written out explicitly by using the Laurent phenomenon to express $x_4, \ldots, x_7$ in terms of x_1, x_2. The flip X^+ has a $cA/3$ singularity along $C^+ = \mathbb{P}^1_{(x_5\colon x_6)}$ at the point $(1\colon 0)$ and is smooth at $(0\colon 1)$. The hyperplane section H_{X^+} is a surface with T-singularity $\frac{1}{3^2\cdot 2}(1, 3\cdot 1\cdot 2 - 1)$ and the corresponding dual intersection diagram of curves in $\widetilde{H}_{X^+}$ is given as follows, where the white node is C^+.

Example 4.4 revisited. Now we explain how to soup up the previous example to obtain something more general. From the role that they play in equations computed in Example 4.4, it is clear that we may as well have assumed $\alpha(u)$, $\beta(u)$, $\gamma(u)$ to be variables in their own right. Moreover, all monomials in z that appear in the equations are products of $(z^{-s_k}, z^{-s_{k+1}}) = (z^3, z)$, and so we could also replace these with arbitrary parameters (A, B). Doing this we obtain the equations

$$x_0x_2 = \alpha x_1^4 + \beta A^7 B^{12} x_1^2 + A^{24}B^{14} \qquad x_3x_5 = x_4^2 + \alpha^2\gamma B$$
$$x_1x_3 = \gamma x_2^2 + A^4B^7 \qquad x_4x_6 = A^2x_5^4 + \alpha^3\beta\gamma^2 A x_5^2 + \alpha^7\gamma^4$$
$$x_2x_4 = x_3^4 + \beta AB^2x_3^2 + \alpha A^2B^4 \qquad x_5x_7 = Bx_6^2 + \alpha^{12}\gamma^7$$

which define a cluster algebra $\mathcal{A}$ over a bigger coefficient ring $S = \mathbb{C}[\alpha, \beta, \gamma, A, B]$. Now let $V = \operatorname{Spec}\mathcal{U}$ be the corresponding cluster variety, which comes with a natural projection $\pi\colon V \to \mathbb{A}^5 = \operatorname{Spec} S$. By observation, the equations are graded with respect to the rank 2 torus action $(\mathbb{C}^*)^2 \curvearrowright V$ given by the following weights.

	x_0	x_1	x_2	x_3	x_4	x_5	x_6	x_7	A	B	α	β	γ
w_A	24	7	4	1	0	−1	0	1	2	0	0	0	0
w_B	41	12	7	2	1	0	−1	0	0	2	0	0	0

We can obtain the affine cover of our flip by taking a *regular pullback* $W = V \times_{\mathbb{A}^5} \mathbb{A}^2$

$$\begin{array}{ccc} W & \longrightarrow & V \\ \downarrow & \square & \downarrow \pi \\ \mathbb{A}^2 & \xrightarrow[\phi]{} & \mathbb{A}^5 \end{array}$$

where $\phi^*\colon S \to \mathbb{C}[u, z]$ is $(\alpha, \beta, \gamma, A, B) \mapsto (\alpha(u), \beta(u), \gamma(u), z^3, z)$. By taking $\mathbb{C}[u, z]$ to be graded with u, z of degree $0, 2$, then ϕ is equivariant with respect to $3w_A + w_B$ and W inherits its $\mathbb{C}^*$-action from V. The main strength of this point of view is that now, by varying the entries of the morphism ϕ (chosen equivariantly with respect to some torus weight $aw_A + bw_B$), we can construct and deform flips in families.

5 Diptych Varieties

In the same spirit as the generalisation of Example 4.4, diptych varieties were introduced by Brown and Reid [BR13] in order to be used as key varieties for constructing $k2A$ flips. They restrict their attention to the special case of $k2A$ flips starting from equations of the form

$$x_0x_2 = u^{\alpha_1}x_1^{d_1} + z^{r_1}, \quad x_1x_3 = u^{\alpha_2}x_2^{d_2} + z^{r_2} \tag{5.1}$$

or, in other words, those with binomial exchange relations (which also happens to be the original cluster algebra format introduced by Fomin and Zelevinsky). In this section, we follow [BR13] but, as before, we warn the reader that our notation is not always chosen compatibly.

5.1 Smoothing a Tent

The tent. A diptych variety is an affine 6-fold, defined to be a special deformation of a reducible toric surface T called a *tent*. A tent is a surface of the form

$$T = \bigcup_{i=1}^{4} S_i = \mathbb{A}^2 \cup \tfrac{1}{r}(1,\alpha) \cup \mathbb{A}^2 \cup \tfrac{1}{s}(1,\beta)$$

where we glue S_i to S_{i+1} for $i = 1,2,3$ and glue S_4 to S_1 by identifying toric boundary strata. We can obtain an affine embedding $T \subset \mathbb{A}^{k+l+4}_{u_0,\dots,u_{k+1},v_0,\dots,v_{l+1}}$ by writing down the Hirzebruch–Jung continued fraction expansions

$$\frac{r}{r-\alpha} = [a_1,\dots,a_k], \qquad \frac{s}{s-\beta} = [b_1,\dots,b_l]$$

and considering the ideal of equations

$$I(T) = \begin{pmatrix} u_{i-1}u_{i+1} = u_i^{a_i} & 1 \le i \le k \\ v_{j-1}v_{j+1} = v_j^{b_j} & 1 \le j \le l \\ u_i v_j = 0 & (i,j) \neq (0,0),(k+1,l+1) \end{pmatrix}$$

where $S_1 = \mathbb{A}^2_{u_0,v_0}$, $S_3 = \mathbb{A}^2_{u_{k+1},v_{l+1}}$, the u_i are coordinates on $S_2 = \frac{1}{r}(1,\alpha)$ and the v_j are coordinates on $S_4 = \frac{1}{s}(1,\beta)$.

Two deformations of T. Starting at the 'bottom' with the coordinate axes $\mathbb{A}^1_{u_0} \cup \mathbb{A}^1_{v_0} \subset T$, we can make a toric deformation of T as follows. For some $a_0, b_0 \in \mathbb{Z}_{\ge 0}$ and two deformation parameters A, B we can replace the equations $u_1v_0 = u_0v_1 = 0$ by $u_1v_0 = Au_0^{a_0}$ and $u_0v_1 = Bv_0^{b_0}$. As long as a_0 and b_0 were chosen to be large enough, this extends to a deformation of T, giving a toric 4-fold $V_{AB} \subset \mathbb{A}^{k+l+4} \times \mathbb{A}^2$. Similarly we can also obtain V_{LM}, deforming the 'top' of T along the coordinate axes $\mathbb{A}^1_{u_{k+1}} \cup \mathbb{A}^1_{v_{l+1}}$, by writing down appropriate $a_{k+1}, b_{l+1} \in \mathbb{Z}^2_{\ge 0}$, two deformation parameters L, M and the equations $u_kv_l = Lu_{k+1}^{a_{k+1}}$ and $u_{k+1}v_l = Mv_{l+1}^{b_{k+1}}$.

The diptych variety. Under the right combinatorial conditions on (r,α), (s,β) and a_0, b_0, a_{k+1}, b_{l+1}, the deformations V_{AB} and V_{LM} are actually *smoothings* of T along the corresponding coordinate strata Figure 5.3. In this case, we would like to 'join' these two smoothings together to obtain a 6-fold V, which is a (total) smoothing of T over $R = \mathbb{C}[A,B,L,M]$. For the moment, let us just consider how we might do this for the two equations $u_1v_0 = u_0v_1 = 0$. We know that they deform

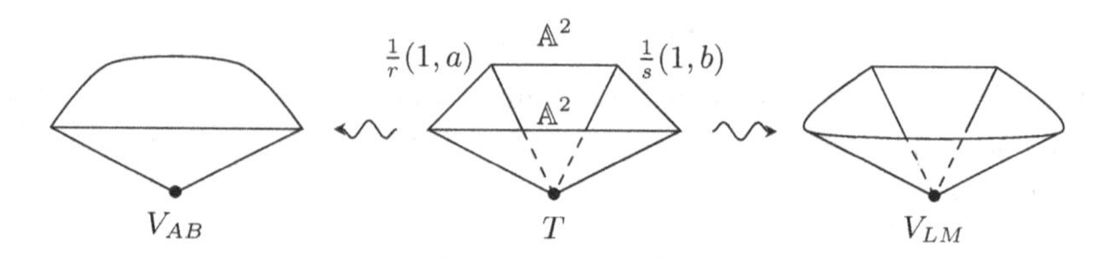

Figure 5.3 Deforming T to obtain V_{AB} and V_{LM}.

to $u_1v_0 = Au_0^{a_0}$ and $u_0v_1 = Bv_0^{b_0}$ over $\mathbb{C}[A,B]$. Moreover they must also deform to some binomial equations $u_1v_0 = L^\gamma M^\delta X$ and $u_0v_1 = L^\epsilon M^\zeta Y$ over $\mathbb{C}[L,M]$, for some monomials $X, Y \in \mathbb{C}[u_i, v_j]$ and some exponents $\gamma, \delta, \epsilon, \zeta \in \mathbb{Z}_{\geq 0}$. Naively, we could stick these binomial equations together to get two trinomial equations

$$u_1v_0 = Au_0^{a_0} + L^\gamma M^\delta X, \qquad u_0v_1 = Bv_0^{b_0} + L^\epsilon M^\zeta Y$$

and hope that these equations can be extended to a deformation V of T over $\mathbb{C}[A,B,L,M]$. Such a V, if it exists, is called a *diptych variety.*

5.2 Classification of Diptych Varieties

We now explain what the 'right combinatorial conditions' are on (r,α), (s,β) and the exponents a_0, b_0, a_{k+1}, b_{l+1}, which are required for the existence of V.

A 2-periodic product of matrices. In order for V_{AB} to smooth the bottom axes of T, by [BR13, Theorem 2.10] it turns out that there must exist $a, b \in \mathbb{Z}_{\geq 0}$ such that

$$a \equiv \alpha \mod r, \quad b \equiv \beta \mod s \quad \text{and} \quad \begin{pmatrix} r & a \\ b & s \end{pmatrix} \in \mathrm{SL}(2,\mathbb{Z})$$

which is necessary data needed to define the fan of V_{AB}. Looking from the other end of T, the two cyclic quotient singularities down the sides of T are of the form $\frac{1}{r}(\alpha^{-1},1)$, $\frac{1}{s}(\beta^{-1},1)$. Therefore, in order for V_{LM} to smooth the top axes of T, there must also exist $l, m \in \mathbb{Z}_{\geq 0}$ such that

$$m \equiv \alpha^{-1} \mod r, \quad l \equiv \beta^{-1} \mod s \quad \text{and} \quad \begin{pmatrix} r & m \\ l & s \end{pmatrix} \in \mathrm{SL}(2,\mathbb{Z}).$$

This system of congruences and the equations $ab = lm = rs - 1$ turn out only to have solutions of the following special form [BR13, Theorem 3.3].

There must exist two integers $d_1, d_2 \geq 0$ such that $b + m = d_1 s$, $a + l = d_2 r$, and for some $k \geq 4$

$$\begin{pmatrix} r & m \\ l & s \end{pmatrix} = \begin{pmatrix} d_1 & -1 \\ 1 & 0 \end{pmatrix} \begin{pmatrix} d_2 & -1 \\ 1 & 0 \end{pmatrix} \cdots \begin{pmatrix} d_{k-3} & -1 \\ 1 & 0 \end{pmatrix} \begin{pmatrix} 0 & 1 \\ -1 & 0 \end{pmatrix} \tag{5.2}$$

where $d_1, d_2, \ldots, d_{k-3}$ is a 2-periodic integer sequence (i.e. $d_i = d_{i+2}$ for all $i \in \mathbb{Z}$). All other solutions listed in [BR13, Theorem 3.3] are obtained either by switching $(r, \alpha) \leftrightarrow (s, \beta)$ (which flips the tent T left-to-right) or by switching $(r, \alpha) \leftrightarrow (r, \alpha^{-1})$ and $(s, \beta) \leftrightarrow (s, \beta^{-1})$ (which flips T top-to-bottom). If $d_1 d_2 < 4$ then, for all integers to be nonnegative, the length of this sequence of matrices is constrained by $k \leq 4, 5, 6, 8$ for $d_1 d_2 = 0, 1, 2, 3$ (which we note is equal to period of the cluster variables in the corresponding cluster algebra of finite type), but k can be arbitrary for $d_1 d_2 \geq 4$.

The monomials from V_{AB} and V_{LM}. It follows from our chosen solution (5.2) to the existence of the diptych variety, that $(d_1 - 1)l < r < d_1 l$ and $(d_1 - 1)s < m < d_1 s$. From this and the equations $b + m = d_1 s$, $a + l = d_2 r$ we obtain $0 < b < s$ and $(d_2 - 1)r < a < d_2 r$. Now, by [BR13, Lemma 2.7], the first smoothing V_{AB} deforms the two equations $u_0 v_1 = u_1 v_0 = 0$ to

$$u_1 v_0 = A u_0^{\lceil \frac{b}{s} \rceil} = A u_0 \quad \text{and} \quad u_0 v_1 = B v_0^{\lceil \frac{a}{r} \rceil} = B v_0^{d_2},$$

and the second smoothing V_{LM} deforms them to

$$\begin{aligned} u_1 v_0 &= L^r M^m u_0^{\lceil -\frac{m}{s} \rceil} = L^r M^m u_0^{-(d_1-1)} \quad \text{and} \\ u_0 v_1 &= L^l M^s v_0^{\lceil -\frac{l}{r} \rceil} = L^l M^s. \end{aligned}$$

The negative exponent in $L^r M^m u_0^{-(d_1-1)}$ appears problematic at first. However Brown and Reid deal with this problem by observing that on V_{LM} we have

$$\begin{aligned} L^r M^m u_0^{-(d_1-1)} &= L^r M^m v_1^{d_1-1} (u_0 v_1)^{-(d_1-1)} \\ &= L^{r-(d_1-1)l} M^{m-(d_1-1)s} v_1^{d_1-1} \end{aligned}$$

and, since $r > (d_1 - 1)l$ and $m > (d_1 - 1)s$, this can be rewritten as a genuine monomial.

Constructing V by unprojection. Consider the complete intersection $V_0 \subset \mathbb{A}^8$ of codimension two, defined by the two equations

$$u_1 v_0 = A u_0 + L^{r-(d_1-1)l} M^{m-(d_1-1)s} v_1^{d_1-1}, \quad u_0 v_1 = B v_0^{d_2} + L^l M^s. \tag{5.3}$$

Starting from V_0, Brown and Reid prove the existence of V using *serial unprojection.* This is a systematic way of building up the coordinate ring $\mathbb{C}[V]$ by successively reintroducing the variables $u_2, \dots, u_{k+1}$ and $v_2, \dots, v_{l+1}$ as rational functions on V_0. In a nutshell, the variety V that we want to construct is isomorphic to V_0 on the complement of a 'bad' Weil divisor $D_0 = V(u_0, v_0, L^l M^s) \subset V_0$. Since D_0 and V_0 are both projectively Gorenstein the Kustin–Miller unprojection theorem guarantees the existence of a rational function $\phi \in \mathbb{C}[V_0]$ with divisor of poles equal to D_0. Taking V_1 to be the graph of ϕ we obtain a birational map $(D_0 \subset V_0) \dashrightarrow (Q \in V_1)$ which contracts D_0 to a point Q. (The inverse map is projection from Q, hence the terminology.) We can continue this game by unprojecting any new divisor that may appear in V_1 in the locus above D_0. As shown by Brown and Reid, working in this way we can obtain all of the generators and (in principle) all of the relations defining $\mathbb{C}[V]$.

Existence of diptych varieties. The existence of the diptych variety is covered in [BR17] in the case $\Delta = d_1 d_2 - 4 \leq 0$, with a further division of the $\Delta > 0$ case into two subcases: the *main case* $d_1, d_2 \geq 2$ [BR13] and the *redundant generators* case $d_1 = 1$ and $d_2 \geq 5$ [BR]. The reason for this subdivision is simply that the inductive procedure for constructing V runs a little differently in these subcases, and there is no simultaneously convenient notation. From now on we will also assume that we are in the main case, although the other cases should be no harder to prove and would only require simple changes to our arguments.

6 Proof of Theorem 1.1

We now prove Theorem 1.1, which is to first show that the diptych variety is a cluster variety and then to give an explicit generating set for the upper cluster algebra $\mathcal{U}$.

6.1 Diptych Varieties as Cluster Varieties

We rephrase the notation describing the diptych variety in a more convenient form. Let $R = \mathbb{C}[A, B, L, M]$ be a coefficient ring and fix $d_1, d_2 \geq 2$ and $k \geq 4$ such that $\Delta = d_1 d_2 - 4 > 0$. We let $d_i := d_{(i \bmod 2)}$ for $i \in \mathbb{Z}$.

Some integer sequences. Most of the subtlety involved in the proof of Theorem 1.1 requires simple (but fairly tedious) bookkeeping with the following sequences.

(1) We define $l_i, m_i \in \mathbb{Z}_{\geq 0}$ for all $0 \leq i \leq k-3$ by the following 'ascending' matrix product

$$\begin{pmatrix} l_i & m_i \\ l_{i+1} & m_{i+1} \end{pmatrix} = \begin{pmatrix} d_i & -1 \\ 1 & 0 \end{pmatrix} \begin{pmatrix} d_{i+1} & -1 \\ 1 & 0 \end{pmatrix} \cdots \begin{pmatrix} d_{k-3} & -1 \\ 1 & 0 \end{pmatrix} \begin{pmatrix} 0 & 1 \\ -1 & 0 \end{pmatrix}$$

and let $l_{k-2}, l_{k-1}, l_k = 0, -1, 0$ and $m_{k-2}, m_{k-1}, m_k = 1, 0, -1$. The sequence $l_1 > \ldots > l_{k-1}$ is strictly decreasing, satisfies the recurrence $l_{i+1} + l_{i-1} = d_{i-1} l_i$ and $(d_{i-1} - 1) l_i < l_{i-1} < d_{i-1} l_i$ for all i. Similarly for the sequence $m_1, \ldots, m_k$.

(2) We define $a_i, b_i \in \mathbb{Z}_{\geq 0}$ for all $4 \leq i \leq k$ by the following 'descending' matrix product

$$\begin{pmatrix} b_i & a_i \\ b_{i-1} & a_{i-1} \end{pmatrix} = \begin{pmatrix} d_i & -1 \\ 1 & 0 \end{pmatrix} \begin{pmatrix} d_{i-1} & -1 \\ 1 & 0 \end{pmatrix} \cdots \begin{pmatrix} d_4 & -1 \\ 1 & 0 \end{pmatrix} \begin{pmatrix} 0 & 1 \\ -1 & 0 \end{pmatrix}$$

and let $a_1, a_2, a_3 = -1, 0, 1$ and $b_1, b_2, b_3 = 0, -1, 0$. The sequence $a_1 < \ldots < a_k$ is strictly increasing, satisfies the recurrence $a_{i+1} + a_{i-1} = d_{i+1} a_i$ and $(d_{i+1} - 1) a_i < a_{i+1} < d_{i+1} a_i$ for all i. Similarly for the sequence $b_2, \ldots, b_k$.

(3) It is also useful to define the following 'interpolating' sequences a_{ij}, b_{ij}, l_{ij}, m_{ij} for all $0 \leq i \leq k+1$ and all $1 \leq j \leq d_{i-1} - 1$.

$$\begin{pmatrix} l_{ij} & m_{ij} \\ l_i & m_i \end{pmatrix} = \begin{pmatrix} j & -1 \\ 1 & 0 \end{pmatrix} \begin{pmatrix} l_i & m_i \\ l_{i+1} & m_{i+1} \end{pmatrix}$$

$$\begin{pmatrix} b_{ij} & a_{ij} \\ b_i & a_i \end{pmatrix} = \begin{pmatrix} j & -1 \\ 1 & 0 \end{pmatrix} \begin{pmatrix} b_i & a_i \\ b_{i-1} & a_{i-1} \end{pmatrix}$$

Lemma 6.1. *We have the following useful relations.*

$$\begin{pmatrix} b_{i+2} & -b_{i+1} \\ a_i & -a_{i-1} \end{pmatrix} = \begin{pmatrix} d_1 & -1 \\ 1 & 0 \end{pmatrix} \cdots \begin{pmatrix} d_{i-2} & -1 \\ 1 & 0 \end{pmatrix} \quad \textit{and}$$

$$\begin{pmatrix} l_{i-1} & m_{i-1} \\ l_i & m_i \end{pmatrix} = \begin{pmatrix} -a_{i-1} & b_{i+1} \\ -a_i & b_{i+2} \end{pmatrix} \begin{pmatrix} l_1 & m_1 \\ l_2 & m_2 \end{pmatrix}.$$

Proof The first statement follows from a simple induction. Note that $\left(\begin{smallmatrix} b_5 & -b_4 \\ a_3 & -a_2 \end{smallmatrix}\right) = \left(\begin{smallmatrix} d_1 & -1 \\ 1 & 0 \end{smallmatrix}\right)$ (since $d_5 = d_1$), so the base case holds. Then

$$\begin{pmatrix} b_{i+2} & -b_{i+1} \\ a_i & -a_{i-1} \end{pmatrix} \begin{pmatrix} d_{i-1} & -1 \\ 1 & 0 \end{pmatrix} = \begin{pmatrix} d_{i-1} b_{i+2} - b_{i+1} & -b_{i+2} \\ d_{i-1} a_i - a_{i-1} & -a_i \end{pmatrix}$$

$$= \begin{pmatrix} b_{i+3} & -b_{i+2} \\ a_{i+1} & -a_i \end{pmatrix}.$$

The second statement follows from the first, since

$$\begin{pmatrix} b_{i+2} & -b_{i+1} \\ a_i & -a_{i-1} \end{pmatrix} \in \mathrm{SL}(2, \mathbb{Z}). \qquad \square$$

The diptych variety as a cluster variety. We now note that the complete intersection V_0 (5.3) is birational to the complete intersection $W_0 \subset \mathbb{A}^4_{x_0,x_1,x_2,x_3} \times \mathbb{A}^4_{ABLM}$ defined by the two equations

$$x_0x_2 = Ax_1^{d_1} + L^{l_1}M^{m_1}, \qquad x_1x_3 = Bx_2^{d_2} + L^{l_2}M^{m_2} \tag{6.1}$$

where $\left(\begin{smallmatrix} r & m \\ l & s \end{smallmatrix}\right) = \left(\begin{smallmatrix} l_1 & m_1 \\ l_2 & m_2 \end{smallmatrix}\right)$ and the change in variables is given by $x_1, x_2, x_3 = u_0, v_0, v_1$ and

$$x_0 = u_0^{d_1-1}u_1 - v_0^{-1}\left(L^{l_2-l_3}M^{m_2-m_3}(Bv_0^{d_2} + L^{l_2}M^{m_2})^{d_1-1} - L^{l_1}M^{m_1}\right).$$

Moreover this is an isomorphism $(V_0 \setminus V(u_0, v_0)) \cong (W_0 \setminus V(x_1, x_2))$. Now, by construction, Brown and Reid's diptych variety $V = \operatorname{Spec} S$ is the unique normal affine variety birational to V_0 such that the locus $V(S \cdot (u_0, v_0))$ has codimension ≥ 2 in V. Therefore V is the unique normal affine variety birational to W_0 such that the locus $V(S \cdot (x_1, x_2))$ has codimension ≥ 2 in V. By Proposition 3.3 and Lemma 3.5, we must have $S = \mathcal{U}$, where $\mathcal{U}$ is the upper cluster algebra of the cluster algebra with initial exchange relations (6.1). We have proved the first part of the statement of Theorem 1.1.

The diptych variety as a key variety for $k2A$ flips. Before moving on to give our explicit generating set for $\mathcal{U}$, we briefly show how the diptych variety can be used as a key variety for $k2A$ neighbourhoods with initial equations of the form (5.1). The diptych variety V comes with a projection $\pi\colon V \to \mathbb{A}^4_{A,B,L,M}$ realising V as a flat family of affine Gorenstein surfaces with central fibre given by the tent $T = \pi^{-1}(0)$. As in §4.4, we can now pick a morphism $\phi\colon \mathbb{A}^2_{u,z} \to \mathbb{A}^4_{A,B,L,M}$, where $\phi^*(A, B, L, M) = (u^{\alpha_1}, u^{\alpha_2}, z^{-s_k}, z^{-s_{k+1}})$ and take the regular pullback $W = V \times_{\mathbb{A}^4} \mathbb{A}^2$ to get the affine cover of a $k2A$ neighbourhood. The torus action on the two toric 4-folds $V_{AB} = \pi^{-1}(\mathbb{A}^2_{A,B})$ and $V_{LM} = \pi^{-1}(\mathbb{A}^2_{L,M})$ extends compatibly to a $(\mathbb{C}^*)^4$-action on V (see §6.5) and, after pulling back and taking the quotient, V_{AB} and V_{LM} become the general hypersurface section H_X and the general elephant E_X respectively.

6.2 An Explicit Generating Set for $\mathcal{U}$

We recall the notation of Definition 3.6, which we use to expand elements of $\mathcal{U}$ as Laurent polynomials in $R[x_1^{\pm 1}, x_2^{\pm 1}]$.

Cluster variables. Starting from (6.1), it is easy to use induction and the happy coincidence that $d_{i+1} = d_{i-1}$ to check that the cluster variables for $i = 3, \ldots, k$ satisfy the following exchange relations.

$$x_{i-1}x_{i+1} = \begin{cases} x_i^{d_i} + A^{a_i}B^{b_i}L^{l_i}M^{m_i} & 3 \le i \le k-2 \\ Lx_{k-1}^{d_{k-1}} + A^{a_{k-1}}B^{b_{k-1}} & i = k-1 \\ Mx_k^{d_k} + A^{a_k}B^{b_k} & i = k \end{cases}$$

Clearly $x_i \in \mathcal{U}$ for all i. We will now describe the rest of the generators needed to generate $\mathcal{U}$.

Outline of the proof. The proof will follow these five steps.

(1) Describe the subset $S \subset \mathcal{U}$ which we claim generates $\mathcal{U}$.
(2) Define a set $\Theta = \{\theta_{(a,b)} \colon (a,b) \in \mathbb{Z}^2\} \subset R[S]$, indexed by $\mathbb{Z}^2$, of well-chosen monomials in the elements of S.
(3) Define a homogeneous $\mathbb{Z}^4$-grading w on $\mathcal{U}$ and, for any homogeneous element $z \in \mathcal{U}$, show that $w(z) \ge w(\theta_{(a,b)})$ for all $(a,b) \in \operatorname{supp}(z)$.
(4) Using w, find some element $\theta = \sum_{a,b>0} r_{(a,b)}\theta_{(a,b)} \in R[S]$ such that $z' := z - \theta$ has $\operatorname{supp}(z') \cap \mathbb{R}^2_{>0} = \emptyset$.
(5) Show $z' \in R[x_0, x_1, x_2, x_3] \subseteq R[S]$ and hence $z = z' + \theta \in R[S]$.

6.3 Step 1: The Missing Generators

We describe the missing variables y_{ij} which appear in the construction of the diptych variety. Pick any five consecutive cluster variables $x_{i-2}, x_{i-1}, x_i, x_{i+1}, x_{i+2}$, which between them satisfy three exchange relations. The variables y_{ij} come from *rolling factors relations*, which interpolate between the two exchange relations $x_{i-2}x_i = F_{i+1}(x_{i+1})$ and $x_ix_{i+2} = F_{i-1}(x_{i-1})$.

Definition 6.2. For $0 < j < d_{i-1}$, we define y_{ij} by

$$x_iy_{ij} := A^{a_{ij}}B^{b_{ij}}x_{i-1}^{d_{i-1}-j} + L^{l_{i,d_{i-1}-j}}M^{m_{i,d_{i-1}-j}}x_{i+1}^j.$$

Taking $j = 0$ in the definition of y_{ij} gives $y_{i0} = A^{-a_{i-1}}B^{-b_{i-1}}x_{i-2}$ and taking $j = d_{i-1}$ gives $y_{id_{i-1}} = L^{-l_{i+1}}M^{-m_{i+1}}x_{i+2}$.

Lemma 6.3. *We have $y_{ij} \in \mathcal{U}$ for all $0 < j < d_{i-1}$ and moreover, $y_{i,d_{i-1}-1} = y_{i+1,1}$ for all i.*

Proof Since $x_{i+1} \in U_{i-1}$, it follows that $y_{ij} \in U_{i-1}$. Similarly $y_{ij} \in U_i$. To simplify the notation we let $j' = d_{i-1} - j$. Now we have

$$\begin{aligned} x_{i-1}^j x_i y_{ij} &= A^{a_{ij}} B^{b_{ij}} x_{i-1}^{d_{i-1}} + L^{l_{ij'}} M^{m_{ij'}} (x_{i+1} x_{i-1})^j \\ &= A^{a_{ij}} B^{b_{ij}} \left(x_{i-2} x_i - A^{a_{i-1}} B^{b_{i-1}} L^{l_{i-1}} M^{m_{i-1}} \right) \\ &\qquad + L^{l_{ij'}} M^{m_{ij'}} \left(x_i^{d_i} + A^{a_i} B^{b_i} L^{l_i} M^{m_i} \right)^j \\ &\equiv A^{j a_i} B^{j b_i} L^{l_{ij'} + j l_i} M^{m_{ij'} + j m_i} \\ &\qquad - A^{a_{i-1} + a_{ij}} B^{b_{i-1} + b_{ij}} L^{l_{i-1}} M^{m_{i-1}} \mod x_i \\ &\equiv 0 \mod x_i \end{aligned}$$

which implies that $y_{ij} \in R[x_{i-2}, x_{i-1}^{\pm 1}, x_i]$, and hence that $y_{ij} \in U_{i-2}$. Lemma 3.4 implies $y_{ij} \in \mathcal{U}$. The second statement can be verified by expanding $x_i x_{i+1} y_{i,d_{i-1}-1}$ and $x_i x_{i+1} y_{i+1,1}$ and checking equality using the relations $a_{i,d_{i-1}-1} = (d_{i-1} - 1)a_i - a_{i-1} = a_{i+1} - a_i = a_{i+1,1}$ etc. □

Definition 6.4. Let S be the following set, which we claim generates $\mathcal{U}$

$$S = \{x_i \colon 0 \le i \le k+1\} \cup \{y_{ij} \colon 2 \le i \le k-1,\ 1 \le j \le d_{i-1} - 1\}.$$

6.4 Step 2: The Fan $\mathcal{F}$ and Basis Θ

We define a sequence of vectors v_i for $i = 0, \dots, k+1$ by setting $v_0 = (0,1)$, $v_1 = (-1,0)$, $v_2 = (0,-1)$, $v_3 = (1,0)$, and then by setting $v_{i+1} = d_i v_i - v_{i-1}$ for $3 \le i \le k+1$. We also define the vectors $v_{ij} := j v_{i+1} - v_i$ for $2 \le i \le k-1$ and $1 \le j \le d_{i+1} - 1$. Note that $v_{i,d_{i+1}-1} = v_{i+1,1}$. We let $\rho_i = \mathbb{R}_{\ge 0} v_i$ and $\rho_{ij} = \mathbb{R}_{\ge 0} v_{ij}$ be the rays spanned by these vectors.

Lemma 6.5. *Let $\mathcal{F}$ be the complete fan in $\mathbb{R}^2$ spanned by the set of rays*

$$\{\rho_i \colon 0 \le i \le k+1\} \cup \{\rho_{ij} \colon 2 \le i \le k-1,\ 1 \le j \le d_{i-1} - 1\}.$$

Then every cone of $\mathcal{F}$ is smooth. Moreover $v(x_i) = v_i = (b_{i+1}, a_{i-1})$ and $v(y_{ij}) = v_{ij}$ for all $x_i, y_{ij} \in S$.

Proof It is clear that the cone $\langle v_1, v_2 \rangle$ is smooth (see Figure 5.4). To show that all of the other cones are smooth it suffices to show that between any two adjacent cones, $\langle u_1, u_2 \rangle$ and $\langle u_2, u_3 \rangle$ say, there is a well-defined tag $a \in \mathbb{Z}$ such that $u_1 + u_3 = a u_2$. This provides a change

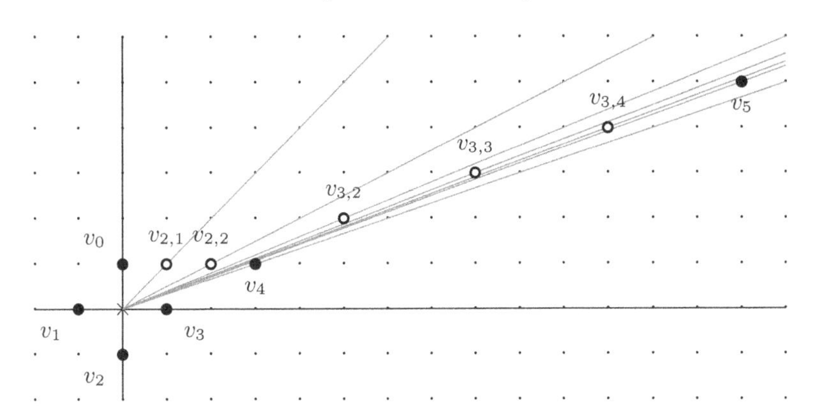

Figure 5.4 The fan $\mathcal{F}$ in the case that $(d_1, d_2) = (3, 5)$ and $k = 4$. The black vertices correspond to the cluster variables x_i and the white vertices correspond to the missing variables y_{ij} (see the 'scissors diagram' [BR13, Figure 4.2]).

of basis $\left(\begin{smallmatrix} a & -1 \\ 1 & 0 \end{smallmatrix}\right) \in \mathrm{SL}(2, \mathbb{Z})$ between cones. By definition we have $v_{i-1} + v_{i+1} = d_i v_i$ for $i \geq 3$. We also see that $v_0 + v_2 = 0$, $v_1 + v_3 = 0$, $v_1 + v_{21} = v_0$ and $v_k + v_{k1} = v_{k+1}$. Finally, it is not hard to check that $v_{i,d_{i+1}-2} + v_{i+1,2} = 3v_{i+1,1}$ and $v_{i,j-1} + v_{i,j+1} = 2v_{ij}$ if $2 \leq j \leq d_{i+1} - 2$.

For the second statement, note that $v(x_i) = v_i$ for $i = 0, 1, 2, 3$. Now

$$v(x_{i+1}) + v(x_{i-1}) = v(x_{i+1}x_{i-1}) = v(x_i^{d_i} + A^{a_i} B^{b_i} L^{l_i} M^{m_i}) = d_i v(x_i).$$

So, by induction $v(x_i) = v_i$ for $i \geq 3$. The explicit formula $v_i = (b_{i+1}, a_{i-1})$ is also easy to check by induction, as in the proof of Lemma 6.1. We have $v(x_{i+1}^j) > v(x_{i-1}^{d_{i-1}-j})$, since $v(x_{i+1}^j) - v(x_{i-1}^{d_{i-1}-j}) =$

$$jv_{i+1} - (d_{i-1} - j)v_{i-1} = jd_i v_i - d_{i-1}v_{i-1} = (jd_i - 1)v_i - v_{i-2} > 0$$

where the last inequality follows from the fact that $jd_i > 1$ and $v_i > v_{i-2}$, since the sequences a_i, b_i are increasing. Therefore

$$v(y_{ij}) = v\left(A^{a_{ij}} B^{b_{ij}} x_{i-1}^{d_{i-1}-j} + L^{l_{i,d_{i-1}-j}} M^{m_{i,d_{i-1}-j}} x_{i+1}^j\right) - v(x_i)$$

which equals $jv_{i+1} - v_i = v_{ij}$. □

Any $(a, b) \in \mathbb{Z}^2$ belongs to some cone $\langle v(s_1), v(s_2)\rangle \subset \mathcal{F}$, for two elements $s_1, s_2 \in S$. Since the cone is smooth we can write $(a, b) = \alpha v(s_1) + \beta v(s_2)$ for some $\alpha, \beta \in \mathbb{Z}_{\geq 0}$.

Definition 6.6. For any $(a, b) \in \mathbb{Z}^2$, we let $\theta_{(a,b)} := s_1^\alpha s_2^\beta$ and we consider the set of all such elements $\Theta = \{\theta_{(a,b)} : (a, b) \in \mathbb{Z}^2\} \subset R[S]$.

6.5 Step 3: The $\mathbb{Z}^4$-Grading w

Let $w = (w_A, w_B, w_L, w_M)$ be a $\mathbb{Z}^4$-grading determined by:

	A	B	L	M	x_0	x_1	x_2	x_3
w_A	d_1	0	0	0	0	-1	0	1
w_B	0	d_2	0	0	1	0	-1	0
w_L	l_1	l_2	1	0	l_1	0	0	l_2
w_M	m_1	m_2	0	1	m_1	0	0	m_2

We observe that the initial exchange relations (6.1) are homogeneously graded with respect to w. It follows that U_0, U_1, U_2 and $\mathcal{U} = U_0 \cap U_1 \cap U_2$ are homogenenously graded with respect to w too.

Lemma 6.7. *Suppose that $z \in \mathcal{U}$ is homogeneous and $v(z) = (a, b)$. Then*

$$w(z) \geq \Big(a,\ b,\ \max(bl_1, al_2),\ \max(bm_1, am_2)\Big)$$

and equality holds if $z = x_i$ or $z = y_{ij}$.

Proof Since $G_z(x_1, x_2)$ is homogeneous with respect to w and is not divisible by x_1, it follows that $w_A(G_z) \geq 0$ and strict inequality holds if and only if A divides G_z. Therefore

$$w_A(z) = w_A(G_z) - w_A(x_1^a x_2^b) = w_A(G_z) + a \geq a$$

and $w_A(z) > a$ if and only if A divides G_z. Similarly for w_B. Now consider $G_z(x_1, 0)$, which must be a nonzero polynomial in x_1. Since

$$z = x_1^{-a} \left(x_0^{-1} F_1(x_1)\right)^{-b} G_z\left(x_1, x_0^{-1} F_1(x_1)\right) \in U_0$$

it follows that $F_1(x_1)^b$ divides $G_z(x_1, 0)$, and so

$$w_L(G_z) = w_L(G_z(x_1, 0)) \geq b w_L(F_1(x_1)) = bl_1.$$

By repeating a similar argument with $G_z(0, x_2)$ it follows that $w_L(G_z) \geq a w_L(F_2(x_2)) = al_2$, proving that $w_L(z) = w_L(G_z) \geq \max(bl_1, al_2)$. Similarly for w_M.

We now show that equality holds for $z = x_i$. Equality for $w_A(x_i)$ follows from the fact that A does not divide G_{x_i} (see proof of Lemma 6.8) and similarly for $w_B(x_i)$. Note that $v(x_i) = (b_{i+1}, a_{i-1})$ by Lemma 6.5, so we are only required to prove that $w_L(x_i) = \max(a_{i-1}l_1,\ b_{i+1}l_2)$, and similarly for w_M. In fact, by Lemma 6.1, $l_{i-1} = b_{i+1}l_2 - a_{i-1}l_1 \geq 0$ for

$i \le k-1$, so we must show that $w_L(x_i) = b_{i+1}l_2$ for $i \le k$. By inspection this holds for $i = 2$ and 3, so by induction we have

$$w_L(x_{i+1}) = w_L(x_i^{d_i}) - w_L(x_{i-1}) = (d_i b_{i+1} - b_i)l_2 = b_{i+2}l_2$$

for $i \le k-1$. Finally for x_{k+1} we have $w_L(x_{k+1}) = w_L(A^{a_k}B^{b_k}) - w_L(x_{k-1}) = a_k l_1 > b_{k+2}l_2$ since $l_k = b_{k+2}l_2 - a_k l_1 = -1$.

For $z = y_{ij}$, we have $v_{ij} = v_{i+1} - jv_i = (jb_{i+2} - b_{i+1}, ja_i - a_{i-1})$. So, we must show that $w_L(y_{ij}) = \max((ja_i - a_{i-1})l_1, (jb_{i+2} - b_{i+1})l_2)$. But now, for all $2 \le i \le k-1$ and all $1 \le j \le d_{i-1} - 1$, we have

$$\begin{aligned} l_{i,d_{i-1}-j} = (d_{i-1} - j)l_i - l_{i+1} = l_{i-1} - jl_i \\ = (ja_i - a_{i-1})l_1 - (jb_{i+2} - b_{i+1})l_2 > 0. \end{aligned}$$

So, we must show that $w_L(y_{ij}) = (ja_i - a_{i-1})l_1$. This follows from

$$\begin{aligned} w_L(y_{ij}) &= w_L(L^{l_{i,d_{i-1}-j}} M^{m_{i,d_{i-1}-j}} x_{i+1}^j) - w_L(x_i) \\ &= l_{i,d_{i-1}-j} + (jb_{i+2} - b_{i+1})l_2 \\ &= (b_{i+1}l_2 - a_{i-1}l_1) - j(b_{i+2}l_2 - a_i l_1) + (jb_{i+2} - b_{i+1})l_2 \\ &= (ja_i - a_{i-1})l_1. \end{aligned}$$

□

6.6 Step 4: Shrinking supp(z)

Suppose that $z \in \mathcal{U}$ is homogeneous. If $\mathrm{supp}(z) \cap \mathbb{R}^2_{>0} = \emptyset$ then we move to Step 5. If not, our goal is to show that we can subtract monomials from Θ off of z until we are left with z' such that $\mathrm{supp}(z') \cap \mathbb{R}^2_{>0} = \emptyset$.

Lemma 6.8. *If* $\mathrm{supp}(z) \cap \mathbb{R}^2_{>0} \neq \emptyset$*, then the boundary of* $\mathrm{supp}(z)$ *contains a vertex* (a, b) *in the halfspace* $\{(a', b') \in \mathbb{Z}^2 \colon a' + b' > 0\}$*. Then there is a monomial* $r \in R$ *with* $\mathrm{supp}(z - r\theta_{(a,b)}) \subseteq (\mathrm{supp}(z) \setminus \{(a, b)\})$.

Proof We first show that the numerator polynomial of any $s \in S$ has constant term $G_s(0,0) = L^{w_L(s)}M^{w_M(s)}$. The constant term must be of the form $G_s(0,0) = \lambda A^{w_A(s)}B^{w_B(s)}L^{w_L(s)}M^{w_M(s)}$ for some $\lambda \in \mathbb{C}$, where the exponents are uniquely determined by the $\mathbb{Z}^4$-grading. Now, since $G_s(x_1, x_2)$ is homogeneous with respect to w_A and w_B, it follows that G_s can be written as a polynomial in $Ax_1^{d_1}$ and $Bx_2^{d_2}$. To show the claim it is enough to show that $s = x_1^{-a}x_2^{-b}$ for some a, b, after setting $A = B = 0$ and $L = M = 1$. Specialising the coefficients this way reduces to the toric surface determined by $x_0x_2 = x_1x_3 = 1$ and

$$x_{i-1}x_{i+1} = x_i^{d_i} \text{ for } 3 \le i \le k,$$
$$x_i y_{ij} = x_{i+1}^j \text{ for } 2 \le i \le k-1,\ 1 \le j \le d_{i-1} - 1.$$

Setting $x_1 = x_2 = 1$ we find that $s = 1$ for all $s \in S$, from which the conclusion that $s = x_1^{-a} x_2^{-b}$ is easy.

Now, turning to the statement of the Lemma, the existence of the vertex $(a, b) \in \operatorname{supp}(z)$ is clear. Let $\lambda \in \mathbb{C}^*$ be the coefficient of the monomial $A^\alpha B^\beta L^\gamma M^\delta x_1^{-a} x_2^{-b}$ in z, where $(\alpha, \beta, \gamma, \delta) = w(x_1^a x_2^b z)$. Now $v(z) = (a', b')$ for some $(a', b') \ge (a, b)$. So, by Lemma 6.7, we have $\gamma = w_L(z) = \max(b'l_1, a'l_2) \ge \max(bl_1, al_2) = w(\theta_{(a,b)})$ and similarly for w_M. Therefore, the monomial $r = \lambda A^\alpha B^\beta L^{\gamma - w_L(\theta_{(a,b)})} M^{\delta - w_M(\theta_{(a,b)})}$ has nonnegative exponents and, since $\operatorname{supp}(\theta_{(a,b)}) \subseteq \operatorname{supp}(z)$, this r satisfies the claim in the statement of the Lemma. □

Reducing to step 5. Let $z_0 := z$ and suppose we have constructed z_i. We replace z_i with the element $z_{i+1} := z_i - r_i \theta_{(a_i, b_i)}$ given by Lemma 6.8. We carry out this procedure until we reach an element z' such that $\operatorname{supp}(z') \cap \{(a', b') \in \mathbb{Z}^2 \colon a' + b' > 0\} = \emptyset$, and so $\operatorname{supp}(z') \cap \mathbb{R}^2_{>0} = \emptyset$. This happens after finitely many steps, since the intersection $\operatorname{supp}(z) \cap \{(a', b') \in \mathbb{Z}^2 \colon a' + b' > 0\}$ can contain only finitely many points.

6.7 Step 5: The Case $\operatorname{supp}(z) \cap \mathbb{R}^2_{>0} = \emptyset$.

Finally we deal with the easy case of $z \in \mathcal{U}$ supported outside of the positive quadrant.

Lemma 6.9. *Suppose that $z \in \mathcal{U}$. For $i = 0, 1, 2$, if $\operatorname{Newt}(z) \subseteq \langle v_i, v_{i+1} \rangle$ then $z \in R[x_i, x_{i+1}]$. In particular, if $\operatorname{supp}(z) \cap \mathbb{R}^2_{>0} = \emptyset$ then $z \in R[x_0, x_1, x_2, x_3]$.*

Proof The statement for $i = 1$ is obvious. Assume that $i = 0$. (The case $i = 2$ is similar.) Thus $v(z) = (-a, b)$ where $a, b \ge 0$. Since $\operatorname{Newt}(z) \subseteq \langle v_0, v_1 \rangle$, any monomial $x_1^\alpha x_2^\beta \in G_z(x_1, x_2)$ satisfies $\beta \le b$. Now, since $G_z(x_1, x_2)$ does not contain any monomial divisible by x_2^{b+1}, the numerator of the expression

$$z = \frac{x_1^a G_z(x_1, x_2)}{x_2^b} = \frac{x_0^b x_1^a G_z\left(x_1, x_0^{-1} F_1(x_1)\right)}{F_1(x_1)^b}$$

is actually a polynomial in $R[x_0, x_1]$. Since $z \in U_0$ we see that $F_1(x_1)^b$ must divide the numerator exactly, and hence $z \in R[x_0, x_1]$.

If $\operatorname{supp}(z) \cap \mathbb{R}^2_{>0} = \emptyset$ then we can write z as $z = z_0 + z_1 + z_2$ where $\operatorname{Newt}(z_i) \subseteq \langle v_i, v_{i+1} \rangle$. Since $z_0, z_1, z_2 \in R[x_0, x_1, x_2, x_3]$ we conclude that $z \in R[x_0, x_1, x_2, x_3]$. □

7 Hacking, Tevelev and Urzúa's Universal Family

Using the cluster algebra interpretation of Mori's algorithm, [HTU17] constructs $k1A$ and $k2A$ neighbourhoods in families. For the application they had in mind they restrict to the case in which the general hyperplane section H_X of X has second Betti number $b_2(H_X) = 1$ (or equivalently the condition that H_X has only *simple* T-singularities, i.e. those of the form $\frac{1}{r_i^2}(1, r_i a_i - 1)$). For $k2A$ neighbourhoods, this means taking functions of the form $f_i(\zeta^{r_i}, u) = \zeta^{r_i} + u^{\alpha_i}$ for $i = 1, 2$ in Theorem 4.1. This case is covered by Brown and Reid's diptych construction by taking $d_1 = d_2$. We now illustrate how the basic idea of their universal family extends to the slightly more general setting of diptych varieties.

7.1 Glueing Together Diptych Varieties

For fixed $d_1, d_2 \geq 2$ and varying $k \in \mathbb{Z}$ we consider the diptych variety V_k defined over the coefficient ring $\mathbb{C}[A, B, L_k, M_k]$, which is birational to $V_k^0 \subset \mathbb{A}^4_{x_0,x_1,x_2,x_3} \times \mathbb{A}^2_{A,B} \times \mathbb{A}^2_{L_k,M_k}$, the initial complete intersection with equations

$$x_0 x_2 = A x_1^{d_1} + L_k^{l_1^{(k)}} M_k^{m_1^{(k)}}, \qquad x_1 x_3 = A x_2^{d_2} + L_k^{l_2^{(k)}} M_k^{m_2^{(k)}},$$

where the exponents are determined by $\begin{pmatrix} l_1^{(k)} & m_1^{(k)} \\ l_2^{(k)} & m_2^{(k)} \end{pmatrix}$ as in (5.2). The idea behind the universal family is that we will glue together diptych varieties V_k and V_{k+1} for all $k \in \mathbb{Z}$. By a simple calculation and using the fact that $d_{k-2} = d_k$, for $k \geq 4$ we find that V_{k+1} is determined by

$$\begin{pmatrix} l_1^{(k+1)} & m_1^{(k+1)} \\ l_2^{(k+1)} & m_2^{(k+1)} \end{pmatrix} = \begin{pmatrix} m_1^{(k)} & d_k m_1^{(k)} - l_1^{(k)} \\ m_2^{(k)} & d_k m_2^{(k)} - l_2^{(k)} \end{pmatrix}.$$

Now, for $i = 1, 2$ we see that

$$L_{k+1}^{l_i^{(k+1)}} M_{k+1}^{m_i^{(k+1)}} = L_{k+1}^{m_i^{(k)}} M_{k+1}^{d_k m_i^{(k)} - l_i^{(k)}} = (M_{k+1}^{-1})^{l_i^{(k)}} (L_{k+1} M_{k+1}^{d_k})^{m_i^{(k)}}$$

and so, identifying $x_0, \ldots, x_3, A, B$ on V_k^0 and V_{k+1}^0, the birational map

$$\phi_{k,k+1} \colon \mathbb{A}^2_{L_k,M_k} \dashrightarrow \mathbb{A}^2_{L_{k+1},M_{k+1}}, \quad \phi_{k,k+1}(L_k, M_k) = (M_{k+1}^{-1}, L_{k+1} M_{k+1}^{d_k})$$

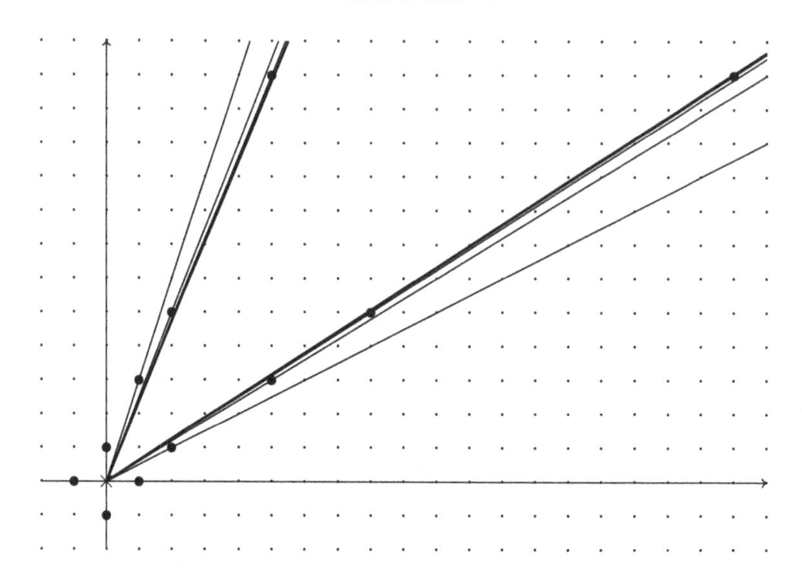

Figure 5.5 The fan $\mathcal{F}_{2,3}$ for the toric surface $\mathcal{M}_{2,3}$.

induces a birational map $V_k^0 \dashrightarrow V_{k+1}^0$, which is an isomorphism over $\mathbb{A}^2_{L_k,M_k} \setminus V(L_k) \to \mathbb{A}^2_{L_{k+1},M_{k+1}} \setminus V(M_{k+1})$. Thinking of the two diptychs as families $V_k \to \mathbb{A}^2_{A,B} \times \mathbb{A}^2_{L_k,M_k}$ and $V_{k+1} \to \mathbb{A}^2_{A,B} \times \mathbb{A}^2_{L_{k+1},M_{k+1}}$, we glue them over the base and consider the limit as $k \to \infty$.

Moreover, it is clearly also possible to glue to V_k to V_{k-1} by maps $\phi_{k,k-1} = \phi^{-1}_{k-1,k}$. Doing this we can also consider the limit as $k \to -\infty$, taking care to see that the appropriate gluing map is $\phi_{k,k-1}(L_k, M_k) = (M_{k-1}, L^{-1}_{k-1})$ if $k = 1, 2$. Since all of these transition maps $\phi_{k,k+1}$ are toric, these copies of $\mathbb{A}^2$ patch together to obtain a toric surface $\mathcal{M}_{d_1,d_2}$ (only locally of finite type) defined by the fan with rays $\{\mathbb{R}_{\geq 0} v_i \colon i \in \mathbb{Z}\}$ (see Figure 5.5), where the v_i are given by the recurrence relation

$$v_0, v_1, v_2, v_3 = (0,1), (-1,0), (0,-1), (1,0)$$
$$v_{i+1} + v_{i-1} = d_i v_i \text{ for all } i \neq 1, 2.$$

We have souped up our diptych variety into a flat family of surfaces $\widehat{V} \to \mathbb{A}^2_{A,B} \times \mathcal{M}_{d_1,d_2}$.

7.2 Hacking, Tevelev and Urzúa's Universal Family

Building the affine cover of flip. Note that the two gradings w_A, w_B defined in §6.5 are independent of k and therefore we have a $(\mathbb{C}^*)^2$-action on $\widehat{V}$ which lifts the standard $(\mathbb{C}^*)^2$-action on $\mathbb{A}^2_{A,B}$. In the same spirit as the calculation of §4.4, to construct the affine cover $W = \operatorname{Spec} R_f$ of a $k2A$ flipping contraction $f\colon X \to Y$ we can pull back $\widehat{V}$ along an appropriately chosen graded morphism $\psi\colon \mathbb{A}^2_{u,z} \to \mathbb{A}^2_{A,B} \times \mathscr{M}_{d_1,d_2}$ which is equivariant with respect to some choice of $\mathbb{C}^*$-action on $\mathbb{A}^2_{A,B}$, where u and z are considered to be of degree 0 and $\delta > 0$ respectively. As a consequence of the grading, ψ is determined by a map $\psi_1\colon \mathbb{A}^2_{u,z} \to \mathbb{A}^2_{A,B}$ and a map $\psi_2\colon \mathbb{A}^1_u \to \mathscr{M}_{d_1,d_2}$. Now ψ_1 must be of the form $\psi_1^*(A,B) = (\alpha(u)z^a, \beta(u)z^b)$ for some $a,b \in \mathbb{Z}$ and $\alpha,\beta \in \mathbb{C}[u]$, with $\alpha(0),\beta(0) \neq 0$ by the conditions of Theorem 4.1. Thus the grading essentially fixes the choice of ψ_1 and it is harmless to reduce to the case $\alpha = \beta = 1$. On the other hand, we have more freedom in allowing ψ_2 to vary, as we will consider below.

The universal family. In the case $d := d_1 = d_2$, Hacking, Tevelev and Urzúa construct $k1A$ and $k2A$ flips by taking pullbacks in essentially exactly the same way as just described. They construct three different flat families of surfaces $\mathbb{Y}$, $\mathbb{U}$ and $\mathbb{X}^+$ over the surface $\mathscr{M}_{d,d}$ and use a morphism $g\colon \mathbb{A}^1 \to \mathcal{M}_{d,d}$ to pullback Y, X and X^+ respectively. The infinite diptych $\widehat{V}$ packages up these three families into one object in the same way that the affine cover $W = \operatorname{Spec} R_f$ packages up the flip.

Smoothing a $k2A$ flip into a $k1A$ flip. We now consider what happens to a $k2A$ flip if we allow the map $\psi_2\colon \mathbb{A}^1_u \to \mathscr{M}_{d_1,d_2}$ to vary. The surface $\mathscr{M}_{d_1,d_2}$ contains an infinite chain of rational curves $\bigcup_{i\in\mathbb{Z}} D_i$ where D_i is the torus invariant curve corresponding to v_i. Note that ψ_2 is determined by the image in any of the open sets $\mathbb{A}^2_{L_i,M_i} \subset \mathscr{M}_{d_1,d_2}$. Since we can build the flip from a regular diptych V_k, there is some $k \geq 4$ such that $\psi_2(0) = (0,0) \in \mathbb{A}^2_{L_k,M_k}$, i.e. $\psi_2(0) = D_k \cap D_{k+1}$ lies in the intersection of two components of the boundary divisor D. A very nice feature of the universal family of [HTU17], which also extends to our setting, is that by perturbing ψ_2 so that the point $\psi_2(0)$ moves along D, we can construct a family of $k1A$ flips degenerating to our $k2A$ flip. Intuitively, moving $\psi_2(0)$ away from $D_k \cap D_{k+1}$ along D_k smooths one of the two cA/r singularities of the $k2A$ neighbourhood and moving $\psi(0)$ along D_{k+1} smooths the other. This initial deformation was already visible from the view of the original diptych variety $V_{d_1,d_2;k}$, since it is defined over $\mathbb{A}^2_{L_k,M_k}$. The advantage of the infinite diptych variety $\widehat{V}$ (or equivalently the universal family of [HTU17]) is that it connects up all of these degenerations of flips into a single family.

7.3 Gross, Hacking and Keel's Deformation of the Vertex

We would like to extend this proof for the finite generation of the diptych variety into a proof of finite generation for the upper cluster algebras of more general rank generalised cluster algebras (i.e. ones with polynomial coefficients). These appear in previous work on extremal contractions (see [Duc15, §7]) and should be able to handle cases in which $f: X \to Y$ is a extremal contraction with a reducible central fibre $C \subset X$. Moreover, this should lead to a much deeper understanding of how extremal contractions behave and deform in families.

Gross, Hacking and Keel [GHK15b] used ideas coming from mirror symmetry to write down deformations of the *vertex*, that is a cycle of 2-planes glued along toric boundary strata $\mathbb{V}_n = \bigcup_{i=1}^n \mathbb{A}^2_{x_i, x_{i+1}} \subset \mathbb{A}^n$ (where $x_{n+1} = x_1$). A *Looijenga pair* (Y, D) is a rational surface with an anticanonical cycle $D = \sum_{i=1}^n D_i \in |-K_Y|$. Given such a pair, they construct a deformation of $\mathbb{V}_n$ in terms of the Gromov–Witten theory of (Y, D). When $D \subset Y$ supports an ample divisor, their construction gives a flat family of affine Gorenstein surfaces $\pi: (\mathbb{V}_n \subset \mathcal{V}) \to (0 \in B)$ which is a deformation of $\mathbb{V}_n$ over the toric variety $B = \operatorname{Spec} \overline{\mathrm{NE}}(Y)$.

Since the tent T at the center of a diptych variety admits a degeneration into a vertex, it should be contained in one of these families. We would like to understand diptych varieties, and eventually the general form of Mori's algorithm, in these terms. In particular, given a 3-fold extremal neighbourhood f, how one can write down a Looijenga pair (Y, D) giving a family $\pi: \mathcal{V} \to B$ and a (rational) map $\phi: \mathbb{A}^2 \dashrightarrow B$ such that the affine cover $W = \operatorname{Spec} R_f$ is given by the pullback $W = \mathcal{V} \times_B \mathbb{A}^2$.

Acknowledgements

This work clearly owes a large intellectual debt to both Shigefumi Mori and Miles Reid, and I thank them, Gavin Brown and Paul Hacking for many useful conversations around the subject of this paper.

References

[BFZ05] Arkady Berenstein, Sergey Fomin, and Andrei Zelevinsky. Cluster algebras. III. Upper bounds and double Bruhat cells. *Duke Math. J.*, 126(1):1–52, 2005.

[BR] Gavin Brown, and Miles Reid. Diptych varieties. III: Redundant generators. In preparation, 10pp.

[BR13] Gavin Brown, and Miles Reid. Diptych varieties, I. *Proc. Lond. Math. Soc. (3)*, 107(6):1353–1394, 2013.

[BR17] Gavin Brown and Miles Reid. Diptych varieties. II: Apolar varieties. *Higher dimensional algebraic geometry—in honour of Professor Yujiro Kawamata's sixtieth birthday*, volume 74 of *Adv. Stud. Pure Math.*, pages 41–72. Math. Soc. Japan, Tokyo, 2017.

[Duc15] Thomas Ducat. *Mori extractions from singular curves in a smooth 3-fold.* PhD thesis, 2015. 126pp.

[GHK15b] Mark Gross, Paul Hacking, and Sean Keel. Mirror symmetry for log Calabi-Yau surfaces I. *Publ. Math. Inst. Hautes Études Sci.*, 122:65–168, 2015.

[HTU17] Paul Hacking, Jenia Tevelev, and Giancarlo Urzúa. Flipping surfaces. *J. Algebraic Geom.*, 26(2):279–345, 2017.

[KM92] János Kollár, and Shigefumi Mori. Classification of three-dimensional flips. *J. Amer. Math. Soc.*, 5(3):533–703, 1992.

[MM15] Jacob P. Matherne, and Greg Muller. Computing upper cluster algebras. *Int. Math. Res. Not. IMRN*, (11):3121–3149, 2015.

[Mor02] Shigefumi Mori. On semistable extremal neighborhoods. In *Higher dimensional birational geometry (Kyoto, 1997)*, volume 35 of *Adv. Stud. Pure Math.*, pages 157–184. Math. Soc. Japan, Tokyo, 2002.

[Rei92] Miles Reid. What is a flip? Scanned notes available at https://homepages.warwick.ac.uk/~masda/3folds/what_flip.pdf, 1992.

The Mirror of the Cubic Surface

Mark Gross, Paul Hacking, Sean Keel and Bernd Siebert

To Miles Reid on the occasion of his 70th birthday

1 Introduction

A number of years ago, one of us (Mark Gross) was giving a lecture at the University of Warwick on the material on scattering diagrams from [GPS10]. Of course, Miles was in the audience, and he asked (paraphrasing as this was many years ago) whether, at some point, the lecturer would come back down to earth. The goal of this note is to show, in fact, we have not left the planet by considering a particularly beautiful example of the mirror symmetry construction of [GHK15b], namely the mirror to a cubic surface.

More precisely, the paper [GHK15b], building on [GS11, GPS10, CPS], constructs mirrors of rational surfaces equipped with anti-canonical cycles of rational curves. Specifically, one begins with the data of a pair (Y, D), where Y is a non-singular projective rational surface over an algebraically closed field $\Bbbk$ of characteristic 0, and $D \in |-K_Y|$ is an effective reduced anti-canonical divisor with at least one node, necessarily then forming a wheel of projective lines. Choose in addition a finitely generated, saturated sub-monoid $P \subset H_2(Y, \mathbb{Z})$ whose only invertible element is 0, such that P contains the class of every effective curve on Y. Let $\mathfrak{m}$ denote the maximal monomial ideal of the monoid ring $\Bbbk[P]$ and $\widehat{\Bbbk[P]}$ denote the completion of $\Bbbk[P]$ with respect to $\mathfrak{m}$. Then the main construction of [GHK15b] produces a family of formal schemes $\mathfrak{X} \to \operatorname{Spf} \widehat{\Bbbk[P]}$ which is interpreted as the mirror family to the pair (Y, D). In the more pleasant case when D supports an ample divisor, the construction is, in fact, algebraic: there is a family $X \to S := \operatorname{Spec} \Bbbk[P]$ of affine surfaces extending the above formal family. In general, if D has $n \geq 3$ components, then X is a closed subscheme of $\mathbb{A}^n_S$, with central fibre a reducible union of n copies of $\mathbb{A}^2$.

[GHK15b, Example 6.13] contains the equation[1] for X in the case that Y is a cubic surface in $\mathbb{P}^3$ and $D = D_1 + D_2 + D_3$ is a triangle of lines. The intent was to include a proof of this in [GHK], which, at the time, was circulated rather narrowly in an extreme rough draft form.

As [GHK] has seen no change for more than five years, and many pieces of it have been cannibalized for other papers or become out-of-date, it seemed that, in the grand tradition of second parts of papers, this paper is unlikely to ever see the light of day. On the other hand, the full details of the cubic surface have not appeared anywhere else, although Lawrence Barrott [Bar20] verifies the given equation for the mirror of the cubic surface. The cubic surface is in particular especially attractive. This is unsurprising, given the rich classical geometry of the cubic (see, for example, [Rei88]). So, we felt that it would be a pity for this construction never to appear. Further, since [GHK] first began to circulate, the technology for understanding the product rule for theta functions on the mirror, and hence the equations for the mirror, has improved at a theoretical level, see [GS18, GS, KY]. Thus in particular it will be possible to give a completely enumerative interpretation for the equations to the mirror cubic. This gives us an opportunity to exposit a number of different viewpoints on the construction here.

Without further ado, here is the main result. Describe the pair (Y, D) as follows. First fix the pair $(\mathbb{P}^2, \overline{D} = \overline{D}_1 + \overline{D}_2 + \overline{D}_3)$ where $\overline{D}$ is a triangle of lines. Let (Y, D) be obtained as the blow-up of two general distinct points on each of the three lines, with D the strict transform of $\overline{D}$. Let E_{ij}, $i = 1, 2, 3$, $j = 1, 2$ be the exceptional curves, with E_{ij} intersecting D_i. For $i = 1, 2$ or 3, denote by L_{ij}, $1 \leq j \leq 8$, the eight lines on the cubic surface not contained in D but intersecting D_i. We note that $\{E_{i1}, E_{i2}\} \subseteq \{L_{ij} \,|\, 1 \leq j \leq 8\}$.

Theorem 1.1. *Taking $P = \mathrm{NE}(Y)$, the cone of effective curves of Y, $S = \operatorname{Spec} \Bbbk[P]$, the mirror family defined over S to the cubic surface $(Y, D = D_1 + D_2 + D_3)$ is given by the equation in $\mathbb{A}^3_S$:*

$$\vartheta_1\vartheta_2\vartheta_3 = \sum_i z^{D_i}\vartheta_i^2 + \sum_i \left(\sum_j z^{L_{ij}}\right) z^{D_i}\vartheta_i + \sum_\pi z^{\pi^* H} + 4z^D.$$

[1] Unfortunately with a sign error!

Here for a curve class C, z^C denotes the corresponding monomial of $\Bbbk[P]$, and $\vartheta_1, \vartheta_2, \vartheta_3$ are the coordinates on the affine 3-space. The sum over π is the sum over all possible birational morphisms $\pi\colon Y \to Y'$ of (Y, D) to a pair (Y', D') isomorphic to $\mathbb{P}^2$ with its toric boundary, with $\pi|_D\colon D \to D'$ an isomorphism and H the class of a line in $\mathbb{P}^2$.

The original guess for the shape of these equations was motivated by the paper [Obl04] which gave a similar equation for a non-commutative cubic surface. Once one knows the shape of the equation, it is not difficult to verify it, as we shall see.

Finally, we note that this paper does not intend to be a complete exposition of the ideas of [GHK15b], but rather, we move quickly to discuss the cubic surface. For a more comprehensive expository account, see the forthcoming work of Argüz [Arg].

2 The Tropicalization of the Cubic Surface

We explain the basic combinatorial data we associate to the pair (Y, D), namely a pair (B, Σ) where:

(1) B is an *integral linear manifold with singularities*;
(2) Σ is a decomposition of B into cones.

First, an integral linear manifold B is a real manifold with coordinate charts $\psi_i\colon U_i \to \mathbb{R}^n$ (where $\{U_i\}$ is an open covering of B) and transition maps $\psi_i \circ \psi_j^{-1} \in \mathrm{GL}_n(\mathbb{Z})$. An *integral linear manifold with singularities* is a manifold B with an open set $B_0 \subseteq B$ and $\Delta = B \setminus B_0$ of codimension at least 2 such that B_0 carries an integral linear structure.

We build B and Σ by pretending that the pair (Y, D) is a toric variety. If it were, we could reconstruct its fan in $\mathbb{R}^2$ (up to $\mathrm{GL}_2(\mathbb{Z})$) knowing the intersection numbers of the irreducible components D_i of D. So, we just start constructing a fan and we will run into trouble when (Y, D) isn't a toric variety. This problem is fixed by introducing a singularity in the linear structure of $\mathbb{R}^2$ at the origin.

Explicitly, for the cubic surface, $D_i^2 = -1$ for $1 \leq i \leq 3$, and we proceed as follows. Take rays in $\mathbb{R}^2$ corresponding to D_1 and D_2 to be $\rho_1 := \mathbb{R}_{\geq 0}(1, 0)$ and $\rho_2 := \mathbb{R}_{\geq 0}(0, 1)$ respectively. See the left-hand picture in Figure 6.1.

Since $D_2^2 = -1$, toric geometry instructs us that the ray corresponding to D_3 would be $\rho_3 := \mathbb{R}_{\geq 0}(-1, 1)$ if (Y, D) were a toric pair. Indeed, if

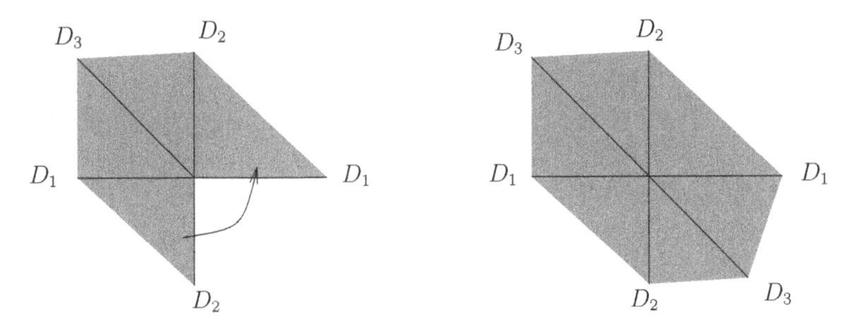

Figure 6.1 The pair (B, Σ).

ρ_1, ρ_2, ρ_3 are successive rays in a two-dimensional fan defining a non-singular complete toric surface, and if n_i is the primitive generator of ρ_i and D_i is the divisor corresponding to ρ_i, we have the relation

$$n_1 + D_2^2 n_2 + n_3 = 0.$$

Thus with $n_1 = (1,0)$ and $n_2 = (0,1)$, n_3 is determined by D_2^2. Since $D_3^2 = -1$, we then need a ray corresponding to D_1 to be $\mathbb{R}_{\geq 0}(-1,0)$, which does not coincide with the ray ρ_1 (telling us that (Y, D) wasn't really a toric pair). If we continue, we obtain a new ray $\mathbb{R}_{\geq 0}(0,-1)$ for D_2 also. Thus we have two cones spanned by the rays corresponding to D_1 and D_2, and there is an integral linear transformation identifying these two cones. In this case, this transformation is $-\mathrm{id}$. After cutting out the fourth quadrant from $\mathbb{R}^2$ and gluing the third and first quadrants via $-\mathrm{id}$, we obtain the integral affine manifold B, along with a decomposition (or fan) Σ into rational polyhedral cones. Here the cones of Σ consist of $\{0\}$, the images of the rays ρ_1, ρ_2, ρ_3, and three two-dimensional cones $\sigma_{i,i+1}$ $i = 1,2,3$, with indices taken mod 3, and $\sigma_{i,i+1}$ having faces ρ_i and ρ_{i+1}. Note the rays correspond to irreducible components of D and the two-dimensional cones to double points of D.

To see the details of this construction in general, and further examples, see [GHK15b, §1.2].

We use the convention that $v_i \in B$ is the primitive integral point on the ray ρ_i, so that any element of $\sigma_{i,i+1}$ can be written as $av_i + bv_{i+1}$ for some $a, b \in \mathbb{R}_{\geq 0}$.

While we have just described the general construction for (B, Σ) as applied to our particular case, in fact, there is a more elegant description for the cubic surface. By continuing to build the fan, we close up to get a fan $\widetilde{\Sigma}$ in $\mathbb{R}^2$. Then $B = \mathbb{R}^2/\langle -\mathrm{id}\rangle$, and $\widetilde{\Sigma}$ descends to Σ on the quotient. See the right-hand side of Figure 6.1.

Remark 2.1. Note that $\widetilde{\Sigma}$ defines a toric variety $\widetilde{Y}$ which is a del Pezzo surface of degree 6. The automorphism $-\mathrm{id}$ of $\widetilde{\Sigma}$ induces an involution $\iota\colon \widetilde{Y} \to \widetilde{Y}$, given on the dense torus orbit as $(z_1, z_2) \mapsto (z_1^{-1}, z_2^{-1})$. This surface can be embedded in $\mathbb{P}^3$ as follows. First, one maps the quotient of the dense torus orbit of $\widetilde{Y}$ to $\mathbb{A}^3$ using the map

$$(z_1, z_2) \mapsto (z_1 + z_1^{-1}, z_2 + z_2^{-1}, z_1^{-1}z_2 + z_1 z_2^{-1}).$$

The image satisfies the equation $x_1x_2x_3 = x_1^2 + x_2^2 + x_3^2 - 4$ which is then projectivized to obtain the Cayley cubic given by the equation

$$x_1x_2x_3 = x_0(x_1^2 + x_2^2 + x_3^2) - 4x_0^3,$$

the unique cubic surface with four ordinary double points, the images of the fixed points of ι. This is, in fact, isomorphic to $\widetilde{Y}/\langle\iota\rangle$. □

Remark 2.2. We write $B_0 = B \setminus \{0\}$. Let $B_0(\mathbb{Z})$ be the subset of B_0 of points with integer coordinates with respect to any integral linear chart. Set $B(\mathbb{Z}) = B_0(\mathbb{Z}) \cup \{0\}$.

The set $B(\mathbb{Z})$ has another natural interpretation, as the *tropicalization* of the log Calabi–Yau manifold $U = Y \setminus D$, see [GHK15a, Definition 1.7]. Here one takes a nowhere vanishing 2-form Ω on U with at worst simple poles along D and we have

$$B(\mathbb{Z}) = \{\text{divisorial discrete valuations } \nu\colon k(U)^* \to \mathbb{Z} \mid \nu(\Omega) < 0\} \cup \{0\}.$$

The advantage of this description is that an automorphism of U which does not extend to an automorphism of Y still induces an automorphism of $B(\mathbb{Z})$, which in general extends to a piecewise linear automorphism of B. □

It will also be useful to consider piecewise linear functions on B with respect to the fan Σ, i.e., continuous functions $F\colon B \to \mathbb{R}$ which restrict to linear functions on each $\sigma \in \Sigma$. Just as in the toric case, there is, in fact, a one-to-one correspondence between such functions with integral slopes and divisors supported on the boundary. Indeed, each boundary divisor D_i defines a piecewise linear function on B, written as $\langle D_i, \cdot\rangle$, uniquely defined by the requirement that

$$\langle D_i, v_j\rangle = \delta_{ij}.$$

Additively, this allows us to obtain a PL function $\langle D', \cdot\rangle$ associated to any divisor D' supported on D. Conversely, given a piecewise linear function $F\colon B \to \mathbb{R}$ with integral slopes, we obtain a divisor $\sum_i F(v_i)D_i$ supported on D.

3 The Scattering Diagram Associated with the Cubic Surface

As (B, Σ) only involves purely combinatorial information about (Y, D), it is insufficient to determine an interesting mirror object. We need to include extra data of a *scattering diagram* on (B, Σ).

Before doing so, we need to select some additional auxiliary data, namely, a monoid $P \subseteq H_2(Y, \mathbb{Z})$ of the form $\sigma_P \cap H_2(Y, \mathbb{Z})$ for $\sigma_P \subseteq H_2(Y, \mathbb{R})$, a strictly convex rational polyhedral cone which contains all effective curve classes. In the case of the cubic surface, we may take σ_P to be the Mori cone, i.e., the cone generated by all effective curve classes. We write $\Bbbk[P]$ for the corresponding monoid ring, and $\mathfrak{m} \subseteq \Bbbk[P]$ for the maximal monomial ideal, generated by $\{z^p \mid p \in P \setminus \{0\}\}$.

Definition 3.1. A *ray* in B is a pair $(\mathfrak{d}, f_\mathfrak{d})$ where:

(1) $\mathfrak{d} \subseteq \sigma_{i,i+1}$ for some i is a ray generated by some $av_i + bv_{i+1} \neq 0$, $a, b \in \mathbb{Z}_{\geq 0}$ relatively prime. We call $\mathfrak{d}$ the *support* of the ray.
(2) $f_\mathfrak{d} = 1 + \sum_{k \geq 1} c_k X_i^{-ak} X_{i+1}^{-bk} \in \Bbbk[P][\![X_i^{-a} X_{i+1}^{-b}]\!]$ with $c_k \in \mathfrak{m}$ for all k, satisfying the property that for any monomial ideal $I \subseteq \Bbbk[P]$ with $\Bbbk[P]/I$ Artinian, (i.e., I is co-Artinian), $f_\mathfrak{d} \mod I$ is a finite sum.

Definition 3.2. A *scattering diagram* $\mathfrak{D}$ for B is a collection of rays with the property that for each co-Artinian monomial ideal $I \subseteq \Bbbk[P]$,

$$\mathfrak{D}_I := \{(\mathfrak{d}, f_\mathfrak{d}) \in \mathfrak{D} \mid f_\mathfrak{d} \not\equiv 1 \mod I\}$$

is finite. We assume further that $\mathfrak{D}$ contains at most one ray with a given support.

The purpose of a scattering diagram is to give a way of building a flat family over $\operatorname{Spec} A_I$, where $A_I := \Bbbk[P]/I$. Explicitly, suppose given a scattering diagram $\mathfrak{D}$ and a co-Artinian ideal I. Assume further that each ρ_i is the support of a ray $(\rho_i, f_i) \in \mathfrak{D}$. This ray is allowed to be trivial, i.e., $f_i = 1$. Now define rings

$$R_{i,I} := A_I[X_{i-1}, X_i^{\pm 1}, X_{i+1}]/(X_{i-1}X_{i+1} - z^{[D_i]} X_i^{-D_i^2} f_i)$$
$$R_{i,i+1,I} := A_I[X_i^{\pm 1}, X_{i+1}^{\pm 1}].$$

Here $z^{[D_i]}$ is the monomial in $\Bbbk[P]$ corresponding to the class of the boundary curve D_i, necessarily lying in P by the assumption that P contains all effective curve classes.

Localizing, note we have canonical isomorphisms

$$(R_{i,I})_{X_{i+1}} \cong R_{i,i+1,I} \text{ and } (R_{i,I})_{X_{i-1}} \cong R_{i-1,i,I}.$$

Set

$$U_{i,I} := \operatorname{Spec} R_{i,I} \text{ and } U_{i,i+1,I} := \operatorname{Spec} R_{i,i+1,I}.$$

Note that if $I = \mathfrak{m}$, then $U_{i,I}$ is the reducible variety $X_{i-1}X_{i+1} = 0$ in $\mathbb{A}^2_{X_{i-1},X_{i+1}} \times (\mathbb{G}_m)_{X_i}$, where the subscripts denote the coordinates on the respective factors. On the other hand, $U_{i,i+1,I} = (\mathbb{G}_m^2)_{X_i,X_{i+1}}$. For more general I, we instead obtain thickenings of these schemes just described.

For any I, we have canonical open immersions $U_{i-1,i,I}, U_{i,i+1,I} \hookrightarrow U_{i,I}$. As $U_{i,I}$ and $U_{i,\mathfrak{m}}$ have the same underlying topological space, we can describe the underlying open sets in $U_{i,\mathfrak{m}}$ of these two open immersions as subsets of $V(X_{i-1}X_{i+1}) \subseteq \mathbb{A}^2 \times \mathbb{G}_m$ as follows. We have $U_{i-1,i,I}$ is given by the open set where $X_{i-1} \neq 0$ (hence $X_{i+1} = 0$) and $U_{i,i+1,I}$ is given by the open set where $X_{i+1} \neq 0$ (hence $X_{i-1} = 0$). Thus in particular the images of these immersions are disjoint. Thus, if for all i we glue $U_{i,I}$ and $U_{i+1,I}$ via the canonically identified copies of $U_{i,i+1,I}$, there is no cocycle gluing condition to check and we obtain a scheme X_I° flat over $\operatorname{Spec} A_I$.

It is easy to describe this if we take $I = \mathfrak{m}$. One obtains in this case that $X_I^\circ = \mathbb{V}_n \setminus \{0\}$, where n is the number of irreducible components of D and, assuming $n \geq 3$,

$$\mathbb{V}_n = \mathbb{A}^2_{x_1,x_2} \cup \cdots \cup \mathbb{A}^2_{x_{n-1},x_n} \cup \mathbb{A}^2_{x_n,x_1} \subseteq \mathbb{A}^n = \operatorname{Spec} \Bbbk[x_1, \ldots, x_n]$$

where $\mathbb{A}^2_{x_i,x_{i+1}}$ denotes the affine coordinate plane in $\mathbb{A}^n$ for which all coordinates but x_i, x_{i+1} are zero. Here $\mathbb{V}_n$ is called the *n-vertex*.

The problem is that for I general, X_I° may be insufficiently well-behaved to extend to a flat deformation of $\mathbb{V}_n$. To do so, we need to perturb the gluings we made above, and the role of the scattering diagram is to provide a data structure for doing so.

Let $\gamma\colon [0,1] \to \operatorname{Int}(\sigma_{i,i+1})$ be a path. We define an automorphism of $R_{i,i+1,I}$ called the *path ordered product*. Assume that whenever γ crosses a ray in $\mathfrak{D}_I$ it passes from one side of the ray to the other. In particular, suppose γ crosses a given ray

$$(\mathfrak{d} = \mathbb{R}_{\geq 0}(av_i + bv_{i+1}), f_{\mathfrak{d}}) \in \mathfrak{D}_I$$

with a and b relatively prime. Define the A_I-algebra homomorphism $\theta_{\gamma,\mathfrak{d}}\colon R_{i,i+1,I} \to R_{i,i+1,I}$ by

$$\theta_{\gamma,\mathfrak{d}}(X_i) = X_i f_{\mathfrak{d}}^{\mp b}$$
$$\theta_{\gamma,\mathfrak{d}}(X_{i+1}) = X_{i+1} f_{\mathfrak{d}}^{\pm a}$$

where the signs are $-b, +a$ if γ passes from the ρ_{i+1} side of $\mathfrak{d}$ to the ρ_i side of $\mathfrak{d}$, and $+b, -a$ if γ crosses in the opposite direction. Note these two choices are inverse automorphisms of $R_{i,i+1,I}$, and $f_{\mathfrak{d}}$ is invertible because $f_{\mathfrak{d}} \equiv 1 \mod \mathfrak{m}$ from Definition 3.1(*2*).

If γ crosses precisely the rays $(\mathfrak{d}_1, f_{\mathfrak{d}_1}), \ldots, (\mathfrak{d}_s, f_{\mathfrak{d}_s}) \in \mathfrak{D}_I$, in that order, then we define the *path ordered product*

$$\theta_{\gamma,\mathfrak{D}} := \theta_{\gamma,\mathfrak{d}_s} \circ \cdots \circ \theta_{\gamma,\mathfrak{d}_1}.$$

Now, for each i, choose a path γ inside $\sigma_{i,i+1}$ which starts near ρ_{i+1} and ends near ρ_i so that it crosses all rays of $\mathfrak{D}_I$ intersecting the interior of $\sigma_{i,i+1}$. Then $\theta_{\gamma,\mathfrak{D}}$ induces an automorphism $\theta_{\gamma,\mathfrak{D}}\colon U_{i,i+1,I} \to U_{i,i+1,I}$, and we can use this to modify our gluing via

$$U_{i,I} \hookleftarrow U_{i,i+1,I} \xrightarrow{\theta_{\gamma,\mathfrak{D}}} U_{i,i+1,I} \hookrightarrow U_{i+1,I}.$$

This produces a new scheme $X^{\circ}_{I,\mathfrak{D}}$, still a flat deformation of $\mathbb{V}_n \setminus \{0\}$ over $\operatorname{Spec} A_I$.

Now comes the key point: we need to make a good choice of $\mathfrak{D}$ in order to be able to construct a partial compactification $X_{I,\mathfrak{D}}$ of $X^{\circ}_{I,\mathfrak{D}}$ such that $X_{I,\mathfrak{D}} \to \operatorname{Spec} A_I$ is a flat deformation of $\mathbb{V}_n$. One of the main ideas of [GHK15b] is the use of results of [GPS10] to write down a good choice of scattering diagram, the *canonical scattering diagram*, in terms of relative Gromov–Witten invariants of the pair (Y, D).

We first discuss the nature of these invariants. Choose a curve class β and a point $v \in B_0(\mathbb{Z})$, say $v = av_i + bv_{i+1}$. We sketch the construction of a Gromov–Witten type invariant N_v^{β} counting what we call $\mathbb{A}^1$-*curves*. Roughly speaking, these are one-pointed stable maps of genus 0, $f\colon (C, p) \to Y$, representing the class β, with $f^{-1}(D) = \{p\}$. Further, f has contact order $\langle D_i, v\rangle$ with D_i at p. Roughly, this contact order is the order of vanishing of the regular function $f^*(t)$ at p, for t a local defining equation for D_i at $f(p)$. However, as stated, this isn't quite right because of standard issues of compactness in relative Gromov–Witten theory. In [GHK15b], these numbers are defined rigorously following [GPS10] by peforming a weighted blow-up of (Y, D) at $D_i \cap D_{i+1}$ determined by $\mathfrak{d}$

and then using relative Gromov–Witten theory. As relative Gromov–Witten theory only works relative to a smooth divisor, one removes all double points of the proper transform of D under this blow-up, and then shows that this doesn't interfere with compactness of the moduli space. We refer to [GHK15b, §3.1] for the precise definition, as we will not need here the subtleties of the general definition. However, we note that in order for such a map to exist, and hence possibly have $N_v^\beta \neq 0$, we must have β an effective curve class and

$$\beta \cdot D_j = \langle D_j, v \rangle.$$

A more modern definition of these invariants, as developed by [GS13, AC14, Che14], is via logarithmic Gromov–Witten theory. Using that theory, one can allow contact orders with multiple divisors simultaneously, and thus do not need to perform the weighted blow-up. It follows from invariance of logarithmic Gromov–Witten theory under toric blow-ups [AW18] and the comparison theorem of relative and logarithmic invariants [AMW14] that these two definitions agree.

Definition 3.3. The *canonical scattering diagram* $\mathfrak{D}_{\mathrm{can}}$ of (Y, D) consists of rays $(\mathfrak{d}, f_{\mathfrak{d}})$ ranging over all possible supports $\mathfrak{d} \subseteq B$ where, if $\mathfrak{d} \subseteq \sigma_{i,i+1}$ with $\mathfrak{d} = \mathbb{R}_{\geq 0}(av_i + bv_{i+1})$ and a, b relatively prime, then

$$f_{\mathfrak{d}} = \exp\left(\sum_{k\geq 1}\sum_{\beta\in H_2(Y,\mathbb{Z})} kN^{\beta}_{akv_i+bkv_{i+1}} z^{\beta}(X_i^{-a}X_{i+1}^{-b})^k\right).$$

We now return to the cubic, where $\mathfrak{D}_{\mathrm{can}}$ is particularly interesting. One might also consider higher degree del Pezzo surfaces. However, del Pezzo surfaces of degree $6, 7, 8$ and 9 are all toric, assuming one takes as D the toric boundary, and they have a trivial scattering diagram (i.e., all $f_{\mathfrak{d}} = 1$ as the invariants N_v^β are always zero). The case of a degree 5 del Pezzo surface was considered as a running example in [GHK15b], see, for example, Example 3.7 there. A degree 4 surface is not that much more complicated, see [Bar20] for details. On the other hand, for the cubic surface, no $f_{\mathfrak{d}}$ is 1, but nevertheless we can essentially determine $f_{\mathfrak{d}}$. On the other hand, the degree 2 del Pezzo surface requires use of a computer to analyze, see [Bar20].

To describe curve classes on the cubic surface Y, we use the description of Y as a blow-up of $\mathbb{P}^2$ given in the introduction, so that $H_2(Y, \mathbb{Z})$ is generated by the classes of the exceptional divisors E_{ij}, $1 \leq i \leq 3$, $1 \leq j \leq 2$, and the class L of a pull-back of a line in $\mathbb{P}^2$.

With this notation, we have:

Proposition 3.4. *The ray* (ρ_i, f_{ρ_i}) *satisfies*

$$f_{\rho_i} = \frac{\prod_{j=1}^{8}(1 + z^{L_{ij}} X_i^{-1})}{(1 - z^{D_k + D_\ell} X_i^{-2})^4},$$

where the L_{ij} *as in the introduction are the lines not contained in* D *but meet* D_i*, and* $\{i, k, \ell\} = \{1, 2, 3\}$.

Proof We take $i = 1$, the other cases following from symmetry. We need to calculate the numbers $N^{\beta}_{kv_1}$. In particular, for β to be represented by an $\mathbb{A}^1$-curve contributing to $N^{\beta}_{kv_1}$, we must have $\beta \cdot D_1 = k$ and $\beta \cdot D_i = 0$ for $i \neq 1$.

We will first consider those curve classes β which may be the curve class of a generically injective map $f\colon \mathbb{P}^1 \to Y$ with the above intersection numbers with the D_i. Write

$$\beta = aL - \sum_{i,j} b_{ij} E_{ij}.$$

Then

$$k = \beta \cdot D_1 = a - b_{11} - b_{12}, 0 = \beta \cdot D_2 = a - b_{21} - b_{22}, 0 = \beta \cdot D_3 = a - b_{31} - b_{32}.$$

Thus

$$a = b_{21} + b_{22} = b_{31} + b_{32}.$$

Further, we must have $p_a(f(C)) \geq 0$, so by adjunction and the fact that $K_Y = -D$,

$$-2 \leq 2p_a(f(C)) - 2 = \beta \cdot (\beta + K_Y) = a^2 - \sum_{i,j} b_{ij}^2 - k. \tag{3.1}$$

Now of course the curve classes E_{11}, E_{12} satisfy the above equalities and inequality, with $k = 1$, while E_{ij}, $i \neq 1$ do not. Then any other class of an irreducible curve which may contribute necessarily has $a > 0$ and $b_{ij} \geq 0$. Let us fix a and k and try to maximize the right-hand side of (3.1) in the hopes that we can make it at least -2. This means in particular that we should try to minimize $b_{i1}^2 + b_{i2}^2$ for $i = 2, 3$.

We split the analysis into two cases. If a is even, then this sum of squares is minimized by taking $b_{i1} = b_{i2} = a/2$. Thus we see that

$$-2 \leq 2p_a(f(C)) - 2 \leq a^2 - k - b_{11}^2 - b_{12}^2 - 4(a^2/4) = -k - b_{11}^2 - b_{12}^2.$$

Since $k \geq 1$, we see we immediately get three possibilities:

(1) $k = 1$, $b_{11} = 1$, $b_{12} = 0$, in which case $a = 2$ and the only possible curve class is $\beta = 2L - E_{11} - E_{21} - \cdots - E_{32}$.
(2) $k = 1$, $b_{11} = 0$, $b_{12} = 1$, in which case $a = 2$ and the only possible curve class is $\beta = 2L - E_{12} - E_{21} - \cdots - E_{32}$.
(3) $k = 2$, $b_{11} = b_{12} = 0$, in which case $a = 2$ and the only possible curve class is $2L - E_{21} - \cdots - E_{32}$.

If a is odd, then we minimize $b_{i1}^2 + b_{i2}^2$ by taking $b_{i1} = (a-1)/2$, $b_{i2} = (a+1)/2$ or vice versa. Thus

$$a^2 - b_{21}^2 - \cdots - b_{32}^2 \leq a^2 - 2\left(\frac{(a-1)^2}{4} + \frac{(a+1)^2}{4}\right) = -1.$$

Again, since $k \geq 1$, the only possibility is $k = 1$, $b_{11} = b_{12} = 0$, and hence $a = 1$, giving the following possible choices for β:

$$L - E_{21} - E_{31}, \quad L - E_{21} - E_{32}, \quad L - E_{22} - E_{31}, \quad L - E_{22} - E_{32}.$$

Note that these four classes, along with E_{11}, E_{12}, and cases (*1*) and (*2*) in the a even case, represent the 8 (-1)-curves in Y which meet D_1 transversally, i.e., the curves L_{1j}. Each of these curve classes is then represented by a unique $\mathbb{A}^1$-curve, and $N_{v_1}^{\beta} = 1$ in these cases.

In the case $a = k = 2$, we consider the curve class

$$\beta = 2L - E_{21} - \cdots - E_{32} \sim D_2 + D_3.$$

The linear system $|D_2 + D_3|$ induces a conic bundle $g\colon Y \to \mathbb{P}^1$, and D_1 is a 2-section of g, i.e., $g|_{D_1}\colon D_1 \to \mathbb{P}^1$ is a double cover, necessarily branched over two points $p_1, p_2 \in \mathbb{P}^1$. Thus the conics $f^{-1}(p_1)$, $f^{-1}(p_2)$ are also $\mathbb{A}^1$-curves, now with contact order 2 with D_1. So, $N_{2v_1}^{\beta} = 2$.[2]

Unfortunately, these are not the only $\mathbb{A}^1$-curves, as there may be stable maps $f\colon (C, p) \to Y$ which are either not generically injective or don't have irreducible image. Indeed, one may have multiple covers of one of the above $\mathbb{A}^1$-curves already considered, provided the cover is totally ramified at the point p of contact with D. However, for general choice of (Y, D), we will now show that there is no possibility of reducible images.

[2] In general, in Gromov–Witten theory, it is not enough to just count the stable maps, as there may be a virtual count. However, in all the cases just considered, the stable map $f\colon C \to Y$ in question is a closed immersion, and hence has no automorphisms as a stable map. Further, the obstruction space to the moduli space of stable maps at the point $[f]$ is $H^1(C, f^*T_Y(-\log D))$ which is seen without much difficulty to vanish. Hence each curve, in fact, contributes 1 to the Gromov–Witten number.

As in [GP10, Lemma 4.2] the image of any $\mathbb{A}^1$-curve must be a union of irreducible curves which intersect the boundary at the same point. In particular, if $f(C) = C_1 \cup \cdots \cup C_n$ is the irreducible decomposition, then $D \cap C_1 \cap \cdots \cap C_n$ consists of one point, necessarily contained in D_1.

Since $C_1, \ldots, C_n$ must be a subset of the 10 curves identified above, the possibilities are as follows. The first is that two of these curves are lines on the cubic surface, and hence we must have three lines (including D_1) intersecting in a common point. Such a point on a cubic surface is called an *Eckardt point*, see [Dol12, §9.1.4]. However, the set of cubic surfaces containing Eckardt points is codimension one in the moduli of all cubic surfaces. Since we may assume (Y, D) is general in moduli (as the Gromov–Witten invariants being calculated are deformation invariant), we may assume Y has no Eckardt points, so this doesn't occur.

On the other hand, one or both of C_1, C_2 could be fibres of the conic bundle induced by $|D_2 + D_3|$. Since two distinct fibres are disjoint, they can't both be fibres of the conic bundle. Further, any line E of the cubic surface intersecting D_1 at one point has $E \cdot (D_2 + D_3) = 0$ and hence is contained in a fibre of the conic bundle g, and thus is again disjoint from a different fibre of g.

We thus come to the conclusion that any stable map contributing to the $\mathbb{A}^1$-curve count must have irreducible image, and hence be a multiple cover of one of the curves discussed above. The moduli space of such multiple covers is always positive dimensional, but happily the virtual count has been calculated in [GPS10, Proposition 6.1]. Degree d covers of a non-singular rational curve which meets D transversally contributes $(-1)^{d-1}/d^2$, whilst degree d covers of a non-singular rational curve which is simply tangent to D is $1/d^2$. Note

$$\exp\left(\sum_{d\geq 1} d \cdot \frac{(-1)^{d-1}}{d^2} z^{d\beta} X_1^{-d}\right) = 1 + z^{\beta} X_1^{-1}$$

and

$$\exp\left(\sum_{d\geq 1} 2d \cdot \frac{1}{d^2} z^{d\beta} X_1^{-2d}\right) = \frac{1}{(1 - z^{\beta} X_1^{-2})^2}.$$

From this, the result follows. □

We now observe that the cubic surface carries sufficient symmetry so that the above computation determines the scattering diagram completely.

Noting that the group $\mathrm{SL}_2(\mathbb{Z})$ acts on $\mathbb{R}^2$ and $-\mathrm{id}$ lies in the center of $\mathrm{SL}_2(\mathbb{Z})$, we obtain an action of $\mathrm{PGL}_2(\mathbb{Z})$ on $B = \mathbb{R}^2/\langle -\mathrm{id}\rangle$. Of course, this action acts transitively on all the rays of rational slope in B, so if we can show that this action preserves the scattering diagram in a certain sense, we will have completely determined the scattering diagram.

We first observe that there is a rotational symmetry. For example, the calculation of f_{ρ_1} equally applies to f_{ρ_2} and f_{ρ_3}, subject to a change of relevant curve classes. More generally, if we know $f_{\mathfrak{d}}$ for

$$\mathfrak{d} = \mathbb{R}_{\geq 0}(av_i + bv_{i+1})$$

then we know it for $S(\mathfrak{d})$, where $S(av_i + bv_{i+1}) = av_{i+1} + bv_{i+2}$, with indices taken modulo 3. Here S is an automorphism of B which lifts to an automorphism of the cover $\mathbb{R}^2$, with $S(1,0) = (0,1)$ and $S(0,1) = (-1,1)$ (so that $S(-1,1) = (-1,0)$, completing the rotation). Thus on the cover, S is represented by $\begin{pmatrix} 0 & -1 \\ 1 & 1 \end{pmatrix}$. We also have an action S^* on $H_2(Y,\mathbb{Z})$ given by $S^*(L) = L$, $S^*(E_{ij}) = E_{i+1,j}$. Then we can write $f_{S(\mathfrak{d})} = S^*(f_{\mathfrak{d}})$, where $X_i^{-ka}X_{i+1}^{-kb} \mapsto X_{i+1}^{-ka}X_{i+2}^{-kb}$ and $z^\beta \mapsto z^{S^*(\beta)}$ give the action of S^* on $f_{\mathfrak{d}}$.

The second symmetry arises from a birational change to the boundary. We may blow-up the point of intersection of D_1 and D_2, and blow-down D_3, to obtain a surface (Y', D') with $Y' \setminus D' = Y \setminus D$. We use the convention that $D' = D_1' + D_2' + D_3'$ with D_1' the strict transform of D_1, D_2' the exceptional curve of the blow-up, and D_3' the strict transform of D_2.

But, in fact, Y' is still a cubic surface and hence we may apply the calculation of Proposition 3.4 with respect to the new divisor D_2'. Because of the way $\mathbb{A}^1$-curve counts are defined, these counts do not depend on toric blow-ups and blow-downs of the boundary. Thus if we know a ray in the scattering diagram for (Y', D'), we have a corresponding ray in the scattering diagram for (Y, D). For example, it is not difficult to check that for $1 \leq j \leq 8$, the curve in the pencil $|D_3 + L_{3j}|$ passing through $D_1 \cap D_2$ has strict transform in (Y', D') a line meeting D_2'. On the other hand, the strict transform of a curve of class $D_1 + D_2 + 2D_3$ on (Y, D) which is cuspidal at $D_1 \cap D_2$ is a conic on Y' which meets D_2' tangentially.

To see this as an action on B, let B' be the integral linear manifold with singularities corresponding to (Y', D'). Then there is a canonical piecewise linear identification of B' with B arising from the description of the tropicalization of Remark 2.2. In particular, this identification sends v_1' to v_1, v_2' with v_1+v_2, and v_3' with v_2. Thus if we know a ray $(\mathfrak{d}', f_{\mathfrak{d}'})$ for (Y', D'), we obtain a ray $(\mathfrak{d}, f_{\mathfrak{d}})$ for (Y, D) under this identification. Instead, we can view this identification as giving an automorphism of B, i.e., consider the automorphism T given by $v_1 \mapsto v_1$, $v_2 \mapsto v_1 + v_2$ and $v_3 \mapsto v_2$. Note this is induced by $\begin{pmatrix} 1 & 1 \\ 0 & 1 \end{pmatrix} \in \mathrm{SL}_2(\mathbb{Z})$.

It is not difficult to work out the action[3] of T^* on $H_2(Y, \mathbb{Z})$. It is

$$\begin{aligned} L &\mapsto 2L - E_{31} - E_{32} \\ E_{1j} &\mapsto E_{1j} \\ E_{2j} &\mapsto L - E_{3j} \\ E_{3j} &\mapsto E_{2j} \end{aligned}$$

for $j = 1, 2$. The symmetry T takes a ray $(\mathfrak{d}, f_{\mathfrak{d}})$ to a ray $(T(\mathfrak{d}), T^*(f_{\mathfrak{d}}))$, where $T^*(f_{\mathfrak{d}})$ does the obvious thing. In particular, one ray in $\mathfrak{D}_{\mathrm{can}}$ is $\mathfrak{d} = \mathbb{R}_{\geq 0}(v_1 + v_2)$ with

$$f_{\mathfrak{d}} := \frac{\prod_{j=1}^{8}(1 + z^{D_3 + L_{3j}} X_1^{-1} X_2^{-1})}{(1 - z^{D_1 + D_2 + 2D_3} X_1^{-2} X_2^{-2})^4} \tag{3.2}$$

Happily, this is the only additional ray we will need to understand other than ρ_1, ρ_2 and ρ_3.

Since S and T generate $\mathrm{SL}_2(\mathbb{Z})$, we have now proved:

Theorem 3.5. *Let $\mathfrak{d} = \mathbb{R}_{\geq 0}(av_i + bv_{i+1})$ for $a, b \in \mathbb{Z}_{\geq 0}$ relatively prime. Then there exists curve classes $\beta_1, \ldots, \beta_9 \in H_2(Y, \mathbb{Z})$ such that*

$$f_{\mathfrak{d}} = \frac{\prod_{j=1}^{8}(1 + z^{\beta_j} X_i^{-a} X_{i+1}^{-b})}{(1 - z^{\beta_9} X_i^{-2a} X_{i+1}^{-2b})^4}.$$

We note that this $\mathrm{SL}_2(\mathbb{Z})$-action has a beautiful explanation in terms of work of Cantat and Loray [CL09]. They describe the $\mathrm{SL}_2(\mathbb{C})$ character variety of the four-punctured sphere $S_4^2 = S^2 \setminus \{p_1, \ldots, p_4\}$, i.e., the variety of $\mathrm{SL}_2(\mathbb{C})$ representations of the fundamental group $\pi_1(S_4^2)$, up to conjugation by elements of $\mathrm{SL}_2(\mathbb{C})$. This character variety is naturally embedded in $\mathbb{A}^7$ with coordinates x, y, z, A, B, C, D and has equation

$$xyz + x^2 + y^2 + z^2 = Ax + By + CZ + D,$$

[3] In fact, this action is not unique: it can always be composed with an automorphism of $H^2(Y, \mathbb{Z})$ preserving the intersection form, permuting the (-1)-curves, and keeping the boundary divisors D_1, D_2, D_3 fixed. We give one possible action.

i.e., is a family of affine cubic surfaces whose natural compactifications in $\mathbb{P}^3$ are then precisely of the form we are considering.

Now S^2 can be viewed as a quotient of a torus $T^2 = \mathbb{R}^2/\mathbb{Z}^2$ by negation, $S^2 = T^2/\langle -\mathrm{id}\rangle$, and the map $T^2 \to S^2$ has four branch points, the two-torsion points of T^2. We take the image of these branch points to be $p_1, \ldots, p_4$, so that any element of $\mathrm{SL}_2(\mathbb{Z})$ acting on T^2 then induces an automorphism of S^2_4, possibly permuting the punctures. Thus we obtain a $\mathrm{PGL}_2(\mathbb{Z})$ action on S^2_4 and hence a $\mathrm{PGL}_2(\mathbb{Z})$ action on the character variety, which, in fact, is compatible with the projection to $\mathbb{A}^4$ with coordinates A, B, C, D. An element of $\mathrm{PGL}_2(\mathbb{Z})$ permutes fibres if it permutes two-torsion points. Thus we obtain an action of $\mathrm{PGL}_2(\mathbb{Z})$ on the "relative" tropicalization of this family of log Calabi–Yau manifolds, i.e., the set of valuations with centers surjecting onto $\mathbb{A}^4$ and with simple poles of the relative holomorphic 2-form. This can be shown to be the same action considered above generated by S and T. We omit the details.

4 Broken Lines, Theta Functions and the Derivation of the Equation

We now explain how to construct theta functions, and what is special about the canonical scattering diagram. We first recall the notion of broken line, fixing here a scattering diagram $\mathfrak{D}$ and a co-Artinian ideal $I \subseteq \Bbbk[P]$.

Definition 4.1. A *broken line* γ in (B, Σ) for $q \in B_0(\mathbb{Z})$ and end-point $Q \in B_0$ is a proper continuous piecewise integral affine map $\gamma\colon (-\infty, 0] \to B_0$, real numbers $t_0 = -\infty < t_1 < \cdots < t_n = 0$, and monomials m_i, $1 \le i \le n$, satisfying the following properties:

(1) $\gamma(0) = Q$.

(2) $\gamma|_{[t_{i-1},t_i]}$ is affine linear for all i, and $\gamma([t_{i-1}, t_i])$ is contained in some two-dimensional cone $\sigma_{j,j+1} \in \Sigma$, where j depends on i. Furthermore, $m_i = c_i X_j^a X_{j+1}^b$ for some $a, b \in \mathbb{Z}$, a, b not both zero, $c_i \in \Bbbk[P]/I$, and $\gamma'(t) = -av_j - bv_{j+1}$ for any $t \in (t_{i-1}, t_i)$.

(3) If $q \in \mathrm{Int}(\sigma_{j,j+1})$ for some j, we can write $q = av_j + bv_{j+1}$ for some $a, b \in \mathbb{Z}_{\ge 0}$ and then $\gamma((-\infty, t_1]) \subset \sigma_{j,j+1}$ and $m_1 = X_j^a X_{j+1}^b$. If $q \in \rho_j$ for some j, then $\gamma((-\infty, t_1])$ is either contained in $\sigma_{j-1,j}$ or $\sigma_{j,j+1}$, and writing $q = av_j$, we have $m_1 = X_j^a$.

(4) If $\gamma(t_i)$ lies in the interior of a maximal cone of Σ then $\gamma(t_i)$ lies in the support of a ray $(\mathfrak{d}, f_\mathfrak{d})$ and γ passes from one side of $\mathfrak{d}$ to the other,

so that $\theta_{\gamma,\mathfrak{d}}$ is defined. Then m_{i+1} is a monomial in $\theta_{\gamma,\mathfrak{d}}(m_i)$. In other words, we expand the expression $\theta_{\gamma,\mathfrak{d}}(m_i)$ into a sum of monomials, and choose m_{i+1} to be one of the terms of this sum.

(5) If $\gamma(t_i) \in \rho_j$ for some j, then γ passes from $\sigma_{j-1,j}$ to $\sigma_{j,j+1}$, and $m_i = c_i X_{j-1}^a X_j^b$, then m_{i+1} is a monomial in the expression

$$c_i(z^{[D_j]} X_j^{-D_j^2} f_{\rho_j} X_{j+1}^{-1})^a X_j^b.$$

If, on the other hand, γ passes from $\sigma_{j,j+1}$ to $\sigma_{j-1,j}$, then, with $m_i = c_i X_j^a X_{j+1}^b$, m_{i+1} is a monomial in the expression

$$c_i X_j^a (z^{[D_j]} X_j^{-D_j^2} f_{\rho_j} X_{j-1}^{-1})^b.$$

In other words, in the first case, the monomial m_i, written in the variables X_{j-1}, X_j, is rewritten, using the defining equation of the ring $R_{j,I}$, in the variables X_j, X_{j+1}. The second case is similar.

Definition 4.2. Let $q \in B_0(\mathbb{Z})$ and $Q \in B_0$ be a point with irrational coordinates. Then we define

$$\vartheta_{q,Q} = \sum_{\gamma} \mathrm{Mono}(\gamma)$$

where the sum is over all broken lines for q with endpoint Q, and $\mathrm{Mono}(\gamma)$ denotes the last monomial attached to γ.

We extend this definition to $q = 0 \in B(\mathbb{Z}) \setminus B_0(\mathbb{Z})$ by setting

$$\vartheta_{0,Q} = \vartheta_0 = 1.$$

It follows from the definition of broken line that if $Q \in \sigma_{i,i+1}$ then $\vartheta_{q,Q} \in R_{i,i+1,I}$.

Definition 4.3. We say $\mathfrak{D}$ is *consistent* if for all $q \in B_0(\mathbb{Z})$ and co-Artinian ideals I:

(1) If $Q, Q' \in \sigma_{i,i+1}$ are points with irrational coordinates and γ is a path in $\sigma_{i,i+1}$ joining Q to Q', then $\theta_{\gamma,\mathfrak{D}}(\vartheta_{q,Q}) = \vartheta_{q,Q'}$.
(2) If $Q \in \sigma_{i-1,i}$, $Q' \in \sigma_{i,i+1}$ are chosen sufficiently close to ρ_i such that there is no non-trivial ray of $\mathfrak{D}_I$ between Q and ρ_i or between Q' and ρ_i, then there exists an element $\vartheta_{q,\rho_i} \in R_{i,I}$ whose images in $R_{i-1,i,I}$ and $R_{i,i+1,I}$ are $\vartheta_{q,Q}$ and $\vartheta_{q,Q'}$ respectively.

One of the main theorems of [GHK15b], namely Theorem 3.8, states that $\mathfrak{D}_{\mathrm{can}}$ is a consistent scattering diagram.

The benefit of a consistent scattering diagram is that the $\vartheta_{q,Q}$ for various Q can then be glued to give a global function $\vartheta_q \in \Gamma(X^\circ_{I,\mathfrak{D}}, \mathcal{O}_{X^\circ_{I,\mathfrak{D}}})$. This allows us to construct a partial compactification $X_{I,\mathfrak{D}}$ of $X^\circ_{I,\mathfrak{D}}$ by

$$X_{I,\mathfrak{D}} := \operatorname{Spec}\Gamma(X^\circ_{I,\mathfrak{D}}, \mathcal{O}_{X^\circ_{I,\mathfrak{D}}}),$$

and the existence of the theta functions ϑ_q guarantees that this produces a flat deformation of $\mathbb{V}_n$ over $\operatorname{Spec} A_I$, see [GHK15b, §2.3] for details. Morally, another way to view this is that we are embedding $X^\circ_{I,\mathfrak{D}}$ in $\mathbb{A}^n_{A_I}$ using the theta functions $\vartheta_{v_1}, \ldots, \vartheta_{v_n}$, and then taking the closure.

Example 4.4. Unfortunately, in the cubic surface example, it is very difficult to write down expressions for theta functions. While for any fixed ideal I, $\vartheta_{q,Q}$ is a finite sum of monomials, in fact, if we take the limit over all I we obtain an infinite sum. Here are some very simple examples of this. Take $Q = \alpha v_1 + \beta v_2$ for some irrational $\alpha, \beta \in \mathbb{R}_{>0}$, and take $q = v_1$. We give examples of broken lines for q ending at Q. Consider a ray $\mathfrak{d} = \mathbb{R}_{\geq 0}(av_1 + bv_2)$ with $(a-1)/b < \alpha/\beta < a/b$. Note that given the choice of Q, there are an infinite number of choices of relatively prime a, b satisfying this condition.

We will construct a broken line as depicted in Figure 6.2. The monomial attached to the segment coming in from infinity is X_1. If we bend along the ray $\mathfrak{d}$, we apply $\theta_{\gamma,\mathfrak{d}}$ to X_1. By Theorem 3.5, $f_{\mathfrak{d}}$ contains a non-zero term $cX_1^{-a}X_2^{-b}$ for some $c \in \Bbbk[P]$, so $\theta_{\gamma,\mathfrak{d}}(X_1)$ contains a term $c'X_1^{1-a}X_2^{-b}$. Choosing this monomial, we now proceed in the direction $(a-1, b)$. In particular, take the bending point to be

$$Q' := \left(\frac{a}{b}((1-a)\beta + b\alpha), (1-a)\beta + b\alpha\right)$$

which lies on $\mathfrak{d}$ because $(1-a)\beta + b\alpha > 0$ by the assumption that $(a-1)/b < \alpha/\beta$. Then

$$Q - Q' = (\frac{a}{b}\beta - \alpha)(a-1, b).$$

Thus the broken line reaches Q, as depicted. So, we have indeed constructed a broken line and there are an infinite number of such broken lines (albeit only a finite number modulo any co-Artinian ideal I).

Of course, here we are using only one term from the infinite power series expansion of $f_{\mathfrak{d}}$ and only considering one possible bend, and we already have an infinite number of broken lines. We believe it would be extremely difficult to get a useful description of all broken lines and hence broken lines provide a useful theoretical, but not practical, description of theta functions.

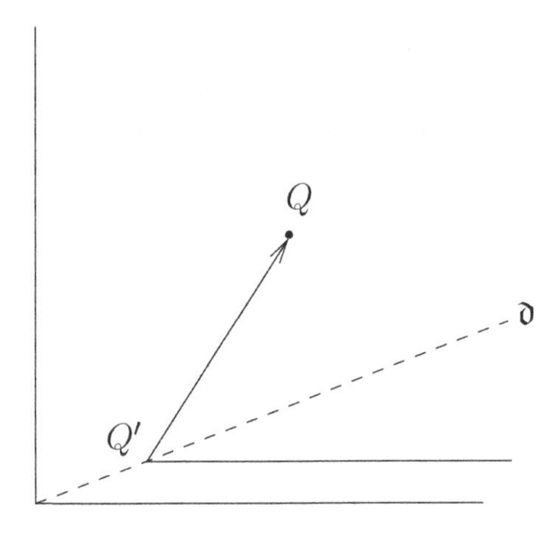

Figure 6.2 The broken line of Example 4.4.

This shows that if we take the limit over all I and obtain a formal scheme $\widehat{\mathfrak{X}} \to \mathrm{Spf}\, \widehat{\Bbbk[P]}$, there is no hope to express the theta functions as algebraic expressions. Thus it is perhaps a bit of a surprise that often the relations satisfied by these theta functions are much simpler, so that we can extend the construction over $\mathrm{Spec}\, \Bbbk[P]$. In the case of the cubic surface, we will see this explicitly by using the product rule for theta functions [GHK15b, Theorem 2.34]:

Theorem 4.5. *Let $p_1, p_2 \in B(\mathbb{Z})$. In the canonical expansion*

$$\vartheta_{p_1} \cdot \vartheta_{p_2} = \sum_{r \in B(\mathbb{Z})} \alpha_{p_1 p_2 r} \vartheta_r$$

where $\alpha_{p_1 p_2 r} \in \Bbbk[P]/I$ for each q, we have

$$\alpha_{p_1 p_2 r} = \sum_{\gamma_1, \gamma_2} c(\gamma_1) c(\gamma_2)$$

where the sum and notation is as follows. We fix $z \in B_0$ a point very close to r contained in the interior of a cone $\sigma_{i,i+1}$ for some i. We then sum over all broken lines γ_1, γ_2 for p_1, p_2 satisfying:

(1) Both broken lines have endpoint z.
(2) If $\mathrm{Mono}(\gamma_j) = c(\gamma_j) X_i^{a_j} X_{i+1}^{b_j}$ with $c(\gamma_j) \in \Bbbk[P]/I$, $j = 1, 2$, then $r = (a_1 + a_2) v_i + (b_1 + b_2) v_{i+1}$.

We shall see that in the case of the cubic surface, only a very small part of $\mathfrak{D}_{\mathrm{can}}$ is necessary to find the equation of the mirror.

There is one more ingredient for the calculation of the equation, namely, the notion of a *min-convex function* in the context of a scattering diagram. Let $F\colon B \to \mathbb{R}$ be a piecewise linear function on B. If γ is a broken line, we obtain a (generally discontinuous) function on $(-\infty, 0]$, the domain of γ, written as $t \mapsto dF(\gamma'(t))$. This means that at a time t, provided F is linear at $\gamma(t)$, we evaluate the differential dF at $\gamma(t)$ on the tangent vector $\gamma'(t)$. Thus $dF(\gamma'(\cdot))$ is a piecewise constant function.

We say F is *min-convex* if for any broken line γ, $dF(\gamma'(\cdot))$ is a decreasing function: see [GHKK18, Definition 8.2] where the definition is given in a slightly different context. The use of such a function is that [GHKK18, Lemma 8.4] applies, so that F is *decreasing* in the sense of [GHKK18, Definition 8.3], i.e., if the coefficient $\alpha_{p_1p_2r} \neq 0$, then

$$F(r) \geq F(p_1) + F(p_2). \tag{4.1}$$

Indeed, suppose that γ_1, γ_2 are broken lines for p_1, p_2 respectively contributing to the expression for $\alpha_{p_1p_2r}$. Note that $F(p_i) = -dF(\gamma_i'(t))$ for $t \ll 0$, while $r = -\gamma_1'(0) - \gamma_2'(0)$. Thus for $t \ll 0$,

$$F(r) - F(p_1) - F(p_2) = \sum_{i=1}^{2} (dF(\gamma_i'(t)) - dF(\gamma_i'(0))) \tag{4.2}$$

which is positive under the decreasing assumption.

We note [Bar20] also makes use of such a function (with the opposite sign convention). Barrott, however, used a computer program to enumerate all contributions to the products, as his main goal was to find the mirror to a degree 2 del Pezzo, which has a considerably more complex equation than the mirror to the cubic. In the case of the cubic surface, the products can be computed by hand.

In our case, we may take $F = \langle K_Y, \cdot \rangle$. Note this pulls back to the PL function on the cover $\mathbb{R}^2 \to B = \mathbb{R}^2/\langle -\mathrm{id}\rangle$ which corresponds to $K_{\tilde{Y}}$.

It is easy to check that, in fact, $dF(\gamma'(\cdot))$ decreases whenever a broken line crosses one of the rays ρ_i (this is just local convexity of F) or when a broken line bends, and hence F is decreasing. However, it will be important to quantify by how much $dF(\gamma'(\cdot))$ changes with each of these occurences. For example, suppose a broken line passes from $\sigma_{1,2}$ into $\sigma_{2,3}$ without bending, with tangent direction $av_1 + bv_2$ where necessarily $a < 0$. Then via parallel transport of this tangent vector into $\sigma_{2,3}$, we can rewrite the vector using the relation $v_1 + v_3 = -D_2^2 v_2$, i.e., $av_1 + bv_2$ is rewritten as $(a+b)v_2 - av_3$. Thus $dF(\gamma'(\cdot))$ takes the value $-(a+b)$ before crossing ρ_2 and the value $-(a+b) + a$ after crossing ρ_2, hence decreasing as $a < 0$.

If γ bends in, say, $\sigma_{i,i+1}$, then γ' changes by some $av_i + bv_{i+1} \neq 0$ for $a, b \geq 0$. But $dF(av_i + bv_{i+1}) = -a - b$, so $dF(\gamma'(\cdot))$ changes by $-(a+b) < 0$.

Note that if γ bends when it crosses ρ_i, in fact, $dF(\gamma'(t))$ decreases by at least 2. These observations will be crucial for bounding the search for possible broken lines contributing to the product.

We now calculate the key products necessary to prove the main theorem.

Lemma 4.6. *We have the following products:*

$$\vartheta_{v_i}^2 = \vartheta_{2v_i} + 2z^{D_j+D_k}, \quad \{i,j,k\} = \{1,2,3\}$$

$$\vartheta_{v_1}\vartheta_{v_2} = \vartheta_{v_1+v_2} + z^{D_3}\vartheta_{v_3} + \sum_{j=1}^{8} z^{D_3+L_{3j}}$$

$$\vartheta_{v_1+v_2}\vartheta_{v_3} = z^{D_1}\vartheta_{2v_1} + z^{D_2}\vartheta_{2v_2} + \vartheta_{v_1}\sum_j z^{D_1+L_{1j}} + \vartheta_{v_2}\sum_j z^{D_2+L_{2j}} + \sum_\pi z^{\pi^*H} + 8z^{D_1+D_2+D_3}.$$

Proof We consider first $\vartheta_{v_i}^2$. By symmetry, we can take $i = 1$. Since $F(v_1) = -1$, if ϑ_r contributes to this product, then we must have $-2 \leq F(r) \leq 0$, the first inequality from (4.1) and the second since F is non-positive. Let γ_1, γ_2 be broken lines contributing to $\alpha_{v_1v_1r}$ as in Theorem 4.5. It follows immediately from (4.2) that if $F(r) = -2$, then γ_i neither bends nor crosses a wall. It is then obvious the only possible r in this case is $r = 2v_1$. Fixing $z \in \mathrm{Int}(\sigma_{1,2})$ near $2v_1$, we obtain the contribution from two broken lines as in the left in Figure 6.3, this is responsible for the ϑ_{2v_1} term.

There are only three points $r \in B(\mathbb{Z})$ with $F(r) = -1$, namely $r = v_i$, $i = 1, 2, 3$. Now if a pair of broken lines γ_1, γ_2 contributes to $\alpha_{v_1v_1r}$, then one of the γ_j either bends or crosses one of the ρ_k. Such a possibility can be ruled out, however. It is easiest to work on the cover $\mathbb{R}^2 \to B = \mathbb{R}^2/\langle -\mathrm{id}\rangle$, bearing in mind that there are two possible initial directions for the lifting of a broken line γ_i, namely it can come in parallel to $\mathbb{R}_{\geq 0}(1,0)$ or parallel to $\mathbb{R}_{\geq 0}(-1,0)$. If $r = v_2$ or v_3, we can fix z in the interior of $\sigma_{2,3}$, and then both broken lines must cross rays to reach z. If $r = v_1$, we may take z in the interior of $\sigma_{1,2}$, and then the only possibility is that one of the γ_i bends. However, if γ_i bends at any ray of $\mathfrak{D}_{\mathrm{can}}$ not supported on one of the ρ_i, then dF decreases by at least 2,

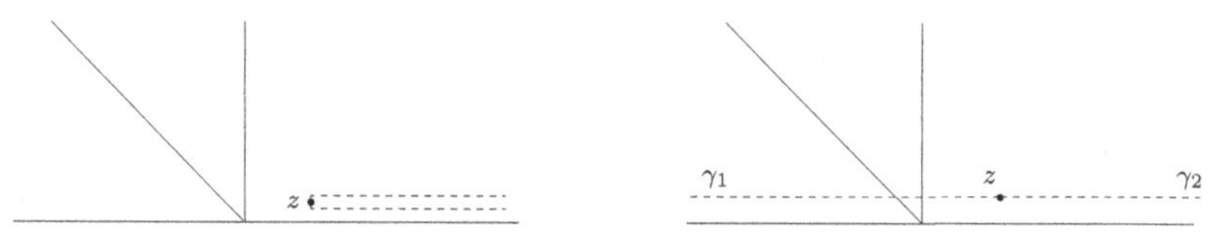

Figure 6.3 The contributions to the product $\vartheta_{v_1}^2$. In the left-hand picture, the two broken lines, in fact, lie on top of each other, but we depict them as distinct lines with endpoint z.

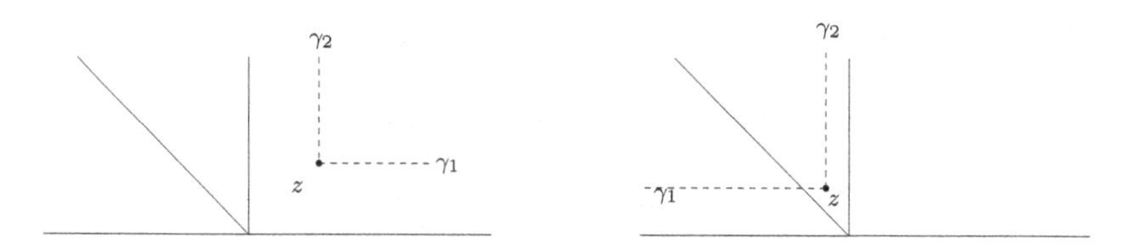

Figure 6.4 Two contributions to the product $\vartheta_{v_1} \cdot \vartheta_{v_2}$.

ruling out this possibility. Thus we can rule out the case $F(r) = -1$. We shall omit this kind of analysis in the sequel, as it is straightforward.

Finally, if $F(r) = 0$, then $r = 0$. Taking z in $\sigma_{1,2}$ close to the origin, we obtain the possibility shown on the right-hand side of Figure 6.3. This actually represents two possibilities, as the labels γ_1 and γ_2 can be interchanged. Each such pair of broken lines contributes $z^{D_2+D_3}\vartheta_0$, recalling that $\vartheta_0 = 1$. One checks easily that there are no possibilities where one of the broken lines bends. This gives the claimed description of $\vartheta_{v_1}^2$.

Turning to $\vartheta_{v_1} \cdot \vartheta_{v_2}$, if $F(r) = -2$, then again broken lines can't bend or cross walls. In this case, the only possibility is as depicted on the left in Figure 6.4, contributing the term $\vartheta_{v_1+v_2}$.

If $F(r) = -1$, then $r = v_1, v_2$ or v_3. By putting the endpoint z in $\sigma_{1,2}$, $\sigma_{2,3}$ or $\sigma_{2,3}$ respectively, a quick analysis shows the only possible contribution is from the right-hand picture in Figure 6.4, contributing $z^{D_3}\vartheta_{v_3}$.

Finally, if $F(r) = 0$, again $r = 0$. Taking z near ρ_1 and the origin in the interior of $\sigma_{1,2}$, we now obtain the possibility of γ_2 bending along the ray $\mathfrak{d} = \mathbb{R}_{\geq 0}(1,1)$ as depicted in Figure 6.5. The bend on γ_2 is calcluated by seeing how $\theta_{\mathfrak{d},\gamma_2}$ acts on the initial monomial X_2, i.e., $X_2 \mapsto X_2 f_{\mathfrak{d}}$. By the form given for $f_{\mathfrak{d}}$ in (3.2), we get the given expression for $\vartheta_{v_1} \cdot \vartheta_{v_2}$.

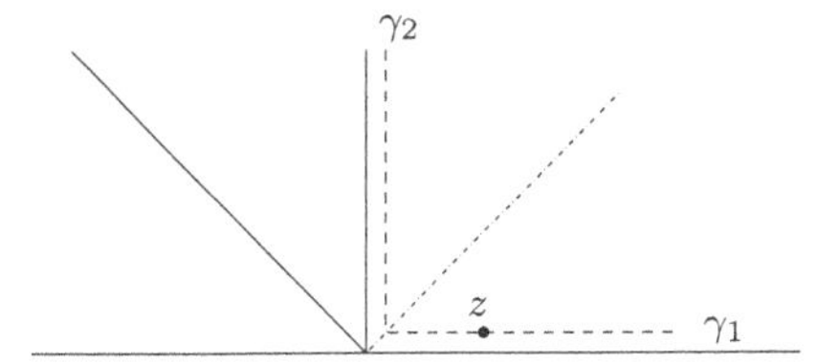

Figure 6.5 An additional contribution to the product $\vartheta_{v_1} \cdot \vartheta_{v_2}$.

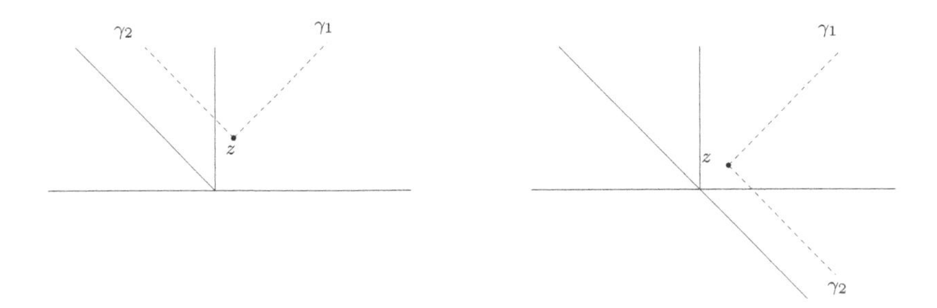

Figure 6.6 Two contributions to the product $\vartheta_{v_1+v_2} \cdot \vartheta_{v_3}$.

Turning to $\vartheta_{v_1+v_2} \cdot \vartheta_{v_3}$ we have the range $-3 \leq F(r) \leq 0$. However, $F(r) = -3$ is impossible, as this does not allow either γ_i to cross a ray ρ_j, and necessarily γ_1 and γ_2 come in from infinity in different cones.

If $F(r) = -2$, as at least one of the γ_i crosses a ray ρ_j, no bends are possible. One then sees the two possibilities in Figure 6.6. These give rise to the contributions $z^{D_2}\vartheta_{2v_2}$ and $z^{D_1}\vartheta_{2v_1}$ respectively.

If $F(r) = -1$, then a bend is also permitted, and $r = v_1, v_2$ or v_3. By placing z in the interiors of $\sigma_{1,2}$, $\sigma_{1,2}$ or $\sigma_{2,3}$ respectively, near ρ_1, ρ_2, or ρ_3, one rules out v_3 as a possibility and has as remaining possibilities as in Figure 6.7. These contribute $\vartheta_{v_1}\sum_j z^{D_1+L_{1j}}$ and $\vartheta_{v_2}\sum_j z^{D_2+L_{2j}}$ respectively.

Finally, we have $F(r) = 0$, i.e., $r = 0$. We put z in the interior of $\sigma_{1,2}$ near the origin, close to ρ_1. Then γ_1 stays in the interior of $\sigma_{1,2}$, and therefore can only bend at a ray of $\mathfrak{D}_{\mathrm{can}}$ intersecting the interior of $\sigma_{1,2}$. However, if it bends at any ray other than $\mathfrak{d} = \mathbb{R}_{\geq 0}(1,1)$, dF decreases by at least 3, while γ_2 crosses some ρ_i, which would require $F(r) \geq 1$. On the other hand, as γ_1 is initially parallel to $\mathfrak{d}$, it cannot cross $\mathfrak{d}$, and hence γ_1 does not bend. This leaves only the possibility depicted in Figure 6.8. This involves rewriting X_3 using the equation $X_2X_3 = f_{\rho_1}z^{[D_1]}X_1$, that is, $X_3 = f_{\rho_1}z^{[D_1]}X_1X_2^{-1}$ and choosing a monomial of

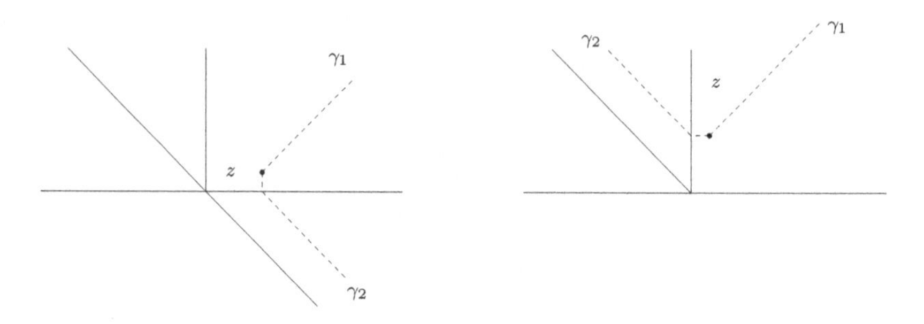

Figure 6.7 Additional contributions to the product $\vartheta_{v_1+v_2} \cdot \vartheta_{v_3}$.

the form $cX_1^{-1}X_2^{-1}$ from this expression. We thus need to consider the coefficient of X_1^{-2} in f_{ρ_1}, and this is $\sum_{1\le j<j'\le 8} z^{L_{1j}+L_{1j'}} + 4z^{D_2+D_3}$. Thus we get a contribution of

$$\sum_{1\le j<j'\le 8} z^{D_1+L_{1j}+L_{1j'}} + 4z^{D_1+D_2+D_3} \tag{4.3}$$

to the product. We can give a clearer description of this expression, however. Consider the class $L_{1j} + L_{1j'}$. If $L_{1j} \cap L_{1j'} = \emptyset$, then L_{1j} and $L_{1j'}$ can be simultaneously contracted. One can easily check that given this choice of j, j', there are unique pairs $L_{2k}, L_{2k'}$ and $L_{3\ell}, L_{3\ell'}$ such that all six of these curves can be simultaneously contracted to give a morphism $\pi\colon (Y,D) \to (Y',D')$, where $Y' \cong \mathbb{P}^2$ and D' is the image of D. This morphism is, in fact, induced by the two-dimensional linear system $|D_1 + L_{1j} + L_{1j'}|$. In particular, $D_1 + L_{1j} + L_{1j'} = \pi^*H$ where H is the class of a line on Y'.

On the other hand, a plane in $\mathbb{P}^3$ containing both D_1 and L_{1j} contains a third line $L_{1j'}$ for some j' with $L_{1j} \cap L_{1j'}$ a point. Thus the set $\{L_{1j}\}$ is partitioned into four pairs, with two lines L_{1j}, $L_{1j'}$ in the same pair if $L_{1j} \cap L_{1j'} \neq \emptyset$, in which case $D_1 + L_{1j} + L_{1j'} \sim D_1 + D_2 + D_3$. Thus we can express (4.3) as

$$\sum_{\pi} z^{\pi^*H} + 8z^{D_1+D_2+D_3}.$$

This is responsible for the last contribution to $\vartheta_{v_1+v_2}\vartheta_{v_3}$. □

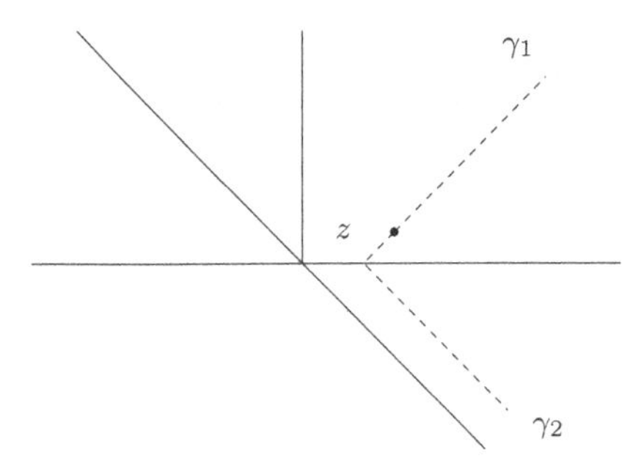

Figure 6.8 The final contribution to the product $\vartheta_{v_1+v_2} \cdot \vartheta_{v_3}$.

Proof of Theorem 1.1. Using the lemma, we calculate

$$\begin{aligned}
\vartheta_{v_1}\vartheta_{v_2}\vartheta_{v_3} &= \left(\vartheta_{v_1+v_2} + z^{D_3}\vartheta_{v_3} + \sum_{j=1}^{8} z^{D_3+L_{3j}}\right)\vartheta_{v_3} \\
&= z^{D_1}\vartheta_{2v_1} + z^{D_2}\vartheta_{2v_2} + z^{D_3}\vartheta_{v_3}^2 + \\
&\qquad \sum_i \left(\sum_j z^{L_{ij}}\right) z^{D_i}\vartheta_{v_i} + \sum_\pi z^{\pi^* H} + 8z^{D_1+D_2+D_3} \\
&= z^{D_1}\vartheta_{v_1}^2 + z^{D_2}\vartheta_{v_2}^2 + z^{D_3}\vartheta_{v_3}^2 + \\
&\qquad \sum_i \left(\sum_j z^{L_{ij}}\right) z^{D_i}\vartheta_{v_i} + \sum_\pi z^{\pi^* H} + 4z^{D_1+D_2+D_3}
\end{aligned}$$

as desired. □

Remark 4.7. We have constructed a family of cubic surfaces over $S = \operatorname{Spec} \Bbbk[P]$ where $P = \mathrm{NE}(Y)$, where Y is a non-singular cubic surface. There is an intriguing slice of this family related to the Cayley cubic which has already made its appearance in Remark 2.1.

We may obtain the Cayley cubic as follows. Take four general lines $L_1, \ldots, L_4$ in $\mathbb{P}^2$, giving 6 pairwise intersection points. By blowing up these six points, we obtain a surface Y. Note that the cone of effective curves of Y is different than that of a general cubic surface because it contains some (-2)-curves. However, for this discussion it is convenient to keep P to be the cone of effective curves of a general cubic surface.

The strict transforms of the four lines become disjoint (-2)-curves which may be contracted, giving a cubic surface Y' with four ordinary double points: this is the Cayley cubic.

If we take $\overline{D}_1$ to be the line joining $L_1 \cap L_2$ and $L_3 \cap L_4$, $\overline{D}_2$ the line joining $L_1 \cap L_3$ and $L_2 \cap L_4$, and $\overline{D}_3$ the line joining $L_1 \cap L_4$ and $L_2 \cap L_3$, and let D_i be the strict transform of $\overline{D}_i$ in Y, we obtain a log Calabi–Yau pair $(Y, D = D_1 + D_2 + D_3)$ as usual. With suitable labelling of the exceptional curves, we can write the classes F_i of the strict transforms of the L_i as

$$\begin{aligned} F_1 &:= L - E_{11} - E_{21} - E_{31} \\ F_2 &:= L - E_{11} - E_{22} - E_{32} \\ F_3 &:= L - E_{12} - E_{21} - E_{32} \\ F_4 &:= L - E_{12} - E_{22} - E_{31} \end{aligned}$$

Now consider the big torus $S^\circ = \operatorname{Spec} \Bbbk[P^{\mathrm{gp}}] \subset S = \operatorname{Spec} \Bbbk[P]$, and consider further the subscheme $T \subseteq S^\circ$ defined by the equations $z^{F_i} = 1$, $i = 1, \ldots, 4$. Then $T = \operatorname{Spec} \Bbbk[P^{\mathrm{gp}}/\mathbf{F}]$, where $\mathbf{F}$ is the subgroup of P^{gp} generated by the curve classes F_i. However, it is not difficult to see that this quotient has two-torsion. Indeed,

$$2E_{11} - 2E_{12} = F_3 + F_4 - F_1 - F_2 \in \mathbf{F},$$

while $E_{11} - E_{12} \notin \mathbf{F}$. We also have

$$E_{11} - E_{12} \equiv E_{21} - E_{22} \equiv E_{31} - E_{32} \mod \mathbf{F}.$$

In fact, T has two connected components: one containing the identity element in the torus S°, the other satisfying the equations $z^{E_{i1}} = -z^{E_{i2}}$ for $i = 1, 2, 3$. Let T' denote this latter component, and restrict the family of cubic surfaces given over S in Theorem 1.1 to T'. One may check that the equation becomes

$$\vartheta_1 \vartheta_2 \vartheta_3 = \sum_i z^{D_i} \vartheta_i^2 - 4z^{D_1 + D_2 + D_3} \tag{4.4}$$

and that this is, in fact, a family of Cayley cubics.

Somewhat more directly, one may also consider the restriction of the scattering diagram to T'. For example, consider the ray (ρ_1, f_{ρ_1}) described in Proposition 3.4. We note that the set of eight lines $\mathbf{L} := \{L_{1i}\}$ split into two groups of four,

$$\begin{aligned} \mathbf{L}_1 := \{&E_{11}, L - E_{21} - E_{31}, L - E_{22} - E_{32}, \\ &2L - E_{11} - E_{21} - E_{22} - E_{31} - E_{32}\} \\ \mathbf{L}_2 := \{&E_{12}, L - E_{21} - E_{32}, L - E_{22} - E_{31}, \\ &2L - E_{12} - E_{21} - E_{22} - E_{31} - E_{32}\} \end{aligned}$$

such that if $L_1, L_2 \in \mathbf{L}$ lie in the same $\mathbf{L}_i$ then $L_1 - L_2 \in \mathbf{F}$, so $z^{L_1} = z^{L_2}$ on T'. If they do not lie in the same $\mathbf{L}_i$, then $2(L_1 - L_2) \in \mathbf{F}$ and $z^{L_1} = -z^{L_2}$ on T'. Further, if $L \in \mathbf{L}$, then $2L - D_2 - D_3 \in \mathbf{F}$.

Thus we obtain, after restriction to T', that, for any choice of $L \in \mathbf{L}$,

$$f_{\rho_1} = \frac{[(1 + z^L X_1^{-1})(1 - z^L X_1^{-1})]^4}{(1 - z^{2L} X_1^{-2})^4} = 1.$$

So, the ray becomes trivial after restriction to T'. Following the argument of §2, one then sees all rays of $\mathfrak{D}_{\mathrm{can}}$ become trivial after restriction to T'. Thus (4.4) may, in fact, be obtained using broken lines as carried out in this section, but this time all broken lines involved are straight.

It is unusual that the trivial scattering diagram is consistent when the affine manifold B has a singularity. In the K3 case, there is a similar situation arising when B is an affine two-sphere arising as a quotient of $\mathbb{R}^2/\mathbb{Z}^2$ via negation.

It is also intriguing that the relevant subtorus $T' \subseteq S^\circ$ is translated, i.e., does not pass through the origin; we speculate that this might have an explanation in terms of orbifold Gromov–Witten invariants of the Cayley cubic.

5 The enumerative interpretation of the equation

There is also a much more recent interpretation of the multiplication law of Theorem 4.5 which gives a Gromov–Witten interpretation for the $\alpha_{p_1 p_2 r} \in \mathbb{k}[P]/I$, writing

$$\alpha_{p_1 p_2 r} = \sum_{\beta \in P} N^{\beta}_{p_1 p_2 r} z^{\beta}$$

with $N^{\beta}_{p_1 p_2 r} \in \mathbb{Q}$. In [GS18], Gross and Siebert explain how to associate certain Gromov–Witten numbers to the data β, p_1, p_2 and r. The details are given in [GS]. The paper [KY] gives a different approach for the same ideas. In general, the construction of these invariants is quite subtle, but happily in the case at hand, all invariants will be easy to calculate. Roughly put, the numbers are defined as follows.

The choice of r defines a choice of stratum $Z_r \subseteq Y$. Indeed, if $\sigma \in \Sigma$ is the minimal cone containing r, then σ corresponds to a stratum of Y, i.e., if $r = 0$, $Z_r = Y$, if $r \in \mathrm{Int}(\rho_i)$ then $Z_r = D_i$, and if $r \in \mathrm{Int}(\sigma_{i,i+1})$ then $Z_r = D_i \cap D_{i+1}$. We choose a general point $z \in Z_r$.

Then $N^{\beta}_{p_1p_2r}$ is a count of the number of stable logarithmic maps $f\colon (C, x_1, x_2, x_{\text{out}}) \to Y$ such that:

(1) the order of tangency of f at x_j with D_k is $\langle D_k, p_j\rangle$, $j = 1, 2$.
(2) $f(x_{\text{out}}) = z$ and the order of tangency of f at x_{out} with D_k is $-\langle D_k, r\rangle$.

Note this involves negative orders of tangency at x_{out}, and defining this is subtle. See [GS18, ACGS20] for more details for how this notion is defined. Here, usually $r = 0$, so we will only have a couple of cases where we have to accept the possibility of a negative order of tangency. For the complete technically correct definition of the above invariant, see [GS, §3].

[GS18, Proposition 2.4] is quite useful in telling us when the relevant moduli space is empty. In particular, that proposition tells us that $N^{\beta}_{p_1p_2r} \neq 0$ implies that if D' is any divisor supported on D, then

$$\beta \cdot D' = \langle D', p_1\rangle + \langle D', p_2\rangle - \langle D', r\rangle. \tag{5.1}$$

In particular, if we take $D' = D = -K_Y$, we obtain

$$\langle D, p_1\rangle + \langle D, p_2\rangle = \beta \cdot D + \langle D, r\rangle.$$

As D is ample in the case of the cubic surface, $\beta \cdot D \geq 0$ for any effective curve class, so we get the stronger result that if $N^{\beta}_{p_1p_2r} \neq 0$, then

$$\langle D, p_1\rangle + \langle D, p_2\rangle \geq \langle D, r\rangle \geq 0,$$

compare with (4.1) and (4.2). In fact, $F = \langle K_Y, \cdot\rangle$, and the above formulae play the same role as those of §4.

We now revisit the calculation of Lemma 4.6. The arguments which follow are necessarily sketchy as we have not given a full definition of the invariants here. We trust the arguments should be sufficiently plausible.

For example, reconsider the product $\vartheta_{v_1}^2$. We see that if $N^{\beta}_{v_1v_1r} \neq 0$ then $\langle D, r\rangle \leq 2$ with equality if and only if $\beta = 0$. Thus if we do have equality, then any map $f\colon (C, x_1, x_2, x_{\text{out}}) \to Y$ contributing to $N^{\beta}_{v_1v_1r}$ is constant. This is discussed in [GS, Lemma 1.15] where it is shown that if $N^{0}_{p_1p_2r} \neq 0$, then p_1, p_2 lie in the same cone and $r = p_1 + p_2$. Further, $N^{0}_{p_1p_2r} = 1$ in this case. In particular, $N^{0}_{v_1v_1,2v_1} = 1$. This gives the contribution ϑ_{2v_1} to $\vartheta_{v_1}^2$.

If $\beta \cdot D = 1$, then the only possibilities for r are v_1, v_2 or v_3. Suppose $r = v_1$. As $\beta \cdot D = 1$, β is the class of a line on Y. Since we choose $z \in Z_r$ general, none of the lines L_{ij} pass through z and thus the image of f may not be L_{ij}. If the image of f is D_2, then f has non-trivial contact with D_3, which is not allowed. Similarly, the image of f may not be D_3.

Finally, if the image of f is D_1, (5.1) yields a contradiction if one takes $D' = D_1$. Thus we eliminate this case. The cases that $r = v_2, v_3$ are similarly ruled out.

We have one remaining case, when $\beta \cdot D = 2$ and $r = 0$. Thus we consider conics which meet D_1 transversally at two points (labelled x_1, x_2), are disjoint from D_2 and D_3, and have a third point x_{out} which coincides with a fixed general point $z \in Y$. It is easy to see that any such conic must be in the linear system $|D_2 + D_3|$, and there is one such conic passing through z. However, as the labels of the intersection points of the conic with D_1 can be interchanged, in fact, $N^{D_2+D_3}_{v_1 v_2 0} = 2$. This gives the second term in the product $\vartheta_{v_1}^2$.

We now move onto $\vartheta_{v_1} \cdot \vartheta_{v_2}$. A similar analysis with the possible degree of the class β leads to the following choices. First, we may have $\beta = 0$, and so $r = v_1 + v_2$ and $N^0_{v_1, v_2, v_1+v_2} = 1$, giving the first contribution to the product.

Next, if $\beta \cdot D = 1$, then $r = v_1, v_2$ or v_3. As before, β must be the class of a line, and as before, we must have $\beta = D_i$ for some i as otherwise the image of f will not contain z. If $\beta = D_3$, we can identify C with D_3, taking x_1 to be the intersection of D_1 and D_3 and x_2 to be the intersection of D_2 with D_3. Since $\beta \cdot D_3 = -1$, (5.1) tells us that $r = v_3$. After fixing $z \in D_3$, we take $x_{\text{out}} = z$. One can show that $N^{D_3}_{v_1 v_2 v_3} = 1$.[4] This contributes the term $z^{D_3} \vartheta_{v_3}$ to the product. On the other hand, if $\beta = D_1$, taking $D' = D_1$ in (5.1) results in a contradiction regardless of the choice of $r = v_k$, and the same holds if $\beta = D_2$. Thus there are no further choices.

Finally, if $\beta \cdot D = 2$, $r = 0$, we fix $z \in Y$ general. We now need to consider conics which meet both D_1 and D_2 transversally, pass through z, and are disjoint from D_3. There are a total of 27 conic bundles on Y: for E the class of a line on Y, $|D - E|$ is a pencil of conics. Thus one easily checks that only eight of these have the correct intersection properties with D, precisely conics of classes $D_3 + L_{3j}$, $1 \leq j \leq 8$. For each j, there is precisely one conic in the pencil $|D_3 + L_{3j}|$ passing through z. This is responsible for the last term in the product $\vartheta_{v_1} \cdot \vartheta_{v_2}$.

[4] We note that the full verification of this statement is somewhat involved, as one must construct the unique punctured curve in the relevant moduli space and show that it is unobstructed. However, this is fairly routine for those familiar with log Gromov–Witten theory, and we omit the details here as it would involve introducing a lot of additional technology into this survey.

We now turn to the product $\vartheta_{v_1+v_2} \cdot \vartheta_{v_3}$. As $v_1 + v_2$ and v_3 do not lie in a common cone of Σ, constant maps cannot occur. Thus we are faced with the possibilities $1 \leq \beta \cdot D \leq 3$.

If $\beta \cdot D = 1$, the same arguments as before reduce to the possibilities that $\beta = D_1$, D_2 or D_3. First $\beta = D_3$ is impossible: any curve with contact order at a point given by $v_1 + v_2$ must pass through $D_1 \cap D_2$. However, in each of the other cases, there is exactly one allowable map. For example, in case $\beta = D_1$, we take $z \in D_1$ general, identify C with D_1, take x_1 to be the intersection point of D_1 and D_2, x_2 the intersection point of D_1 and D_3, take $x_{\mathrm{out}} = z$, and take $r = 2v_1$. Again it is possible to show that these curves exist as punctured logarithmic curves and $N^{D_i}_{v_1+v_2,v_3,2v_i} = 1$ for $i = 1, 2$. This gives the first two terms in the product.

If $\beta \cdot D = 2$, then $r = v_i$ for some i, and we must consider conics which pass through $D_1 \cap D_2$, are transversal to D_3, and pass through an additional point $z \in D_i$. We may now show the image of any punctured map contributing to $N^{\beta}_{v_1+v_2,v_3,v_i}$ is reducible. If the image is an irreducible conic, that conic must pass through $D_1 \cap D_2$, intersect D_3 in at least one point, and pass through the generally chosen point $z \in D_i$. This implies that $\beta \cdot (D_1 + D_2 + D_3) \geq 3$. Since $D_1 + D_2 + D_3$ is the class of a hyperplane section of the cubic surface, this contradicts β being a degree 2 class. If, on the other hand, the image of the punctured map is a line (hence the punctured map is a double cover), this line must be D_1 or D_2, being the only lines passing through $D_1 \cap D_2$. Thus $\beta = 2D_1$ or $2D_2$. However, this case is ruled out via an application of (5.1).

Thus necessarily the image of the punctured map is a union of two lines. The only lines passing through $D_1 \cap D_2$ are D_1 and D_2, and thus $\beta = D_i + L$ for $i = 1$ or 2 and L some other line. As the image of f must be connected, this only leaves the option of $\beta = D_1 + L_{1j}$, $\beta = D_2 + L_{2j}$, or $\beta = D_j + D_k$. The third case can be ruled out from (5.1), and for the first two cases, one can show that $N^{\beta}_{v_1+v_2,v_3,v_i} = 1$. This gives the third and fourth terms in the expression for $\vartheta_{v_1+v_2} \cdot \vartheta_{v_3}$.

Finally, we consider the case of $\beta \cdot D = 3$, so that β is a cubic. There are two choices. Either β is the class of a twisted cubic, i.e., $-2 = \beta \cdot (\beta + K_Y)$, or β is the class of an elliptic curve, i.e., $0 = \beta \cdot (\beta + K_Y)$. Now if β is the class of a twisted cubic, it is easy to see that the linear system $|\beta|$ is two-dimensional and induces a morphism $\pi \colon Y \to Y' \cong \mathbb{P}^2$. If in addition, $\beta \cdot D_i = 1$ for each i (which follows from (5.1)), π maps D_1, D_2 and D_3 to lines in Y'. Hence there is a one-to-one correspondence between such classes β and morphisms $\pi \colon Y \to Y'$ as before.

Given such a morphism, $\beta = \pi^*H$, and there is a unique twisted cubic in the linear system $|\beta|$ passing through both z and $D_1 \cap D_2$. Thus $N^{\beta}_{v_1+v_2,v_3,0} = 1$. This gives the fifth term in the expression for $\vartheta_{v_1+v_2} \cdot \vartheta_{v_3}$.

Finally, if β is the class of an elliptic curve of degree 3, it is necessarily planar, and hence $\beta = D$. We now calculate $N^{\beta}_{v_1+v_2,v_3,0}$. First, there is a pencil of plane cubics passing through $D_1 \cap D_2$ and z. If $\ell \subseteq \mathbb{P}^3$ denotes the line joining these points, then each element of the pencil is of the form $H \cap Y$ for $H \subseteq \mathbb{P}^3$ a plane containing ℓ. To study this pencil, we may blow-up its basepoints, which are the three points of $\ell \cap Y$. This gives a rational elliptic surface $g\colon \tilde{Y} \to \mathbb{P}^1$. Via a standard Euler characteristic computation, such a surface is expected to have 12 singular fibres. However, note that if H contains D_i, $i = 1$ or 2, then $H \cap Y$ is a union $D_i \cup C$ of a line and a conic. In general, C intersects D_i in two points. By normalizing one of these two nodes, we obtain a stable map to Y. However, none of these maps can be equipped with the structure of a stable log map because the point of normalization on the conic maps into D and has non-zero contact order with D, yet it is not a marked point.

Since we have just seen that two of the fibres of this elliptic fibration are of Kodaira type I_2, this leaves 8 additional nodal elliptic curves. By normalizing the node, one obtains a genus zero stable map with the desired intersection behaviour with D. This yields the last term in the description of $\vartheta_{v_1+v_2} \cdot \vartheta_{v_3}$.

We close by noting that the Frobenius structure conjecture (see the first arXiv version of [GHK15b, Conjecture 0.9] or [Man, KY]) gives us another explanation for the constant term (i.e., the coefficient of ϑ_0) $\sum_{\pi} z^{\pi^*H} + 10z^{D_1+D_2+D_3}$ in the equation defining the mirror to the cubic surface. Here we write 10 rather than 4 as we rewrite the equation for the mirror in terms of ϑ_{2v_i} instead of $\vartheta^2_{v_i}$.

The Frobenius conjecture implies that we may calculate the constant term in the triple product $\vartheta_{v_1}\vartheta_{v_2}\vartheta_{v_3}$ as $\sum_{\beta} N^{\beta}_{v_1v_2v_30} z^{\beta}$ where, roughly, $N^{\beta}_{v_1v_2v_30}$ is a count defined as follows. Fix $z \in Y$ general and $\lambda \in \overline{\mathcal{M}}_{0,4}$. We count four-pointed stable log maps $f\colon (C, x_1, x_2, x_3, x_{\mathrm{out}}) \to Y$ such that f meets D_i transversally at x_i, $f(x_{\mathrm{out}}) = z$, and the modulus of the stabilization of C is λ. This can be viewed as fixing the cross-ratio of the four points $x_1, x_2, x_3, x_{\mathrm{out}}$ to be λ. This part of the Frobenius conjecture is shown in [GS, KY], and see also [Man] for related results.

The class β of such a curve C must satisfy $\beta \cdot D = 3$, so β is either a twisted cubic or a plane cubic. In the former case, one immediately recovers $\sum_\pi z^{\pi^* H}$. Indeed, if one fixes $z \in \mathbb{P}^2$ and a cross-ratio λ, there is a unique line H in $\mathbb{P}^2$ passing through z such that the cross-ratio of z and the three points of intersection of H with the boundary divisor is λ.

The count of plane cubics is more subtle. In this case, it is easiest to fix the modulus of the stabilization of C by insisting the stabilization is a singular curve, with x_2, x_3 on one irreducible component and x_1, x_{out} on the other. There are the following possibilities.

(1) The image of f is a union of three lines. This cannot occur, as such a curve does not pass through a general $z \in Y$.

(2) The image of f is the union of a line and a conic, $E \cup Q$. Suppose $E \neq D_i$ for any i. Then E meets D at one point and is rigid, hence does not pass through z. Thus three of the four marked points of C must lie in Q. This contradicts the choice of modulus. Thus $E = D_i$ for some i, and $Q \in |D - D_i|$. In particular, as Q is irreducible, Q is disjoint from D_j, D_k for $\{i, j, k\} = \{1, 2, 3\}$. Thus D_i must contain those marked points mapping to D_j and D_k, so necessarily $D_i = D_1$. In particular, C is the normalization of $D_1 \cup Q$ at one of the two nodes, and the marked point x_1 is the point of Q mapping to the chosen node. Note that this marking is what allows us to count this curve, as opposed to the same curve considered in the contribution to the constant term in $\vartheta_{v_1+v_2} \cdot \vartheta_{v_3}$. Because of the choice of nodes, this gives two curves of class D.

(3) The image of f is an irreducible nodal cubic. In order for the domain to have the given modulus, x_2 and x_3 must lie on a contracted component of C, i.e., $C = C_1 \cup C_2$ with $x_2, x_3 \in C_1$, $x_1, x_{\text{out}} \in C_2$, $f|_{C_1}$ constant with image $D_2 \cap D_3$, and $f(C_2)$ a nodal cubic. The count is now exactly the same as in the case of the contribution of nodal cubics to $\vartheta_{v_1+v_2} \cdot \vartheta_{v_3}$ and we have 8 such nodal cubics.

This explains the term $10z^{D_1+D_2+D_3}$.

Acknowledgements

We would like to thank L. Barrott, A. Neitzke, A. Oblomkov and Y. Zhang for useful discussions. M.G. was supported by EPSRC grant EP/N03189X/1 and a Royal Society Wolfson Research Merit Award. P.H. was supported by NSF grant DMS-1601065 and DMS-1901970. S.K. was supported by NSF grant DMS-1561632.

References

[AC14] Dan Abramovich and Qile Chen. Stable logarithmic maps to Deligne-Faltings pairs II. *Asian J. Math.*, 18(3):465–488, 2014.

[ACGS20] Dan Abramovich, Qile Chen, Mark Gross and Bernd Siebert. Punctured logarithmic maps. arXiv:2009.07720, 2020.

[AMW14] Dan Abramovich, Steffen Marcus and Jonathan Wise. Comparison theorems for Gromov-Witten invariants of smooth pairs and of degenerations. *Ann. Inst. Fourier (Grenoble)*, 64(4):1611–1667, 2014.

[Arg] Hulya Argüz. Canonical scattering for log Calabi–Yau surfaces. In preparation.

[AW18] Dan Abramovich and Jonathan Wise. Birational invariance in logarithmic Gromov-Witten theory. *Compos. Math.*, 154(3):595–620, 2018.

[Bar20] Lawrence Jack Barrott. Explicit equations for mirror families to log Calabi-Yau surfaces. *Bull. Korean Math. Soc.*, 57(1):139–165, 2020.

[Che14] Qile Chen. Stable logarithmic maps to Deligne-Faltings pairs I. *Ann. of Math. (2)*, 180(2):455–521, 2014.

[CL09] Serge Cantat and Frank Loray. Dynamics on character varieties and Malgrange irreducibility of Painlevé VI equation. *Ann. Inst. Fourier (Grenoble)*, 59(7):2927–2978, 2009.

[CPS] Michael Carl, Max Pumperla and Bernd Siebert. A tropical view of Landau-Ginzburg models. Available at http://www.math .uni-hamburg.de/home/siebert/preprints/LGtrop.pdf.

[Dol12] Igor V. Dolgachev. *Classical algebraic geometry*. Cambridge University Press, Cambridge, 2012. A modern view.

[GHK] Mark Gross, Paul Hacking and Sean Keel. Mirror symmetry for log Calabi–Yau surfaces II. Draft.

[GHK15a] Mark Gross, Paul Hacking and Sean Keel. Birational geometry of cluster algebras. *Algebr. Geom.*, 2(2):137–175, 2015.

[GHK15b] Mark Gross, Paul Hacking and Sean Keel. Mirror symmetry for log Calabi–Yau surfaces I. *Publ. Math. Inst. Hautes Études Sci.*, 122:65–168, 2015.

[GHKK18] Mark Gross, Paul Hacking, Sean Keel and Maxim Kontsevich. Canonical bases for cluster algebras. *J. Amer. Math. Soc.*, 31(2):497–608, 2018.

[GP10] Mark Gross and Rahul Pandharipande. Quivers, curves, and the tropical vertex. *Port. Math.*, 67(2):211–259, 2010.

[GPS10] Mark Gross, Rahul Pandharipande, and Bernd Siebert. The tropical vertex. *Duke Math. J.*, 153(2):297–362, 2010.

[GS] Mark Gross and Bernd Siebert. Intrinsic mirror symmetry. arXiv:1909.07649, 2019.

[GS11] Mark Gross and Bernd Siebert. From real affine geometry to complex geometry. *Ann. of Math. (2)*, 174(3):1301–1428, 2011.

[GS13] Mark Gross and Bernd Siebert. Logarithmic Gromov-Witten invariants. *J. Amer. Math. Soc.*, 26(2):451–510, 2013.

[GS18] Mark Gross and Bernd Siebert. Intrinsic mirror symmetry and punctured Gromov-Witten invariants. In *Algebraic geometry: Salt Lake City 2015*, volume 97 of *Proc. Sympos. Pure Math.*, pages 199–230. Amer. Math. Soc., Providence, RI, 2018.

[KY] Sean Keel and Tony Yue Yu. The Frobenius structure theorem for affine log Calabi–Yau varieties containing a torus. arXiv:1908.09861, 2019.

[Man] Travis Mandel. Theta bases and log Gromov-Witten invariants of cluster varieties. arXiv:1903.03042, 2019.

[Obl04] Alexei Oblomkov. Double affine Hecke algebras of rank 1 and affine cubic surfaces. *Int. Math. Res. Not.*, (18):877–912, 2004.

[Rei88] Miles Reid. *Undergraduate algebraic geometry*, volume 12 of *London Mathematical Society Student Texts*. Cambridge University Press, Cambridge, 1988.

Semi-Orthogonal Decomposition of a Derived Category of a 3-Fold With an Ordinary Double Point

Yujiro Kawamata

Dedicated to Professor Miles Reid for his seventieth birthday.

Abstract

We consider semi-orthogonal decompositions of derived categories for 3-dimensional projective varieties in the case when the varieties have ordinary double points.

1 Introduction

The bounded derived category of coherent sheaves $D^b(\mathrm{coh}(X))$ of an algebraic variety X reflects important properties of the variety X. The abelian category of coherent sheaves $\mathrm{coh}(X)$ uniquely determines X itself, but $D^b(\mathrm{coh}(X))$ has more symmetries and the investigation of its structure may reveal something more fundamental. If X is a smooth projective variety, then $D^b(\mathrm{coh}(X))$ satisfies finiteness properties such as having finite homological dimension, Hom-finiteness, and saturatedness [BvdB03]. A birational map like blowing up is reflected to a semi-orthogonal decomposition [BO95]. Even singular varieties with quotient singularities can be considered in the same way by replacing them by smooth Deligne–Mumford stacks [Kaw02, Kaw18a], and we observe the parallelism between the minimal model program (MMP) and semi-orthogonal decompositions.

2010 Mathematics Subject Classification. 14F05, 14F17, 14E05.

Recently we found that the category $D^b(\mathrm{coh}(X))$ for a singular variety X may have a nice semi-orthogonal decomposition when the singularities are not so bad [Kaw18b, Kuz17, KKS18] at least in the case of surfaces. The paper [KKS18] shows that the structure of $D^b(\mathrm{coh}(X))$ is quite interesting at least in the case of a rational surface with cyclic quotient singularities.

In this paper, we calculate the case of dimension 3. We expect that there are still richer structures in dimension 3. We mainly consider 3-folds with an ordinary double point (ODP). There are two cases; $\mathbb{Q}$-factorial and non-$\mathbb{Q}$-factorial. We will see the difference in the following. We note that an ordinary double point is $\mathbb{Q}$-factorial if and only if it is factorial. We also note that an ordinary double point on a rational surface is never factorial, but there are rational 3-folds with factorial and non-factorial ODP's.

We first recall some definitions and theorems on generators of categories in §2 and the triangulated category of singularities $D_{\mathrm{sg}}(X)$ of a variety X in §3. We calculate triangulated categories of singularities in the case of ODP in §4. Then we prove the following in §5:

Theorem 1.1 (= Theorem 5.9). *Let X be a Gorenstein projective variety, let L be a maximally Cohen-Macaulay coherent sheaf on X which generates $D_{\mathrm{sg}}(X)$, and let F be a coherent sheaf which is constructed from L by a flat non-commutative (NC) deformation over the endomorphism algebra $R = \mathrm{End}(F)$ such that $\mathrm{Hom}(F, F[p]) = 0$ for all $p \neq 0$. Then $D^b(\mathrm{coh}(X))$ has a semi-orthogonal decomposition into the triangulated subcategory generated by F, which is equivalent to $D^b(\text{mod-}R)$, and its right orthogonal complement which is saturated, i.e., the complement is similar to the case of a smooth projective variety.*

We apply the theorem to the case where X is a 3-fold with only one ODP which is not $\mathbb{Q}$-factorial and calculate the structure of $D^b(\mathrm{coh}(X))$ in §6:

Theorem 1.2 (= Theorem 6.1). *Let X be a 3-dimensional projective variety with only one ODP which is not $\mathbb{Q}$-factorial. Assume that there are reflexive sheaves L_1, L_2 of rank 1 on X such that*

$$\dim H^0(X, \mathcal{H}om(L_i, L_j)) = \delta_{ij} \text{ and } H^p(X, \mathcal{H}om(L_i, L_j)) = 0 \text{ for } p > 0.$$

Assume moreover that L_1, L_2 generate the triangulated category of singularities $D_{\mathrm{sg}}(X)$. Then L_1, L_2 generate an admissible subcategory of $D^b(\mathrm{coh}(X))$, which is equivalent to $D^b(\text{mod-}R)$ for a 4-dimensional non-

commutative algebra R, such that its right orthogonal complement is a saturated category.

In §7, we calculate some examples. We give two examples where the assumptions of Theorem 6.1 are satisfied and then we consider examples where the singularities are factorial ODP's. In the latter case, non-commutative (NC) deformations [Kaw18b] of L do not terminate and do not yield suitable coherent sheaf F. Indeed, the versal deformation becomes a quasi-coherent sheaf corresponding to an infinite chain of coherent sheaves. In the appendix, we make a correction to an error in a cited paper [Kaw18b] on NC deformations.

We assume that the base field k is algebraically closed of characteristic 0 in this paper.

I would like to dedicate this paper to Professor Miles Reid for the long lasting friendship since he was a Postdoc and I was a graduate student in Tokyo (he kindly corrected my English in my master's thesis at that time).

The author would like to thank Professor Keiji Oguiso for his help in Example 7.7, and Professor Alexander Kuznetsov for the correction in Appendix. The author would also like to thank the anonymous referee for many improvements.

This work was partly done while the author stayed at Korea Advanced Institute of Science and Technology (KAIST) and National Taiwan University. The author would like to thank Professor Yongnam Lee, Professor Jungkai Chen, Department of Mathematical Sciences of KAIST and National Center for Theoretical Sciences of Taiwan for the hospitality and excellent working conditions.

This work was partly supported by Grant-in-Aid for Scientific Research (A) 16H02141. This was also partly supported by the National Science Foundation Grant DMS-1440140 while the author stayed at the Mathematical Sciences Research Institute in Berkeley during the Spring 2019 semester.

2 Generators

We collect some definitions and results concerning generators of categories and representability of functors. In this section, T denotes a triangulated category.

A set of objects $E \subset T$ is said to be a *generator* if the right orthogonal complement defined by $E^{\perp} = \{A \in T \mid \mathrm{Hom}_T(E, A[p]) = 0 \ \forall p\}$ is zero:

$E^\perp = 0$. E is said to be a *classical generator* if T coincides with the smallest triangulated subcategory which contains E and closed under direct summands. E is saids to be a *strong generator* if there exists a number n such that T coincides with the full subcategory obtained from E by taking finite direct sums, direct summands, shifts and at most $n-1$ cones.

T is said to be *Karoubian* if every projector splits. The *idempotent completion* or the *Karoubian envelope* of a triangulated category is defined to be the category consisting of all kernels of all projectors.

A triangulated full subcategory B (resp. C) of T is called *right (resp. left) admissible* if the natural functor $F\colon B \to T$ (resp. $F'\colon C \to T$) has a right (resp. left) adjoint functor $G\colon T \to B$ (resp. $G'\colon T \to C$). An expression

$$T = \langle C, B \rangle$$

is said to be a *semi-orthogonal decomposition* if B, C are triangulated full subcategories such that $\mathrm{Hom}_T(b, c) = 0$ for any $b \in B$ and $c \in C$ and such that, for any $a \in T$, there exists a distinguished triangle $b \to a \to c \to b[1]$ for some $b \in B$ and $c \in C$. In this case, B (resp. C) is a right (resp. left) admissible subcategory. Conversely, if B (resp. C) is a right (resp. left) admissible subcategory, then there is a semi-orthogonal decomposition $T = \langle C, B \rangle$ for $C = B^\perp$ (resp. $B = {}^\perp C$) [Bon89].

Let Mod-k (resp. mod-k) be the abelian category of all (resp. finite dimensional) k-vector spaces. A k-linear triangulated category T is said to be *Hom-finite* if $\sum_i \dim \mathrm{Hom}(A, B[i]) < \infty$ for any objects $A, B \in T$. A cohomological functor $H : T^{\mathrm{op}} \to$ Mod-k is said to be of *finite type* if $\sum_i \dim H(A[i]) < \infty$ for any object $A \in T$. A Hom-finite triangulated category T is said to be *right saturated* if any cohomological functor H of finite type is representable by some object $B \in T$. A right saturated full subcategory of a Hom-finite triangulated category is automatically right admissible and yields a semi-orthogonal decomposition [Bon89]. In a similar way, we define homological functor of finite type and a left saturated category which is automatically left admissible. If T is Hom-finite, has a strong generator, and is Karoubian, then T is right saturated [BvdB03, Theorem 1.3].

Let T be a triangulated category which has arbitrary coproducts (e.g. infinite direct sums). An object $A \in T$ is said to be *compact* if

$$\coprod_\lambda \mathrm{Hom}(A, B_\lambda) \cong \mathrm{Hom}(A, \coprod_\lambda B_\lambda)$$

for all coproducts $\coprod_\lambda B_\lambda$. We denote by T^c the full subcategory of T consisting of all compact objects. If X is a quasi-separated and quasi-compact scheme and $T = D(\mathrm{Qcoh}(X))$, then $T^c = \mathrm{Perf}(X)$, the triangulated subcategory of perfect complexes [BvdB03, Theorem 3.1.1].

T is said to be *compactly generated* if $(T^c)^\perp = 0$. If T is compactly generated, then $E \in T^c$ classically generates T^c if and only if E generates T [Nee92], [BvdB03, Theorem 2.1.2]. The following is used in the proof of Theorem 5.9:

Theorem 2.1 (Brown representability theorem [Nee96, Theorem 4.1]). *Let T be a compactly generated triangulated category and T' be another triangulated category. Let $F\colon T \to T'$ be an exact functor which commutes with coproducts:*

$$\coprod_\lambda F(A_\lambda) \cong F(\coprod_\lambda A_\lambda).$$

Then there exists a right adjoint functor $G\colon T' \to T$:

$$\mathrm{Hom}_{T'}(A, G(B)) \cong \mathrm{Hom}_T(F(A), B).$$

3 Orlov's Triangulated Category of Singularities

We recall Orlov's theory of the triangulated category of singularities. Let X be a separated noetherian scheme of finite Krull dimension whose category of coherent sheaves has enough locally free sheaves, e.g., a quasi-projective variety. Orlov defined a *triangulated category of singularities* $D_{\mathrm{sg}}(X)$ as the quotient category of the bounded derived category of coherent sheaves $D^b(\mathrm{coh}(X))$ by the category of perfect complexes $\mathrm{Perf}(X)$: $D_{\mathrm{sg}}(X) = D^b(\mathrm{coh}(X))/\,\mathrm{Perf}(X)$.

The triangulated category of singularities behaves well when X is Gorenstein:

Proposition 3.1 ([Orl04, Proposition 1.23]). *Assume that X is Gorenstein. Then any object in $D_{\mathrm{sg}}(X)$ is isomorphic to the image of a coherent sheaf F such that $\mathcal{H}om(F, \mathcal{O}_X[i]) = 0$ for all $i > 0$.*

If X is Gorenstein, then the natural morphism

$$F \longrightarrow R\mathcal{H}om(R\mathcal{H}om(F, \mathcal{O}_X), \mathcal{O}_X)$$

is an isomorphism. If (R, M) is a Gorenstein complete local ring of dimension d, then the local duality theorem says that

$$\operatorname{Ext}^i_R(F, R) \cong \operatorname{Hom}_R(H^{d-i}_M(F), E)$$

for any R-module F, where E is an injective hull of $k = R/M$. Thus the condition $\mathcal{H}om(F, \mathcal{O}_X[i]) = 0$ for all $i > 0$ is equivalent to saying that F is a maximally Cohen-Macaulay (MCM) sheaf.

Theorem 3.2 ([Orl04, Proposition 1.21]). *Assume that X is Gorenstein. Let F be coherent sheaf such that $\mathcal{H}om(F, \mathcal{O}_X[i]) = 0$ for all $i > 0$. Let $G \in D^b(\operatorname{coh}(X))$ and let N be an integer such that $\operatorname{Hom}(P, G[i]) = 0$ for all locally free sheaves P and all $i > N$, e.g., $N = 0$ if G is a sheaf and X is affine. Then*

$$\operatorname{Hom}_{D_{\rm sg}(X)}(F, G[N]) \cong \operatorname{Hom}_{D^b(\operatorname{coh}(X))}(F, G[N])/R$$

where R is the subspace of morphisms which factor through locally free sheaves P such as $F \to P \to G[N]$.

Proof [Orl04, Proposition 1.21] assumes that G is a sheaf. But this assumption is not used in the proof. We note that N depends on G. □

Theorem 3.3 (Knörrer periodicity [Orl04, Theorem 2.1]). *Let V be a separated regular noetherian scheme of finite Krull dimension (e.g., a smooth quasi-projective variety) and let $f : V \to \mathbb{A}^1$ be a flat morphism. Let $W = V \times \mathbb{A}^2$ and let $g = f + xy : W \to \mathbb{A}^1$ for coordinates (x, y) on $\mathbb{A}^2$. Let X (resp. Y) be the fiber of f (resp. g) above 0. Let $Z = \{f = x = 0\} \subset Y$, and let $i : Z \to Y$ and $q : Z \to X$ be natural morphisms. Then $\Phi = Ri_*q^* \colon D^b(\operatorname{coh}(X)) \to D^b(\operatorname{coh}(Y))$ induces an equivalence $\bar{\Phi} \colon D_{\rm sg}(X) \to D_{\rm sg}(Y)$.*

Theorem 3.4 ([Orl11, Theorem 2.10]). *Let X and X' be quasi-projective varieties. Assume that the formal completions $\hat{X}$ and $\hat{X}'$ along their singular loci are isomorphic. Then the idempotent completions of the triangulated categories of singularities $\overline{D_{\rm sg}(X)}$ and $\overline{D_{\rm sg}(X')}$ are equivalent.*

4 Triangulated Categories of Singularities for ODPs

We calculate triangulated categories of singularities for varieties with only ordinary double points. An *ordinary double point* is a singularity which is analytically isomorphic to an isolated hypersurface singularity defined by an equation $x_0^2 + \cdots + x_n^2 = 0$ in $\mathbb{A}^{n+1}$.

In general, we have $M \cong M[2]$ for $M \in D_{\mathrm{sg}}(X)$ when X is a hypersurface or a complete intersection in a regular local ring [Eis80].

Example 4.1 ([Orl04, §3.3]). Let $X_n = \{x_0^2 + \cdots + x_n^2 = 0\} \subset \mathbb{A}^{n+1}$ be an ordinary double point of dimension n. Then $D_{\mathrm{sg}}(X_n) \cong D_{\mathrm{sg}}(X_{n+2})$ by the Knörrer periodicity (Theorem 3.3 = [Orl04, Theorem 2.1]).

We consider X_0. Let $A = k[z]/z^2$. Then any object of $D_{\mathrm{sg}}(X_0)$ is represented by a finite A-module. Therefore it is a direct sum of $V_1 = k = A/(z)$.

A natural exact sequence $0 \to (z) \to A \to A/(z) \to 0$ yields an exact triangle $V_1 \to A \to V_1 \to V_1[1]$ in $D^b(\mathrm{coh}(X_0))$, hence an isomorphism $V_1 \cong V_1[1]$ in $D_{\mathrm{sg}}(X_0)$.

Therefore we have $\mathrm{Hom}_{D_{\mathrm{sg}}(X_0)}(V_1, V_1) \cong k$ which is generated by the identity Id. It follows that $D_{\mathrm{sg}}(X_0)$ is already idempotent complete.

The translation functor takes $V_1 \mapsto V_1$ and $\mathrm{Id} \mapsto \mathrm{Id}$. The exact triangles are only trivial ones.

Example 4.2. We consider X_1.

Let $B = k[z, w]/(zw)$. Then any object of $D_{\mathrm{sg}}(X_1)$ is represented by a finite B-module M such that $\mathrm{Hom}(M, B[i]) = 0$ for all $i > 0$. Therefore it is a direct sum of $M_z = B/(w)$ and $M_w = B/(z)$.

A natural exact sequence $0 \to zB \to B \to B/(z) \to 0$ yields an exact triangle $M_z \to B \to M_w \to M_z[1]$ in $D^b(\mathrm{coh}(X_1))$, hence an isomorphism $M_w \cong M_z[1]$ in $D_{\mathrm{sg}}(X_1)$. We also have $M_z \cong M_w[1]$ in $D_{\mathrm{sg}}(X_1)$.

We have $\mathrm{Hom}_{D^b(\mathrm{coh}(X_1))}(M_z, M_z) \cong k[z]$. Since the multiplication map $z : M_z \to M_z$ is factored as $M_z \cong zB \subset B \to B/(w)$, we have $z \in R$ in the notation of Theorem 3.2. Therefore $\mathrm{Hom}_{D_{\mathrm{sg}}(X_1)}(M_z, M_z) \cong k$ which is generated by the identity Id_z by Theorem 3.2.

Since $\mathrm{Hom}_{D^b(\mathrm{coh}(X_1))}(M_z, M_w) = 0$, so $\mathrm{Hom}_{D_{\mathrm{sg}}(X_1)}(M_z, M_w) = 0$. It follows that $D_{\mathrm{sg}}(X_1)$ is already idempotent complete.

The translation functor takes $M_z \mapsto M_w$ and $\mathrm{Id}_z \mapsto \mathrm{Id}_w$. The exact triangle are only trivial ones.

Example 4.3. We consider another 1-dimensional scheme Y_1 whose singularity is analytically isomorphic to that of X_1 but not algebraically.

Let $C = k[z, w]/(z^2 + z^3 + w^2)$. The completion of $Y_1 = \mathrm{Spec}(C)$ at the singularity is isomorphic to that of X_1, hence $\overline{D_{\mathrm{sg}}(Y_1)} \cong \overline{D_{\mathrm{sg}}(X_1)}$ by Theorem 3.4. But we will see that $D_{\mathrm{sg}}(Y_1)$ is not equivalent to $D_{\mathrm{sg}}(X_1)$ [Orl11, Introduction].

Let $C' \to C$ be the normalization. Then $C' \cong k[t]$ with $z = -t^2 - 1$ and $w = -t^3 - t$.

Any object of $D_{\mathrm{sg}}(Y_1)$ is represented by a finite C-module N such that $\mathrm{Hom}(N, C[i]) = 0$ for all $i > 0$. Therefore it is a direct sum of C'.

There are 2 points of $Y_1' = \mathrm{Spec}(C') \cong \mathbb{A}^1$ above the singular point of Y; we have $(z, w)C' = P \cap Q$ for prime ideals $P = (t + \sqrt{-1})$ and $Q = (t - \sqrt{-1})$ of C'. There is a surjective homomorphism $C^{\oplus 2} \to C'$ given by $(a, b) \mapsto a - bt$. The kernel of this homomorphism is equal to $(w, z)C'$, which is isomorphic to C' as a C-module. Indeed, we have $w - zt = 0$ and $wt = -z^2 - z$, etc. Therefore we have an exact sequence

$$0 \to (w, z)C' \to C^{\oplus 2} \to C' \to 0$$

yielding an exact triangle $C' \to C^{\oplus 2} \to C' \to C'[1]$ in $D^b(\mathrm{coh}(Y_1))$, hence an isomorphism $C' \cong C'[1]$ in $D_{\mathrm{sg}}(Y_1)$.

We have $\mathrm{Hom}_{D^b(\mathrm{coh}(Y_1))}(C', C') \cong C' \cong k[t]$ as C-modules, where a C-module homomorphism on the left hand side is mapped to the image of 1 by the homomorphism. We note that a C-homomorphism is determined by the image of 1 because $C' \to C$ is birational. Since the multiplication map $z \colon C' \to C'$ is factored as $C' \cong zC' \subset C \to C'$, we have $z \in R$ in the notation of Theorem 3.2. Therefore we have $\mathrm{Hom}_{D_{\mathrm{sg}}(Y_1)}(C', C') \cong k[t]/(t^2 + 1)$ by by Theorem 3.2.

The translation functor takes $C' \mapsto C'$. The exact triangle are only trivial ones.

There is an idempotent $(1 \pm \sqrt{-1}t)/2 \in \mathrm{Hom}_{D_{\mathrm{sg}}(Y_1)}(C', C')$. But there is no corresponding sheaf on Y_1. Therefore $D_{\mathrm{sg}}(Y_1)$ is not idempotent complete. The corresponding idempotent completion is equivalent to $D_{\mathrm{sg}}(X_1)$.

Example 4.4. We consider X_2.

We rewrite the equation of X_2 as $xy + z^2 = 0$. There is an equivalence $\Phi_1 \colon D_{\mathrm{sg}}(X_0) \cong D_{\mathrm{sg}}(X_2)$ (Theorem 3.3) which is given as follows. Let $Z_1 = \mathrm{Spec}(k[y, z]/z^2)$. Then there are natural morphisms $q_1 \colon Z_1 \to X_0$ and $i_1 \colon Z_1 \subset X_2$. The equivalence is given by $\Phi_1 = Ri_{1*}q_1^*$.

Let $L = \{x = z = 0\}$ be a line on the surface X_2 through the origin. L is a prime divisor which is not a Cartier divisor, but $2L$ is a Cartier divisor.

Since $V_1 = A/(z)$, we have

$$\Phi_1(V_1) = Ri_{1*}(k[y, z]/(z)) = k[x, y, z]/(x, z) = \mathcal{O}_L.$$

We have an exact sequence

$$0 \to \mathcal{O}_{X_2}(-L) \to \mathcal{O}_{X_2} \to \mathcal{O}_L \to 0.$$

Thus $\mathcal{O}_L \cong \mathcal{O}_{X_2}(-L)[1]$ in $D_{\mathrm{sg}}(X_2)$. Therefore $D_{\mathrm{sg}}(X_2)$ is spanned by a reflexive sheaf $\mathcal{O}_{X_2}(-L)$ of rank 1, i.e., any object is isomorphic to a direct sum of $\mathcal{O}_{X_2}(-L)$.

We note that $\mathcal{O}_L$ is a torsion sheaf, but $\mathcal{O}_{X_2}(-L)$ is a Cohen-Macaulay sheaf.

There is an exact sequence

$$0 \to \mathcal{O}_{X_2}(-L) \to F \to \mathcal{O}_{X_2}(-L) \to 0$$

for a locally free sheaf F of rank 2 [Kaw18b, Example 5.5]. Therefore $\mathcal{O}_{X_2}(-L) \cong \mathcal{O}_{X_2}(-L)[1]$ in $D_{\mathrm{sg}}(X_2)$.

Example 4.5. We consider X_3. This is the case of a non-$\mathbb{Q}$-factorial 3-fold.

We rewrite the equation of X_3 as $xy + zw = 0$. There is an equivalence $\Phi_2 \colon D_{\mathrm{sg}}(X_1) \cong D_{\mathrm{sg}}(X_3)$ (Theorem 3.3) which is given as follows. Let $Z_2 = \mathrm{Spec}(k[y,z,w]/zw)$. There are natural morphisms $q_2 \colon Z_2 \to X_1$ and $i_2 \colon Z_2 \subset X_3$. The equivalence is given by $\Phi_2 = Ri_{2*}q_2^*$.

Let $L = \{x = w = 0\}$ and $L' = \{x = z = 0\}$ be planes on X_3 through the origin. They are prime divisors which are not $\mathbb{Q}$-Cartier divisors.

Since $M_z = B/(w)$ and $M_w = B/(z)$, we have

$$\Phi_2(M_z) = Ri_{2*}(k[y,z,w]/(w)) = k[x,y,z,w]/(x,w) = \mathcal{O}_L$$

and

$$\Phi_2(M_w) = Ri_{2*}(k[y,z,w]/(z)) = k[x,y,z,w]/(x,z) = \mathcal{O}_{L'}.$$

We have an exact sequence

$$0 \to \mathcal{O}_{X_3}(-L) \to \mathcal{O}_{X_3} \to \mathcal{O}_L \to 0.$$

Thus $\mathcal{O}_L \cong \mathcal{O}_{X_3}(-L)[1]$ in $D_{\mathrm{sg}}(X_3)$. We also have $\mathcal{O}_{L'} \cong \mathcal{O}_{X_3}(-L')[1]$ in $D_{\mathrm{sg}}(X_3)$. Therefore $D_{\mathrm{sg}}(X_3)$ is spanned by reflexive sheaves $\mathcal{O}_{X_3}(-L)$ and $\mathcal{O}_{X_3}(-L')$ of rank 1, i.e., any object is isomorphic to a direct sum of these objects.

We note that $\mathcal{O}_L$ and $\mathcal{O}_{L'}$ are torsion sheaves, but $\mathcal{O}_{X_3}(-L)$ and $\mathcal{O}_{X_3}(-L')$ are Cohen-Macaulay sheaves.

There is an exact sequence

$$0 \to \mathcal{O}_{X_3}(-L) \to F \to \mathcal{O}_{X_3}(-L') \to 0$$

for a locally free sheaf F of rank 2 [Kaw18b, Example 5.6]. Therefore $\mathcal{O}_{X_3}(-L') \cong \mathcal{O}_{X_3}(-L)[1]$ in $D_{\rm sg}(X_3)$. We also have $\mathcal{O}_{X_3}(-L) \cong \mathcal{O}_{X_3}(-L')[1]$ in $D_{\rm sg}(X_3)$.

Example 4.6. We consider a 3-dimensional scheme Y_3 which is analytically isomorphic to X_3 at the singular points but not algebraically. Y_3 is $\mathbb{Q}$-factorial, hence factorial, because the fundamental group of the punctured neighborhood of the singularity is trivial.

Let Y_3 be a 3-fold defined by an equation $xy + z^2 + z^3 + w^2 = 0$. This has an ordinary double point which is $\mathbb{Q}$-factorial.

There is an equivalence $\Phi'_2\colon D_{\rm sg}(Y_1) \cong D_{\rm sg}(Y_3)$ (Theorem 3.3) which is given as follows. Let $Z'_2 = \mathrm{Spec}(k[y,z,w]/(z^2 + z^3 + w^2))$. Then there are natural morphisms $q'_2\colon Z'_2 \to Y_1$ and $i'_2\colon Z'_2 \subset Y_3$. The equivalence is given by $\Phi'_2 = Ri'_{2*}(q'_2)^*$.

Let $D = \{x = z^2 + z^3 + w^2 = 0\}$ be a prime divisor on Y_3, which is a Cartier divisor. Let $D' \to D$ be the normalization. On Y_1, we have $C' = k[t]$ with $z = -t^2 - 1$ and $w = -t^3 - t$. On Y_3, we have $\mathcal{O}_{D'} = k[y,t]$. Thus we have $\Phi'_2(C') = Ri_{2*}(k[y,t]) = k[y,t] = \mathcal{O}_{D'}$.

There are surjective homomorphisms $\mathcal{O}_{Y_3}^{\oplus 2} \to \mathcal{O}_D^{\oplus 2} \to \mathcal{O}_{D'}$, hence an exact sequence

$$0 \to F \to \mathcal{O}_{Y_3}^{\oplus 2} \to \mathcal{O}_{D'} \to 0$$

defining a coherent sheaf F. Then $D_{\rm sg}(Y_3)$ is spanned by F, i.e., any object is isomorphic to a direct sum of F. The completion of Y_3 at the singular point is isomorphic to that of X_3, and the completion of F corresponds to that of $\mathcal{O}_{X_3}(-L) \oplus \mathcal{O}_{X_3}(-L')$.

5 Semi-Orthogonal Decomposition for a Gorenstein Variety

The following is the main result of this section:

Theorem 5.1. *Let X be a Gorenstein projective variety, let L be a maximally Cohen-Macaulay sheaf on X, let F be a coherent sheaf which is a perfect complex on X and let $R = \mathrm{End}_X(F)$ be the endomorphism ring. Assume the following conditions:*

(a) $\mathrm{Hom}_{D^b(\mathrm{coh}(X))}(F, F[p]) = 0$ *for* $p \ne 0$.
(b) F *is flat over* R.

(c) L generates the triangulated category of singularities

$$D_{\text{sg}}(X) = D^b(\text{coh}(X))/\operatorname{Perf}(X)$$

in the sense that $\operatorname{Hom}(L, A) \neq 0$ *for any* $A \not\cong 0$ *in* $D_{\text{sg}}(X)$ *(note that there is no shift of* A*).*

(d) L belongs to the triangulated subcategory T of $D^b(\text{coh}(X))$ generated by F. More precisely, T is the image of the functor

$$\Phi_0 \colon D^b(\textit{mod-}R) \to D^b(\text{coh}(X))$$

defined by $\Phi_0(A) = A \otimes_R F$, where $D^b(\textit{mod-}R)$ is the bounded derived category of finitely generated right R-modules.

Denote by $T^\perp$ the right orthogonal complement of T in $D^b(\text{coh}(X))$, that is, the full subcategory consisting of objects A such that $\operatorname{Hom}_{D^b(\text{coh}(X))}(F, A[p]) \cong 0$ *for all p. Then the following hold.*

(1) There is an equivalence $T \cong D^b(\textit{mod-}R)$, the bounded derived category of finitely generated right R-modules.

(2) There is a semi-orthogonal decomposition

$$D^b(\text{coh}(X)) = \langle T^\perp, T \rangle.$$

(3) $T^\perp \subset \operatorname{Perf}(X)$.

(4) $T^\perp$ is two-sided saturated.

(5) $T^\perp$ has a Serre functor, i.e., an auto-equivalence $S : T^\perp \to T^\perp$ with bifunctorial isomorphism $\operatorname{Hom}(\bullet_1, \bullet_2) \cong \operatorname{Hom}(\bullet_2, S(\bullet_1))^*$.

Proof (*1*) and (*2*). (see [TU10, Lemma 3.3].) We define two functors $\Phi \colon D(\text{Mod-}R) \to D(\text{Qcoh}(X))$ and $\Psi \colon D(\text{Qcoh}(X)) \to D(\text{Mod-}R)$ between unbounded triangulated categories, where $D(\text{Mod-}R)$ denotes the unbounded derived category of right R-modules which are not necessarily finitely generated. We set $\Phi(\bullet) = \bullet \otimes_R^L F$, where lower R stands for the tensor product over R and the upper L for the left derived functor, and $\Psi(\bullet) = R\operatorname{Hom}_X(F, \bullet)$. That is, we define $\Phi(A) = P_* \otimes_R F$ as complexes for a K-projective resolution $P_* \to A$ in $D(\text{Mod-}R)$, and $\Psi(B) = \operatorname{Hom}_X(F, I_*)$ as complexes for a K-injective resolution $B \to I_*$ in $D(\text{Qcoh}(X))$. Since F is flat over R and perfect on X, these functors induce functors $\Phi_0 \colon D^b(\text{mod-}R) \to D^b(\text{coh}(X))$ and $\Psi_0 \colon D^b(\text{coh}(X)) \to D^b(\text{mod-}R)$, where $D^b(\text{mod-}R)$ denotes the bounded derived category of right R-modules which are finitely generated.

We have

$$\begin{aligned}\operatorname{Hom}_X(\Phi(A), B) &\cong \operatorname{Hom}_X(P_* \otimes_R F, I_*) \\ &\cong \operatorname{Hom}_R(P_*, \operatorname{Hom}_X(F, I_*)) \cong \operatorname{Hom}_R(A, \Psi(B)).\end{aligned}$$

Therefore Φ and Ψ are adjoints.

We have

$$\Psi\Phi(A) = R\operatorname{Hom}_X(F, P_* \otimes_R F) \cong P_* \otimes_R R\operatorname{Hom}(F, F) \cong P_* \cong A.$$

Thus adjunction morphism $A \to \Psi\Phi(A)$ is an isomorphism and Φ is fully faithful. Let T' be the image of Φ, i.e., the triangulated subcategory of $D(\operatorname{Qcoh}(X))$ generated by F. Then we conclude that there is a semi-orthogonal decomposition $D(\operatorname{Qcoh}(X)) = \langle (T')^\perp, T' \rangle$.

By restriction, we deduce that Φ_0 is fully faithful and Ψ_0 is its right adjoint. Therefore we have *(1)* and *(2)*. We needed the unbounded version of the statement for the following proof of the assertion *(4)*.

(3). Let $G \in D^b(\operatorname{coh}(X))$. If $G \notin \operatorname{Perf}(X)$, then $\bar{G}[N] \not\cong 0 \in D_{\mathrm{sg}}(X)$, where N is as in Theorem 3.2 and $\bar{G}$ denotes the object G in $D_{\mathrm{sg}}(X)$. Then $\operatorname{Hom}_{D_{\mathrm{sg}}(X)}(\bar{L}, \bar{G}[N]) \neq 0$, hence $\operatorname{Hom}_{D^b(\operatorname{coh}(X))}(L, G[N]) \neq 0$ by Theorem 3.2. Thus $G \notin T^\perp$. Therefore $T^\perp \subset \operatorname{Perf}(X)$.

(4) and *(5)*. We first prove that $T^\perp$ is right saturated. We modify the proof of [BvdB03, Theorem A.1].

$(T')^\perp$ has arbitrary coproduct, and it is compactly generated by $T^\perp = (T')^\perp \cap \operatorname{Perf}(X)$. Indeed, for any object $0 \not\cong B \in (T')^\perp$, there is an $A \in \operatorname{Perf}(X)$ such that $\operatorname{Hom}_X(A, B) \neq 0$. By *(2)*, there is a left adjoint functor $\Xi : D(\operatorname{Qcoh}(X)) \to (T')^\perp$ of the inclusion functor $\Theta : (T')^\perp \to D(\operatorname{Qcoh}(X))$ which induces a left adjoint functor $\Xi_0 : D^b(\operatorname{coh}(X)) \to T^\perp$ of the inclusion functor $\Theta_0 : T^\perp \to D^b(\operatorname{coh}(X))$. Then we have $\operatorname{Hom}_X(\Xi_0(A), B) \neq 0$ because $\operatorname{Hom}_X(F, B) = 0$.

We use [CKN01, Lemma 2.14]. Let $H : (T^\perp)^{\mathrm{op}} \to \text{mod-}k$ be any cohomological functor of finite type. We define $G = DH : T^\perp \to \text{mod-}k$ using the duality functor $D : (\text{mod-}k)^{\mathrm{op}} \to \text{mod-}k$ given by $D(V) = \operatorname{Hom}(V, k)$. Let $G' : (T')^\perp \to \text{Mod-}k$ be the Kan extension of G given by $G'(B) = \operatorname{colim}_{C \to B} G(C)$, where the colimit is taken for all morphisms from all compact objects $C \in T^\perp$. DG' is represented by an object $A \in (T')^\perp$ by the Brown representability theorem, Theorem 2.1. Since $DDH = H$ on $T^\perp$, we deduce that H is represented by A.

We have to prove that $A \in T^\perp = (T')^\perp \cap D^b(\operatorname{coh}(X))$. We take an embedding $p : X \to \mathbb{P}^n$, and let $H' = H\Xi_0 p^* : D^b(\operatorname{coh}(\mathbb{P}^n))^{\mathrm{op}} \to \text{mod-}k$ be the induced cohomological functor. By Beilinson's theorem [Bei78], there is an equivalence $t' : D(\text{Mod-}S) \cong D(\text{Qcoh}(\mathbb{P}^n))$ which induces an equivalence $t : D^b(\text{mod-}S) \cong D^b(\text{coh}(\mathbb{P}^n))$ for some finite dimensional associative ring S. Let $H'' = H't : D^b(\text{mod-}S)^{\text{op}} \to \text{mod-}k$ and $A' = (t')^{-1}p_*\Theta(A) \in D(\text{Mod-}S)$. Then we have

$$H''(B) = \text{Hom}_X(\Xi_0 p^* t(B), A) = \text{Hom}(B, A')$$

for any $B \in D^b(\text{mod-}S)$, i.e., H'' is represented by A'. Therefore our assertion is reduced to showing that $A' \in D^b(\text{mod-}S)$. We have

$$\sum \dim \text{Hom}_S(S[n], A') = \sum H''(S[n]) < \infty,$$

hence $A' \in D^b(\text{mod-}S)$ and we are done.

Next we prove the existence of a Serre functor S. We note that $T^\perp$ is Hom-finite because it is contained in $\text{Perf}(X)$. $\text{Perf}(X)$ has also a Serre functor S' defined by $S'(\bullet) = \bullet \otimes \omega_X[\dim X]$; we have $\text{Hom}_{\text{Perf}(X)}(a, b) \cong \text{Hom}_{\text{Perf}(X)}(b, S'(a))^*$. Since $T^\perp$ is right saturated, the inclusion functor $i_* \colon T^\perp \to \text{Perf}(X)$ has a right adjoint functor $i^! \colon \text{Perf}(X) \to T^\perp$. If we define $S = i^! S' i_*$, then we have for $a, b \in T^\perp$:

$$\begin{aligned} \text{Hom}_{T^\perp}(a, b) &\cong \text{Hom}_{\text{Perf}(X)}(i_* a, i_* b) \\ &\cong \text{Hom}_{\text{Perf}(X)}(i_* b, S' i_*(a))^* \cong \text{Hom}_{T^\perp}(b, S(a))^*. \end{aligned}$$

Thus S is a Serre functor of $T^\perp$.

Finally, we prove that $T^\perp$ is also left saturated. Let $H' \colon T^\perp \to \text{mod-}k$ be any homological functor of finite type. Then $H \colon (T^\perp)^{\text{op}} \to \text{mod-}k$ defined by $H(a) = (H'(a))^*$ is a cohomological functor of finite type, hence is representable. Thus there is an object $b \in T^\perp$ such that

$$H'(a) = H(a)^* \cong \text{Hom}(a, b)^* \cong \text{Hom}(S^{-1}(b), a).$$

Therefore H' is represented by an object $S^{-1}(b)$. □

Remark 5.2. We will use the theorem in the case where F is obtained as a versal non-commutative deformation of a simple collection L as described in [Kaw18b]. In this case F is flat over the parameter algebra $R = \text{End}_X(F)$.

We do not know how to generalize the theorem in the case where F is not necessarily flat over its endomorphism ring. Indeed, we do not know how to prove that the functor $\Phi \colon D^-(\text{mod-}R) \to D^-(\text{coh}(X))$ defined by $\Phi(\bullet) = \bullet \otimes^L_R F$ is bounded, i.e., Φ sends $D^b(\text{mod-}R)$ to $D^b(\text{coh}(X))$, though its right adjoint functor $\Psi \colon D^-(\text{coh}(X)) \to D^-(\text{mod-}R)$ defined by $\Psi(\bullet) = R\,\text{Hom}(F, \bullet)$ is bounded.

6 Non-commutative Deformation on a 3-Fold with a Non-$\mathbb{Q}$-factorial ODP

We will apply Theorem 5.9 to a 3-fold with a non-$\mathbb{Q}$-factorial ordinary double point.

Let (X, P) be a not $\mathbb{Q}$-factorial 3-dimensional ordinary double point. Then we have two $\mathbb{Q}$-factorializations of X in the following way. Since it is analytically isomorphic to a hypersurface singularity defined by $x_1^2+\cdots+x_4^2=0$ in $\mathbb{A}^4$, the blowing up $\mu: V \to X$ at the singularity P is a desingularization. The exceptional locus of μ is a prime divisor E which is isomorphic to $\mathbb{P}^1 \times \mathbb{P}^1$. We have $\mathcal{O}_E(E) \cong \mathcal{O}_E(K_X) \cong \mathcal{O}_E(-1,-1)$. Since X is not $\mathbb{Q}$-factorial, there is a prime divisor $\bar{D}$ on X which is not a $\mathbb{Q}$-Cartier divisor. Let D be the strict transform of $\bar{D}$ on V. Then we have $O_E(D) \cong O_E(a,b)$ for some integers a, b such that $a \neq b$. Assuming that $a > b$, we apply the Base Point Free Theorem to a divisor $L = D + bE$ on V over X [KMM87, Theorem 3.1.1]. Since L is μ-nef and $L - K_X$ is μ-nef and μ-big, we deduce that there is a positive integer m_0 such that mL is relatively base point free over X for any integer $m \geq m_0$. We obtain a corresponding contraction morphism $\alpha: V \to Y$ to a smooth variety Y over X which contracts E along the second ruling to a curve $l \cong \mathbb{P}^1$ with normal bundle $N_{l/Y} \cong \mathcal{O}_l(-1) \oplus \mathcal{O}_l(-1)$ (if we take $L = -D - aE$, then we obtain a contraction morphism along another ruling of E). The induced morphism $f: Y \to X$ is called a $\mathbb{Q}$-factorialization of X. Since Y is analytically isomorphic to the normal bundle $N_{l/Y}$ along l, we can calculate that $R^1 f_* \mathcal{O}_Y(D) = 0$ if $(D, l) \geq -1$ for a divisor D.

The following theorem says that the assumptions of Theorem 5.9 are satisfied under some additional assumptions:

Theorem 6.1. *Let X be a projective variety of dimension 3. Assume that there is only one singular point P which is a non-$\mathbb{Q}$-factorial ordinary double point. Then there is a $\mathbb{Q}$-factorialization $f: Y \to X$, a projective birational morphism from a smooth projective variety whose exceptional locus l is a smooth rational curve. Assume that there are divisors D_1, D_2 on Y such that, for $L_i = f_* \mathcal{O}_Y(-D_i)$, the following conditions are satisfied:*

(a) $(D_1, l) = 1$ *and* $(D_2, l) = -1$.

(b) (L_1, L_2) *is a simple collection, i.e.,* $\dim \operatorname{Hom}(L_i, L_j) = \delta_{ij}$.

(c) $H^p(X, f_* \mathcal{O}_Y(-D_i + D_j)) = 0$ *for all* $p > 0$ *and all* i, j.

Then there are locally free sheaves F_1, F_2 of rank 2 on X given by non-trivial extensions

$$0 \to L_2 \to F_1 \to L_1 \to 0$$
$$0 \to L_1 \to F_2 \to L_2 \to 0$$

such that, for $L = L_1 \oplus L_2$ and $F = F_1 \oplus F_2$, the following assertions hold:

(1) $\mathrm{Ext}^p(F, F) = 0$ *for* $p > 0$.

(2) *F is flat over*

$$R := \mathrm{End}(F) \cong \begin{pmatrix} k & kt \\ kt & k \end{pmatrix} \mod t^2$$

where the multiplication of R is defined according to the matrix rule.

(3) L is a Cohen-Macaulay sheaf which generates the triangulated category of singularities $D_{\mathrm{sg}}(X)$.

(4) L belongs to the triangulated subcategory of $D^b(\mathrm{coh}(X))$ generated by F in the sense of Theorem 5.9.

Proof We consider 2-pointed non-commutative (NC) deformations of a simple collection (L_1, L_2) [Kaw18b]. There is a spectral sequence

$$E_2^{p,q} = H^p(X, \mathcal{E}xt^q(L_i, L_j)) \Rightarrow \mathrm{Ext}^{p+q}(L_i, L_j)$$

for any i, j. Since $\mathcal{H}om(L_i, L_j) = f_*\mathcal{O}_Y(-D_i + D_j)$, the condition (c) says that $E_2^{p,0} = 0$ for all $p > 0$ and all i, j. Hence

$$\mathrm{Ext}^p(L_i, L_j) \cong H^0(X, \mathcal{E}xt^p(L_i, L_j))$$

for all p.

A neighborhood of the singular point $P \in X$ is analytically isomorphic to that of the vertex of the cone over $\mathbb{P}^1 \times \mathbb{P}^1$ in $\mathbb{P}^4$ considered in [Kaw18b, Example 5.6] (see also Section 7.1). Since the sheaves L_1 and L_2 here correspond to the sheaves $\mathcal{O}_X(0, -1)$ and $\mathcal{O}_X(-1, 0)$ there, the extension space $\mathcal{E}xt^1(L_1, L_2)$ is isomorphic to $\mathcal{E}xt^1(\mathcal{O}_X(0, -1), \mathcal{O}_X(-1, 0))$ there. By the argument there, there exists a locally free sheaf F_1 in an analytic neighborhood of P with the extension sequence as stated in the theorem, and similarly F_2. Since $\mathrm{Ext}^1(L_i, L_j) \cong H^0(X, \mathcal{E}xt^1(L_i, L_j))$, we deduce that these extensions exist globally on X.

We need to calculate $\mathcal{E}xt^p(L_i, L_j)$ for all p at P. For this purpose we need the following calculation:

Lemma 6.2. *Let X' be the cone over $\mathbb{P}^1 \times \mathbb{P}^1$ in $\mathbb{P}^4$ as in [Kaw18b, Example 5.6] (we use the notation X' in order to avoid a confusion). Let G_1, G_2 be locally free sheaves of rank 2 on X' defined by non-trivial extensions:*

$$0 \to \mathcal{O}_{X'}(-1,0) \to G_1 \to \mathcal{O}_{X'}(0,-1) \to 0$$
$$0 \to \mathcal{O}_{X'}(0,-1) \to G_2 \to \mathcal{O}_{X'}(-1,0) \to 0.$$

Then the following hold:

(1) $H^p(X', \mathcal{O}_{X'}(-1,0)) = H^p(X', \mathcal{O}_{X'}(-2,0)) = H^p(X', \mathcal{O}_{X'}(-1,1)) = 0, \quad \forall p.$

(2) $H^p(\mathcal{O}_{X'}(0,-1)) = H^p(X', \mathcal{O}_{X'}(0,-2)) = H^p(X', \mathcal{O}_{X'}(1,-1)) = 0, \quad \forall p.$

(3) $\mathrm{Ext}^p(G_1, \mathcal{O}_{X'}(-1,0)) = \mathrm{Ext}^p(G_1, \mathcal{O}_{X'}(0,-1)) = 0, \quad \forall p > 0.$

(4) $\mathrm{Ext}^p(G_2, \mathcal{O}_{X'}(-1,0)) = \mathrm{Ext}^p(G_2, \mathcal{O}_{X'}(0,-1)) = 0, \quad \forall p > 0.$

(5) $\dim \mathrm{Hom}(G_1, \mathcal{O}_{X'}(0,-1)) = \dim \mathrm{Hom}(G_2, \mathcal{O}_{X'}(-1,0)) = 1.$

(6) $\mathrm{Hom}(G_1, \mathcal{O}_{X'}(-1,0)) = \mathrm{Hom}(G_2, \mathcal{O}_{X'}(0,-1)) = 0.$

(7) $\mathrm{Ext}^p(\mathcal{O}_{X'}(0,-1), \mathcal{O}_{X'}(0,-1)) = \mathrm{Ext}^p(\mathcal{O}_{X'}(-1,0), \mathcal{O}_{X'}(-1,0)) = 0,\ p > 0,\ p \equiv 1 \pmod 2.$

(8) $\mathrm{Ext}^p(\mathcal{O}_{X'}(0,-1), \mathcal{O}_{X'}(-1,0)) = \mathrm{Ext}^p(\mathcal{O}_{X'}(-1,0), \mathcal{O}_{X'}(0,-1)) = 0,\ p > 0,\ p \equiv 0 \pmod 2.$

(9) $\dim \mathrm{Ext}^p(\mathcal{O}_{X'}(0,-1), \mathcal{O}_{X'}(0,-1)) = \dim \mathrm{Ext}^p(\mathcal{O}_{X'}(-1,0), \mathcal{O}_{X'}(-1,0)) = 1,\ p > 0,\ p \equiv 0 \pmod 2.$

(10) $\dim \mathrm{Ext}^p(\mathcal{O}_{X'}(0,-1), \mathcal{O}_{X'}(-1,0)) = \dim \mathrm{Ext}^p(\mathcal{O}_{X'}(-1,0), \mathcal{O}_{X'}(0,-1)) = 1,\ p > 0, p \equiv 1 \pmod 2.$

Proof (1) and (2). Let $D \cong \mathbb{P}^2$ be a prime divisor on X' corresponding to $\mathcal{O}_{X'}(1,0)$ such that there is an exact sequence

$$0 \to \mathcal{O}_{X'}(-1,0) \to \mathcal{O}_{X'} \to \mathcal{O}_D \to 0.$$

Since $H^p(X', \mathcal{O}_{X'}) \cong H^p(D, \mathcal{O}_D)$ for all p, so $H^p(X', \mathcal{O}_{X'}(-1,0)) = 0$ for all p. We have an exact sequence

$$0 \to \mathcal{O}_{X'}(-2,0) \to \mathcal{O}_{X'}(-1,0) \to \mathcal{O}_D(-P) \to 0$$

where $\mathcal{O}_D(-P)$ is the ideal sheaf of P on D:

$$0 \to \mathcal{O}_D(-P) \to \mathcal{O}_D \to \mathcal{O}_P \to 0.$$

Since $H^p(\mathcal{O}_D) \cong H^p(\mathcal{O}_P)$ for all p, we have $H^p(\mathcal{O}_D(-P)) = 0$ for all p. Then we deduce that $H^p(X', \mathcal{O}_{X'}(-2,0)) = 0$ for all p.

Let $S \cong \mathbb{P}^1 \times \mathbb{P}^1$ be a prime divisor on X' corresponding to $\mathcal{O}_{X'}(1,1)$ such that there is an exact sequence

$$0 \to \mathcal{O}_{X'}(-1,-1) \to \mathcal{O}_{X'} \to \mathcal{O}_S \to 0.$$

Then we have

$$0 \to \mathcal{O}_{X'}(-2,0) \to \mathcal{O}_{X'}(-1,1) \to \mathcal{O}_S(-1,1) \to 0.$$

Since $H^p(S, \mathcal{O}_S(-1,1)) = 0$ for all p, $H^p(X', \mathcal{O}_{X'}(-1,1)) = 0$ for all p.

The second assertion follows by symmetry.

(3) and *(4)*. There are spectral sequences

$$\begin{aligned} E_2^{p,q} &= H^p(X, \mathcal{E}xt^q(\mathcal{O}_{X'}(0,-1), \mathcal{O}_{X'}(-1,0))) \\ &\qquad\qquad \Longrightarrow \operatorname{Ext}^{p+q}(\mathcal{O}_{X'}(0,-1), \mathcal{O}_{X'}(-1,0)) \\ E_2^{p,q} &= H^p(X, \mathcal{E}xt^q(\mathcal{O}_{X'}(1,0), \mathcal{O}_{X'}(-1,0))) \\ &\qquad\qquad \Longrightarrow \operatorname{Ext}^{p+q}(\mathcal{O}_{X'}(1,0), \mathcal{O}_{X'}(-1,0)). \end{aligned}$$

Then by *(1)*, we obtain natural isomorphisms

$$\begin{aligned} \operatorname{Ext}^p(\mathcal{O}_{X'}(0,-1), \mathcal{O}_{X'}(-1,0)) &\cong H^0(\mathcal{E}xt^p(\mathcal{O}_{X'}(0,-1), \mathcal{O}_{X'}(-1,0)) \\ &\cong \operatorname{Ext}^p(\mathcal{O}_{X'}(1,0), \mathcal{O}_{X'}(-1,0)) \end{aligned}$$

for all $p > 0$.

We have a commutative diagram of exact sequences

$$\begin{array}{ccccccccc} 0 & \longrightarrow & \mathcal{O}_{X'}(-1,0) & \longrightarrow & G_1 & \longrightarrow & \mathcal{O}_{X'}(0,-1) & \longrightarrow & 0 \\ & & \Big\downarrow{=} & & \Big\downarrow & & \Big\downarrow & & \\ 0 & \longrightarrow & \mathcal{O}_{X'}(-1,0) & \longrightarrow & \mathcal{O}_{X'}^2 & \longrightarrow & \mathcal{O}_{X'}(1,0) & \longrightarrow & 0 \end{array}$$

where the cokernels of the middle and right vertical arrows are isomorphic to $\mathcal{O}_S(1,0)$. By *(1)* we have isomorphisms

$$\operatorname{Ext}^p(G_1, \mathcal{O}_{X'}(-1,0)) \cong \operatorname{Ext}^{p+1}(\mathcal{O}_S(1,0), \mathcal{O}_{X'}(-1,0)) \cong 0$$

for all $p > 0$.

Since $H^p(X', \mathcal{O}_{X'}) = 0$ for $p > 0$ and $H^p(X', \mathcal{O}_{X'}(-1,-1)) = 0$ for all p, we also have

$$\operatorname{Ext}^p(G_1, \mathcal{O}_{X'}(0,-1)) \cong \operatorname{Ext}^{p+1}(\mathcal{O}_S(1,0), \mathcal{O}_{X'}(0,-1)) \cong 0$$

for all $p > 0$.

The second assertion follows by symmetry.

(5–10). We have a long exact sequence

$$\begin{aligned}
0 &\to \operatorname{Hom}(\mathcal{O}_{X'}(0,-1), \mathcal{O}_{X'}(0,-1)) \\
&\quad \to \operatorname{Hom}(G_1, \mathcal{O}_{X'}(0,-1)) \to \operatorname{Hom}(\mathcal{O}_{X'}(-1,0), \mathcal{O}_{X'}(0,-1)) \\
&\to \operatorname{Ext}^1(\mathcal{O}_{X'}(0,-1), \mathcal{O}_{X'}(0,-1)) \\
&\quad \to \operatorname{Ext}^1(G_1, \mathcal{O}_{X'}(0,-1)) \to \operatorname{Ext}^1(\mathcal{O}_{X'}(-1,0), \mathcal{O}_{X'}(0,-1)) \to \dots
\end{aligned}$$

Since $\operatorname{Hom}(\mathcal{O}_{X'}(-1,0), \mathcal{O}_{X'}(0,-1)) = 0$ and $\operatorname{Ext}^p(G_1, \mathcal{O}_{X'}(0,-1)) = 0$ for $p > 0$, we deduce (for $p > 0$)

$$\begin{aligned}
&\dim \operatorname{Hom}(G_1, \mathcal{O}_{X'}(0,-1)) = 1 \\
&\operatorname{Ext}^1(\mathcal{O}_{X'}(0,-1), \mathcal{O}_{X'}(0,-1)) = 0 \\
&\operatorname{Ext}^p(\mathcal{O}_{X'}(-1,0), \mathcal{O}_{X'}(0,-1)) \cong \operatorname{Ext}^{p+1}(\mathcal{O}_{X'}(0,-1), \mathcal{O}_{X'}(0,-1)).
\end{aligned} \tag{6.1}$$

We have a long exact sequence

$$\begin{aligned}
0 &\to \operatorname{Hom}(\mathcal{O}_{X'}(0,-1), \mathcal{O}_{X'}(-1,0)) \\
&\quad \to \operatorname{Hom}(G_1, \mathcal{O}_{X'}(-1,0)) \to \operatorname{Hom}(\mathcal{O}_{X'}(-1,0), \mathcal{O}_{X'}(-1,0)) \\
&\to \operatorname{Ext}^1(\mathcal{O}_{X'}(0,-1), \mathcal{O}_{X'}(-1,0)) \\
&\quad \to \operatorname{Ext}^1(G_1, \mathcal{O}_{X'}(-1,0)) \to \operatorname{Ext}^1(\mathcal{O}_{X'}(-1,0), \mathcal{O}_{X'}(-1,0))) \to \dots
\end{aligned}$$

By construction, the homomorphism

$$k \cong \operatorname{Hom}(\mathcal{O}_{X'}(-1,0), \mathcal{O}_{X'}(-1,0)) \to \operatorname{Ext}^1(\mathcal{O}_{X'}(0,-1), \mathcal{O}_{X'}(-1,0))$$

is injective. Hence we deduce (for $p > 0$)

$$\begin{aligned}
&\operatorname{Hom}(G_1, \mathcal{O}_{X'}(-1,0)) = 0 \\
&\dim \operatorname{Ext}^1(\mathcal{O}_{X'}(0,-1), \mathcal{O}_{X'}(-1,0)) = 1 \\
&\operatorname{Ext}^p(\mathcal{O}_{X'}(-1,0), \mathcal{O}_{X'}(-1,0)) \cong \operatorname{Ext}^{p+1}(\mathcal{O}_{X'}(0,-1), \mathcal{O}_{X'}(-1,0)).
\end{aligned} \tag{6.2}$$

By symmetry, we also obtain (for $p > 0$)

$$\begin{aligned}
&\operatorname{Hom}(G_2, \mathcal{O}_{X'}(0,-1)) = 0 \\
&\dim \operatorname{Hom}(G_2, \mathcal{O}_{X'}(-1,0)) = 1 \\
&\operatorname{Ext}^1(\mathcal{O}_{X'}(-1,0), \mathcal{O}_{X'}(-1,0)) = 0 \\
&\dim \operatorname{Ext}^1(\mathcal{O}_{X'}(-1,0), \mathcal{O}_{X'}(0,-1)) = 1 \\
&\operatorname{Ext}^p(\mathcal{O}_{X'}(0,-1), \mathcal{O}_{X'}(-1,0)) \cong \operatorname{Ext}^{p+1}(\mathcal{O}_{X'}(-1,0), \mathcal{O}_{X'}(-1,0)) \\
&\operatorname{Ext}^p(\mathcal{O}_{X'}(0,-1), \mathcal{O}_{X'}(0,-1)) \cong \operatorname{Ext}^{p+1}(\mathcal{O}_{X'}(-1,0), \mathcal{O}_{X'}(0,-1)).
\end{aligned} \tag{6.3}$$

Combining equations (6.1), (6.2) and (6.3), we obtain

$$\begin{aligned}0 = \operatorname{Ext}^1(\mathcal{O}_{X'}(0,-1), \mathcal{O}_{X'}(0,-1)) &\cong \operatorname{Ext}^2(\mathcal{O}_{X'}(-1,0), \mathcal{O}_{X'}(0,-1)) \\ &\cong \operatorname{Ext}^3(\mathcal{O}_{X'}(0,-1), \mathcal{O}_{X'}(0,-1)) \\ &\cong \operatorname{Ext}^4(\mathcal{O}_{X'}(-1,0), \mathcal{O}_{X'}(0,-1)) \cong \dots \\ 1 = \dim \operatorname{Ext}^1(\mathcal{O}_{X'}(-1,0), \mathcal{O}_{X'}(0,-1)) & \\ &= \dim \operatorname{Ext}^2(\mathcal{O}_{X'}(0,-1), \mathcal{O}_{X'}(0,-1)) \\ &= \dim \operatorname{Ext}^3(\mathcal{O}_{X'}(-1,0), \mathcal{O}_{X'}(0,-1)) \\ &= \operatorname{Ext}^4(\mathcal{O}_{X'}(0,-1), \mathcal{O}_{X'}(0,-1)) = \dots\end{aligned}$$

hence our remaining results. □

We go back to our original situation:

Corollary 6.3.
(1) $\operatorname{Ext}^p(L_1, L_1) = \operatorname{Ext}^p(L_2, L_2) = 0, \quad p > 0,\ p \equiv 1 \pmod 2.$
(2) $\operatorname{Ext}^p(L_1, L_2) = \operatorname{Ext}^p(L_2, L_1) = 0, \quad p > 0,\ p \equiv 0 \pmod 2.$
(3) $\dim \operatorname{Ext}^p(L_1, L_1) = \dim \operatorname{Ext}^p(L_2, L_2) = 1, \quad p > 0,\ p \equiv 0 \pmod 2.$
(4) $\dim \operatorname{Ext}^p(L_1, L_2) = \dim \operatorname{Ext}^p(L_2, L_1) = 1, \quad p > 0,\ p \equiv 1 \pmod 2.$
(5) $\dim \operatorname{Hom}(F_i, L_i) = 1$ *for all* i, $\quad \operatorname{Hom}(F_i, L_j) = 0 \quad (i \neq j)$.
(6) $\operatorname{Ext}^p(F_i, L_j) = 0 \quad p > 0$, *for all* i, j.
(7) $\operatorname{Ext}^p(F_i, F_j) = 0 \quad p > 0$, *for all* i, j.

Proof Since $P \in X$ is analytically isomorphic to the singular cone point of Lemma 6.2, we have isomorphisms between internal extension sheaves at the singular points. By the spectral sequence arguments, we obtain global assertions (*1*) through (*4*).

We have exact sequences

$$\begin{aligned}0 &\to \operatorname{Hom}(L_1, L_1) \to \operatorname{Hom}(F_1, L_1) \to \operatorname{Hom}(L_2, L_1) \\ &\to \operatorname{Ext}^1(L_1, L_1) \to \operatorname{Ext}^1(F_1, L_1) \to \operatorname{Ext}^1(L_2, L_1) \to \dots \\ 0 &\to \operatorname{Hom}(L_1, L_2) \to \operatorname{Hom}(F_1, L_2) \to \operatorname{Hom}(L_2, L_2) \\ &\to \operatorname{Ext}^1(L_1, L_2) \to \operatorname{Ext}^1(F_1, L_2) \to \operatorname{Ext}^1(L_2, L_2) \to \dots\end{aligned}$$

Since $\operatorname{Hom}(L_2, L_1) = 0$, we have $\dim \operatorname{Hom}(F_1, L_1) = 1$. The homomorphism $\operatorname{Hom}(L_2, L_2) \to \operatorname{Ext}^1(L_1, L_2)$ is non-trivial because the extension is non-trivial. Hence $\operatorname{Hom}(F_1, L_2) = 0$.

We have $\operatorname{Ext}^1(L_1, L_1) = 0$. For all $p > 0$, we have a commutative diagram

$$\begin{array}{ccccc} \mathrm{Ext}^p(\mathcal{O}_{X'}(-1,0),\mathcal{O}_{X'}(0,-1)) & \overset{\cong}{\to} & H^0(X,\mathcal{E}xt^p(L_2,L_1)) & \overset{\cong}{\to} & \mathrm{Ext}^p(L_2,L_1) \\ \downarrow & & \downarrow & & \downarrow \\ \mathrm{Ext}^{p+1}(\mathcal{O}_{X'}(0,-1),\mathcal{O}_{X'}(0,-1)) & \overset{\cong}{\to} & H^0(X,\mathcal{E}xt^{p+1}(L_1,L_1)) & \overset{\cong}{\to} & \mathrm{Ext}^{p+1}(L_1,L_1) \end{array}$$

Therefore the homomorphisms $\mathrm{Ext}^p(L_2, L_1) \to \mathrm{Ext}^{p+1}(L_1, L_1)$ are bijective. Thus we obtain $\mathrm{Ext}^p(F_1, L_1) = 0$ for all $p > 0$.

In a similar way, we deduce that the homomorphisms $\mathrm{Ext}^p(L_2, L_2) \to \mathrm{Ext}^{p+1}(L_1, L_2)$ are bijective, and we obtain $\mathrm{Ext}^p(F_1, L_2) = 0$ for all $p > 0$. The assertions for F_2 are obtained by symmetry.

(7) follows from exact sequences

$$\mathrm{Ext}^p(F_i, L_{j'}) \to \mathrm{Ext}^p(F_i, F_j) \to \mathrm{Ext}^p(F_i, L_j)$$

for all i, j and $j' \neq j$. □

Remark 6.4. The 2-periodicity is a consequence of the equalities $\bar{L}_1 = \bar{L}_2[1] = \bar{L}_1[2] \in D_{\mathrm{sg}}(X)$.

We go back to the proof of the theorem. *(7)* of the above corollary implies our assertion *(1)*.

The assertion *(2)* is a consequence of the fact that F is an NC deformation of L. Then we can construct a functor $\Phi_0 \colon D^b(\text{mod-}R) \to D^b(\mathrm{coh}(X))$ as in Theorem 5.9*(4)*. The L_i are images of the simple R-modules by Φ_0, hence the assertion *(4)*.

For the assertion *(3)*, we consider the following exact sequences of local cohomologies:

$$H^{p-1}_P(L_i) \to H^p_P(L_j) \to H^p_P(F_i)$$

for $i \neq j$. Since the L_i are reflexive, we have $H^p_P(L_j) = 0$ for $p < 2$. Since X is Cohen-Macaulay, we have $H^p_P(F_i) = 0$ for $p < 3$. Therefore we deduce that $H^p_P(L_i) = 0$ for $p < 3$, i.e., the L_i are maximally Cohen-Macaulay sheaves. They generate $D_{\mathrm{sg}}(X)$ by the condition (a) with the help of Example 4.5. Thus we complete the proof of the theorem. □

Remark 6.5. There is an exact sequence

$$0 \to \mathcal{O}_{\mathbb{P}^1 \times \mathbb{P}^1}(-m, 0) \to \mathcal{O}^2_{\mathbb{P}^1 \times \mathbb{P}^1} \to \mathcal{O}_{\mathbb{P}^1 \times \mathbb{P}^1}(m, 0) \to 0$$

on $\mathbb{P}^1 \times \mathbb{P}^1$ for any positive integer m. But the corresponding sequence

$$0 \to \mathcal{O}_{X_3}(-m, 0) \to \mathcal{O}^2_{X_3} \to \mathcal{O}_{X_3}(m, 0) \to 0$$

on the cone X_3 over $\mathbb{P}^1 \times \mathbb{P}^1$ in $\mathbb{P}^4$ is not exact if $m \geq 2$. This follows from the fact that the fiber $\mathcal{O}_{X_3}(m, 0) \otimes \mathcal{O}_P \cong \mathcal{O}_{X_3}(0, -m) \otimes \mathcal{O}_P$ at the singular point P has $m+1$ generators. Indeed if $xy+zw=0$ is the equation of X_3 at P, then $\mathcal{O}_{X_3}(0,-m)=(x^m, x^{m-1}z, \dots, z^m)$. Therefore the condition (a) of the theorem is necessary for the conclusion (*3*).

Remark 6.6. Our construction may be generalized to higher dimensions X_n with $n \geq 4$ by using spinor bundles. Let $n' = n-1$. On n'-dimensional smooth quadric $\mathbb{Q}$, there are locally free sheaves Σ_Q (resp. Σ_Q^+ and Σ_Q^-) of rank 2^{m-1} for $n' = 2m-1$ (resp. rank 2^{m-1} for $n' = 2m$) called *spinor bundles.* There are semi-orthogonal decompositions [Kap88]:

$$D^b(\mathrm{coh}(Q)) = \begin{cases} \langle \Sigma_Q(-n'), \mathcal{O}_Q(-n'+1), \dots, \mathcal{O}_Q(-1), \mathcal{O}_Q \rangle, & n' \text{ odd} \\ \langle \Sigma_Q^+(-n'), \Sigma_Q^-(-n'), \\ \qquad \mathcal{O}_Q(-n'+1), \dots, \mathcal{O}_Q(-1), \mathcal{O}_Q \rangle, & n' \text{ even.} \end{cases}$$

By [Ott88], there are exact sequences

$$\begin{aligned} &0 \to \Sigma_Q(-1) \to \mathcal{O}_Q^{2^m} \to \Sigma_Q \to 0, && n' = 2m-1 \\ &0 \to \Sigma_Q^+(-1) \to \mathcal{O}_Q^{2^m} \to \Sigma_Q^- \to 0, && n' = 2m \\ &0 \to \Sigma_Q^-(-1) \to \mathcal{O}_Q^{2^m} \to \Sigma_Q^+ \to 0, && n' = 2m. \end{aligned}$$

Correspondingly, there are Cohen-Macaulay sheaves Σ_X (resp. Σ_X^+ and Σ_X^-) of rank 2^{m-1} for $n = 2m$ (resp. rank 2^{m-1} for $n = 2m+1$) on $X = X_n$, and there are exact sequences

$$\begin{aligned} &0 \to \Sigma_X(-1) \to \mathcal{O}_X^{2^m} \to \Sigma_X \to 0, && n = 2m \\ &0 \to \Sigma_X^+(-1) \to \mathcal{O}_X^{2^m} \to \Sigma_X^- \to 0, && n = 2m+1 \\ &0 \to \Sigma_X^-(-1) \to \mathcal{O}_X^{2^m} \to \Sigma_X^+ \to 0, && n = 2m+1. \end{aligned}$$

As in the case of $n = 2, 3$ [Kaw18b, Examples 5.5–6], if we pull back the sequences by injective homomorphisms $\Sigma_X(-1) \to \Sigma_X$ and $\Sigma_X^{\mp}(-1) \to \Sigma_X^{\pm}$, we obtain non-commutative deformations of $\Sigma_X(-1)$ and $\Sigma_X^{\mp}(-1)$ yielding locally free sheaves of rank 2^m which generate semi-orthogonal components of $D^b(\mathrm{coh}(X))$.

7 Examples

7.1 Quadric Cone

This is an example of a projective variety with a non-$\mathbb{Q}$-factorial ordinary double point considered in [Kaw18b, Example 5.6] and used in the proof of Theorem 6.1.

Let X be a cone over $\mathbb{P}^1 \times \mathbb{P}^1$ in $\mathbb{P}^4$ defined by $xy + zw = 0$ as in Lemma 6.2. X has one ordinary double point P, which is not $\mathbb{Q}$-factorial. Let $\mathcal{O}_X(a,b)$ be reflexive sheaf of rank 1 for integers a, b corresponding to the invertible sheaf $\mathcal{O}_{\mathbb{P}^1\times\mathbb{P}^1}(a,b)$ of bidegree (a,b). The sheaf $\mathcal{O}_X(a,b)$ is invertible if and only if $a = b$. For example, the hyperplane section bundle is $\mathcal{O}_X(1,1)$.

There is an NC deformation $G = G_1 \oplus G_2$ of $\mathcal{O}_X(0,-1) \oplus \mathcal{O}_X(-1,0)$ over $R = \mathrm{End}(G)$, which is given in Theorem 6.1(*2*). There is a semi-orthogonal decomposition

$$D^b(X) = \langle \mathcal{O}(-2,-2), \mathcal{O}(-1,-1), G, \mathcal{O}\rangle \cong \langle D^b(k), D^b(k), D^b(R), D^b(k)\rangle$$

where G in the middle term stands for the subcategory generated by G in the sense of Theorem 5.9(*4*). The referee suggested that it might be better to write instead in the following way:

$$D^b(X) = \langle \mathcal{O}(-2,-2), \mathcal{O}(-1,-1), \langle \mathcal{O}_X(0,-1), \mathcal{O}_X(-1,0)\rangle, \mathcal{O}\rangle$$

or even

$$\begin{aligned} D^b(X) &= \langle \mathcal{O}(-3,-3), \mathcal{O}(-2,-2), \mathcal{O}(-1,-1), \langle \mathcal{O}_X(0,-1), \mathcal{O}_X(-1,0)\rangle\rangle \\ &\cong \langle D^b(k), D^b(k), D^b(k), D^b(R)\rangle. \end{aligned}$$

7.2 2 Point Blow Up of $\mathbb{P}^3$

This is another example of a projective variety with a non-$\mathbb{Q}$-factorial ordinary double point.

Let $g\colon Y \to \mathbb{P}^3$ be a blowing up at 2 distinct points P_1, P_2, with exceptional divisors E_1, E_2. Let l be the strict transform of the line connecting P_1, P_2. Let $f\colon Y \to X$ be the contraction of l to a point.

Let H be the strict transform of a general plane on $\mathbb{P}^3$ to Y. We have $K_Y = -4H + 2E_1 + 2E_2$, and f is given by the anti-canonical linear system $|-K_Y|$ which is nef and big. By the contraction theorem [KMM87], if $(D, l) = 0$ for a Cartier divisor D on Y, then $\mathcal{O}_Y(D) = f^*\mathcal{O}_X(f_*D)$ for a Cartier divisor f_*D on X. By [Bei78, BO95], $D^b(Y)$ is classically generated by a full exceptional collection

$$\begin{aligned} D^b(Y) = \langle \mathcal{O}_{E_1}(2E_1), \mathcal{O}_{E_2}(2E_2), &\mathcal{O}_{E_1}(E_1), \mathcal{O}_{E_2}(E_2), \\ &\mathcal{O}(-3H), \mathcal{O}(-2H), \mathcal{O}(-H), \mathcal{O}\rangle. \end{aligned}$$

The following lemma shows that the conditions of Theorem 6.1 are satisfied:

Lemma 7.1. *Let* $D_1 = -H+E_1+E_2$, $D_2 = -E_1$, *and* $L_i = f_*\mathcal{O}_Y(-D_i)$ *for* $i = 1, 2$. *Then the following hold.*

(1) $(D_1, l) = 1$, $(D_2, l) = -1$ *and* $R^p f_* \mathcal{O}_Y(D_i) = 0$ *for* $p > 0$ *and all* i.
(2) (L_1, L_2) *is a simple collection.*
(3) $H^p(X, f_*(D_i - D_j)) = 0$ *for all* $p > 0$ *and all* i, j.

Proof (*1*) is clear. (*2*) Since the L_i are reflexive sheaves of rank 1, we have $\dim \mathrm{Hom}(L_i, L_i) = 1$. We have $\mathrm{Hom}(L_1, L_2) = H^0(Y, D_1 - D_2) = H^0(Y, -H + 2E_1 + E_2) = 0$. We have also

$$\mathrm{Hom}(L_2, L_1) = H^0(Y, -D_1 + D_2) = H^0(Y, H - 2E_1 - E_2) = 0.$$

(*3*) If $i = j$, then $H^p(X, f_*\mathcal{O}_X) = H^p(Y, \mathcal{O}_Y) = 0$ for $p > 0$. We consider the case $i \neq j$. There is a commutative diagram of exact sequences

$$\begin{array}{ccccccccc} 0 & \longrightarrow & \mathcal{O}_Y(-D_1 + D_2) & \longrightarrow & \mathcal{O}_Y(H) & \longrightarrow & \mathcal{O}_{2E_1} \oplus \mathcal{O}_{E_2} & \longrightarrow & 0 \\ & & \downarrow & & \downarrow & & \downarrow & & \\ 0 & \longrightarrow & \mathcal{O}_l(-D_1 + D_2) & \longrightarrow & \mathcal{O}_l(H) & \longrightarrow & \mathcal{O}_{2Q_1} \oplus \mathcal{O}_{Q_2} & \longrightarrow & 0 \end{array}$$

where $Q_i = E_i \cap l$. In the associated long exact sequences, we have

$$\begin{aligned} &H^0(\mathcal{O}_Y(-D_1 + D_2)) = H^0(\mathcal{O}_l(-D_1 + D_2)) = 0, \\ &\dim H^0(Y, \mathcal{O}(H)) = 4, \qquad \dim H^0(l, \mathcal{O}_l(H)) = 2, \\ &\dim H^0(\mathcal{O}_{2E_1} \oplus \mathcal{O}_{E_2}) = 5, \qquad \dim H^0(\mathcal{O}_{2Q_1} \oplus \mathcal{O}_{Q_2}) = 3, \text{ and} \\ &H^p(Y, \mathcal{O}(H)) = H^p(l, \mathcal{O}_l(H)) = 0 \text{ for } p > 0. \end{aligned}$$

It follows that $\dim H^1(Y, \mathcal{O}(-D_1 + D_2)) = \dim H^1(\mathcal{O}_l(-D_1 + D_2)) = 1$ and $H^p(Y, \mathcal{O}(-D_1 + D_2)) = 0$ for $p \neq 1$. Moreover, the homomorphisms

$$\begin{aligned} H^0(Y, \mathcal{O}(H)) &\longrightarrow H^0(l, \mathcal{O}_l(H)) \quad \text{and} \\ H^0(\mathcal{O}_{2E_1} \oplus \mathcal{O}_{E_2}) &\longrightarrow H^0(\mathcal{O}_{2Q_1} \oplus \mathcal{O}_{Q_2}) \end{aligned}$$

are surjective. It follows that the homomorphism

$$H^1(Y, \mathcal{O}_Y(-D_1 + D_2)) \longrightarrow H^1(\mathcal{O}_l(-D_1 + D_2))$$

is also surjective. We have an exact sequence

$$\begin{aligned} 0 \to H^1(X, f_*\mathcal{O}_Y(-D_1 + D_2)) &\to H^1(Y, \mathcal{O}_Y(-D_1 + D_2)) \to \\ H^0(X, R^1 f_*\mathcal{O}_Y(-D_1 + D_2)) &\to \\ H^2(X, f_*\mathcal{O}_Y(-D_1 + D_2)) &\to H^2(Y, \mathcal{O}_Y(-D_1 + D_2)). \end{aligned}$$

Since $H^1(Y, \mathcal{O}_Y(-D_1 + D_2)) \to H^1(l, \mathcal{O}_l(-D_1 + D_2))$ is surjective, we conclude that $H^p(X, f_*(-D_1 + D_2)) = 0$ for $p > 0$.

There is an exact sequence

$$0 \to \mathcal{O}(-H) \to \mathcal{O}(D_1 - D_2) \to \mathcal{O}_{2E_1}(2E_1) \oplus \mathcal{O}_{E_2}(E_2) \to 0.$$

We have for all p

$$H^p(Y, \mathcal{O}(-H)) = H^p(\mathcal{O}_{2E_1}(2E_1) \oplus \mathcal{O}_{E_2}(E_2)) = 0.$$

Hence $H^p(Y, \mathcal{O}_Y(D_1 - D_2)) = 0$ for all p. Since $R^p f_* \mathcal{O}_Y(D_1 - D_2) = 0$ for $p > 0$, we conclude that $H^p(X, f_* \mathcal{O}_Y(D_1 - D_2)) = 0$ for $p > 0$. □

Let

$$0 \to L_2 \to F_1 \to L_1 \to 0$$
$$0 \to L_1 \to F_2 \to L_2 \to 0$$

be the universal extensions corresponding to $\mathrm{Ext}^1(L_i, L_j)$ for $i \neq j$. Let $F = F_1 \oplus F_2$. We will calculate the right orthogonal complement $F^\perp$ in the rest of the example.

We denote

$$C'_1 = -3H + 2E_1 + E_2, \; C'_2 = -3H + E_1 + 2E_2$$
$$C'_3 = -2H + E_1 + E_2, \; C'_4 = -H + E_1, \; C'_5 = 0.$$

Then $(C'_i, l) = 0$ for all i. Let $C_i = f_* C'_i$ be Cartier divisors on X such that $C'_i = f^* C_i$.

Lemma 7.2. *The right orthogonal complement $F^\perp$ in $D^b(\mathrm{coh}(X))$ is generated by a strong exceptional collection consisting of the invertible sheaves $(\mathcal{O}_X(C_1), \ldots, \mathcal{O}_X(C_5))$.*

Proof We first prove that the sequence is a strong exceptional collection. Since $H^p(X, \mathcal{O}_X) = 0$ for $p > 0$, the $\mathcal{O}_X(C_i)$ are exceptional objects. We prove that $\mathrm{Hom}(\mathcal{O}(C_i), \mathcal{O}(C_j)[p]) = 0$ for $i > j$ and for all p, and $\mathrm{Hom}(\mathcal{O}(C_i), \mathcal{O}(C_j)[p]) = 0$ for all i, j and $p > 0$. We note that $\mathrm{Hom}_X(\mathcal{O}_X(C_i), \mathcal{O}_X(C_j)[p]) \cong \mathrm{Hom}_Y(\mathcal{O}_Y(C'_i), \mathcal{O}_Y(C'_j)[p])$. We calculate

$$\begin{aligned} H^p(Y, E_1 - E_2) &= H^p(Y, -H + E_1) = H^p(Y, -H + E_2) = \\ &H^p(Y, -2H + E_1 + E_2) = H^p(Y, -2H + 2E_2) = \\ &H^p(Y, -3H + 2E_1 + E_2) = H^p(Y, -3H + E_1 + 2E_2) = 0 \end{aligned}$$

for all p. For example, we have an exact sequence

$$\cdots \to H^p(Y, E_1 - E_2) \to H^p(Y, E_1) \to H^p(E_2, \mathcal{O}_{E_2}) \to \cdots.$$

Since $H^p(Y, E_1) \cong H^p(Y, \mathcal{O}_Y)$, we have $H^p(Y, E_1 - E_2) = 0$ for all p. We have $H^p(Y, -3H + E_1 + 2E_2) = H^p(Y, -3H) = 0$ for all p, etc. We also calculate

$$\begin{aligned}H^p(Y, -E_1+E_2) &= H^p(Y, H-E_1) = H^p(Y, H-E_2) = \\ &H^p(Y, 2H-E_1-E_2) = H^p(Y, 2H-2E_2) = \\ &H^p(Y, 3H-2E_1-E_2) = H^p(Y, 3H-E_1-2E_2) = 0\end{aligned}$$

for $p > 0$. For example, we have an exact sequence

$$\cdots \to H^p(Y, H-E_1) \to H^p(Y, H) \to H^p(E_1, \mathcal{O}_{E_1}) \to \cdots .$$

Since $H^0(Y, H) \to H^0(E_1, \mathcal{O}_{E_1})$ is surjective and $H^p(Y, H) = 0$ for $p > 0$, we have $H^p(Y, H-E_1) = 0$ for $p > 0$, etc.

Now we prove that the $\mathcal{O}(C_i)$ belong to $F^\perp$. By Grothendieck duality,

$$\begin{aligned}\mathrm{Hom}_X(L_i, \mathcal{O}(C_j)[p]) &= \mathrm{Hom}_X(Rf_*\mathcal{O}_Y(D_i), \mathcal{O}_X(C_j)[p]) \\ &\cong \mathrm{Hom}_Y(\mathcal{O}_Y(D_i), \mathcal{O}_Y(C'_j)[p])\end{aligned}$$

because $f^!\mathcal{O}_X(C_j) \cong \mathcal{O}_Y(C'_j)$. Therefore we will prove that

$$H^p(Y, C_j - H + E_1 + E_2) = H^p(Y, C_j - E_1) = 0 \quad \text{for all } j \text{ and } p.$$

Since $K_Y = -4H + 2E_1 + 2E_2$, we have

$$H^p(Y, -4H + 3E_1 + 2E_2) = H^{3-p}(Y, -E_1)^* = 0$$

for all p. We also have

$$H^p(Y, -3H + E_1 + E_2) \cong H^p(\mathbb{P}^3, -3H) = 0$$

for all p. Therefore we have $\mathcal{O}(C_1) \in F^\perp$.

We have

$$\begin{aligned}H^p(Y, -4H + 2E_1 + 3E_2) &= H^{3-p}(Y, -E_2)^* = 0 \\ H^p(Y, -3H + 2E_2) &\cong H^p(\mathbb{P}^3, -3H) = 0\end{aligned}$$

for all p, hence we have $\mathcal{O}(C_2) \in F^\perp$.

We have

$$\begin{aligned}H^p(Y, \mathcal{O}(-3H + 2E_1 + 2E_2)) &\cong H^p(\mathbb{P}^3, \mathcal{O}(-3H)) = 0 \\ H^p(Y, \mathcal{O}(-2H + E_2)) &\cong H^p(\mathbb{P}^3, \mathcal{O}(-3H)) = 0 \\ H^p(Y, \mathcal{O}(-2H + 2E_1 + E_2)) &\cong H^p(\mathbb{P}^3, \mathcal{O}(-2H)) = 0 \\ H^p(Y, \mathcal{O}(-H)) &= 0 \\ H^p(Y, \mathcal{O}(-H + E_1 + E_2)) &\cong H^p(\mathbb{P}^3, \mathcal{O}(-H)) = 0 \\ H^p(Y, \mathcal{O}(-E_1)) &= 0\end{aligned}$$

for all p, hence we have $\mathcal{O}_X(C_i) \in F^\perp$ for $i = 3, 4, 5$.

Next, we prove that the $\mathcal{O}_X(C_i)$ and the L_j generate $D^b(\mathrm{coh}(X))$. It follows that the $\mathcal{O}_X(C_i)$ generates $F^\perp$. We denote by C the triangulated subcategory of $D^b(\mathrm{coh}(X))$ generated by the $\mathcal{O}_X(C_i)$ and the L_j.

The linear system $|H-E_1-E_2|$ is a pencil, and its base locus is nothing but l. The image of the natural homomorphism $\mathcal{O}_Y^2 \to \mathcal{O}_Y(H-E_1-E_2)$ is equal to $I_l(H-E_1-E_2)$, where I_l is the ideal sheaf of l. Hence we have an exact sequence

$$0 \to \mathcal{O}_Y(-H+E_1+E_2) \to \mathcal{O}_Y^2 \to \mathcal{O}_Y(H-E_1-E_2) \to \mathcal{O}_l(-1) \to 0.$$

By pushing down to X, we obtain an exact sequence

$$0 \to f_*\mathcal{O}_Y(-H+E_1+E_2) \to \mathcal{O}_X^2 \to L_1 \to 0.$$

Therefore C coincides with the triangulated subcategory generated by the $\mathcal{O}_X(C_j)$, the L_j and $f_*\mathcal{O}_Y(-H+E_1+E_2)$.

We have exact sequences

$$0 \to \mathcal{O}_X \to f_*\mathcal{O}_Y(E_1) \to f_*\mathcal{O}_{E_1}(E_1) \to 0$$
$$0 \to f_*\mathcal{O}_Y(-H+E_2) \to f_*\mathcal{O}_Y(-H+E_1+E_2) \to f_*\mathcal{O}_{E_1}(E_1) \to 0.$$

Thus $f_*\mathcal{O}_Y(-H+E_2)$ can also be included in the set of generators of C. We note that $(-H+E_2, l) = 0$, hence $f_*\mathcal{O}_Y(-H+E_2)$ is an invertible sheaf on X.

We need some lemmas:

Lemma 7.3. *$D^b(\mathrm{coh}(Y))$ has the following full exceptional collection:*

$$\begin{aligned} D^b(\mathrm{coh}(Y)) = \langle &\mathcal{O}_Y(-3H+2E_1+E_2), \mathcal{O}_Y(-3H+E_1+2E_2), \\ &\mathcal{O}_Y(-2H+E_1+E_2), \mathcal{O}_Y(-H+E_1), \mathcal{O}_Y(-H+E_2)), \\ &\mathcal{O}_Y(-H+E_1+E_2), \mathcal{O}_Y, \mathcal{O}_Y(H-E_1-E_2)\rangle. \end{aligned}$$

By definition, these exceptional objects classically generate $D^b(\mathrm{coh}(Y))$. We note that the images by Rf_* of these exceptional objects on Y are exactly the objects considered above as the generators of C.

Proof We first prove that these objects constitute an exceptional collection. Since they are all line bundles, they are exceptional objects. We check their semi-orthogonality. We have $H^p(Y, E_1-E_2) = H^p(Y, -H+E_i) = H^p(Y, -2H+E_1+E_2) = H^p(Y, -2H+2E_i) = 0$ for all p and all i, hence the first 5 are semi-orthogonal. We have $H^p(Y, -2H+2E_1+2E_2) = H^p(-H+E_1+E_2) = 0$ for all p, hence the latter 3 are also semi-orthogonal.

We have $H^p(Y, -2H + E_i) = H^p(Y, -H) = H^p(Y, -E_i) = 0$ and $H^p(Y, -3H+2E_i+E_j) = H^p(Y, -2H+2E_i+E_j) = 0$ for all p and $i \ne j$. By Serre duality, $H^p(Y, -4H + 3E_i + 2E_j)$ is dual to $H^{3-p}(Y, -E_i) = 0$ for $i \ne j$. Hence the first 5 and the latter 3 are semi-orthogonal, and these 8 objects make an exceptional collection.

We prove that these objects classically generate $D^b(\mathrm{coh}(Y))$. Let T be the full triangulated subcategory of $D^b(\mathrm{coh}(Y))$ classically generated by the above exceptional collection. By the exact sequences

$$0 \to \mathcal{O}_Y(-H + E_i) \to \mathcal{O}_Y(-H + E_1 + E_2) \to \mathcal{O}_{E_j}(E_j) \to 0$$
$$0 \to \mathcal{O}_Y(-H) \to \mathcal{O}_Y(-H + E_i) \to \mathcal{O}_{E_i}(E_i) \to 0$$
$$0 \to \mathcal{O}_Y(-2H) \to \mathcal{O}_Y(-2H + E_1 + E_2) \to \mathcal{O}_{E_1}(E_1) \oplus \mathcal{O}_{E_2}(E_2) \to 0$$

for $i \ne j$, we deduce that the objects $\mathcal{O}_{E_i}(E_i)$ for $i = 1, 2$, $\mathcal{O}_Y(-H)$ and $\mathcal{O}_Y(-2H)$ can be included in the set of classical generators of T.

From the exact sequences

$$0 \to \mathcal{O}(-H + E_1 + E_2) \to \mathcal{O}^2 \to \mathcal{O}(H - E_1 - E_2) \to \mathcal{O}_l(-1) \to 0$$
$$0 \to \mathcal{O}(-3H + 2E_1 + 2E_2) \to \mathcal{O}(-2H + E_1 + E_2)^2 \to \mathcal{O}(-H) \to \mathcal{O}_L(-1) \to 0$$

we deduce that $\mathcal{O}_Y(-3H + 2E_1 + 2E_2)$ can also be included. From

$$0 \to \mathcal{O}_Y(-3H + E_i + 2E_j) \to \mathcal{O}_Y(-3H + 2E_1+2E_2) \to \mathcal{O}_{E_i}(2E_i) \to 0$$

for $i \ne j$, we deduce that $\mathcal{O}_{E_i}(2E_i)$ can be included for $i = 1, 2$. From

$$0 \to \mathcal{O}_Y(-3H) \to \mathcal{O}_Y(-3H + 2E_1+2E_2) \to \mathcal{O}_{2E_1}(2E_1) \oplus \mathcal{O}_{2E_2}(2E_2) \to 0$$

we deduce that $\mathcal{O}_Y(-3H)$ can be included. Thus $T = D^b(\mathrm{coh}(Y))$. □

The following lemma says that the direct image functor Rf_* for a birational morphism f is almost surjective for the bounded derived categories of coherent sheaves if Rf_* preserves the structure sheaves:

Lemma 7.4. *Let $f : Y \to X$ be a birational morphism of projective varieties. Assume that $Rf_*\mathcal{O}_Y \cong \mathcal{O}_X$. Then the Karoubian envelope of the image of the functor $Rf_* : D^b(\mathrm{coh}(Y)) \to D^b(\mathrm{coh}(X))$ coincides with $D^b(\mathrm{coh}(X))$.*

Proof By the assumption, we have $Rf_*Lf^* \cong \mathrm{Id}$ on $D(\mathrm{Qcoh}(X))$. Let $A \in D^b(\mathrm{coh}(X))$ be any object. Then we have $Lf^*A \in D^-(\mathrm{coh}(Y))$. We take a large integer m and consider a natural distinguished triangle for truncations:

$$\tau_{<-m}Lf^*A \xrightarrow{h} Lf^*A \longrightarrow \tau_{\geq -m}Lf^*A \longrightarrow (\tau_{<-m}Lf^*A)[1].$$

We have a morphism $Rf_*(h)\colon Rf_*(\tau_{<-m}Lf^*A) \to Rf_*Lf^*A \cong A$. Since Rf_* is bounded and A is bounded, $Rf_*(h) = 0$ for sufficiently large m. It follows that A is a direct summand of $Rf_*(\tau_{\geq -m}Lf^*A)$ in $D(\mathrm{Qcoh}(X))$ because $D(\mathrm{Qcoh}(X))$ is Karoubian by [BN93]:

$$Rf_*(\tau_{\geq -m}Lf^*A) \cong Rf_*(\tau_{<-m}Lf^*A)[1] \oplus A.$$

Since $Rf_*(\tau_{\geq -m}Lf^*A) \in D^b(\mathrm{coh}(X))$, we conclude that A belongs to the Karoubian completion of the image of Rf_*. □

Let G be the set of exceptional objects which classically generates $D^b(\mathrm{coh}(Y))$ in Lemma 7.3. We prove that Rf_*G generates $D^b(\mathrm{coh}(X))$. Let $A \in D^b(\mathrm{coh}(X))$ be any object. If $\mathrm{Hom}(Rf_*G, A[p]) = 0$ for all p, we will prove $A \cong 0$.

We know that, for some object $B \in D^b(\mathrm{coh}(Y))$, A is a direct summand of Rf_*B. Since G classically generates $D^b(\mathrm{coh}(Y))$, we deduce that $\mathrm{Hom}(Rf_*B, A[p]) = 0$ for all B and all p. It follows that $A \cong 0$. □

Remark 7.5. It is interesting to calculate the derived categories of varieties which are obtained by blowing up $\mathbb{P}^3$ at more than 2 points. Especially, if we blow up 6 or more points, then the blown-up varieties have moduli. In this case, it is interesting to determine the semi-orthogonal complement of a trivial factor consisting of a non-full but maximal exceptional collection.

7.3 Factorial Case

We consider an example of a 3-fold with a $\mathbb{Q}$-factorial ordinary double point. We will see that similar arguments to the non-$\mathbb{Q}$-factorial case do not work because NC deformations do not terminate.

We start with an example of a singular curve:

Example 7.6. Let X be a nodal cubic curve defined by an equation $(x^2 + y^2)z + y^3 = 0$ in $\mathbb{P}^2$.

Let $P \in X$ be the singular point and let $\nu : X' \to X$ be the normalization. Then $X' \cong \mathbb{P}^1$ and $H^1(\mathcal{O}_X) = k$. We consider non-commutative deformations of a Cohen-Macaulay sheaf $L = \nu_*\mathcal{O}_{X'}$ which generates $D_{\mathrm{sg}}(X)$ (see Example 4.3).

From an exact sequence $0 \to \mathcal{O}_X \to L \to \mathcal{O}_P \to 0$, we consider the associated long exact sequence to obtain

$$\mathcal{H}om_{\mathcal{O}_X}(L, L) \cong \mathcal{H}om_{\mathcal{O}_X}(\mathcal{O}_X, L) \cong L$$
$$\mathcal{E}xt^p_{\mathcal{O}_X}(\mathcal{O}_P, L) \cong \mathcal{E}xt^p_{\mathcal{O}_X}(L, L), \; p > 0.$$

Since X is Gorenstein, we have

$$\mathcal{E}xt^p_{\mathcal{O}_X}(\mathcal{O}_P, \mathcal{O}_X) \cong \begin{cases} \mathcal{O}_P, & p = 1 \\ 0, & p \neq 1. \end{cases}$$

Therefore from another associated long exact sequence, we obtain

$$\mathcal{E}xt^p_{\mathcal{O}_X}(\mathcal{O}_P, L) \cong \mathcal{E}xt^p_{\mathcal{O}_X}(\mathcal{O}_P, \mathcal{O}_P)$$

for all $p > 0$. In particular, we have $\mathcal{E}xt^1_{\mathcal{O}_X}(L, L) \cong \mathcal{E}xt^1_{\mathcal{O}_X}(\mathcal{O}_P, \mathcal{O}_P) \cong k^2$.

The versal NC deformation of $\mathcal{O}_P$ on X is given by the completion of X at P. Its parameter algebra is given by $k[[x, y]]/(xy)$. The versal NC deformation of L is given by an infinite chain of smooth rational curves. It is the inverse limit of the sheaves $L_{i,j}$ for $i, j \to \infty$ defined in the following way. $L_{i,j}$ is the direct image sheaf of an invertible sheaf $L'_{i,j}$ on a chain of smooth rational curves of type A_{i+j+1}, where the degree of $L_{i,j}$ on the m-th component is equal to 0 for $m = i + 1$, and to 1 otherwise. The parameter algebra for $L_{i,j}$ is given by $k[x, y]/(xy, x^{i+1}, y^{j+1})$. In particular, NC deformations of L do not stop after finitely many steps.

We consider a factorial surface:

Example 7.7. Let $\pi : Y_2 \to \mathbb{P}^2$ be a double cover whose ramification divisor is a generic curve of degree 6 with one node at $Q \in \mathbb{P}^2$. Then $P = \pi^{-1}(Q) \in Y_2$ is the only singular point of Y_2.

We claim that $P \in Y_2$ is a factorial ordinary double point. This construction and the following argument is communicated by Keiji Oguiso. Let $Y' \to Y_2$ be a minimal resolution with an exceptional divisor E. Then Y' is a K3 surface and the Neron–Severi lattice is given by $NS(Y') = \mathbf{Z}H \oplus \mathbf{Z}E$, where H is the pull-back of a line on $\mathbb{P}^2$, due to the genericity of the ramification divisor. We have $(H^2) = 2$, $(E^2) = -2$ and $(H, E) = 0$, hence $P \in Y_2$ is factorial.

We note that an ordinary double point on a rational surface, say S, is always non-factorial (though 2-factorial). Indeed, the Neron–Severi lattice of its resolution $S' \to S$ is always unimodular since $H^2(\mathcal{O}_{S'}) = 0$. Hence there is a curve on S' whose intersection number with the exceptional curve is odd, and its image on S is not a Cartier divisor.

Let l be a generic line in $\mathbb{P}^2$ through Q, and let $C = \pi^{-1}(l)$. Then C is an irreducible curve of genus 1 with a node. Let $\nu\colon C' \to C$ be the normalization, and let $L_{C'}$ be an invertible sheaf on C' of degree 2. Then there is a surjective homomorphism $\mathcal{O}_C^{\oplus 2} \to \nu_* L_{C'}$. Let L be the kernel of the composite homomorphism $\mathcal{O}_{Y_2}^{\oplus 2} \to \mathcal{O}_C^{\oplus 2} \to \nu_* L_{C'}$:

$$0 \to L \to \mathcal{O}_{Y_2}^{\oplus 2} \to \nu_* L_{C'} \to 0.$$

L is a reflexive sheaf of rank 2 which is locally free except at P. We obtain $H^p(L) = 0$ for all p from a long exact sequence. C has two analytic branches at P, and L is analytically isomorphic to a direct sum of reflexive sheaves of rank 1 near P.

We have exact sequences

$$0 \to L^{\oplus 2} \to \mathcal{H}om(L, L) \to \mathcal{E}xt^1(\nu_* L_{C'}, L) \to 0$$
$$0 \to \mathcal{H}om(\nu_* L_{C'}, \nu_* L_{C'}) \to \mathcal{E}xt^1(\nu_* L_{C'}, L) \to \mathcal{E}xt^1(\nu_* L_{C'}, \mathcal{O}_{Y_2}^{\oplus 2}).$$

Since Y_2 is Gorenstein with $\omega_{Y_2} \cong \mathcal{O}_{Y_2}$, we have by Grothendieck duality

$$\mathcal{E}xt^1(\nu_* L_{C'}, \mathcal{O}_{Y_2}) \cong \nu_* L_{C'}^{-1}.$$

Since $H^0(\nu_* L_{C'}^{-1}) = 0$, we have

$$\mathrm{Hom}(L, L) \cong H^0(\mathcal{E}xt^1(\nu_* L_{C'}, L)) \cong \mathrm{Hom}(\nu_* L_{C'}, \nu_* L_{C'}) \cong k.$$

Thus L is a simple sheaf on Y_2.

The NC deformations of L do not stop after finitely many steps. Indeed, the two analytic components extend in an infinite chain as in the previous example.

Now we consider a 3-fold with a $\mathbb{Q}$-factorial, hence factorial, ordinary double point:

Example 7.8. Let X_0 be a cubic 3-fold in $\mathbb{P}^4$ defined, in coordinates (x, y, z, w, t), by an equation

$$x^3 + 3xy^2 + w^3 - 3t(xy + z^2 + w^2) = 0.$$

The singular locus of X_0 consists of two points $P = (0, 0, 0, 0, 1)$ and $P' = (0, -1, 0, 0, 1)$ which are ordinary double points. Let $g\colon X \to X_0$ be the blowing up at P' with the exceptional divisor $E \cong \mathbb{P}^1 \times \mathbb{P}^1$.

The local equation of X at P can be written as

$$x(x^2 + 3y^2 + y) + z^2 + w^2 + w^3 = 0.$$

Hence X is a projective variety with one $\mathbb{Q}$-factorial ODP.

Let D_0 be a prime divisor on X_0 defined by $x = 0$. Then D_0 has equation $w^3 - 3t(z^2 + w^2) = 0$ in $\mathbb{P}^3$ with coordinates (y, z, w, t). The divisor D_0 is a cone over a nodal cubic curve. The vertex of the cone is at $Q = (0, 1, 0, 0, 0)$. The singular locus of D_0 is a line defined by $z = w = 0$. Let $\nu_0 \colon D_0' \to D_0$ be the normalization. D_0' is the cone over a normal rational curve of degree 3.

g induces a blowing up $g_D \colon D \to D_0$ at P'. The exceptional locus of g_D consists of two lines m_1, m_2 on $\mathbb{P}^1 \times \mathbb{P}^1$. The singular locus of D is a smooth rational curve which is the strict transform of the line $\{z = w = 0\}$. Let $\nu \colon D' \to D$ be the normalization.

Let l be a generic line on D' through the vertex. Then there is a surjective homomorphism $\mathcal{O}_D^{\oplus 2} \to \nu_* \mathcal{O}_{D'}(l)$. Let L be the kernel of the composition $\mathcal{O}_X^{\oplus 2} \to \mathcal{O}_D^{\oplus 2} \to \nu_* \mathcal{O}_{D'}(l)$. Thus we have an exact sequence

$$0 \to L \to \mathcal{O}_X^{\oplus 2} \to \mu_* \mathcal{O}_{D'}(l) \to 0.$$

There is an exact sequence

$$\cdots \to H_P^p(L) \to H_P^p(\mathcal{O}_X^2) \to H_P^p(\nu_* \mathcal{O}_{D'}(l)) \to \dots$$

Since $\mathcal{O}_X$ and $\nu_* \mathcal{O}_{D'}(l)$ have depth 3 and 2, respectively, L is a maximally Cohen-Macaulay sheaf of rank 2 on X which is locally free except at P. L generates $D_{\mathrm{sg}}(X)$.

We will prove that L is a simple sheaf, i.e., $\mathrm{End}(L) \cong k$. We have $\dim H^0(\mathcal{O}_X) = 1$, $H^p(\mathcal{O}_X) = 0$ for $p > 0$, $\dim H^0(\mathcal{O}_{D'}(l)) = 2$ and $H^p(\mathcal{O}_{D'}(l)) = 0$ for $p > 0$. Therefore we have $H^p(L) = 0$ for all p. We have an exact sequence

$$0 \to L^{\oplus 2} \to \mathcal{H}om(L, L) \to \mathcal{E}xt^1(\mu_* \mathcal{O}_{D'}(l), L) \to 0$$

and $\mathcal{E}xt^p(L, L) \cong \mathcal{E}xt^{p+1}(\mu_* \mathcal{O}_{D'}(l), L)$ for $p > 0$.

Since X is Gorenstein, we have

$$\mathcal{E}xt^p(\mu_* \mathcal{O}_{D'}(l), \mathcal{O}_X) \cong \begin{cases} \mu_* \omega_{D'/X}(-l), & p = 1 \\ 0, & p \neq 1 \end{cases}$$

by Grothendieck duality. From a long exact sequence, we deduce

$$\begin{aligned} 0 \to &\mathcal{H}om(\mu_* \mathcal{O}_{D'}(l), \mu_* \mathcal{O}_{D'}(l)) \to \mathcal{E}xt^1(\mu_* \mathcal{O}_{D'}(l), L) \to \mu_* \omega_{D'/X}(-l)^{\oplus 2} \\ &\to \mathcal{E}xt^1(\mu_* \mathcal{O}_{D'}(l), \mu_* \mathcal{O}_{D'}(l)) \to \mathcal{E}xt^2(\mu_* \mathcal{O}_{D'}(l), L) \to 0 \end{aligned}$$

and $\mathcal{E}xt^p(\mu_* \mathcal{O}_{D'}(l), \mu_* \mathcal{O}_{D'}(l)) \cong \mathcal{E}xt^{p+1}(\mu_* \mathcal{O}_{D'}(l), L)$ for $p > 1$.

Since $D \sim \mathcal{O}_X(1) - E$ on X, we calculate

$$\omega_{D'/X}(-l) \cong \mathcal{O}_{D'}(3l - (m_1 + m_2) - (l - m_1) - (l - m_2) - l) \cong \mathcal{O}_{D'}.$$

Thus $H^0(\mu_*\omega_{D'/X}(-l)^{\oplus 2}) \cong k^2$. L is a locally free sheaf outside P, and $\mu_*\mathcal{O}_{D'}(l)$ is an invertible sheaf on the smooth locus of a Cartier divisor D. Moreover, $\mu_*\mathcal{O}_{D'}(l)$ is analytically isomorphic to the sum of invertible sheaves on two Cartier divisors along the double locus of D except at P and Q. Therefore we have $\mathcal{E}xt^2(\mu_*\mathcal{O}_{D'}(l), L)$ is supported at $\{P, Q\}$. On the other hand, $\mathcal{E}xt^1(\mu_*\mathcal{O}_{D'}(l), \mu_*\mathcal{O}_{D'}(l))$ is an invertible sheaf on the smooth locus of D and has higher rank along the double locus of D. It follows that the homomorphism

$$H^0(\mu_*\omega_{D'/X}(-l)^{\oplus 2}) \to H^0(\mathcal{E}xt^1(\mu_*\mathcal{O}_{D'}(l), \mu_*\mathcal{O}_{D'}(l)))$$

is injective. Therefore

$$H^0(\mathcal{E}xt^1(\mu_*\mathcal{O}_{D'}(l), L)) \cong H^0(\mathcal{H}om(\mu_*\mathcal{O}_{D'}(l), \mu_*\mathcal{O}_{D'}(l))) \cong k$$

hence L is a simple sheaf.

We consider NC deformations of L. The successive extensions of L become an infinite chain as in the previous examples so that they do not stop after finitely many steps. Thus we do not obtain a semi-orthogonal decomposition unlike the case of non-$\mathbb{Q}$-factorial ODP.

Note that there is an extension F of L by L which is locally free. But F still has an extension by L; the successive extensions by L do not stop.

8 Appendix: Correction to [Kaw18b]

We correct an error in [Kaw18b]. In [Kaw18b, Example 5.7], it is claimed that a collection $(\mathcal{O}_X(-d), G, \mathcal{O}_X)$ yields a semi-orthogonal decomposition of $D^b(\text{coh}(X))$. But it is not the case because the semi-orthogonality fails: $R\,\text{Hom}(G, \mathcal{O}_X(-d)) \neq 0$.

The correct collection is $(G(-d), \mathcal{O}_X(-d), \mathcal{O}_X)$. Then $D^b(\text{coh}(X)) \cong \langle D^b(R), D^b(k), D^b(k)\rangle$. It remains to prove the semi-orthogonality. But we have $R\,\text{Hom}(\mathcal{O}_X(-d), G(-d)) \cong RH(X, G) = 0$.

There is more information in [Kuz17, KKS18].

References

[Bei78] A. A. Beilinson. Coherent sheaves on $\mathbf{P}^n$ and problems in linear algebra. *Funktsional. Anal. i Prilozhen.*, 12(3):68–69, 1978.

[BN93] Marcel Bökstedt and Amnon Neeman. Homotopy limits in triangulated categories. *Compositio Math.*, 86(2):209–234, 1993.

[BO95] A. Bondal and D. Orlov. Semiorthogonal decompositions for algebraic varieties. 1995.

[Bon89] A. I. Bondal. Representations of associative algebras and coherent sheaves. *Izv. Akad. Nauk SSSR Ser. Mat.*, 53(1):25–44, 1989.

[BvdB03] A. Bondal and M. van den Bergh. Generators and representability of functors in commutative and noncommutative geometry. *Mosc. Math. J.*, 3(1):1–36, 258, 2003.

[CKN01] J. Daniel Christensen, Bernhard Keller, and Amnon Neeman. Failure of Brown representability in derived categories. *Topology*, 40(6):1339–1361, 2001.

[Eis80] David Eisenbud. Homological algebra on a complete intersection, with an application to group representations. *Trans. Amer. Math. Soc.*, 260(1):35–64, 1980.

[Kap88] M. M. Kapranov. On the derived categories of coherent sheaves on some homogeneous spaces. *Invent. Math.*, 92(3):479–508, 1988.

[Kaw02] Yujiro Kawamata. D-equivalence and K-equivalence. *J. Differential Geom.*, 61(1):147–171, 2002.

[Kaw18a] Yujiro Kawamata. Birational geometry and derived categories. *Surveys in differential geometry 2017. Celebrating the 50th anniversary of the Journal of Differential Geometry*, volume 22 of *Surv. Differ. Geom.*, pages 291–317. Int. Press, Somerville, MA, 2018.

[Kaw18b] Yujiro Kawamata. On multi-pointed non-commutative deformations and Calabi-Yau threefolds. *Compos. Math.*, 154(9):1815–1842, 2018.

[KKS18] J. Karmazyn, A. Kuznetsov, and E. Shinder. Derived categories of singular surfaces. 2018.

[KMM87] Yujiro Kawamata, Katsumi Matsuda, and Kenji Matsuki. Introduction to the minimal model problem. *Algebraic geometry, Sendai, 1985*, volume 10 of *Adv. Stud. Pure Math.*, pages 283–360. North-Holland, Amsterdam, 1987.

[Kuz17] A. Kuznetsov. Derived categories of families of sextic del Pezzo surfaces. 2017.

[Nee92] Amnon Neeman. The connection between the K-theory localization theorem of Thomason, Trobaugh and Yao and the smashing subcategories of Bousfield and Ravenel. *Ann. Sci. École Norm. Sup. (4)*, 25(5):547–566, 1992.

[Nee96] Amnon Neeman. The Grothendieck duality theorem via Bousfield's techniques and Brown representability. *J. Amer. Math. Soc.*, 9(1):205–236, 1996.

[Orl04] D. O. Orlov. Triangulated categories of singularities and D-branes in Landau-Ginzburg models. *Tr. Mat. Inst. Steklova*, 246(Algebr. Geom. Metody, Svyazi i Prilozh.):240–262, 2004.

[Orl11] Dmitri Orlov. Formal completions and idempotent completions of triangulated categories of singularities. *Adv. Math.*, 226(1):206–217, 2011.

[Ott88] Giorgio Ottaviani. Spinor bundles on quadrics. *Trans. Amer. Math. Soc.*, 307(1):301–316, 1988.

[TU10] Yukinobu Toda and Hokuto Uehara. Tilting generators via ample line bundles. *Adv. Math.*, 223(1):1–29, 2010.

Duality and Normalization, Variations on a Theme of Serre and Reid

János Kollár
with an Appendix by Hailong Dao

To Miles Reid

Abstract

We discuss the naive duality theory of coherent, torsion free, S_2 sheaves on schemes.

1 Introduction

On a normal algebraic variety the most important coherent sheaf is its canonical sheaf, also called dualizing sheaf. One can define a dualizing sheaf on more general schemes, but frequently the definitions show neither that the object is canonical nor that it has anything to do with duality; see for example, [BH93, 3.3.1] or [Sta18, Tag 0A7B]. I started to contemplate this while re-reading [Ser59, Chapter IV] and [Rei94, §§2–3]. These notes are the result of a subsequent attempt to generalize the naive duality of reflexive sheaves from normal varieties to schemes. As the reader will see, the essential ideas are in the works of my predecessors, especially [Ser59, Rei94, Har07]. However, I hope that the formulation and the generality of some of the results may be new and of interest.

On a normal scheme X the reflexive hull of a coherent sheaf F is given by

$$F^{**} := \mathcal{H}om_X\big(\mathcal{H}om_X(F, \mathcal{O}_X), \mathcal{O}_X\big).$$

While this definition makes sense over any integral scheme (see [Sta18, Tag 0AUY] and [Sta18, Tag 0AVT]), it does not seem to have many of the good properties of the normal case.

On a normal scheme a coherent sheaf is reflexive iff it is torsion free and S_2; see Definitions 2.5–2.7. Our claim is that, on non-normal schemes, the correct analogs of reflexive sheaves and reflexive hulls are

- torsion free, S_2 sheaves, abbreviated as TfS_2 sheaves, and
- torsion free, S_2-hulls, abbreviated as TfS_2-hulls.

Our first result, proved at the end of Section 2, says that the theory of torsion free and S_2 sheaves on noetherian schemes can be reduced to the study of S_2 schemes; see [HH94, Section 2] for related local results.

Theorem 1.1. *Let X be a noetherian scheme. Then there is a unique noetherian, S_2 scheme X^H and a finite morphism $\pi\colon X^H \to X$ such that $F \mapsto \pi_*F$ establishes an equivalence between the categories of coherent, torsion free, S_2 sheaves on X^H and coherent, torsion free, S_2 sheaves on X.*

This X^H is called the *torsion free, S_2-hull* or *TfS_2-hull* of X.

Note that $\pi\colon X^H \to X$ need not be surjective. For example, if X is of finite type over a field then π is birational if X is pure dimensional. Otherwise, π maps birationally onto the union of those irreducible components of X that do not intersect any larger dimensional irreducible component. The same holds for excellent schemes. There are, however, 2-dimensional, noetherian, integral schemes X where the sole coherent, torsion free, S_2 sheaf is the zero sheaf; see (4.8.2). For these $X^H = \emptyset$.

In general, the most useful dualizing object on a scheme is Grothendieck's dualizing complex [Sta18, Tag 0A7B]. However, the existence of a dualizing complex is a difficult question in general and for our purposes it is an overkill. If X has a dualizing complex $\omega_X^{\bullet}$ then $\tilde{\omega}_X := \mathcal{H}^{-\dim X}(\omega_X^{\bullet})$ is a TfS_2-dualizing sheaf, but it turns out that if one aims to get duality only for torsion free S_2 sheaves, then the required "dualizing sheaf" exists in greater generality. The following is a special case; a necessary and sufficient condition is given in 4.6.

Theorem 1.2. *Let X be a noetherian, S_2 scheme such that the normalization of its underlying reduced scheme $\pi\colon \bar{X} \to \operatorname{red} X$ is finite. Then there is a coherent, torsion free, S_2 sheaf $\tilde{\omega}_X$ such that the torsion free, S_2-hull (see 2.7) of any coherent sheaf F is given by*

$$F^H := \mathcal{H}om_X\big(\mathcal{H}om_X(F, \tilde{\omega}_X), \tilde{\omega}_X\big).$$

*Such an $\tilde{\omega}_X$ is called a TfS_2-*dualizing sheaf *on X.*

In particular, if F itself is torsion free and S_2 then

$$F = \mathcal{H}om_X\big(\mathcal{H}om_X(F, \tilde{\omega}_X), \tilde{\omega}_X\big).$$

The finiteness of $\pi\colon \bar{X} \to \operatorname{red} X$ is called condition N-1; this is the minimal assumption needed for Theorems 1.3 and 1.4 to make sense. If X is excellent, or universally Japanese, or Nagata then this condition holds; see [Sta18, Tag 0BI1] and [Sta18, Tag 033R] for the definitions and their basic properties.

In the literature, duality is usually stated either in the derived category of coherent sheaves as in [Sta18, Tag 0A7C], or for maximal CM modules over CM rings as in [BH93, 3.3.10] and [Eis95, Section 2.19].

After establishing these results, we revisit the '$n_Q = 2\delta_Q$ theorem' of [Ser59, Section IV.11] (who credits earlier works of Severi, Kodaira, Samuel and Gorenstein) and [Rei94, Section 3] (who credits Serre). Although our statements are more general than the usual forms, the gist of the proof is classical.

Theorem 1.3. *Let X be a noetherian, reduced, S_2 scheme. Assume that the normalization $\pi\colon \bar{X} \to X$ is finite with conductors $D \subset X$ and $\bar{D} \subset \bar{X}$ (see 5.1). Then*

(1) $\pi_*[\bar{D}] \geq 2[D]$ and
(2) equality holds iff the semilocal ring $\mathcal{O}_{D,X}$ is Gorenstein.

The proof is given in Section 5 where we also establish the following characterization of seminormal schemes.

Theorem 1.4. *Let X be a noetherian, reduced, S_2 scheme whose normalization $\pi\colon \bar{X} \to X$ is finite with conductors $D \subset X$ and $\bar{D} \subset \bar{X}$. Let $\tilde{\omega}_X$ be a TfS_2-dualizing sheaf on X and let*

$$\tilde{\omega}_{\bar{X}} := \pi^! \tilde{\omega}_X = \mathcal{H}om_X\big(\pi_*\mathcal{O}_{\bar{X}}, \tilde{\omega}_X\big)$$

be the corresponding TfS_2-dualizing sheaf on $\bar{X}$ (see 5.3). The following are equivalent.

(1) X is seminormal.
(2) $\bar{D}$ is reduced.
(3) $\tilde{\omega}_X \subset \pi_*\tilde{\omega}_{\bar{X}}(\operatorname{red} \bar{D})$.

In both of these results, the key step is to understand the dualizing sheaf of a one-dimensional scheme. As a byproduct, we prove the following in Section 6.

Proposition 1.5. *Let X be a noetherian, one-dimensional, S_1 scheme. Then X has a dualizing sheaf iff the following hold.*

(1) The local ring $\mathcal{O}_{x,X}$ has a dualizing module for every point $x \in X$.
(2) There is an open and dense subset $U \subset X$ such that $\operatorname{red} U$ *is Gorenstein.*

The standard definition of a dualizing module ω_R over a CM local ring (R, m) (see, for instance, [BH93, page 107]) requires the vanishing of $\operatorname{Ext}^i_R(R/m, \omega_R)$ for all $i \neq \dim R$. In Section 7, we discuss how to get by if we know vanishing only for $i < \dim R$, or even without any vanishing. See 7.5 and 7.6 for the complete statements.

1.6 (CM-dualizing sheaf). Let X be a CM scheme. If M is a torsion free CM sheaf then so is $\mathcal{H}om_X(M, \omega_X)$, see for example [BH93, 3.3.10]. The Appendix by Hailong Dao shows that if $\dim X \geq 3$ then ω_X is essentially the only coherent sheaf with this property.

This is also related to a question posed by Hochster in a lecture in 1972 whether the set $\{L \in \operatorname{Cl}(X) \colon L \text{ is CM}\}$ finite?

For cones this is proved in [Kar09, Theorem 6.11]; we outline a more geometric argument in 7.15. The case of three-dimensional isolated hypersurface singularities is settled in [DK16, Cor.4.8]; various special examples were treated earlier by [Knö87] and [EP03].

Assumptions. Unless otherwise specified, from now on all schemes and rings are assumed to be noetherian.

2 Torsion Free, S_2 Sheaves

We recall some well-known definitions and results; see [Kol13b, 2.58–63] or [Sta18, Tag 033P] for details.

Definition 2.1. A coherent sheaf F satisfies Serre's propery S_1 if the following equivalent conditions hold.

(1) $\operatorname{depth}_x F_x \geq \min\{1, \operatorname{codim}(x \in \operatorname{supp} F)\}$ for every $x \in X$.
(2) F has no embedded associated primes.
(3) If $\operatorname{codim}(x \in \operatorname{supp} F) \geq 1$ then $H^0_x(F) = H^0_x(F_x) = 0$.

Definition 2.2. A coherent sheaf F satisfies Serre's propery S_2 if the following equivalent conditions hold.

(1) $\operatorname{depth}_x F_x \geq \min\{2, \operatorname{codim}(x \in \operatorname{supp} F)\}$ for every $x \in X$.
(2) F is S_1 and $F|_D$ is also S_1 whenever $D \subset U \subset X$ is a Cartier divisor in an open subset U and D does not contain any associated prime of F.
(3) If $\operatorname{codim}(x \in \operatorname{supp} F) \geq 1$ then $H^0_x(F_x) = 0$. If $\operatorname{codim}(x \in \operatorname{supp} F) \geq 2$ then $H^1_x(F_x) = 0$.
(4) An exact sequence $0 \to F \to F' \to Q \to 0$ splits whenever the intersection $\operatorname{supp} Q \cap \operatorname{supp} F$ has codimension ≥ 2 in $\operatorname{supp} F$.
(5) F is S_1 and for every open $U \subset X$ and every closed $Z \subset U$ of codimension ≥ 2, the restriction map $H^0(U, F|_U) \to H^0(U \setminus Z, F|_{U \setminus Z})$ is an isomorphism.

The following is easiest to prove using 2.2(*3*).

Lemma 2.3. *Let $0 \to F_1 \to F_2 \to F_3 \to 0$ be an exact sequence of coherent sheaves.*

(1) If F_1, F_3 are S_2, so is F_2.
(2) Assume that F_2 is S_2 and if $Z \subset \operatorname{supp} F_1$ has codimension ≥ 2 in $\operatorname{supp} F_1$ then Z is not an associated point of F_3. Then F_1 is S_2. □

We need the following form of Grothendieck's dévissage, see for instance [Sta18, Tag 01YC] or [Kol13a, Section 10.3].

Lemma 2.4. *Let F be a coherent sheaf with $X_j \subset X$ the associated subschemes. Then F has an increasing filtration $0 = F_0 \subset F_1 \subset \cdots$ such that each F_{i+1}/F_i is a coherent, torsion free, rank 1 sheaf over some X_j. Moreover, we can choose $\operatorname{supp}(F_1/F_0)$ arbitrarily and if F is S_2 then the F_{i+1}/F_i are also S_2.* □

Definition 2.5 (Torsion free sheaves). A coherent sheaf T on X is called a *torsion sheaf* if $\operatorname{supp} T$ is nowhere dense in X. Every coherent sheaf F has a largest torsion subsheaf, denoted by $\mathrm{t}(F)$. F is called *torsion free* if $\mathrm{t}(F) = 0$, equivalently, if every associated point of F is a generic point of X.

Warning. Many authors define torsion and torsion free only over integral schemes, see for example [Sta18, Tag 0549] and [Sta18, Tag 0AVR].

Definition 2.6 (*TfS*$_2$ sheaves). We will be especially interested in coherent sheaves that are torsion free and S_2, abbreviated as *TfS*$_2$. Note that if X is normal, these are exactly the *reflexive sheaves* on X. We claim that, on non-normal schemes, *TfS*$_2$ sheaves are the correct analogs of reflexive sheaves.

Definition 2.7 (Torsion free S_2-hull). Let F be a coherent sheaf on X. Its *torsion free S_2-hull* or *TfS_2-hull* is a coherent sheaf F^H with map $q\colon F \to F^H$ such that

(1) F^H is TfS_2,
(2) $\ker q = \mathrm{t}(F)$,
(3) $\operatorname{supp} F^H = \operatorname{supp}\bigl(F/\mathrm{t}(F)\bigr)$ and
(4) $\operatorname{codim}_X \operatorname{supp}(F^H/F) \geq 2$.

It is clear that a torsion free S_2-hull is unique and its existence is a local question on X. Furthermore, $F^H = \bigl(F/\mathrm{t}(F)\bigr)^H$.

Warning. If X is pure dimensional and F is torsion free, then this agrees with every notion of hull that I know of. If F is torsion, then by our definition $F^H = 0$.

Another notion of S_2-hull is defined in [Kol08, Kol17] where TfS_2 in (*1*) is replaced by S_2 and $\mathrm{t}(F)$ in (*2–3*) is replaced by $\mathrm{Emb}(F)$.

I could not think of a better notation for the torsion free S_2-hull than F^H, though this is the same as used in [Kol17, Kol13a]. F^h looks like the Henselization, F^{**} or $F^{[**]}$ like the reflexive hull and F^{TfS_2H} is way too cumbersome.

Example 2.8. The following examples show that in some cases there are fewer coherent, S_2 sheaves than one might expect.

(2.8.1) Let $X \subset \mathbb{A}^3$ be the union of a plane P and a line L meeting at a point p. Let G be a nonzero, coherent, S_2 sheaf on X. Then $\operatorname{supp} G$ is the union of P and of finitely many points in $L \setminus \{p\}$.

Indeed, let $G_L \subset G$ be the largest subsheaf supported on L. If G is S_2 then $H^1_p(G) = 0$, hence $H^1_p(G_L) = 0$. But p has codimension 1 in L, so

$$H^1_p(G_L) = H^0\bigl(L \setminus \{p\}, G_L|_{L\setminus\{p\}}\bigr)/H^0\bigl(L, G_L\bigr)$$

is infinite, unless $p \notin \operatorname{supp} G_L$. In particular, there is no coherent S_2 or TfS_2 sheaf on X whose support equals X.

(2.8.2) Let $\pi\colon \bar{X} \to X$ denote the normalization of the previous example. Then $\mathcal{O}_{\bar{X}}$ is S_2 and TfS_2 but $\pi_*\mathcal{O}_{\bar{X}}$ is neither.

(2.8.3) Set $K := k(x_i\colon i \in \mathbb{N})$. In $\mathbb{A}^2_K$ with coordinates y_1, y_2 let p_1 be the point $(1,1)$ and p_2 the generic point of $(y_2 = 0)$. Then $k(p_1) \cong$

$K = k(x_i \colon i \in \mathbb{N})$ and $k(p_2) \cong K(y_1) = k(y_1, x_i \colon i \in \mathbb{N})$. Choose an isomorphism $\phi \colon k(p_1) \cong k(p_2)$ and set

$$R := \{f \in K[y_1, y_2] \colon \phi(f(p_1)) = f(p_2)\}.$$

Set $X := \operatorname{Spec}_k R$. Then $\mathbb{A}^2_K$ is the normalization of X, the normalization map $\mathbb{A}^2_K \to X$ is finite but there is no coherent S_2 or TfS_2 sheaf on X whose support equals X. Note that ϕ is not K-linear, so X is not a K-scheme, not even of finite type over any field.

2.9 (Construction of TfS_2-hulls). There are several well-known constructions of hulls.

(2.9.1) Let F be a coherent sheaf on X. Assume that there is a dense, open subset $j \colon U \subset X$ such that $F|_U$ is TfS_2 and $\operatorname{codim}_X(X \setminus U) \geq 2$. Then F has a TfS_2-hull iff $j_*(F|_U)$ is coherent, and then it is the TfS_2-hull of F. This follows from 2.2(*5*).

(2.9.2) Let G be a coherent TfS_2 sheaf on X and $F \subset G$ a coherent subsheaf. Then F has a TfS_2-hull and, using 2.3(*2*), it can be constructed as follows.

Let $T \subset F/G$ denote the largest subsheaf whose support has codimension ≥ 2 in X. Then F^H is the preimage of T in G.

(2.9.3) Let F be a coherent sheaf on X. Assume that there is a dense, open subset $U \subset X$ such that $F|_U$ is TfS_2. Then F^H can be constructed as follows.

As we noted in 2.7, we may assume that X is affine. Let g be an equation of $X \setminus U$. Let $W \subset X \setminus U$ denote the closure of the union of those associated points of F/gF that have codimension ≥ 2 in X. Let $j \colon X \setminus W \hookrightarrow X$ be the open embedding. Then $F|_{X \setminus W}$ is TfS_2 and F has a TfS_2-hull iff $F^H := j_*(F|_{X \setminus W})$ is coherent, and then it is the TfS_2-hull of F.

(2.9.4) Let $0 \to F_1 \to F_2 \to F_3$ be an exact sequence of coherent sheaves. If F_1, F_3 have a TfS_2-hull then so does F_2.

To see this note that, after killing the torsion parts, by 2.7(*4*) there is a dense, open subset $j \colon U \subset X$ such that $F_1|_U, F_3|_U$ are TfS_2 and $\operatorname{codim}_X(X \setminus U) \geq 2$. Using (2.9.1), we get an exact sequence

$$0 \to F_1^H \to F_2^H \to F_3^H.$$

Before stating the main equivalence theorem, we need a definition.

Definition 2.10. A scheme X is *formally* S_2 at a point $x \in X$ iff the completion $\hat{X}_x$ does not have associated primes of dimension ≤ 1. (This is probably not the best terminology but it is at least short. If $\hat{X}_x$ is S_2 then it is also formally S_2 (see the argument in (2.8.1)) but the converse does not hold.)

Theorem 2.11. *For a noetherian scheme X the following are equivalent.*

(1) Every coherent sheaf has a TfS_2-hull.

(2) $\mathcal{O}_X$ has a TfS_2-hull.

(3) $\mathcal{O}_{\mathrm{red}\,X}$ has a TfS_2-hull.

(4) There is a finite, surjective morphisms $p\colon Y \to X$ such that Y is S_2 and if $\operatorname{codim}_Y y = 0$ (respectively $= 1$) then $\operatorname{codim}_X p(y) = 0$ (respectively $= 1$).

(5) There is a TfS_2 sheaf whose support equals X.

(6) The TfS_2 locus of any coherent sheaf F contains an open dense set and $\mathrm{red}\,X$ is formally S_2 at every $x \in X$ of codimension ≥ 2.

(7) The TfS_2 locus of any S_1 coherent sheaf F is an open, dense set whose complement $Z(F)$ has codimension ≥ 2 and $\mathrm{red}\,X$ is formally S_2 at every $x \in Z(F)$.

(8) The S_2 locus of X contains an open dense set and $\mathrm{red}\,X$ is formally S_2 at every $x \in X$ of codimension ≥ 2.

(9) The S_2 locus of $\mathrm{red}\,X$ is an open, dense set whose complement $Z(\mathrm{red}\,X)$ has codimension ≥ 2 and $\mathrm{red}\,X$ is formally S_2 at every $x \in Z(\mathrm{red}\,X)$.

Proof It is clear that (*1*) implies (*2*) and (*3*). For (*4*) one can then take either $Y := \operatorname{Spec}_X \mathcal{O}_X^H$ or $Y := \operatorname{Spec}_X \mathcal{O}_{\mathrm{red}\,X}^H$. If (*4*) holds then $p_*\mathcal{O}_Y$ is TfS_2 by 2.12 and we get (*5*).

If F has an TfS_2-hull $q\colon F \to F^H$ then $Z(F) = \operatorname{supp}(F^H/q(F))$, thus (*1*) and 2.14 imply (*6–9*).

The claims (*6–9*) are local, so we may assume that X is affine. Let F be a coherent, S_1 sheaf and $U \subset X$ an open, dense subset such that $F|_U$ is TfS_2. Pick $x \in X \setminus U$ and let g be a local equation of $X \setminus U$. Then F is not S_2 at x iff x is an asociated prime of F/gF. There are only finitely many such points, let $W \subset X$ be the union of their closures. Then the S_2-locus of F is $X \setminus W$, hence (*6*) implies (*7*). Similarly (*8*) implies (*9*) and (*6*) $\Rightarrow$ (*8*) and (*7*) $\Rightarrow$ (*9*) are clear.

If (*9*) holds then let U denote the S_2 locus and $j\colon U \hookrightarrow X$ the natural embedding. Then $\mathcal{O}_X^H = j_*(\mathcal{O}_U)$ is coherent by 2.14, hence (*9*) implies (*2*).

It remains to prove that (*5*) implies (*1*). Let X_i be the irreducible components of X. By 2.4 every $\operatorname{red} X_i$ supports a rank 1 *TfS*$_2$ sheaf L_i. We claim that every rank 1 sheaf M_i on X_i has a *TfS*$_2$-hull. To see this, cover $\operatorname{red} X_i$ with open affine subsets U_{ik}. For every k, we can realize $M_i|_{U_{ik}}$ as a subsheaf of $L_i|_{U_{ik}}$. Thus $M_i|_{U_{ik}}$ has a *TfS*$_2$-hull by (2.9.2) and so does M_i. Finally, using 2.4 and (2.9.4), we obtain that every torsion free coherent sheaf has a *TfS*$_2$-hull, and so does every coherent sheaf by 2.7. □

Lemma 2.12. *Let $p\colon X \to Y$ be a finite morphism and F a coherent, TfS$_2$ sheaf on X. The following are equivalent.*

(1) The push-forward p_*F is also a TfS$_2$ sheaf on Y.

(2) For every $x \in \operatorname{supp} F$, if $\operatorname{codim}_X x = 0$ (respectively $= 1$) then $\operatorname{codim}_Y p(x) = 0$ (respectively $= 1$). □

Corollary 2.13. *Let $p\colon X \to Y$ be a finite morphism that maps generic points to generic points. If Y satisfies the conditions of 2.11, then p satisfies 2.12(2).*

Proof The codimension 0 case holds by the birationality assumption. Assume that $\operatorname{codim}_X x = 1$. Then X is a 1-dimensional irreducible component of the local scheme $Y_{p(x)}$ and the same holds after completion. Thus 2.11(*4*) implies that $\dim Y_{p(x)} = 1$ hence $\operatorname{codim}_Y p(x) = 1$. □

We have used the following result on the coherence of push-forwards, which is a sharpening of [Gro60, IV.5.11.1].

Proposition 2.14. *[Kol17, Theorem 2] Let X be a scheme, $Z \subset X$ a closed subset of codimension ≥ 2 and $j\colon X \setminus Z \hookrightarrow X$ the open embedding. Assume that there is a TfS$_2$ sheaf on $X \setminus Z$ whose support is $X \setminus Z$. The following are equivalent.*

(1) j_*F is coherent for every TfS$_2$ sheaf F on $X \setminus Z$.

(2) There is a TfS$_2$ sheaf on X whose support is X.

(3) X is formally S_2 at every $x \in Z$. □

Corollary 2.15. *Let X be a scheme satisfying the conditions 2.11. Let $j\colon U \hookrightarrow X$ be an open subscheme and G a TfS$_2$ sheaf on U.*

(1) G can be extended to a TfS$_2$ sheaf on X.

(2) If $\operatorname{codim}_X(X \setminus U) \geq 2$ *then* j_*G *is the unique* TfS_2 *extension.*

Proof Let G_X be any coherent extension, then $(G_X)^H$ is a TfS_2 extension. (*2*) was already noted in (2.9.1). □

Definition 2.16 (S_2-hull of a scheme). Let X be a Noetherian scheme. Let $\{x_j : j \in J\}$ be the set of those associated points of $\mathcal{O}_X$ that are not associated points of any TfS_2 sheaf on X. Let $I_X \subset \mathcal{O}_X$ be the largest ideal sheaf whose associated points are $\{x_j : j \in J\}$.

Let $X_1 \subset X$ be the subscheme defined by I_X. Then $X^H := \operatorname{Spec}_X \mathcal{O}^H_{X_1}$ is called the *TfS*$_2$*-hull* of X. By construction, X^H is S_2 and the natural map $\pi : X^H \to X_1$ is finite and birational. We prove next that X^H is the TfS_2-hull as required by Theorem 1.1.

Warning. The construction of X^H is local on X_1 but not on X. If $U \subset X$ is an open subset that is disjoint from X_1 then there may well be a nontrivial TfS_2 sheaf on U that does not extend to a TfS_2 sheaf on X.

Proof of Theorem 1.1 Let X be a noetherian scheme and $X_1 \subset X$ as in 2.16. Let F be a TfS_2 sheaf on X. Then $I_X \cdot F = 0$, thus every TfS_2 sheaf on X is the push forward of a TfS_2 sheaf on X_1. After replacing X by X_1 we may as well assume that X satisfies the equivalent conditions 2.11(*1–9*). The theorem then amounts to saying that every TfS_2 sheaf on X has a natural structure as an $\mathcal{O}^H_X$-module. This is a local question, so we may assume that X is affine. Let $s \in H^0(X, \mathcal{O}^H_X)$. By 2.7(*4*) and 2.15(*2*) there is open subset $U \subset X$ such that $X \setminus U$ has codimension ≥ 2 and $H^0(U, \mathcal{O}_U) = H^0(X, \mathcal{O}^H_X)$. Let F be a TfS_2 sheaf on X. Given $s \in H^0(X, \mathcal{O}^H_X)$ and $\sigma \in H^0(X, F)$, the product $(s|_U) \cdot (\sigma|_U)$ is a section of $H^0(U, F_U)$. By 2.2(*5*), it uniquely extends to a section $s(\sigma) \in H^0(X, F)$. This defines the $\mathcal{O}^H_X$-module structure on F. □

3 Duality for Torsion Free, S_2 Sheaves

Definition 3.1. Let X be a scheme. A *TfS*$_2$*-dualizing sheaf* on X is a TfS_2 sheaf $\tilde{\omega}_X$ such that for every TfS_2 sheaf F the natural map

$$j_F : F \to \mathcal{H}om_X\big(\mathcal{H}om_X(F, \tilde{\omega}_X), \tilde{\omega}_X\big) \tag{3.1}$$

is an isomorphism. If $X = \operatorname{Spec} R$ is affine and $\tilde{\omega}_X$ is the sheaf corresponding to an R-module $\tilde{\omega}_R$, then the latter is called a *TfS*$_2$*-dualizing module* of R.

Another way is to define duality as an anti-equivalence $F \mapsto D(F)$ of the category of TfS_2-sheaves such that $D(D(F)) = F$; for this approach, see [Eis95, §21.1]. Then $\mathcal{H}om_X(F_1, F_2) = \mathcal{H}om_X\big(D(F_2), D(F_1)\big)$. In particular

$$\mathcal{H}om_X\big(F, D(\mathcal{O}_X)\big) = \mathcal{H}om_X\big(\mathcal{O}_X, D(F)\big) = D(F).$$

Thus $D(\mathcal{O}_X)$ is a TfS_2-dualizing sheaf on X. Conversely, if $\tilde{\omega}_X$ is TfS_2-dualizing then $D(F) := \mathcal{H}om_X(F, \tilde{\omega}_X)$ is a duality on the category of TfS_2-sheaves.

By this definition, if the 0 sheaf is the only TfS_2 sheaf on X then it is also a TfS_2-dualizing sheaf. This causes uninteresting exceptions in several statements. Thus we focus on S_2 schemes from now on. The following observation shows that this is not a restriction on the generality.

3.2. Let X be a scheme and $\pi\colon X^H \to X$ its S_2-hull. If X^H has a TfS_2-dualizing sheaf $\tilde{\omega}_{X^H}$ then, as an immediate consequence of Theorem 1.1 we get that

$$\tilde{\omega}_X := \pi_* \tilde{\omega}_{X^H} \tag{3.2}$$

is a TfS_2-dualizing sheaf on X. Thus in studying TfS_2-duality, we may as well restrict ourselves to S_2 schemes. If X is S_2 then applying (3.1) to $F = \mathcal{O}_X$ gives that

$$\mathcal{H}om_X(\tilde{\omega}_X, \tilde{\omega}_X) \cong \mathcal{O}_X. \tag{3.3}$$

A map of finite modules over a local ring is an isomorphism iff its completion is an isomorphism. Thus, once we have a candidate for a TfS_2-dualizing module, we can check it formally.

Lemma 3.3. *Let (R, m) be a local ring and M a finite R-module. If $\hat{M}$ is a TfS_2-dualizing module over $\hat{R}$ then M is a TfS_2-dualizing module over R.* □

The converse of 3.3 holds if $\dim R = 1$ by 3.13 and 7.4. There are, however, 2-dimensional normal rings R whose completions are not Gorenstein at their generic points [FR70]. For these R is a TfS_2-dualizing module over R by (3.5.1), but $\hat{R}$ is not a TfS_2-dualizing module over $\hat{R}$ by 3.12. This is in marked contrast with the dualizing complex, which is preserved by completion [Sta18, Tag 0DWD].

Lemma 3.4. *Let X be an S_2 scheme. A coherent, TfS_2 sheaf $\tilde{\omega}_X$ is TfS_2-dualizing over X iff its pull-back to the localization X_x is TfS_2-dualizing for all points $x \in X$ of codimension ≤ 1.*

Proof By 2.2(5) a map between coherent *TfS*$_2$ sheaves is an isomorphism iff it is an isomorphism at all points of codimension ≤ 1. The converse is established in 3.7. □

Example 3.5. The basic examples are the following.

(3.5.1) If X is normal then $\mathcal{O}_X$ is a *TfS*$_2$-dualizing sheaf on X.

(3.5.2) If X is CM then a dualizing sheaf 7.4 is also a *TfS*$_2$-dualizing sheaf; this follows from [BH93, 3.3.10] and 3.4. More generally, if an arbitrary scheme X has a dualizing complex $\omega_X^{\bullet}$ then $\tilde{\omega}_X := \mathcal{H}^{-\dim X}(\omega_X^{\bullet})$ is a *TfS*$_2$-dualizing sheaf. This follows from [Sta18, Tag 0A7C]. While these are the main examples, we try to work out the theory of *TfS*$_2$-dualizing sheaves without using the general theory of dualizing complexes.

(3.5.3) Note that usually there are many non-isomorphic *TfS*$_2$-dualizing sheaves. If $\tilde{\omega}_X$ is a *TfS*$_2$-dualizing sheaf and L is invertible then $L \otimes \tilde{\omega}_X$ is also a *TfS*$_2$-dualizing sheaf since

$$
\begin{aligned}
&\mathcal{H}om_X\bigl(\mathcal{H}om_X(F, L\otimes\tilde{\omega}_X), L\otimes\tilde{\omega}_X\bigr)\\
&\quad= \mathcal{H}om_X\bigl(\mathcal{H}om_X(F,\tilde{\omega}_X)\otimes L, \tilde{\omega}_X\bigr)\otimes L\\
&\quad= \mathcal{H}om_X\bigl(\mathcal{H}om_X(F,\tilde{\omega}_X), \tilde{\omega}_X\bigr)\otimes L^{-1}\otimes L\\
&\quad= \mathcal{H}om_X\bigl(\mathcal{H}om_X(F,\tilde{\omega}_X), \tilde{\omega}_X\bigr).
\end{aligned}
$$

We will give a precise characterization in 4.2.

(3.5.4) We see in 4.3 that on a regular scheme, *TfS*$_2$-dualizing = invertible.

(3.5.5) [FR70] constructs a 1-dimensional integral scheme X over $\mathbb{C}$ that has no dualizing sheaf. By 3.13, this implies that X has no *TfS*$_2$-dualizing sheaf either.

(3.5.6) As far as I can tell, *TfS*$_2$-dualizing modules are not closely related to the semidualizing modules considered in [Fox72, SW07].

Lemma 3.6. *Let X be an S_2-scheme or, more generally, a scheme satisfying the conditions 2.11. Let $\tilde{\omega}_X$ be a TfS$_2$-dualizing sheaf and $j\colon U \to X$ an open embedding. Then $j^*\tilde{\omega}_X$ is a TfS$_2$-dualizing sheaf on U.*

Proof Let F_U be a *TfS*$_2$ sheaf on U. By 2.15, we can extend F_U to a *TfS*$_2$ sheaf F_X on X. Thus

$$\mathcal{H}om_U\big(\mathcal{H}om_U(F_U, j^*\tilde{\omega}_X), j^*\tilde{\omega}_X\big) \cong j^* \mathcal{H}om_X\big(\mathcal{H}om_X(F_X, \tilde{\omega}_X), \tilde{\omega}_X\big)$$
$$\cong j^* F_X \cong F_U. \qquad \square$$

By passing to the direct limit, we get the following consequence.

Corollary 3.7. *Let X be an S_2-scheme or, more generally, a scheme satisfying the conditions 2.11, with a TfS_2-dualizing sheaf $\tilde{\omega}_X$. Let $j\colon W \to X$ be a direct limit of open embeddings. Then $j^*\tilde{\omega}_X$ is a TfS_2-dualizing sheaf on W.* $\square$

For ease of reference, we recall the following from [Har77, Example.III.6.10].

Lemma 3.8. *Let $p\colon X \to Y$ be a finite morphism and G a coherent sheaf on Y. Then*

(1) The formula $p^!G := \mathcal{H}om_Y(p_*\mathcal{O}_X, G)$ gives a coherent sheaf on X.
(2) There is a trace map $\operatorname{tr}_{X/Y}\colon p_*(p^!G) \to G$ sending a section $\sigma \in H^0(X, p^!G) = \operatorname{Hom}_Y(p_*\mathcal{O}_X, G)$ to $\operatorname{tr}_{X/Y}(\sigma) := \sigma(1) \in H^0(Y, G)$.
(3) For any coherent sheaf F on X, there is a natural isomorphism

$$p_* \mathcal{H}om_X(F, p^!G) = \mathcal{H}om_Y(p_*F, G).$$

(4) For any coherent sheaf F on X there are natural maps

$$\phi_i\colon p_* \mathcal{E}xt^i_X(F, p^!G) \to \mathcal{E}xt^i_Y(p_*F, G).$$

(5) If p is also flat then the ϕ_i are isomorphisms. $\square$

Proposition 3.9. *Let $p\colon X \to Y$ be a finite morphism which satisfies 2.12(2). Let $\tilde{\omega}_Y$ be a TfS_2-dualizing sheaf on Y. Then $\tilde{\omega}_X := p^!\tilde{\omega}_Y$ is a TfS_2-dualizing sheaf on X.*

Proof Applying 3.8(*3*) twice we get that

$$p_* \mathcal{H}om_X\big(\mathcal{H}om_X(F, \tilde{\omega}_X), \tilde{\omega}_X\big) \cong \mathcal{H}om_Y\big(p_* \mathcal{H}om_X(F, \tilde{\omega}_X), \tilde{\omega}_Y\big)$$
$$\cong \mathcal{H}om_Y\big(\mathcal{H}om_Y(p_*F, \tilde{\omega}_Y), \tilde{\omega}_Y\big) \cong p_*F.$$

Since p is affine, this implies that $\mathcal{H}om_X\big(\mathcal{H}om_X(F, \tilde{\omega}_X), \tilde{\omega}_X\big) \cong F$. $\square$

Corollary 3.10. *Let X be a quasi-projective scheme. Then X has a TfS_2-dualizing sheaf.*

Proof By (3.2), we may assume that X is S_2 and then its connected components are pure dimensional by 3.11. If X is projective and pure dimensional, Noether's normalization theorem gives a finite morphism $p\colon X \to \mathbb{P}^{\dim X}$ that maps generic points to generic points. Thus X has a TfS_2-dualizing sheaf by (3.5.2) and 3.9. The quasi-projective case is now implied by 3.7. □

The following is a slightly stronger formulation of [Har62].

Lemma 3.11. *Let X be a connected, S_2 scheme with a dimension function. Then X is pure dimensional and connected in codimension 1.* □

Low Dimensions

Over Artin schemes every coherent sheaf is S_2 and TfS_2-duality is the same as Matlis duality; see, for instance, [BH93, Section 3.2] or [Eis95, Section 21.1].

Lemma 3.12. *Let (A, m) be an Artin, local ring, $k := A/m$ and Ω a finite A-module. The following are equivalent.*

(1) Ω is a TfS_2-dualizing module.
(2) $\Omega \cong E(k)$, the injective hull of k.
(3) $\operatorname{Hom}_A(k, \Omega) \cong k$ and $\operatorname{Ext}^1_A(k, \Omega) = 0$.
(4) $\operatorname{Hom}_A(k, \Omega) \cong k$ and $\operatorname{Ext}^i_A(k, \Omega) = 0$ for $i > 0$.
(5) $\operatorname{Hom}_A(k, \Omega) \cong k$ and $\operatorname{length}(\Omega) = \operatorname{length}(A)$.
(6) $\operatorname{Hom}_A(k, \Omega) \cong k$ and $\operatorname{length}(\Omega) \geq \operatorname{length}(A)$. □

Theorem 3.13. *Let (R, m) be a 1-dimensional, S_1 local ring, $k := R/m$ and Ω a finite R-module. The following are equivalent.*

(1) Ω is a TfS_2-dualizing module.
(2) $\operatorname{Ext}^i_R(k, \Omega) = \delta_{i,1} \cdot k$ for every i.

Note that (*2*) is one of the usual definitions of a dualizing module; see 7.4.

Proof Assume (*1*) and write $A^* := \operatorname{Hom}_R(A, \Omega)$. From any exact sequence of torsion free modules

$$0 \to A \to B \to C \to 0,$$

we get an $A^+ \subset A^*$ such that

$$0 \to C^* \to B^* \to A^+ \to 0$$

is exact. Applying duality again gives an exact sequence

$$0 \to (A^+)^* \to B \to C \to 0.$$

Thus $(A^+)^* = A$. Hence $A^+ = A^*$, which proves the following.

Claim 3.14. Duality sends exact sequences of torsion free modules to exact sequences of their duals. □

Thus if we have an exact sequence of torsion free modules

$$0 \to \Omega \to B \to C \to 0$$

then its dual is

$$0 \to C^* \to B^* \to R \to 0,$$

hence they both split. That is, $\mathrm{Ext}^1_R(C, \Omega) = 0$ for any torsion free module C. Chasing through a projective resolution of a module M we conclude the following.

Claim 3.15. Let M be a finitely generated R-module. Then for $i \geq 2$, $\mathrm{Ext}^i_R(M, \Omega) = 0$. If M is torsion free then $\mathrm{Ext}^i_R(M, \Omega) = 0$ for $i \geq 1$. □

Claim 3.16. Let $Q_1 \subset Q_2$ be torsion free R-modules, and suppose that $\mathrm{length}(Q_2/Q_1)$ is finite. Then $\mathrm{length}(Q_2/Q_1) = \mathrm{length}(Q_1^*/Q_2^*)$.

Proof $M \to M^*$ gives a one-to-one correspondence between intermediate modules $Q_1 \subset M \subset Q_2$ and $Q_2^* \subset M^* \subset Q_1^*$. □

Let T be a torsion R-module and write it as

$$0 \to K \to R^n \to T \to 0.$$

Duality gives

$$0 \to \Omega^n \to K^* \to \mathrm{Ext}^1(T, \tilde{\omega}) \to 0.$$

Combining with 3.16, we get that

$$\mathrm{length}\big(\mathrm{Ext}^1_R(T, \Omega)\big) = \mathrm{length}(T).$$

In particular, $\mathrm{Ext}^1_R(R/m, \Omega) \cong R/m$. This completes the proof of (*2*).

The implication (*2*) ⇒ (*1*) was already noted in (3.5.2). We did not find any shortcuts to the proofs given in the references and there is no point repeating what is there. □

Proposition 3.17. *Let X be a 1-dimensional, S_1 scheme and $\tilde{\omega}_X$ a TfS_2-dualizing sheaf. Then every TfS_2-dualizing sheaf is of the form $L \otimes \tilde{\omega}_X$ for some line bundle L.*

Proof This is a special case of [Sta18, Tag 0A7F] or [BH93, 3.3.4], but here is a direct proof using the above computations.

Let $\tilde{\Omega}_X$ be another TfS_2-dualizing sheaf, and set $L := \mathcal{H}om_X(\tilde{\omega}_X, \tilde{\Omega}_X)$. If the claim holds for all localizations of X then L is a line bundle by (3.3), and then $\tilde{\Omega}_X \cong L \otimes \tilde{\omega}_X$. Thus it is sufficient to prove the claim when X is local.

We keep * for $\tilde{\omega}_X$-duality. Write $\tilde{\Omega}_X^*$ as

$$0 \to K \to \mathcal{O}_X^n \to \tilde{\Omega}_X^* \to 0,$$

we get

$$0 \to \tilde{\Omega}_X \to \tilde{\omega}_X^n \to K^* \to 0.$$

Since $\tilde{\Omega}_X$ is *TfS*$_2$-dualizing, these sequences split by 3.15. So, $\tilde{\Omega}_X^* = \mathcal{H}om_X(\tilde{\Omega}_X, \tilde{\omega}_X)$ is projective, hence isomorphic to $\mathcal{O}_X$. Thus

$$\begin{aligned}\tilde{\Omega}_X \cong (\tilde{\Omega}_X^*)^* &\cong \mathcal{H}om_X\big(\mathcal{H}om_X(\tilde{\Omega}_X, \tilde{\omega}_X), \tilde{\omega}_X\big) \\ &\cong \mathcal{H}om_X(\mathcal{O}_X, \tilde{\omega}_X) \cong \tilde{\omega}_X. \qquad \square\end{aligned}$$

We have not yet discussed the existence of *TfS*$_2$-dualizing sheaves and modules. By 3.13, in the 1-dimensional case this is equivalent to the existence of dualizing sheaves. We recall the main results about dualizing sheaves on 1-dimensional schemes in Section 6.

4 Existence of *TfS*$_2$-Dualizing Sheaves

We start with the uniqueness question for *TfS*$_2$-dualizing sheaves and then prove the main existence theorem. At the end, we give a series of examples of noetherian rings without *TfS*$_2$-dualizing modules.

Definition 4.1. A coherent sheaf L on a scheme X is called *mostly invertible* if it is S_2 and there is an open subset $j: U \hookrightarrow X$ such that $\operatorname{codim}_X(X \setminus U) \geq 2$ and $L|_U$ is invertible. Equivalently, if L is invertible at all points of codimension ≤ 1. Thus L is also *TfS*$_2$.

If F is a *TfS*$_2$ sheaf on X then we set $L \hat{\otimes} F := j_*\big(L|_U \otimes F|_U\big)$. If L_1, L_2 are mostly invertible then so is $L_1 \hat{\otimes} L_2$ and $L_1^{[-1]} := j_*\big((L|_U)^{-1}\big)$.

Theorem 4.2. *Let X be an S_2 scheme with a TfS$_2$-dualizing sheaf $\tilde{\omega}_X$.*

(1) If L is mostly invertible then $L \hat{\otimes} \tilde{\omega}_X$ is also a TfS$_2$-dualizing sheaf.
(2) Every TfS$_2$-dualizing sheaf is obtained this way.

Proof By assumption there is an open subset $j: U \hookrightarrow X$ such that $\operatorname{codim}_X(X \setminus U) \geq 2$ and $L|_U$ is a line bundle. Thus

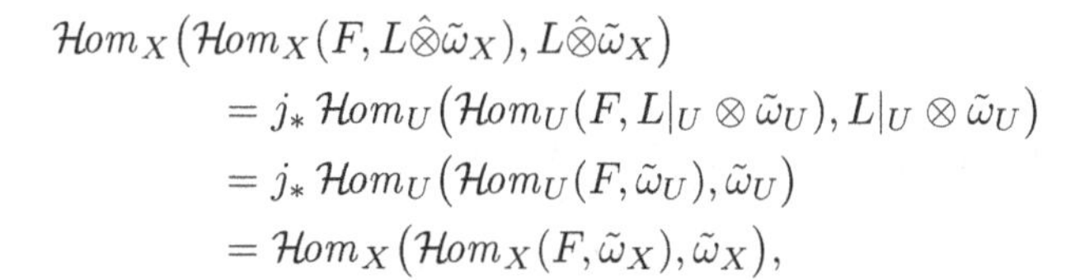

$$\begin{aligned}
&\mathcal{H}om_X\big(\mathcal{H}om_X(F, L\hat{\otimes}\tilde{\omega}_X), L\hat{\otimes}\tilde{\omega}_X\big)\\
&\qquad = j_*\,\mathcal{H}om_U\big(\mathcal{H}om_U(F, L|_U \otimes \tilde{\omega}_U), L|_U \otimes \tilde{\omega}_U\big)\\
&\qquad = j_*\,\mathcal{H}om_U\big(\mathcal{H}om_U(F, \tilde{\omega}_U), \tilde{\omega}_U\big)\\
&\qquad = \mathcal{H}om_X\big(\mathcal{H}om_X(F, \tilde{\omega}_X), \tilde{\omega}_X\big),
\end{aligned}$$

where we used (3.5.3) in the middle and 2.15(*2*) at the ends.

Let $\tilde{\Omega}_X$ be another *TfS*$_2$-dualizing sheaf and set $L := \mathcal{H}om_X(\tilde{\omega}_X, \tilde{\Omega}_X)$. If L is mostly invertible then $\tilde{\Omega}_X \cong L\hat{\otimes}\tilde{\omega}_X$. Thus we need to show that L is invertible at all points of codimension ≤ 1. This can be done after localization. The codimension 0 case follows from 3.12 and the codimension 1 case from 3.13. □

Corollary 4.3. *Let X be a regular scheme. Then a coherent sheaf is TfS$_2$-dualizing iff it is invertible.* □

Next we show that *TfS*$_2$-dualizing sheaves exist for Nagata schemes, in a formulation that is slightly more general than Theorem 1.2.

Theorem 4.4. *Let X be a noetherian scheme whose normalization $\pi\colon \bar{X} \to X$ is finite. Then X has a TfS$_2$-dualizing sheaf.*

Note that we do not assume that X is reduced, thus $\bar{X} = \overline{\operatorname{red} X}$.

Proof By 3.2 it is enough to prove this for the S_2-hull of X. Thus we may as well assume that X is S_2

By 4.5 there is a dense open subset $U \subset X$ with a *TfS*$_2$-dualizing sheaf $\tilde{\omega}_U$. Let $p_i \in X \setminus U$ be a generic point that has codimension one in X. By assumption the normalization $\pi\colon \bar{X}_{p_i} \to X_{p_i}$ is finite. If X is reduced, then X_{p_i} has a dualizing sheaf $\tilde{\omega}_i$ by 5.6. In the non-reduced case, we can use the more general 6.2.

For each generic point of X_{p_i}, the restrictions of $\tilde{\omega}_U$ and of $\tilde{\omega}_i$ are isomorphic. After fixing these isomorphisms, we can identify $\tilde{\omega}_i$ with a subsheaf of $(j_*\tilde{\omega}_U)_{p_i}$. We can now choose a coherent subsheaf $G \subset j_*\tilde{\omega}_U$ such that $G|_U = \tilde{\omega}_U$ and $G_{p_i} = \tilde{\omega}_i$ for every i. By 2.11, G has a *TfS*$_2$-hull G^H and $\tilde{\Omega}_X := G^H$ is a *TfS*$_2$-dualizing sheaf on X by 3.4. □

Lemma 4.5. *Let X be a scheme and $\tilde{\Omega}_X$ a coherent sheaf on X. Assume that* $\operatorname{red} X$ *is normal and $\tilde{\Omega}_{X_g}$ is dualizing for every generic point $g \in X$. Then there is a dense open subset $U \subset X$ such that $\tilde{\Omega}_X|_U$ is TfS$_2$-dualizing.*

Warning. A similar assertion does not hold for the dualizing sheaf; see (4.8.2).

Proof Set $U_1 := X \setminus \operatorname{supp}\big(\mathcal{E}xt^1_X(\mathcal{O}_{\operatorname{red} X}, \tilde{\Omega}_X)\big) \setminus \operatorname{supp}\big(\mathrm{t}(\mathcal{O}_X)\big)$ and then let $U \subset U_1$ be the open set where $\operatorname{socle}(\tilde{\Omega}_X)$ is locally free over $\operatorname{red} X$.

Let F be a coherent, TfS_2 sheaf on U. By assumption

$$j_F \colon F \to \mathcal{H}om_U\big(\mathcal{H}om_U(F, \tilde{\Omega}_U), \tilde{\Omega}_U\big)$$

is an isomorphism at the generic points. It remains to prove that it is also an isomorphism at codimension ≤ 1 points. By 3.4, this can be checked after localizing at codimension ≤ 1 points. At codimension 0 points we get a dualizing sheaf by assumption, at codimension 1 points we get a dualizing sheaf by 6.4. □

The following existence result is a direct generalization of Proposition 1.5; it can be proved exactly as 4.4 and the proof of 1.5 in Section 6.

Theorem 4.6. *Let X be an S_2 scheme. Then X has a TfS_2-dualizing sheaf iff the following hold.*

(1) $\mathcal{O}_{x,X}$ has a dualizing module for every codimension 1 point $x \in X$.
(2) There is an open and dense subset $U \subset X$ such that $\operatorname{red}(X_x)$ *is Gorenstein for every codimension 1 point $x \in U$.* □

Note that, by 6.1, assumption (*1*) can be reformulated as

(*1′*) For every codimension 1 point $x \in X$, the generic fibers of the map from the completion of the localization $\hat{X}_x \to X_x$ are Gorenstein.

Next we give some examples of schemes that do not have a TfS_2-dualizing sheaf, though all localizations have a dualizing sheaf. We use the following general construction, modeled on [Nag62, Appendix, Eg.1].

Proposition 4.7. *Let k be a field, I an arbitrary set and suppose $\{(R_i, m_i) \colon i \in I\}$ is a set of integral, essentially of finitely type k-algebras such that $k \cong R_i/m_i$. Then there is a noetherian, integral k-algebra $R = R(I, R_i, m_i)$ with the following properties.*

(1) The maximal ideals of R can be naturally indexed as $\{M_i \colon i \in I\}$.
(2) $R_{M_i} \cong R_i \otimes_k K_i'$ for some fields $K_i' \supset k$.
(3) Every nonzero ideal of F is contained in only finitely many maximal ideals.

Proof For any finite subset $J \subset I$ set $R_J := \bigotimes_{j \in J} R_j$. If $J_1 \subset J_2$ then using the natural injections $k \hookrightarrow R_j$, we get injections $R_{J_1} \hookrightarrow R_{J_2}$. Let R_I denote the direct limit of the $\{R_J\}$. Usually R_I is not noetherian. Somewhat sloppily we identify R_J with its image in R_I.

This also defines R_{I_1} for any subset $I_1 \subset I$. We use the notation $R'_J := R_{I\setminus J}$ and let K'_J be the quotient field of R'_J. Note that

$$R_I/(R_I m_i) \cong (R_i/m_i) \otimes_k R'_i,$$

hence the $R_I m_i$ are prime ideals. We obtain R from R_I by inverting every element in $R_I \setminus \cup_i R_I m_i$. Set $M_i := R m_i$. By construction the M_i are the maximal ideals of R.

Pick any $r \in R_I$. Then $r \in R_J$ for some finite subset $J \subset I$. If $J' \subset I$ is disjoint from J and $s \in R_{J'} \setminus \{0\}$ then $r+s$ is not in any $R_I m_i$, hence it is invertible in R. Since $r + s \equiv s \mod (r)R_I$ we see that

$$1 \equiv \frac{s}{r+s} \mod R(r).$$

Thus $R/(r) \cong \big(R_J/(r)\big) \otimes_k K'_J$. In particular, $R/(r)$ is Noetherian for every r and so is R. □

Example 4.8. Depending on the choice of the R_i in 4.7, we get many examples of noetherian domains with unexpected behavior.

(4.8.1) A 1-dimensional integral domain without a dualizing module. Pick an infinite set I and for $i \in I$ let R_i be the localization of $k[t^3, t^4, t^5]$ at the origin. Note that R_i is not Gorenstein. The resulting R has a dense set of non-Gorenstein points, so it does not have a TfS_2-dualizing module though all of its localizations at maximal ideals have one.

(4.8.2) A 2-dimensional normal ring without a dualizing module. Pick an infinite set I and for $i \in I$ let R_i be the localization of

$$S := \langle x^a y^b \colon 3 \mid a+b \rangle \subset k[x,y].$$

Note that S is also the ring of invariants $k[x,y]/\frac{1}{3}(1,1)$. Its dualizing module is not free, but isomorphic to the module

$$\omega_S \cong \langle x^a y^b \colon 3 \mid a+b-1 \rangle \subset k[x,y].$$

The resulting R has a dense set of non-Gorenstein points, so it does not have a dualizing module though all of its localizations at maximal ideals have one. By contrast, R has plenty of TfS_2-dualizing modules, for example, R itself.

(4.8.3) A 2-dimensional integral domain without a finite, torsion free, S_2 module. Pick an infinite set I and for $i \in I$ let R_i be the localization of

$$S := \langle x^a y^b \colon a+b \geq 2 \rangle \subset k[x,y].$$

Note that S is not normal and not S_2. The resulting R has a dense set of non-S_2 points, so it does not have a nonzero, finite, torsion free, S_2 module though all of its localizations at maximal ideals have one.

5 Conductors

Definition 5.1. Let X be a reduced scheme whose normalization $\pi\colon \bar{X} \to X$ is finite. Its *conductor ideal sheaf* is defined as

$$\operatorname{cond}_{\bar{X}/X} := \mathcal{H}om_X\bigl(\pi_*\mathcal{O}_{\bar{X}}, \mathcal{O}_X\bigr).$$

It is the largest ideal sheaf on X that is also an ideal sheaf on $\bar{X}$. We define the *conductor subschemes* as

$$D := \operatorname{Spec}_X\bigl(\mathcal{O}_X/\operatorname{cond}_{\bar{X}/X}\bigr) \quad \text{and} \quad \bar{D} := \operatorname{Spec}_{\bar{X}}\bigl(\mathcal{O}_{\bar{X}}/\operatorname{cond}_{\bar{X}/X}\bigr).$$

Since 1 is a local section of $\operatorname{cond}_{\bar{X}/X}$ over $U \subset X$ iff π is an isomorphism over U, we see that

$$\operatorname{supp} D = \pi(\operatorname{supp} \bar{D}) = \operatorname{supp}(\pi_*\mathcal{O}_{\bar{X}}/\mathcal{O}_X).$$

If X is S_2 then $D \subset X$ and $\bar{D} \subset \bar{X}$ are S_1 and of pure codimension 1. Indeed, let $q\colon \mathcal{O}_{\bar{X}} \to \mathcal{O}_{\bar{D}}$ be the quotient map and $T \subset \pi_*\mathcal{O}_{\bar{D}}$ the largest subsheaf whose support has codimension ≥ 2 in X. Assume that $T \neq 0$. Note that $q^{-1}(T) \subset \mathcal{O}_{\bar{X}}$ is an ideal sheaf, so it is not contained in $\mathcal{O}_X$ by the maximality of the conductor. Then $\langle \mathcal{O}_X, q^{-1}(T)\rangle \supset \mathcal{O}_X$ is a nontrivial extension whose cosupport has codimension ≥ 2, contradicting the S_2 condition for X.

Note also that $\pi_*\mathcal{O}_{\bar{X}} \subset \mathcal{H}om_X(\operatorname{cond}_{\bar{X}/X}, \mathcal{O}_X)$, thus the conductor is a coherent ideal sheaf iff π is finite.

Definition 5.2. Let X be a scheme and F a coherent sheaf on X such that $\operatorname{supp} F$ has codimension ≥ 1. The *divisorial support* of F is

$$[F] := \textstyle\sum_x \operatorname{length}_{k(x)}(F_x) \cdot [\bar{x}],$$

where the summation is over all codimension 1 points $x \in X$ and $[\bar{x}]$ denotes the Weil divisor defined by the closure of x. If $Z \subset X$ is a subscheme then we set $[Z] := [\mathcal{O}_Z]$.

5.3 (Normalization and dualizing sheaf). Let X be a reduced scheme with finite normalization $\pi\colon \bar{X} \to X$ satisfying 2.12(*2*). If X has a *TfS*$_2$-dualizing sheaf $\tilde{\omega}_X$ then we always choose

$$\tilde{\omega}_{\bar{X}} := \pi^!\tilde{\omega}_X = \mathcal{H}om_X\bigl(\pi_*\mathcal{O}_{\bar{X}}, \tilde{\omega}_X\bigr) \tag{5.1}$$

as our *TfS*$_2$-dualizing sheaf on $\bar{X}$.

We can view sections of $\tilde{\omega}_X$ as rational sections of $\tilde{\omega}_{\bar{X}}$ with poles along $\bar{D}$. Thus $\tilde{\omega}_X \subset \pi_*\tilde{\omega}_{\bar{X}}(m\bar{D})$ for some $m > 0$.

We can also view $\tilde{\omega}_X$ as a *non-coherent* subsheaf of $\tilde{\omega}_{\bar{X}}(m\bar{D})$, but we need to be careful since $\tilde{\omega}_X$ is not even a sheaf of $\mathcal{O}_{\bar{X}}$-modules.

There is, however, a smallest coherent subsheaf of $\tilde{\omega}_{\bar{X}}(m\bar{D})$ that contains $\tilde{\omega}_X$, we denote it by $\mathcal{O}_{\bar{X}} \cdot \tilde{\omega}_X \subset \tilde{\omega}_{\bar{X}}(m\bar{D})$.

The following duality is quite useful.

Lemma 5.4. *Let $p\colon X \to Y$ be a finite morphism between S_2 schemes that maps generic points to generic points. Let $\tilde{\omega}_Y$ be a TfS_2-dualizing sheaf on Y and $\tilde{\omega}_X := p^!\tilde{\omega}_Y$. Then*

$$\mathcal{H}om_Y(p_*\mathcal{O}_X, \mathcal{O}_Y) \cong \mathcal{H}om_Y(\tilde{\omega}_Y, p_*\tilde{\omega}_X).$$

Proof Note that $\tilde{\omega}_X$ is a TfS_2-dualizing sheaf by 3.9 and 2.13. The claim follows from the isomorphisms

$$\begin{aligned} \mathcal{H}om_Y(\tilde{\omega}_Y, p_*\tilde{\omega}_X) &= \mathcal{H}om_Y\big(\tilde{\omega}_Y, \mathcal{H}om_Y(p_*\mathcal{O}_X, \tilde{\omega}_Y)\big) \\ &= \mathcal{H}om_Y\big(\tilde{\omega}_Y \otimes p_*\mathcal{O}_X, \tilde{\omega}_Y\big) \\ &= \mathcal{H}om_Y\big(p_*\mathcal{O}_X, \mathcal{H}om_Y(\tilde{\omega}_Y, \tilde{\omega}_Y)\big) \\ &= \mathcal{H}om_Y\big(p_*\mathcal{O}_X, \mathcal{O}_Y\big), \end{aligned}$$

where at the end we used (3.3). □

The next result is closely related to [Rei94, Theorem 3.2].

Lemma 5.5. *Let X be a reduced, S_2 scheme whose normalization $\pi\colon \bar{X} \to X$ is finite with conductors $D \subset X$ and $\bar{D} \subset \bar{X}$. Then the following equivalent claims hold.*

(1) $\bar{D} = \inf\{E\colon \tilde{\omega}_X \subset \pi_*\big(\tilde{\omega}_{\bar{X}}(E)\big)\}$.
(2) $\mathcal{O}_{\bar{X}} \cdot \tilde{\omega}_X \subset \tilde{\omega}_{\bar{X}}(\bar{D})$ *and the support of the quotient has codimension* ≥ 2.

Proof Note that $\mathcal{O}_X(-D) = \mathcal{H}om_X\big(\pi_*\mathcal{O}_{\bar{X}}, \mathcal{O}_X\big)$. Thus by 5.4 we get that $\mathcal{O}_X(-D) \cdot \tilde{\omega}_X \subset \pi_*\tilde{\omega}_{\bar{X}}$. Since $\mathcal{O}_X(-D) = \pi_*\mathcal{O}_{\bar{X}}(-\bar{D})$, this implies that $\tilde{\omega}_X \subset \tilde{\omega}_{\bar{X}}(\bar{D})$.

Conversely, assume that $\tilde{\omega}_X \subset \tilde{\omega}_{\bar{X}}(E)$. Then $\mathcal{O}_{\bar{X}}(-E) \cdot \tilde{\omega}_X \subset \tilde{\omega}_{\bar{X}}$, thus 5.4 shows that $\mathcal{O}_{\bar{X}}(-E) \cdot \mathcal{O}_{\bar{X}} \subset \mathcal{O}_X$. So, $\mathcal{O}_{\bar{X}}(-E) \subset \mathcal{O}_{\bar{X}}(-\bar{D})$ and hence $E \geq \bar{D}$. □

As a consequence we get the characterization of seminormal, S_2 schemes.

Proof of Theorem 1.4 The equivalence of (*2*) and (*3*) is a special case of 5.5 and (*1*) $\iff$ (*2*) follows from the equality

$$X^{\mathrm{sn}} = \operatorname{Spec}_X \langle \mathcal{O}_X, \pi_* \mathcal{O}_{\bar{X}}(-\operatorname{red} \bar{D}) \rangle,$$

which is a rewriting of the last formula on [Kol96, page 85]. □

Next we focus on the 1-dimensional case. Then the dualizing sheaf usually exists.

Proposition 5.6. *Let* $(0 \in C)$ *be a local, 1-dimensional, reduced scheme whose normalization* $\pi \colon \bar{C} \to C$ *is finite. Then* C *has a dualizing sheaf* ω_C.

One can view the above claim as a very special converse to 3.9. (I do not know if its higher dimensional versions are true or not.) We give two proofs. The first, in 5.11 gives a concrete construction of the dualizing sheaf. More general results on 1-dimensional rings are treated in 6.2 and 6.1.

The following is taken from [Rei94, page 714].

Proposition 5.7. *Let* $(0 \in C)$ *be a local, 1-dimensional, reduced scheme whose normalization* $\pi \colon (\bar{0} \in \bar{C}) \to (0 \in C)$ *is finite with conductors* $D \subset C$ *and* $\bar{D} \subset \bar{C}$. *Assume that the residue field* $k(0)$ *is infinite and let* $\sigma \in \omega_C$ *be a general section. Then*

$$\mathcal{O}_C(-D) \cdot \sigma = \pi_* \omega_{\bar{C}} \subset \omega_C. \tag{5.2}$$

Proof Let $\bar{c}_i \in \bar{C}$ be the preimages of $0 \in C$. By 5.5 for every i there is a section $\sigma_i \in \omega_C$ such that σ_i generates $\omega_{\bar{C}}(\bar{D})$ at c_i. If $k(0)$ is infinite then a general linear combination $\sigma := \sum_i \lambda_i \sigma_i$ generates $\omega_{\bar{C}}(\bar{D})$ everywhere.

We can now compute $\mathcal{O}_C(-D) \cdot \sigma$ on $\bar{C}$ as

$$\mathcal{O}_{\bar{C}}(-\bar{D}) \cdot \sigma = \mathcal{O}_{\bar{C}}(-\bar{D}) \cdot \mathcal{O}_{\bar{C}} \cdot \sigma = \mathcal{O}_{\bar{C}}(-\bar{D}) \cdot \omega_{\bar{C}}(\bar{D}) = \omega_{\bar{C}}. \quad \square$$

Corollary 5.8. *Let* $(0 \in C)$ *be a local, 1-dimensional, reduced scheme whose normalization* $\pi \colon (\bar{0} \in \bar{C}) \to (0 \in C)$ *is finite with conductors* $D \subset C$ *and* $\bar{D} \subset \bar{C}$. *Then*

(*1*) $\operatorname{length}(\omega_C / \pi_* \omega_{\bar{C}}) \geq \operatorname{length}(\mathcal{O}_D)$ *and*
(*2*) *equality holds iff* ω_C *is free.*

Proof If $k(0)$ is infinite then 5.7 gives an embedding $\mathcal{O}_D \cdot \sigma \hookrightarrow \omega_C / \pi_* \omega_{\bar{C}}$ and equality holds iff σ generates $\omega_C / \pi_* \omega_{\bar{C}}$. By (5.2) $\pi_* \omega_{\bar{C}}$ is contained in $\mathcal{O}_C \cdot \sigma$, thus equality holds iff σ generates ω_C.

If $k(0)$ is finite, then first we take $\mathbb{A}^1_C$ and localize at the generic point of the fiber over $0 \in C$. The residue field is now $k(0)(t)$, hence infinite, and the lengths are unchanged. $\square$

Lemma 5.9. *Let $(0 \in C)$ be a local, 1-dimensional, reduced scheme whose normalization $\pi\colon (\bar{0} \in \bar{C}) \to (0 \in C)$ is finite with conductors $D \subset C$ and $\bar{D} \subset \bar{C}$. Then*

$$\operatorname{length}(\omega_C/\pi_*\omega_{\bar{C}}) = \operatorname{length}(\pi_*\mathcal{O}_{\bar{D}}) - \operatorname{length}(\mathcal{O}_D).$$

Proof By (5.1), $\pi_*\omega_{\bar{C}}$ is the dual of $\pi_*\mathcal{O}_{\bar{C}}$. Therefore 3.16 says that $\operatorname{length}(\omega_C/\pi_*\omega_{\bar{C}}) = \operatorname{length}(\pi_*\mathcal{O}_{\bar{C}}/\mathcal{O}_C)$ and $\pi_*\mathcal{O}_{\bar{C}}/\mathcal{O}_C \cong \pi_*\mathcal{O}_{\bar{D}}/\mathcal{O}_D$. $\square$

We can now prove the '$n_Q = 2\delta_Q$ theorem.'

Proof of Theorem 1.3 The claim can be checked after localizing at various generic points of D. As we noted in 5.1, $D \subset X$ has pure codimension 1. Thus we may assume that $C := X$ is local and $\dim C = 1$. Since the normalization $\pi\colon \bar{C} \to C$ is finite, C has a dualizing sheaf ω_C by 5.6

Combining 5.8 and 5.9., we get that

$$\operatorname{length}(\pi_*\mathcal{O}_{\bar{D}}) - \operatorname{length}(\mathcal{O}_D) = \operatorname{length}(\omega_C/\pi_*\omega_{\bar{C}}) \geq \operatorname{length}(\mathcal{O}_D),$$

and equality holds iff C is Gorenstein. $\square$

Let us state another variant of [Rei94, 3.2.I].

Proposition 5.10. *Let X be a reduced, S_2 scheme with finite normalization $\pi\colon \bar{X} \to X$ and conductors $D \subset X$ and $\bar{D} \subset \bar{X}$. Let $\tilde{\omega}_X$ be a TfS_2-dualizing sheaf on X and set $\tilde{\omega}_{\bar{X}} := \pi^!\tilde{\omega}_X$. Then there is an exact sequence*

$$0 \to \tilde{\omega}_X \to \pi_*\tilde{\omega}_{\bar{X}}(\bar{D}) \xrightarrow{r_D} \mathcal{E}xt^1_X(\mathcal{O}_D, \tilde{\omega}_X) \to 0. \tag{5.3}$$

If X is CM and $\tilde{\omega}_X = \omega_X$ is a dualizing sheaf then the sequence becomes

$$0 \to \omega_X \to \pi_*\omega_{\bar{X}}(\bar{D}) \xrightarrow{r_D} \omega_D \to 0. \tag{5.4}$$

Proof Start with the exact sequence

$$0 \to \mathcal{O}_X(-D) \to \mathcal{O}_X \to \mathcal{O}_D \to 0.$$

Take $\mathcal{H}om_X(\ ,\tilde{\omega}_X)$ to get

$$\begin{aligned} 0 \to \tilde{\omega}_X \to \mathcal{H}om_X(\mathcal{O}_X(-D), \tilde{\omega}_X) &\to \mathcal{E}xt^1_X(\mathcal{O}_D, \tilde{\omega}_X) \\ &\to \mathcal{E}xt^1_X(\mathcal{O}_X, \tilde{\omega}_X) = 0. \end{aligned}$$

By 3.8(*3*), we have the following which proves (5.3):

$$\begin{aligned}\mathcal{H}om_X\big(\mathcal{O}_X(-D),\tilde{\omega}_X\big) &= \mathcal{H}om_X\big(\pi_*\mathcal{O}_{\bar{X}}(-\bar{D}),\tilde{\omega}_X\big)\\ &= \mathcal{H}om_{\bar{X}}\big(\mathcal{O}_{\bar{X}}(-\bar{D}),\tilde{\omega}_{\bar{X}}\big) = \tilde{\omega}_{\bar{X}}(\bar{D}).\end{aligned}$$

If X is CM then general duality shows that $\mathcal{E}xt^1_X(\mathcal{O}_D,\omega_X)\cong\omega_D$; we will not use this part, see [BH93, 3.3.7] or [Sta18, Tag 0AX0] for proofs.

Note finally that the map r_D can be written as the composite of the Poincaré residue map $\Re_{\bar{X}/\bar{D}}$, the map ϕ_1 in 3.8(*4*) and of the trace map $\operatorname{tr}_{\bar{D}/D}$

$$\begin{aligned}r_D\colon \tilde{\omega}_{\bar{X}}(\bar{D}) \xrightarrow{\Re} \mathcal{E}xt^1_{\bar{X}}(\mathcal{O}_{\bar{D}},\tilde{\omega}_{\bar{X}}) &\xrightarrow{\phi} \mathcal{E}xt^1_X(\pi_*\mathcal{O}_{\bar{D}},\tilde{\omega}_X)\\ &\xrightarrow{\operatorname{tr}} \mathcal{E}xt^1_X(\mathcal{O}_D,\tilde{\omega}_X).\end{aligned}$$

We will write this down very explicitly in 5.11. □

5.11 (Construction of ω_C from $\omega_{\bar{C}}$). Let $(0\in C)$ be a local, 1-dimensional, reduced scheme whose normalization $\pi\colon\bar{C}\to C$ is finite. Since $\bar{C}$ is regular, $\omega_{\bar{C}}$ exists, for example we can choose $\omega_{\bar{C}}=\mathcal{O}_{\bar{C}}$. We give a construction of ω_C starting with $\omega_{\bar{C}}$.

If ω_C exists then it sits in the exact sequence (5.4). The other 2 sheaves in (5.4) are $\pi_*\omega_{\bar{D}}$ and ω_D, whose existence is already known. Thus it is natural to set

$$\Omega := \ker\big[\pi_*\omega_{\bar{C}}(\bar{D}) \xrightarrow{\mathcal{R}} \pi_*\omega_{\bar{D}} \xrightarrow{\operatorname{tr}} \omega_D\big] \tag{5.5}$$

and aim to show the following.

Claim 5.12. Ω is a dualizing sheaf over C.

Note that for curves over a field the Poincaré residue map $\mathcal{R}_{\bar{C}/\bar{D}}$ and the trace map $\operatorname{tr}_{\bar{D}/D}$ are both canonical. Over a general scheme the dualizing sheaf is defined only up to tensoring with a line bundle, so we believe that, depending on the choices we make, either $\mathcal{R}_{\bar{C}/\bar{D}}$ or $\operatorname{tr}_{\bar{D}/D}$ in (5.5) is not canonical. This is not a problem since ω_C is not unique as a subsheaf of $\pi_*\omega_{\bar{C}}$; we can multiply it by any section of $\mathcal{O}^*_{\bar{C}}$.

Since π is birational, Ω is dualizing at the generic points of C, thus, by 7.7, the following implies 5.12.

Claim 5.13. $\mathcal{E}xt^1_C(k,\Omega)\cong k$.

Proof To fix our notation, we have $\pi\colon(P\subset\bar{C})\to(0\in C)$ where $P=\{p_i\colon i\in I\}=\pi^{-1}(0)$ and with conductors $D\subset C$ and $\bar{D}\subset\bar{C}$. Set

$k := k(0)$ and $k_i := k(p_i)$. The conductor can be written as $\bar{D} = \sum r_i[p_i]$ for some $r_i \geq 1$ and $P = \operatorname{supp} \bar{D}$. We can indentify

$$\omega_{\bar{D}} \cong \omega_{\bar{C}}(\bar{D})/\omega_{\bar{C}} \quad \text{and} \quad \operatorname{socle}(\omega_{\bar{D}}) \cong \omega_{\bar{C}}(P)/\omega_{\bar{C}}.$$

The residue map gives an isomorphism

$$\mathcal{R}\colon \operatorname{socle}(\omega_{\bar{D}}) \cong \textstyle\sum_i k_i, \tag{5.6}$$

which is, however, not canonical.

Another way to represent $\omega_{\bar{D}}$ is using the isomorphism

$$\omega_{\bar{D}} \cong \mathcal{H}om_D(\mathcal{O}_{\bar{D}}, \omega_D).$$

Using that $\operatorname{socle}(\omega_D) \cong k$, under this isomorphism the socle is represented as

$$\operatorname{socle}(\omega_{\bar{D}}) \cong \textstyle\sum_i \operatorname{Hom}_k(k_i, k). \tag{5.7}$$

Using the form (5.7), the socle of the trace map

$$\operatorname{tr}\colon \operatorname{socle}(\omega_{\bar{D}}) \to \operatorname{socle}(\omega_D) \cong k$$

sends $\{(\phi_i : k_i \to k) : i \in I\}$ to $\sum_i \phi_i(1)$. Since the trace is non-degenerate, using the form (5.6) we get instead the representation

$$\{(x_i \in k_i) : i \in I\} \mapsto \textstyle\sum_i \operatorname{tr}_{k_i/k}(c_i x_i) = 0 \quad \text{for some} \quad c_i \in k_i^*,$$

where the c_i arise from the unknown isomorphism $k_i \cong \operatorname{Hom}_k(k_i, k)$. (Over a field, the correct choices lead to $c_i = 1$ for every i; see [Ser59, Section IV.9–10].) We can summarize these discussions as follows.

Claim 5.14. $\Omega \cap H^0(\bar{C}, \omega_{\bar{C}}(P)) \subset H^0(\bar{C}, \omega_{\bar{C}}(P))$ is a k-hypersurface defined by the equation

$$\textstyle\sum_i \operatorname{tr}_{k_i/k}\big(c_i \cdot \mathcal{R}_{p_i}(\sigma)\big) = 0, \tag{5.8}$$

that is, a section σ of $H^0(\bar{C}, \omega_{\bar{C}}(P))$ is in Ω iff it satisfies (5.8).

We now return to the proof of 5.13. For any nonsplit extension $0 \to \Omega \to \Omega' \to k \to 0$, we can view Ω' uniquely as a subsheaf of $\omega_{\bar{C}}(\bar{D} + P)$. Pick a section σ of Ω' mapping to $1 \in k$. At the points $p_i \in \bar{C}$ we can write $\sigma = v_i x_i^{-r_i - 1}$ where x_i is a local parameter and v_i a local section of $\omega_{\bar{C}}$ at p_i.

Since $m_0\sigma \in \Omega$ and $\mathcal{O}_C(-D) \subset m_0$, we see that $\mathcal{O}_{\bar{C}}(-\bar{D}) \cdot \sigma \in \Omega$. For any $a_i \in k_i$ there is a $g \in \mathcal{O}_{\bar{C}}(-\bar{D})$ such that $g = u_i x_i^{r_i}$ where u_i is a local section of $\mathcal{O}_{\bar{C}}$ at p_i such that $u_i(p_i) = a_i \in k_i$ Thus $g\sigma \in \Omega$ and $g\sigma = u_i v_i x_i^{-1}$ at p_i for every i. By (5.8), we get the equation

$$\textstyle\sum_i \operatorname{tr}_{k_i/k}\big(a_i c_i \mathcal{R}_{p_i}(v_i x_i^{-1})\big) = 0 \quad \text{for every} \quad a_i \in k_i.$$

Since the trace is non-degenerate, we conclude that $\mathcal{R}_{p_i}(v_i x_i^{-1}) = 0$ for every i. That is, $v_i x_i^{-1}$ is a local section of $\omega_{\bar{C}}$ at p_i. Therefore $\sigma = (v_i x_i^{-1}) x_i^{-r}$ is a local section of $\omega_{\bar{C}}(\bar{D})$ at each p_i. Thus $\Omega' \subset \omega_{\bar{C}}(\bar{D})$ and we get a map $\Omega' \to k \to \omega_D$. The image of k is then the socle of ω_D, hence Ω' is the preimage of $\operatorname{socle}(\omega_D) \cong k$. Thus Ω' is unique and so $\mathcal{E}xt^1_C(k, \Omega) \cong k$. This proves 5.13 and hence also 5.12. □

6 Duality for 1-Dimensional Schemes

One can give a complete characterization of those 1-dimensional schemes that have a dualizing sheaf. This is probably known but I did not find a complete reference. The key results are the next local characterization and the global statement Proposition 1.5.

Theorem 6.1. *[FFGR75, 5.3] Let (R, m) be a 1-dimensional local ring. Then R has a dualizing module iff the generic fibers of $R \to \hat{R}$ are Gorenstein.* □

Aside. By [Kaw00, 1.4], an arbitrary local ring has a dualizing complex iff it is a quotient of a local, Gorenstein ring, but this is very hard to use in practice. See [BH93, 3.3.6] for the simpler CM version.

Note that if the normalization $\bar{R}$ is finite over R then the completion of $R/\sqrt{0}$ is reduced by [Kru30, Satz 9] (see also [Kol07, 1.101]). Therefore the generic fibers of $R \to \hat{R}$ are sums of fields, hence Gorenstein. Thus 6.1 is a much stronger existence result than 5.6, though the latter gives ω_R in a more concrete form.

For sake of completeness, we outline the standard proof of the following special case, which starts with complete local rings and descends from there.

Proposition 6.2. *Let (R, m) be a 1-dimensional, S_1 local ring such that the completion of $R/\sqrt{0}$ is reduced. Then R has a dualizing module.*

Proof Set $k = R/m$ and assume first that $\operatorname{char} R = \operatorname{char} k$. If R is complete, then we can view R as a k-algebra, cf. [Sta18, Tag 0323]. Let $y_1, \dots, y_n \in m$ be a system of parameters, where $n = \dim R$. Then R is finite over the power series ring $k[[y_1, \dots, y_n]]$. Since $k[[y_1, \dots, y_n]]$ is a dualizing module over itself by 3.5, R has a TfS_2-dualizing module by 3.9.

In the non-complete case, let $Q(R)$ be the total ring of quotients of R, $Q(\hat{R})$ the total ring of quotients of $\hat{R}$. Let $\omega_{Q(R)}$ and $\omega_{Q(\hat{R})}$ be the corresponding dualizing modules. We check below that

$$Q(\hat{R}) \otimes_{Q(R)} \omega_{Q(R)} \cong \omega_{Q(\hat{R})}. \tag{6.1}$$

If this holds, let Ω_R be a finite R-module such that $Q(R) \otimes_R \Omega_R \cong \omega_{Q(R)}$.

By 6.3 we can realize $\omega_{\hat{R}}$ as a submodule of $\hat{\Omega}_R$ with finite quotient. So, there is a submodule $\omega_R \subset \Omega_R$ such that $\hat{\omega}_R = \omega_{\hat{R}}$. Thus ω_R is a dualizing module by 3.3.

In order to prove (6.1), we check the conditions of 3.12(*5*). We know that $\operatorname{length}(\omega_{Q(R)}) = \operatorname{length}(Q(R))$ and this is preserved when we tensor by $Q(\hat{R})$. We also know that

$$\operatorname{Hom}_{Q(\hat{R})}\big(Q(\hat{R}) \otimes_{Q(R)} K(R), Q(\hat{R}) \otimes_{Q(R)} \omega_{Q(R)}\big) \cong Q(\hat{R}) \otimes_{Q(R)} K(R).$$

If the completion of $R/\sqrt{0}$ is reduced, then the generic fibers of $R \to \hat{R}$ are sums of fields, hence $K(\hat{R}) = Q(\hat{R}) \otimes_{Q(R)} K(R)$ and we are done. (A very similar argument proves 6.1. One needs to argue that the socle of $Q(\hat{R}) \otimes_{Q(R)} \omega_{Q(R)}$ is the socle of $Q(\hat{R}) \otimes_{Q(R)} K(R)$ and the latter is a sum of the residue fields iff $Q(\hat{R}) \otimes_{Q(R)} K(R)$ is Gorenstein. See also [BH93, 3.3.14] or [Sta18, Tag 0E4D].)

We do not know a similarly elementary proof in the mixed characteristic case. Here one writes R as a quotient of a power series ring $S := \Lambda[[x_1, \dots, x_r]]$ where Λ is a complete DVR and then $\operatorname{Ext}_S^{\dim S - \dim R}(R, S)$ is a *TfS*$_2$-dualizing module; see [BH93, 3.3.7] or [Eis95, Section 21.6]. It is also a very special case of general duality theory as in [Sta18, Tag 0AX0]. The rest goes as before. □

Lemma 6.3. *Let R be a 1-dimensional local ring and $Q(R)$ its total ring of quotients. For finite R-modules M and N the following are equivalent.*

(1) $Q(R) \otimes_R M \cong Q(R) \otimes_R N$.
(2) There is a map $\phi\colon M \to N$ whose kernel and cokernel are torsion.

□

For the global existence, we need the following criterion for the dualizing module over a nonreduced ring.

Lemma 6.4. *Let R be a 1-dimensional, S_1, local ring and set $S := R/\sqrt{0}$. Let Ω_R be a finite R-module such that $\operatorname{socle}(\Omega_R) \cong \omega_S$ and $\operatorname{Ext}^1_R(S, \Omega_R) = 0$. Then Ω_R is a dualizing R-module.*

Proof By 3.12, Ω_R is dualizing at the generic points. For a maximal ideal $m \subset S$ with residue field k, duality gives the exact sequence

$$0 \to \mathrm{Hom}_R(S, \Omega_R) \to \mathrm{Hom}_R(m, \Omega_R) \to \mathrm{Ext}^1_R(k, \Omega_R) \to 0. \qquad (6.2)$$

The image of an R-homomorphism from an S-module to Ω_R is contained in the socle, which is ω_S. Thus (6.2) can be rewritten as

$$0 \to \mathrm{Hom}_S(S, \omega_S) \to \mathrm{Hom}_S(m, \omega_S) \to \mathrm{Ext}^1_R(k, \Omega_R) \to 0.$$

This shows that $\mathrm{Ext}^1_R(k, \Omega_R) \cong \mathrm{Ext}^1_S(k, \omega_S) \cong k$ and so, Ω_R is a dualizing module by 7.5. □

Proof of Proposition 1.5 Let ω_X be a dualizing sheaf. Then every localization has a dualizing module by 3.7, and by 3.9 the sheaf $\omega_{\mathrm{red}\, X} := \mathcal{H}om_X(\mathcal{O}_{\mathrm{red}\, X}, \omega_X)$ is a dualizing sheaf of $\mathrm{red}\, X$. It is a coherent, rank 1, torsion free sheaf, hence locally free over a dense, open subset, proving 1.5(*2*).

Conversely, let Ω be a coherent, torsion free sheaf on X that is dualizing at all generic points of X. This implies that the support of $\mathcal{E}xt^1_X(\mathcal{O}_{\mathrm{red}\, X}, \Omega)$ is nowhere dense. Let $U \subset X$ be a dense open subset such that $\mathrm{red}\, U$ is Gorenstein, $\mathrm{socle}(\Omega)$ is invertible on $\mathrm{red}\, U$ and $\mathcal{E}xt^1_U(\mathcal{O}_{\mathrm{red}\, U}, \Omega|_U) = 0$. Then $\Omega|_U$ is dualizing by 6.4.

For each $x \in X \setminus U$ let ω_{X_x} be a dualizing sheaf on X_x. Over the generic points $g_x \in X_x$ we can fix isomorphisms of $\Omega|_{g_x}$ and $\omega_{X_x}|_{g_x}$ and glue the sheaves Ω and ω_{X_x} together. We get a dualizing sheaf on X. □

7 Dualizing Module of CM Rings

We start with two observations that allow us to reduce various questions about CM modules to 1-dimensional CM modules. Let (R, m) be a local ring. Recall that a finite R-module N is *Cohen-Macaulay* or *CM* for short, if there is a system of parameters $x_1, \dots, x_n \in m$ such that x_{i+1} is a non-zerodivisor on $N/(x_1, \dots, x_i)N$ for $i = 0, \dots, \dim N - 1$. A CM module N is called *maximal CM* if $\dim N = \dim R$.

7.1 (CM modules and dimension reduction). Let (R, m) be a local ring and M a finite R-module.

Assume that $\dim M \geq 2$ and $x \in m$ is not contained in any of the positive dimensional associated primes of M. Then M is CM iff M/xM is. Using this inductively we get the following.

Claim 7.2. Let (R, m) be a local ring, $x_1, \dots, x_n$ a system of parameters and M a finite R-module of dimension d. Then for general $x'_i \equiv x_i \mod m^2$, the module M is CM iff $M/(x'_1, \dots, x'_{d-1})M$ is CM. □

Next note that, by 2.4, M admits a filtration where each successive quotient G_j is a rank 1 torsion free module over R/P_j for some prime ideal P_j. There is thus a non-zerodivisor $g \in m$ such that each $(G_j)_g$ is free over $(R/P_j)_g$. Choose $x \in m$ such that g is not contained in any of the minimal primes of M/xM. Let P be a minimal associated prime of M and Q a minimal associated prime of M/xM that contains P. Then

$$\operatorname{length}_Q(M/xM)_Q = \operatorname{length}_Q\bigl(R_P/xR_P\bigr) \cdot \operatorname{length}_P(M_P).$$

Using this inductively, we obtain the following.

Claim 7.3. Let (R, m) be a local ring, $x_1, \dots, x_n$ a system of parameters and M, N finite R-modules of dimension d. Assume also that $\operatorname{length}_P(M_P) \geq \operatorname{length}_P(N_P)$ for every d-dimensional prime P. Then, for general $x'_i \equiv x_i \mod m^2$,

$$\operatorname{length}_Q\bigl(M/(x'_1, \dots, x'_{d-1})M\bigr) \geq \operatorname{length}_Q\bigl(N/(x'_1, \dots, x'_{d-1})N\bigr)$$

for every 1-dimensional prime Q.

There are several standard definitions of a dualizing/canonical module.

Definition 7.4 (Dualizing or canonical module). Let (R, m) be a local, CM ring of dimension n, $k := R/m$ and M a finite R-module. Then M is a *dualizing module* or a *canonical module* iff any of the following equivalent conditions hold.

(7.4.1) Ext version as in [BH93, page 107].

$$\operatorname{Ext}^i_R(k, M) \cong \begin{cases} k & \text{if } i = n \\ 0 & \text{if } i \neq n. \end{cases}$$

(7.4.2) Inductive version as in [Eis95, Section 21.3]. Let $x_1, \dots, x_n \in m$ be a system of parameters. Then M is dualizing iff it is maximal CM and $M/(x_1, \dots, x_n)M \cong E(k)$, the injective hull of k over $R/(x_1, \dots, x_n)$. (Note that the assumption that M be maximal CM is missing in [Eis95, Section 21.3].) (The equivalence can be seen using 3.12 and 7.14.)

(7.4.3) Endomorphism version as in [Eis95, theorem 21.8] or [Sta18, Tag 0A7B]. M is dualizing iff it is maximal CM, has finite injective dimension and $\operatorname{End}_R(M) \cong R$.

Note that $\mathrm{Ext}^i_R(k, M) = 0$ for all $i < \dim R$ iff M is maximal CM. In some sense the key condition is $\mathrm{Ext}^{\dim R}_R(k, M) \cong k$. The vanishing of the higher Ext groups may be harder to see. The following result says that they can be replaced by other conditions that could be easier to check.

Theorem 7.5. *Let (R, m) be a local CM ring of dimension n and $k := R/m$. Let Ω be a finite R-module. The following are equivalent.*

(1) *Ω is dualizing.*
(2) $\mathrm{Ext}^i_R(k, \Omega) = \delta_{in} \cdot k$ *for every i.*
(3) $\mathrm{Ext}^i_R(k, \Omega) = \delta_{in} \cdot k$ *for $0 \le i \le n$ and Ω_P is dualizing over R_P for every minimal prime $P \subset R$.*
(4) $\mathrm{Ext}^i_R(k, \Omega) = \delta_{in} \cdot k$ *for $0 \le i \le n$ and* $\mathrm{length}_P\, \Omega_P = \mathrm{length}_P\, R_P$ *for every minimal prime $P \subset R$.*
(5) $\mathrm{Ext}^i_R(k, \Omega) = \delta_{in} \cdot k$ *for $0 \le i \le n$ and* $\mathrm{length}_P\, \Omega_P \ge \mathrm{length}_P\, R_P$ *for every minimal prime $P \subset R$.*

Proof The first two claims are equivalent by our definition (7.4.1) and it is clear that each assertion implies the next one. Thus it remains to prove that *(5)* $\Rightarrow$ *(2)*.

If $n = 0$ then the first part of *(5)* says that $\mathrm{socle}(\Omega) \cong k$, so we can realize Ω as a submodule of $E(k)$, the injective hull of k. Thus $\mathrm{length}\, \Omega \le \mathrm{length}\, E(k) = \mathrm{length}\, R$, where the last equality holds by 3.12. The second part of *(5)* says that $\mathrm{length}\, \Omega \ge \mathrm{length}\, R$. Thus $\Omega = E(k)$ is dualizing.

If $n = 1$ then let $x \in m$ be a non-zerodivisor. A special case of 7.14 says that $\mathrm{socle}(\Omega/x\Omega) \cong k$. By 7.13 $\mathrm{length}_0(\Omega/x\Omega) \ge \mathrm{length}_0(R/xR)$. Thus $\Omega/x\Omega$ is dualizing over R/xR by the already settled $n = 0$ case. In particular,

$$\mathrm{Ext}^i_{R/xR}(k, \Omega/x\Omega) = \delta_{i0} \cdot k \quad \text{for every } i,$$

hence $\mathrm{Ext}^i_R(k, \Omega) = \delta_{i1} \cdot k$ for every i by 7.14.

If $n \ge 2$ then the first part of *(5)* says that Ω is maximal CM. By 7.3, we can choose a system of parameters $x_1, \ldots, x_n$ such that

$$\mathrm{length}_Q\big(\Omega/(x_1, \ldots, x_{n-1})\Omega\big) \ge \mathrm{length}_Q\big(R/(x_1, \ldots, x_{n-1})R\big)$$

for every 1-dimensional prime Q. Applying 7.14 gives that

$$\mathrm{Ext}^i_{R/(x_1,\ldots,x_{n-1})R}\big(k, \Omega/(x_1, \ldots, x_{n-1})\Omega\big) = \delta_{i,n-1} \cdot k \quad \text{for} \quad i \le 1.$$

The $n = 1$ case now gives that $\Omega/(x_1, \dots, x_{n-1})\Omega$ is dualizing over $R/(x_1, \dots, x_{n-1})R$ and applying 7.14 again shows that $\operatorname{Ext}^i_R(k, \Omega) = \delta_{in} \cdot k$ for every i. Thus Ω is dualizing over R. □

The maximal CM condition, that is, the vanishing of $\operatorname{Ext}^i_R(k, \Omega)$ for $0 \le i < n$ can be checked in other ways.

Lemma 7.6. *Let (R, m) be a local ring of dimension n and $x_1, \dots, x_n$ a system of parameters. Let M, N be finite R-modules of dimension d such that* $\operatorname{length}_P M_P \ge \operatorname{length}_P N_P$ *for every d-dimensional prime $P \subset R$. Assume that N is CM. Then*

(1) $\operatorname{length}_0\bigl(M/(x_1, \dots, x_n)M\bigr) \ge \operatorname{length}_0\bigl(N/(x_1, \dots, x_n)\bigr)$ *and*
(2) equality holds iff M is also CM and $\operatorname{length}_P M_P = \operatorname{length}_P N_P$ *for every d-dimensional prime P.*

Note that M, N need not be isomorphic in case (*2*).

Proof There is nothing to prove if $d = 0$. If $d = 1$ then $M' := M/\mathrm{t}(M)$ is CM and, by 7.13,

$$\begin{aligned} \operatorname{length}_0\bigl(M'/x_1M'\bigr) &= \textstyle\sum_P \operatorname{length}_P\bigl(M'_P\bigr) \cdot \operatorname{length}_0\bigl((R/P)/x_1(R/P)\bigr) \\ \operatorname{length}_0\bigl(N/x_1N\bigr) &= \textstyle\sum_P \operatorname{length}_P\bigl(N_P\bigr) \cdot \operatorname{length}_0\bigl((R/P)/x_1(R/P)\bigr), \end{aligned}$$

where the summation is over the 1-dimensional primes of R. Thus

$$\operatorname{length}_0\bigl(M'/x_1M'\bigr) \ge \operatorname{length}_0\bigl(N/x_1N\bigr)$$

and equality holds iff $\operatorname{length}_P M_P = \operatorname{length}_P N_P$ for every 1-dimensional prime P. Therefore

$$\begin{aligned} \operatorname{length}_0\bigl(M/x_1M\bigr) &= \operatorname{length}_0\bigl(M'/x_1M'\bigr) + \operatorname{length}_0\bigl(\mathrm{t}(M)/x_1\mathrm{t}(M)\bigr) \\ &\ge \operatorname{length}_0\bigl(N/x_1N\bigr) + \operatorname{length}_0\bigl(\mathrm{t}(M)/x_1\mathrm{t}(M)\bigr). \end{aligned}$$

Since $\mathrm{t}(M)/x_1\mathrm{t}(M) = 0$ iff $\mathrm{t}(M) = 0$, this settles the $d = 1$ case. As in the proof of 7.5, the $d \ge 2$ case reduces to the above by 7.2 and 7.3. □

The following is an immediate combination of 3.12(*6*), 7.5 and 7.6.

Corollary 7.7. *Let (R, m) be a local, CM ring of dimension n with residue field k and $x_1, \dots, x_n$ a system of parameters. Let Ω be a finite R-module. Then Ω is dualizing iff*

(1) $\operatorname{length}_P \Omega_P \ge \operatorname{length}_P R_P$ *for every minimal prime $P \subset R$ and*
(2) $\operatorname{socle}\bigl(\Omega/(x_1, \dots, x_n)\Omega\bigr) \cong k$. □

The above result and [Eis95, Section 21.3] leads to the following.

Question 7.8. Let (R, m) be a local, CM ring of dimension n with dualizing module ω_R and $x_1, \dots, x_n$ a system of parameters. Let M be a finite R-module such that $M/(x_1, \dots, x_n)M \cong E(k)$, the injective hull of $k := R/m$ over $R/(x_1, \dots, x_n)$. Is then M a quotient of ω_R?

The next example shows that this is not the case.

Example 7.9. Fix $m \geq 3$. Let $R := k[t^i : i \geq m]$ be a monomial ring and set $x = t^m$. Then $R/xR = \langle 1, t^{m+1}, \dots, t^{2m-1}\rangle$, and in R/xR we have $(t^{m+1}, \dots, t^{2m-1})^2 = 0$. We can write

$$\omega_R = \langle t^{-m}, \dots, t^{-2}, 1, t, \dots \rangle \cdot dt.$$

Then

$$\omega_R/x\omega_R = \langle t^{-m}, \dots, t^{-2}, t^{m-1}\rangle \cdot dt.$$

Setting $\sigma_i = t^{-i}dt$, $\Sigma := \langle \sigma_m, \dots, \sigma_2\rangle$ and $s = t^{m-1}dt$, the module structure on $\omega_R/x\omega_R$ is given by $t^i\sigma_j = \delta_{i,j+m-1} \cdot s$.

Claim 7.10. $\dim \operatorname{Ext}_R(\omega_R/x\omega_R, k) = m^2 - m - 1$.

Proof Consider an extension $0 \to k \to M \xrightarrow{c} \omega_R/x\omega_R \to 0$. If we fix a lifting $\bar{s} \in c^{-1}\langle s\rangle$ then we get k-linear maps

$$\bar{x} := x \circ c^{-1} \colon \Sigma \to k \quad \text{and} \quad \tau_i := t^i \circ c^{-1} \colon \Sigma \to c^{-1}\langle s\rangle/\langle \bar{s}\rangle,$$

the latter for $i = m+1, \dots, 2m-1$. These maps can be chosen arbitrarily and they determine the R-module structure of M. □

Note that if $\bar{x} \neq 0$ then $M/xM \cong \omega_R/x\omega_R$. Extensions as above that are quotients of ω_R correspond to maps $\operatorname{Hom}_R(x\omega_R, k) \cong \operatorname{Hom}_R(\Sigma, k) \cong k^{m-1}$. Comparing this with 7.10 gives a negative answer to 7.8.

Corollary 7.11. *For $m \geq 3$ there are Artinian R-modules M such that $M/xM \cong \omega_R/x\omega_R$ but M is not a quotient of ω_R.*

The next computation shows what changes for torsion free R-modules. These are flat over $k[x]$; we determine the first infinitesimal extension.

Claim 7.12. Let M be an extension of $\omega_R/x\omega_R$ by $\omega_R/x\omega_R$ such that $M/xM \cong \omega_R/x\omega_R$. Then $M \cong \omega_R/x^2\omega_R$.

Proof By assumption we have an extension

$$0 \to \omega_R/x\omega_R \xrightarrow{i_x} M \longrightarrow \omega_R/x\omega_R \to 0$$

as a sequence of R/x^2R-modules. Applying duality to it over R/x^2R gives $0 \to R/xR \to M^* \to R/xR \to 0$, noting that

$$\begin{aligned}\operatorname{Hom}_{R/x^2R}(\omega_R/x\omega_R, \omega_R/x^2\omega_R) &= \operatorname{Hom}_{R/x^2R}(\omega_R/x\omega_R, x\omega_R/x^2\omega_R)\\ &\cong \operatorname{Hom}_{R/xR}(\omega_R/x\omega_R, \omega_R/x\omega_R)\\ &= R/xR.\end{aligned}$$

The quotient map $R/x^2R \to R/xR$ lifts to $R/x^2R \to M^*$. Duality now gives $M \to \omega_R/x^2\omega_R$. Since $xM = i_x(\omega_R/x\omega_R)$ by assumption, the map $M \to \omega_R/x^2\omega_R$ is an isomorphism. □

The following is a special case of Herbrand quotients (see [Ful84, A.1]).

Lemma 7.13. *Let (R, m) be a local, 1-dimensional, S_1 ring with minimal primes P_i. Let F be a finite, torsion free R-module and $r \in m$ a non-zerodivisor. Then*

$$\operatorname{length}_0(F/rF) = \textstyle\sum_i \operatorname{length}_{P_i}(F_{P_i}) \cdot \operatorname{length}_0((R/P_i)/r(R/P_i)).$$

Proof Both sides are additive on short exact sequences of finite, torsion free modules. Thus, by 2.4, it is enough to prove the claim when R is integral and F has rank 1. Then we can realize F as an ideal $F \subset R$ such that R/F has finite length. Computing $\operatorname{length}_0(R/rF)$ two ways we get that

$$\operatorname{length}_0(R/rR) + \operatorname{length}_0(rR/rF) = \operatorname{length}_0(R/F) + \operatorname{length}_0(F/rF).$$

Since multiplication by r gives an isomorphism $R/F \cong rR/rF$, we are done. □

We have repeatedly used the following result of [Ree56], see also [BH93, 3.1.16].

Lemma 7.14. *Let R be a ring, N, M finite R-modules on $r \in R$. Assume that r is a non-zerodivisor on R, M and $rN = 0$. Then there are canonical isomorphisms*

$$\operatorname{Ext}^{i+1}_R(N, M) \cong \operatorname{Ext}^i_{R/rR}(N, M/rM) \quad \textit{for every } i \geq 0. \qquad \square$$

7.15 (Cones over surfaces). Let us recall first the basic facts about cones as in [Kol13b, Section 3.1]. Let S be a smooth, projective surface, H an ample line bundle and

$$(0, X) := C(S, L) := \operatorname{Spec} \textstyle\sum_{m\geq 0} H^0(S, H^m)$$

the corresponding affine cone over S. Then X is CM iff $H^1(S, H^m) = 0$ for every $m \in \mathbb{Z}$. If this holds then $\operatorname{Cl}(X) \cong \operatorname{Cl}(S)/[H] = \operatorname{Pic}(S)/[H]$.

If L is a line bundle on S then let $C(L)$ denote the corresponding divisorial sheaf on X. This sheaf is the sheafification of the module $\sum_{i=-\infty}^{\infty} H^0(S, L\otimes H^i)$. Note that $C(L)$ is CM iff $H^1(S, L\otimes H^m) = 0$ for every $m \in \mathbb{Z}$.

Claim 7.16. Only finitely many of the $C(L)$ are CM.

Proof Let L be a line bundle on S and $m := \lfloor (L\cdot H)/(H\cdot H)\rfloor$. Then $C(L) \cong C(L\otimes H^{-m})$ and the intersection number $\big((L\otimes H^{-m})\cdot H\big)$ is between 0 and $(H\cdot H)$. Set

$$\mathrm{Cl}^b(S) := \{[L]\colon 0 \le (L\cdot H) \le (H\cdot H)\} \subset \mathrm{Cl}(S).$$

We have proved that $\mathrm{Cl}^b(S) \to \mathrm{Cl}(X)$ is surjective. It is thus enough to show that $h^1(S, L) = 0$ holds for only finitely many line bundles in $\mathrm{Cl}^b(S)$.

By the Hodge index theorem, $L \mapsto (L\cdot L)$ is a negative definite quadratic function on $\mathrm{Cl}^b(S)$. Thus, by Riemann–Roch, $L \mapsto \chi(S, L)$ is the sum of a negative definite quadratic function and of a linear function.

On the other hand (see [Mat72] or [Kol96, VI.2.15.8]), by the Matsusaka inequality

$$h^0(S, L) \le \frac{(L\cdot H)^2}{(H\cdot H)} + 2,$$

so both $h^0(S, L)$ and $h^2(S, L) = h^0(S, \omega_S\otimes L^{-1})$ are bounded on $\mathrm{Cl}^b(S)$. Thus $L \mapsto h^1(S, L)$ is the sum of a positive definite quadratic function, a linear function and a bounded function on $\mathrm{Cl}^b(S)$. Therefore it has only finitely many zeros. □

8 Appendix by Hailong Dao

Let X be a CM scheme. We say that a coherent sheaf Ω is *CM-dualizing* if it is *TfS*$_2$-dualizing and the duality preserves maximal CM sheaves. That is, if M is torsion free and CM then so is $\mathcal{H}om_X(M, \Omega)$. Note that a dualizing sheaf ω_X is also CM-dualizing, see for example [BH93, 3.3.10]. If $\dim X = 2$ then CM is the same as S_2, thus every *TfS*$_2$-dualizing sheaf is also CM-dualizing. The situation is, however, quite different if $\dim X \ge 3$, as shown by the following result which answers a question of Kollár that was posed in the first version of this paper.

Theorem 8.1. *Let (x, X) be a local, CM scheme of dimension ≥ 3. Let Ω be a torsion free, coherent sheaf on X such that, for every torsion free, CM sheaf M, its dual $\mathcal{H}om_X(M, \Omega)$ is also torsion free and CM. Then Ω is a direct sum of copies of ω_X.*

Proof Note first that Ω is CM since $\Omega = \mathcal{H}om_X(\mathcal{O}_X, \Omega)$. Therefore

$$\mathcal{E}xt^j_X(k, \Omega) = 0 \quad \text{for } j < \dim X. \tag{8.1}$$

Let k be the residue field at $x \in X$ and consider a free resolution of it

$$\cdots \longrightarrow F_2 \xrightarrow{\phi_2} F_1 \xrightarrow{\phi_1} F_0 \xrightarrow{\phi_0} k.$$

Let $K_i := \ker \phi_i$ be the ith syzygy module of k and set $K_0 := k$. Note that the K_i are locally free on $X \setminus \{x\}$, in particular $\mathcal{E}xt^j_X(K_i, M)$ is supported on $\{x\}$ for every $j \geq 1$.

From $0 \to K_{i+1} \to F_{i+1} \to K_i \to 0$ we get that $\mathcal{E}xt^{j-1}_X(k, K_i) \cong \mathcal{E}xt^j_X(k, K_{i+1})$ for every $j \leq \dim X - 1$; in particular K_i is CM for $i \geq \dim X$.

Similarly, we get an exact sequence

$$0 \to \mathcal{H}om_X(K_i, \Omega) \to \mathcal{H}om_X(F_{i+1}, \Omega) \to \mathcal{H}om_X(K_{i+1}, \Omega) \\ \to \mathcal{E}xt^1_X(K_i, \Omega) \to 0$$

and isomorphisms

$$\mathcal{E}xt^j_X(K_{i+1}, \Omega) \cong \mathcal{E}xt^{j+1}_X(K_i, \Omega) \quad \text{for } j \geq 1.$$

Breaking the four term sequence into two short exact sequence gives that if $\mathcal{H}om_X(K_{i+1}, \Omega)$ is CM then

$$\mathcal{E}xt^1_X(K_i, \Omega) \cong H^2_x(X, \mathcal{H}om_X(K_i, \Omega)). \tag{8.2}$$

If $i \geq \dim X$ and $\dim X \geq 3$ then this shows that $\mathcal{E}xt^1_X(K_i, \Omega) = 0$ for $i \geq \dim X$. Using the isomorphisms (8.2), we get that

$$\mathcal{E}xt^j_X(k, \Omega) = 0 \quad \text{for } j > \dim X.$$

Combining this with (8.1) gives that Ω is a direct sum of copies of ω_X by [BH93, 3.3.28]. □

This immediately implies the following.

Corollary 8.2. *Let X be a CM scheme of pure dimension ≥ 3 and $\tilde{\omega}_1, \tilde{\omega}_2$ CM-dualizing sheaves on X. Then $\tilde{\omega}_1 \cong \tilde{\omega}_2 \otimes L$ for some line bundle L on X.* □

Acknowledgements

I thank H. Dao, D. Eisenbud and M. Hochster for helpful comments and references. Partial financial support was provided by the NSF under grant number DMS-1362960.

References

[BH93] Winfried Bruns and Jürgen Herzog. *Cohen-Macaulay rings*, volume 39 of *Cambridge Studies in Advanced Mathematics*. Cambridge University Press, Cambridge, 1993.

[DK16] Hailong Dao and Kazuhiko Kurano. Boundary and shape of Cohen-Macaulay cone. *Math. Ann.*, 364(3-4):713–736, 2016.

[Eis95] David Eisenbud. *Commutative algebra*, volume 150 of *Graduate Texts in Mathematics*. Springer-Verlag, New York, 1995. With a view toward algebraic geometry.

[EP03] Viviana Ene and Dorin Popescu. Rank one maximal Cohen-Macaulay modules over singularities of type $Y_1^3 + Y_2^3 + Y_3^3 + Y_4^3$. *Commutative algebra, singularities and computer algebra (Sinaia, 2002)*, volume 115 of *NATO Sci. Ser. II Math. Phys. Chem.*, pages 141–157. Kluwer Acad. Publ., Dordrecht, 2003.

[FFGR75] Robert Fossum, Hans-Bjørn Foxby, Phillip Griffith, and Idun Reiten. Minimal injective resolutions with applications to dualizing modules and Gorenstein modules. *Inst. Hautes Études Sci. Publ. Math.*, (45):193–215, 1975.

[Fox72] Hans-Bjørn Foxby. Gorenstein modules and related modules. *Math. Scand.*, 31:267–284 (1973), 1972.

[FR70] Daniel Ferrand and Michel Raynaud. Fibres formelles d'un anneau local noethérien. *Ann. Sci. École Norm. Sup. (4)*, 3:295–311, 1970.

[Ful84] William Fulton. *Intersection theory*, volume 2 of *Ergebnisse der Mathematik und ihrer Grenzgebiete (3)*. Springer-Verlag, Berlin, 1984.

[Gro60] A. Grothendieck. Éléments de géométrie algébrique. I–IV. *Inst. Hautes Études Sci. Publ. Math.*, (4,8,11,17,20,24,28,32), 1960.

[Har62] Robin Hartshorne. Complete intersections and connectedness. *Amer. J. Math.*, 84:497–508, 1962.

[Har77] Robin Hartshorne. *Algebraic geometry*. Springer-Verlag, New York-Heidelberg, 1977. Graduate Texts in Mathematics, No. 52.

[Har07] Robin Hartshorne. Generalized divisors and biliaison. *Illinois J. Math.*, 51(1):83–98, 2007.

[HH94] Melvin Hochster and Craig Huneke. Indecomposable canonical modules and connectedness. *Commutative algebra: syzygies, multiplicities, and birational algebra (South Hadley, MA, 1992)*, volume 159 of *Contemp. Math.*, pages 197–208. Amer. Math. Soc., Providence, RI, 1994.

[Kar09] N. Karroum. *MCM-einfache Moduln.* PhD thesis, Ruhr-Uni. Bochum, 2009.

[Kaw00] Takesi Kawasaki. On Macaulayfication of Noetherian schemes. *Trans. Amer. Math. Soc.*, 352(6):2517–2552, 2000.

[Knö87] Horst Knörrer. Cohen-Macaulay modules on hypersurface singularities. I. *Invent. Math.*, 88(1):153–164, 1987.

[Kol96] János Kollár. *Rational curves on algebraic varieties*, volume 32 of *Ergebnisse der Mathematik und ihrer Grenzgebiete. 3. Folge. A Series of Modern Surveys in Mathematics.* Springer-Verlag, Berlin, 1996.

[Kol07] János Kollár. *Lectures on resolution of singularities*, volume 166 of *Annals of Mathematics Studies.* Princeton University Press, Princeton, NJ, 2007.

[Kol08] János Kollár. Hulls and husks. `arXiv:0805.0576`, 2008.

[Kol13a] János Kollár. Moduli of varieties of general type. *Handbook of moduli. Vol. II*, volume 25 of *Adv. Lect. Math. (ALM)*, pages 131–157. Int. Press, Somerville, MA, 2013.

[Kol13b] János Kollár. *Singularities of the minimal model program*, volume 200 of *Cambridge Tracts in Mathematics.* Cambridge University Press, Cambridge, 2013. With a collaboration of Sándor Kovács.

[Kol17] János Kollár. Coherence of local and global hulls. *Methods Appl. Anal.*, 24(1):63–70, 2017.

[Kru30] Wolfgang Krull. Ein Satz über primäre Integritätsbereiche. *Math. Ann.*, 103(1):450–465, 1930.

[Mat72] T. Matsusaka. Polarized varieties with a given Hilbert polynomial. *Amer. J. Math.*, 94:1027–1077, 1972.

[Nag62] Masayoshi Nagata. *Local rings.* Interscience Tracts in Pure and Applied Mathematics, No. 13. Interscience Publishers a division of John Wiley & Sons New York-London, 1962.

[Ree56] D. Rees. A theorem of homological algebra. *Proc. Cambridge Philos. Soc.*, 52:605–610, 1956.

[Rei94] Miles Reid. Nonnormal del Pezzo surfaces. *Publ. Res. Inst. Math. Sci.*, 30(5):695–727, 1994.

[Ser59] Jean-Pierre Serre. *Groupes algébriques et corps de classes.* Publications de l'institut de mathématique de l'université de Nancago, VII. Hermann, Paris, 1959.

[Sta18] The Stacks Project Authors. *Stacks Project.* https://stacks.math.columbia.edu, 2018.

[SW07] Sean Sather-Wagstaff. Semidualizing modules and the divisor class group. *Illinois J. Math.*, 51(1):255–285, 2007.

Rationality of $\mathbb{Q}$-Fano Threefolds of Large Fano Index

Yuri Prokhorov

To Miles Reid on his 70th birthday

Abstract

We prove that $\mathbb{Q}$-Fano threefolds of Fano index ≥ 8 are rational.

1 Introduction

Recall that a projective algebraic variety X called $\mathbb{Q}$-*Fano* if it has only terminal $\mathbb{Q}$-factorial singularities, $\mathrm{Pic}(X) \simeq \mathbb{Z}$, and the anticanonical divisor $-K_X$ is ample. $\mathbb{Q}$-Fano varieties play a very important role in the higher dimensional geometry since they appear naturally in the minimal model program as building blocks in so-called Mori fiber spaces. It is known that $\mathbb{Q}$-Fano varieties of given dimension are bounded, that is, they form an algebraic family [Kaw92, Bir21]. Moreover, the method of [Kaw92] allows to produce a finite but very huge list of numerical candidates (Hilbert series) of $\mathbb{Q}$-Fanos [BK]. In dimension three there are a lot of classificational results of $\mathbb{Q}$-Fanos of special types, see, for example, [San96, Suz04, Tak06, Pro10, BKQ18, PR16], but the full classification is very far from being complete.

An important invariant of a $\mathbb{Q}$-Fano variety X is its $\mathbb{Q}$-*Fano index* $\mathrm{q}_{\mathbb{Q}}(X)$, the maximal integer q such that $-K_X \sim_{\mathbb{Q}} qA$ for some integral Weil divisor A, where $\sim_{\mathbb{Q}}$ defines the $\mathbb{Q}$-linear equivalence. In this paper we prove the following.

Theorem 1.1. *Let X be a $\mathbb{Q}$-Fano threefold with $\mathrm{q}_{\mathbb{Q}}(X) \geq 8$. Then X is rational.*

Note that in some sense our result is optimal: according to [Oka19] a very general weighted hypersurface $X_{14} \subset \mathbb{P}(2,3,4,5,7)$ is a non-rational (and even non-stably rational) $\mathbb{Q}$-Fano threefold with $\mathrm{q}_{\mathbb{Q}}(X) = 7$. On the other hand, the result of Theorem 1.1 can be essentially improved. We hope that non-rational $\mathbb{Q}$-Fano threefolds of large indices admit a reasonable classification.

The structure of the paper is as follows. Section 2 is preliminary. In Section 3, we list certain kinds of $\mathbb{Q}$-Fano threefolds with torsions in the Weil divisor class group $\mathrm{Cl}(X)$. In §4 the main birational construction is introduced. The proof of the main theorem is given in Sections 5–9 by case by case analysis.

2 Preliminaries

We work over the complex number field $\mathbb{C}$, and with notation:

- $\mathrm{Cl}(X)$ denotes the Weil divisors class group of a normal variety;
- $\mathrm{Cl}(X)_{\mathrm{t}}$ denotes the torsion part of $\mathrm{Cl}(X)$;
- $\mathbf{B}(X)$ is the basket of a terminal threefold X [Rei87];
- $\mathrm{r}(X,P)$ is the singularity index of a terminal point $P \in X$;
- $\mathrm{g}(X) := \dim|-K_X| - 1$ is the genus of a $\mathbb{Q}$-Fano threefold X.

For a $\mathbb{Q}$-Fano threefold X we define its *Fano* and *$\mathbb{Q}$-Fano index* by:

$$\mathrm{q_W}(X) := \max\{q \in \mathbb{Z} \mid -K_X \sim qA \text{ with } A \text{ a Weil divisor}\},$$
$$\mathrm{q}_{\mathbb{Q}}(X) := \max\{q \in \mathbb{Z} \mid -K_X \sim_{\mathbb{Q}} qA \text{ with } A \text{ a Weil divisor}\},$$

where $\sim$ (respectively. $\sim_{\mathbb{Q}}$) is the linear (respectively. $\mathbb{Q}$-linear) equivalence. Clearly, $\mathrm{q_W}(X)$ divides $\mathrm{q}_{\mathbb{Q}}(X)$, and $\mathrm{q_W}(X) = \mathrm{q}_{\mathbb{Q}}(X)$ unless $K_X + qA \in \mathrm{Cl}\, X$ is a nontrivial torsion element. Throughout this paper, for a $\mathbb{Q}$-Fano threefold X, by A we denote a Weil divisor such that $-K_X \sim_{\mathbb{Q}} \mathrm{q}_{\mathbb{Q}}(X)A$. If $\mathrm{q}_{\mathbb{Q}}(X) = \mathrm{q_W}(X)$ we take A so that $-K_X \sim \mathrm{q_W}(X)A$.

Theorem 2.1 ([Suz04]). *Let X be a $\mathbb{Q}$-Fano threefold. Then*

$$\mathrm{q}_{\mathbb{Q}}(X) \in \{1, \ldots, 11, 13, 17, 19\} \tag{2.1}$$

and all the possibilities do occur.

The following easy observation will be used freely.

Lemma 2.2 ([Kaw88, Lemma 5.1]). *Let $(X \ni P)$ be a threefold terminal singularity and let $\mathrm{Cl}^{\mathrm{sc}}(X, P)$ be the subgroup of the (analytic) Weil divisor class group consisting of Weil divisor classes which are $\mathbb{Q}$-Cartier. Then the group $\mathrm{Cl}^{\mathrm{sc}}(X, P)$ is cyclic of order $\mathrm{r}(X, P)$ and is generated by the canonical class K_X.*

Lemma 2.3. *Let X be a $\mathbb{Q}$-Fano threefold and let $\mathrm{r}(X)$ be the global Gorenstein index of X. Then the equality $\mathrm{q}_{\mathbb{Q}}(X) = \mathrm{q}_{\mathrm{W}}(X)$ holds if and only if $\mathrm{q}_{\mathbb{Q}}(X)$ and $\mathrm{r}(X)$ are coprime.*

Proof The "only if" part of the statement immediately follows from Lemma 2.2; see [Suz04, Lemma 1.2(3)]. Let us prove the "if" part. Assume that $\gcd(\mathrm{q}_{\mathbb{Q}}(X), \mathrm{r}(X)) = 1$. Put $q := \mathrm{q}_{\mathbb{Q}}(X)$ and write $-K_X \sim_{\mathbb{Q}} qA'$, where A' is a Weil divisor. Then $\Xi := K_X + qA'$ is a torsion element in $\mathrm{Cl}(X)$. Take $A = A' + t\Xi$, $t \in \mathbb{Z}$. Then

$$K_X + qA \sim (1 + qt)\Xi.$$

Since the order of Ξ in $\mathrm{Cl}(X)$ divides $\mathrm{r}(X)$, there exists $t \in \mathbb{Z}$ such that $(1 + qt)\Xi \sim 0$. □

The following proposition a consequence of the classification of $\mathbb{Q}$-Fano threefolds of large degree; see [Pro07, Pro10, Pro13].

Proposition 2.4. *Let X be a $\mathbb{Q}$-Fano threefold with indices $\mathrm{q}_{\mathbb{Q}}(X) = \mathrm{q}_{\mathrm{W}}(X) \geq 3$. Assume that X is not rational. Then X belongs to one of the following classes below.*

		dim $\|kA\|$				
$\mathrm{q}_{\mathbb{Q}}$	$\mathrm{g}(X)$	$\|A\|$	$\|2A\|$	$\|3A\|$	$\|4A\|$	$\|5A\|$
13	4	-1	-1	0	0	0
11	≤ 9	≤ 0	0	0	1	≤ 2
9	4	-1	0	0	1	1
8	≤ 10	≤ 0	0	≤ 1	≤ 2	≤ 3
7	≤ 14	≤ 0	≤ 1	≤ 2	≤ 4	≤ 6
6	≤ 15	≤ 0	≤ 1	≤ 3	≤ 6	≤ 11
5	≤ 18	≤ 1	≤ 3	≤ 7	≤ 12	
4	≤ 21	≤ 1	≤ 5	≤ 12		
3	≤ 20	≤ 2	≤ 9			

Proof Given an integer q, the $\mathbb{Q}$-Fano threefolds X with $\mathrm{q}_{\mathbb{Q}}(X) = q$ and genus $\mathrm{g}(X) \geq \mathrm{g}_q$ are completely described in [Pro10, Pro13, Pro16], where the number g_q is given by the second column in the table. It is easy to see that all these varieties are rational. The rest can be checked by a computer search as in [Suz04], [Pro10, Lemma 3.5] or [PR16, 2.4] (see also [BK]). □

Proposition 2.5 ([Kaw96, Kaw05]). *Let $Y \ni P$ be a threefold terminal point of index $r > 1$ and let*

$$f\colon (\tilde{Y} \supset E) \to (Y \ni P)$$

be a divisorial Mori extraction, where E is the exceptional divisor and $f(E) = P$. Write

$$K_{\tilde{Y}} = f^*K_Y + \alpha E.$$

Then the following assertions hold.

(1) If $Y \ni P$ is cyclic quotient singularity of type $\frac{1}{r}(1, a, r-a)$, then $\alpha = 1/r$ and f is a weighted blowup with weights $(1, a, r-a)$.
(2) If $Y \ni P$ is a point of type other than cA/r and $r > 2$, then $\alpha = 1/r$.
(3) If $Y \ni P$ is of type cA/r and its basket $\mathbf{B}(Y, P)$ consists of m points of index r, then $\alpha = a/r$, where $m \equiv 0 \mod a$.

3 $\mathbb{Q}$-Fano Threefolds with Torsion in the Class Group

3.1. Let X be a $\mathbb{Q}$-Fano threefold and let $\Xi \in \mathrm{Cl}(X)_{\mathrm{t}}$ be a non-trivial torsion element of order n. Then Ξ defines a finite étale in codimension two cover $\pi\colon X' \to X$ such that X' has only terminal singularities, $K_{X'} = \pi^*K_X$ and $\pi^*\Xi = 0$; see [Rei87, 3.6]. Clearly, X' is a Fano variety. However, in general, we cannot say that X' is $\mathbb{Q}$-factorial neither $\mathrm{Pic}(X') \simeq \mathbb{Z}$. Let $q := \mathrm{q}_{\mathbb{Q}}(X)$. Take A so that $-K_X \sim_{\mathbb{Q}} qA$ and let $A' := \pi^*A$. Then $-K_{X'} \sim_{\mathbb{Q}} qA'$. Hence, $\mathrm{q}_{\mathbb{Q}}(X')$ is divisible by q.

Remark 3.2. In the above notation, suppose $q \geq 5$. Run the MMP on X'. On each step the relation $-K_{X'} \sim_{\mathbb{Q}} qA'$ is preserved. Therefore, at the end, we obtain a $\mathbb{Q}$-Fano threefold X'' such that $-K_{X''} \sim_{\mathbb{Q}} qA''$ where $q \geq 5$. Then by (2.1), we have $\mathrm{q}_{\mathbb{Q}}(X'') = q$ and so $\mathrm{q}_{\mathbb{Q}}(X') = q$. Moreover,

$$\mathrm{g}(X'') \geq \mathrm{g}(X').$$

Proposition 3.3. *With notation as in 3.1, assume that $q \geq 3$ and $q \neq \mathrm{q_W}(X)$. Take $\Xi := K_X + qA$. Then*

$$(q, n) = (3, 3) \quad or \quad (4, 2). \tag{3.1}$$

Proof As in Proposition 2.4 we use a computer search. In this case the algorithm is modified as follows, see [Car08]. We denote $r_P := \mathrm{r}(X, P)$. Let $r := \operatorname{lcm}(\{r_P\})$ be the global Gorenstein index of X.

Step 1. By [Kaw92] we have the inequality

$$0 < -K_X \cdot c_2(X) = 24 - \sum_{P \in \mathbf{B}} \frac{r_P - 1}{r_P}.$$

This produces a finite (but huge) number of possibilities for the basket $\mathbf{B}(X)$ and the number $-K_X \cdot c_2(X)$.

Step 2. (2.1) implies that $q \in \{3, \dots, 11, 13, 17, 19\}$. In each case, we compute A^3 by the formula (see [Suz04])

$$A^3 = \frac{12}{(q-1)(q-2)} \Big(1 - \frac{A \cdot c_2(X)}{12} + \sum_{P \in B} c_P(-A)\Big)$$

where c_P is the correction term in the plurigenus formula [Rei87]. The number rA^3 must be a positive integer [Suz04, Lemma 1.2].

Step 3. Next, by [Suz04, Proposition 2.2] the Bogomolov–Miyaoka inequality (see [Kaw92]) implies that $(4q^2 - 3q)\, A^3 \le -4K_X \cdot c_2(X)$.

Step 4. In a neighborhood of each $P \in X$ we can write $A \sim l_P K_X$ by Lemma 2.2, where $0 \le l_P < r_P$. There is a finite number of possibilities for the collection $\{(l_P)\}$.

Step 5. The number n is determined as minimal positive such that $\chi(n\Xi) = 1$ (by the Kawamata–Viehweg vanishing). Hence, n can be computed by using orbifold Riemann–Roch.

Step 6. Finally, applying Kawamata–Viehweg vanishing we obtain

$$\chi(tA + s\Xi) = h^0(tA + s\Xi) = 0$$

for $-q < t < 0$ and $0 \le s < n$. Again, we check this condition using orbifold Riemann–Roch.

To run this algorithm the author used the computer algebra system PARI/GP [TPG08]. As the result, we get a short list from which one can see that (3.1) holds. □

Proposition 3.4. *Notation as in 3.1. Assume that $q \ge 5$ and $\mathrm{Cl}(X)_{\mathrm{t}}$ contains an element Ξ of order $n \ge 2$. Then $n \le 3$, $\mathrm{q}_{\mathbb{Q}}(X) = q$, and one of the cases listed in Table 9.1 holds. Moreover, the group $\mathrm{Cl}(X)_{\mathrm{t}}$ is cyclic and generated by Ξ.*

Table 9.1 *The possible cases of Proposition 3.4*

	n	q	$\mathrm{g}(X)$	$\mathbf{B}(X)$	A^3	$\mathbf{k}$	$\mathbf{B}(X')$	$\mathrm{g}(X')$
1°	2	5	2	$(2,4,14)$	$1/28$	$(1,0,7)$	$(4^2,7)$	4
2°	3	5	3	$(2,9,9)$	$1/18$	$(0,3,6)$	$(2^3,3^2)$	10
3°	2	5	5	$(4,4,12)$	$1/12$	$(0,2,6)$	$(2,4^2,6)$	10
4°	2	5	7	$(2,2,3,14)$	$5/42$	$(0,1,0,7)$	$(2^2,3^2,7)$	14
5°	2	5	10	$(2,3,4,12)$	$1/6$	$(0,0,2,6)$	$(2^3,3^2,6)$	20
6°	2	5	8	$(2,2,4,8)$	$1/8$	$(1,1,2,4)$	$(2,4)$	16
7°	2	5	11	$(2,4,4,6)$	$1/6$	$(1,2,2,3)$	$(2^2,3)$	21
8°	2	7	6	$(2,6,10)$	$1/30$	$(0,3,5)$	$(2^2,3,5)$	11
9°	2	7	7	$(2,2,3,4,8)$	$1/24$	$(1,1,0,2,4)$	$(2,3^2,4)$	14

Proof Similar to Proposition 3.3. But in this case, $\mathrm{q}_{\mathbb{Q}}(X) = \mathrm{q}_{\mathrm{W}}(X)$ and we have to modify one step:

Step 4′. In this case, $\gcd(q,r) = 1$ by Lemma 2.3. Since $K_X + qA \sim 0$, the numbers l_P are uniquely determined by $1 + ql_P \equiv 0 \mod r_P$. But for Ξ there are several choices. Again, near each point $P \in X$ we can write $\Xi \sim k_P K_X$ by Lemma 2.2, where for the collection $\mathbf{k} = (\{k_P\})$ there are only a finite number of possibilities.

We obtain a list $\{(n, q, \mathbf{B}(X), \mathrm{g}(X), A^3, \mathbf{k})\}$. In each case we compute the basket $\mathbf{B}(X')$ of a (terminal) Fano threefold X' with $A'^3 = nA^3$. By Remark 3.2 we have $\mathrm{q}_{\mathbb{Q}}(X') = q$. Then we can compute $\mathrm{g}(X')$ by orbifold Riemann–Roch. At the end, we get the list in the table and several extra possibilities which do not occur because $\mathrm{g}(X') \le 32$ in the case $\mathrm{q}_{\mathbb{Q}}(X') = 5$ by [Pro13, Theorem. 1.2(v)] and Remark 3.2. □

We do not assert that all the possibilities in Proposition 3.4 occur. We are able only to provide several examples for 2°, 6°-9°.

Example 3.5. The following quotient of weighted hypersurfaces are $\mathbb{Q}$-Fano threefolds as in 2°, 6°-9°.

2° $\{x_1^6 + x_2^3 + x_2'^3 + x_3x_3' = 0\} \subset \mathbb{P}(1, 2^2, 3^2)/\boldsymbol{\mu}_3(0, 1, -1, 1, -1)$

6° $\{x_1^6 + x_1'^6 + x_2x_4 + x_3^2 = 0\} \subset \mathbb{P}(1^2, 2, 3, 4)/\boldsymbol{\mu}_2(0, 1, 1, 1, 1)$

7° $\{x_1^4 + x_1'^4 + x_1x_3 + x_2^2 + x_2'^2 = 0\} \subset \mathbb{P}(1^2, 2^2, 3)/\boldsymbol{\mu}_2(0, 1, 1, 1, 0)$

8° $\{x_1^8 + x_2^4 + x_3x_5 + x_4^2 = 0\} \subset \mathbb{P}(1, 2, 3, 4, 5)/\boldsymbol{\mu}_2(0, 1, 1, 1, 1)$

9° $\{x_1^6 + x_2x_4 + x_3^2 + x_3'^2 = 0\} \subset \mathbb{P}(1, 2, 3^2, 4)/\boldsymbol{\mu}_2(0, 1, 0, 1, 1)$

One can expect also that the variety 1° is a quotient of a codimension four $\mathbb{Q}$-Fano (see [BK, No. 41418] and [CD20, Section 5.4]).

We compute dimensions of linear systems by orbifold Riemann–Roch:

Corollary 3.6. *In the cases 8° and 9° of Proposition 3.4 the dimension of the linear systems $|kA+s\Xi|$ are as follows*

	8°							9°						
k	1	2	3	4	5	6	7	1	2	3	4	5	6	7
$\dim\lvert kA\rvert$	0	0	0	1	2	4	6	−1	0	1	2	3	5	7
$\dim\lvert kA+\Xi\rvert$	−1	0	1	2	3	4	5	0	0	1	2	3	5	7

Combining 3.4 and 2.5 we obtain.

Corollary 3.7. *Let Y be a $\mathbb{Q}$-Fano threefold with $\mathrm{q}_{\mathbb{Q}}(X)\geq 5$. Assume that $\mathrm{Cl}(Y)_{\mathrm{t}}\neq 0$. Let $P\in Y$ be a non-Gorenstein point and let f be a divisorial Mori extraction of P. Then for the discrepancy α of the exceptional divisor $E\subset\tilde{Y}$ we have*

$$\alpha\leq\begin{cases}1 & \text{if } \mathrm{Cl}(Y)_{\mathrm{t}} \text{ is of order } 2,\\ 2/9 & \text{if } \mathrm{Cl}(Y)_{\mathrm{t}} \text{ is of order } 3.\end{cases}$$

4 Main Construction

4.1. Let X be a $\mathbb{Q}$-Fano threefold. For simplicity, we assume that the group $\mathrm{Cl}(X)$ is torsion free (this is the only case that we need). Denote $q=\mathrm{q}_{\mathbb{Q}}(X)=\mathrm{q}_{\mathrm{W}}(X)$. Thus $-K_X\sim qA$ and A is the ample generator of the group $\mathrm{Cl}(X)\simeq\mathbb{Z}$.

Consider a nonempty linear system $\mathscr{M}$ on X without fixed components. Let $c=\mathrm{ct}(X,\mathscr{M})$ be the canonical threshold of the pair $(X,\mathscr{M})$. Consider a log crepant blowup $f\colon\tilde{X}\to X$ with respect to $K_X+c\mathscr{M}$. One can choose f so that $\tilde{X}$ has only terminal $\mathbb{Q}$-factorial singularities, that is, f is a divisorial extraction in the Mori category (see [Cor95, Ale94]). Let E be the exceptional divisor. For α, $\beta\in\mathbb{Q}_{\geq 0}$, write

$$\begin{aligned} K_{\tilde{X}} &\sim_{\mathbb{Q}} f^*K_X+\alpha E,\\ \tilde{\mathscr{M}} &\sim_{\mathbb{Q}} f^*\mathscr{M}-\beta E \end{aligned} \tag{4.1}$$

where $\tilde{\mathscr{M}}$ is the birational transform of $\mathscr{M}$. Then $c=\alpha/\beta$.

Lemma 4.2 ([Pro10, Lemma 4.2]). *Let $P\in X$ be a point of index $r>1$. In a neighborhood of P we can write $\mathscr{M}\sim -tK_X$, where $0<t<r$. Then $c\leq 1/t$ and so $\beta\geq t\alpha$.*

Assume that the log divisor $-(K_X+c\mathscr{M})$ is ample. Run the log minimal model program with respect to $K_{\tilde{X}}+c\tilde{\mathscr{M}}$. We obtain the following diagram (Sarkisov link, see [Ale94, Pro10, Pro16])

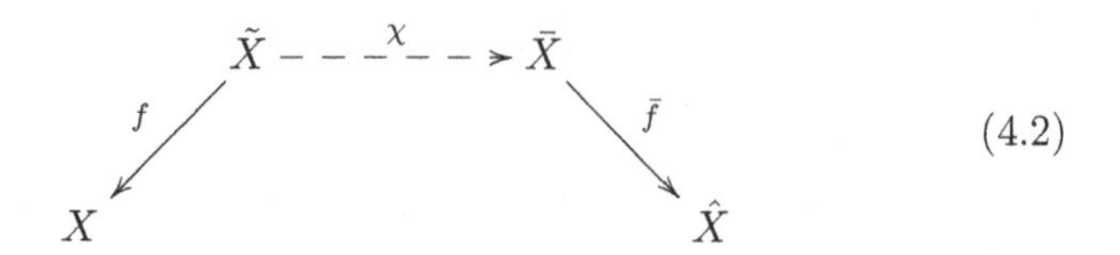

(4.2)

In the diagram, χ is a composition of $K_{\tilde{X}} + c\tilde{\mathscr{M}}$-log flips, the variety $\bar{X}$ has only terminal $\mathbb{Q}$-factorial singularities, $\rho(\bar{X}) = 2$, $\rho(\hat{X}) = 1$ and $\bar{f}\colon \bar{X} \to \hat{X}$ is an extremal $K_{\bar{X}}$-negative Mori contraction. In what follows, for the divisor (or linear system) D on X by $\tilde{D}$ and $\bar{D}$ we denote proper transforms of D on $\tilde{X}$ and $\bar{X}$ respectively.

If $|kA| \neq \emptyset$, we put $\mathscr{M}_k := |kA|$ (is it possible that $\mathscr{M}_k$ has fixed components in general). If $\dim \mathscr{M}_k = 0$, then by M_k we denote a unique effective divisor $M_k \in \mathscr{M}_k$. As in (4.1), we write

$$\tilde{\mathscr{M}}_k \sim_{\mathbb{Q}} f^*\mathscr{M}_k - \beta_k E. \tag{4.3}$$

4.3. Assume that the contraction $\bar{f}$ is birational. Then $\hat{X}$ is a $\mathbb{Q}$-Fano threefold. In this case, denote by $\bar{F}$ the $\bar{f}$-exceptional divisor, by $\tilde{F} \subset \tilde{X}$ its proper transform, $F := f(\tilde{F})$, and $\hat{q} := \mathrm{q}_{\mathbb{Q}}(\hat{X})$. Again we denote by $\hat{D}$ the proper (birational) transform of an object D (resp. $\tilde{D}$, $\bar{D}$) on X (resp. $\tilde{X}$, $\bar{X}$). Let Θ be an ample Weil divisor on $\hat{X}$ generating $\mathrm{Cl}(\hat{X})/\mathrm{Cl}(\hat{X})_{\mathrm{t}}$. Write

$$\hat{E} \sim_{\mathbb{Q}} e\Theta, \qquad \hat{\mathscr{M}}_k \sim_{\mathbb{Q}} s_k\Theta,$$

where $e \in \mathbb{Z}_{>0}$, $s_k \in \mathbb{Z}_{\geq 0}$. If $\dim \mathscr{M}_k = 0$ and $\bar{M}_k = \bar{F}$ (that is, a unique element M_k of the linear system $\bar{\mathscr{M}}_k$ is the $\bar{f}$-exceptional divisor), we put $s_k = 0$.

Lemma 4.4. *If in the above notation $\alpha < 1$, then* $\mathrm{g}(\hat{X}) \geq \mathrm{g}(X)$.

Proof We have $a(E, |-K_X|) < 1$. On the other hand, $0 = K_X + |-K_X|$ is Cartier. Hence, $a(E, |-K_X|) \leq 0$ and $K_{\tilde{X}} + f_*^{-1}|-K_X|$ is linearly equivalent to a non-positive multiple of E. Therefore we have the inclusion $f_*^{-1}|-K_X| \subset |-K_{\tilde{X}}|$ and so

$$\dim|-K_{\hat{X}}| \geq \dim|-K_{\bar{X}}| = \dim|-K_{\tilde{X}}| \geq \dim|-K_X|. \qquad \square$$

Note that in general, the group $\mathrm{Cl}(\hat{X})$ can have torsions:

Lemma 4.5 ([Pro10, Lemma 4.12]). *Write $F \sim dA$. Then*

$$\mathrm{Cl}(\hat{X})_{\mathrm{t}} \simeq \mathbb{Z}/n\mathbb{Z} \qquad \textit{where } n = d/e.$$

4.6. Assume that the contraction $\bar{f}$ is not birational. In this case, $\mathrm{Cl}(\hat{X})$ has no torsion. Therefore, $\mathrm{Cl}(\hat{X}) \simeq \mathbb{Z}$. Denote by Θ the ample generator of $\mathrm{Cl}(\hat{X})$ and by $\bar{F}$ a general geometric fiber. Then $\bar{F}$ is either a smooth rational curve or a del Pezzo surface. The image of the restriction map $\mathrm{Cl}(\bar{X}) \to \mathrm{Pic}(\bar{F})$ is isomorphic to $\mathbb{Z}$. Let Λ be its ample generator. As above, we can write

$$-K_{\bar{X}}|_{\bar{F}} = -K_{\bar{F}} \sim \hat{q}\Lambda, \qquad \bar{E}|_{\bar{F}} \sim e\Lambda, \qquad \bar{\mathscr{M}}_k|_{\bar{F}} \sim s_k\Lambda,$$

where $\hat{q}$, $e \in \mathbb{Z}_{>0}$, $s_k \in \mathbb{Z}_{\geq 0}$.

If $\hat{X}$ is a curve, then $\hat{q} \leq 3$ and $\hat{X} \simeq \mathbb{P}^1$. If $\hat{X}$ is a surface, then $\hat{q} \leq 2$. In this case, the variety $\hat{X}$ can have only Du Val singularities of type $\mathrm{A_n}$ (see [MP08, Theorem 1.2.7]).

Lemma 4.7. *If the contraction $\bar{f}$ is not birational and $\hat{q} > 1$, then X is rational.*

Proof Indeed, if $\hat{X}$ is a curve and $\hat{q} \geq 2$, then a general fiber $\bar{F}$ is a del Pezzo surface with divisible canonical class. Then $\bar{F}$ is either a projective plane or a quadric. Clearly, $\bar{X}$ is rational in this case. Similarly, if $\hat{X}$ is a surface and $\hat{q} = 2$, then there is a divisor which is a generically section of $\bar{f}$ and $\bar{X}$ is again rational. □

4.8. Since the group $\mathrm{Cl}(\bar{X})$ has no torsion, the numerical equivalence of Weil divisors on $\bar{X}$ coincides with linear one. Hence the relations (4.1) and (4.3) give us

$$kK_{\tilde{X}} + q\tilde{\mathscr{M}}_k \sim f^*(kK_X + q\mathscr{M}_k) + (k\alpha - q\beta_k)E \sim (k\alpha - q\beta_k)E$$

where $k\alpha - q\beta_k \in \mathbb{Z}$. From this we obtain the following important equality which will be used throughout this paper:

$$k\hat{q} = qs_k + (q\beta_k - k\alpha)e. \tag{4.4}$$

4.9. Suppose that the morphism $\bar{f}$ is birational. Then

$$K_{\bar{X}} \sim_{\mathbb{Q}} \bar{f}^*K_{\hat{X}} + b\bar{F}, \quad \bar{\mathscr{M}}_k \sim_{\mathbb{Q}} \bar{f}^*\hat{\mathscr{M}}_k - \gamma_k\bar{F}, \quad \bar{E} \sim_{\mathbb{Q}} \bar{f}^*\hat{E} - \delta\bar{F}$$

as (4.1) and (4.3). This gives us

$$s_kK_{\bar{X}} + \hat{q}\bar{\mathscr{M}}_k \sim_{\mathbb{Q}} (bs_k - \hat{q}\gamma_k)\bar{F},$$
$$eK_{\bar{X}} + \hat{q}\bar{E} \sim_{\mathbb{Q}} (be - \hat{q}\delta)\bar{F}.$$

Taking proper transforms of these relations to X, we obtain

$$-qs_k + \hat{q}k = ne(bs_k - \hat{q}\gamma_k), \tag{4.5}$$
$$-q = n(be - \hat{q}\delta). \tag{4.6}$$

Corollary 4.10. *If, in the above notation,* $\gcd(n, q) = 1$, *then* $\bar{f}(\bar{F})$ *is a point on* $\hat{X}$ *whose index is divisible by* n.

Proof Indeed, either the discrepancy b of $\bar{F}$ or the multiplicity δ is fractional and its denominator is divisible by n according to (4.6). □

5 $\mathbb{Q}$-Fano Threefolds of Fano Index 7 and Large Genus

Now we apply the techniques outlined in the previous section to $\mathbb{Q}$-Fano threefolds of indices ≥ 7.

Proposition 5.1. *Let* X *be a* $\mathbb{Q}$*-Fano threefold with* $\mathrm{q}_{\mathbb{Q}}(X) = 7$ *and* $\mathrm{g}(X) \geq 11$. *Then* X *is rational.*

Proof By Proposition 3.4, the group $\mathrm{Cl}(X)$ is torsion free. Assume that X is not rational. By [Pro16, Theorem 1.2, Proposition 2.1], we have

$$\mathbf{B}(X) = (2, 2, 3, r), \tag{5.1}$$

where for r there are only two possibilities:

$$r = 5, \quad A^3 = 1/15, \quad \mathrm{g}(X) = 11;$$
$$r = 12, \quad A^3 = 1/12, \quad \mathrm{g}(X) = 13.$$

In particular, X has only cyclic quotient singularities. By the orbifold Riemann–Roch in both cases we have

$$\dim |kA| = k - 1 \quad \text{for } k = 1,\ 2,\ 3.$$

Hence the linear system $|A|$ contains a unique irreducible surface M_1 and $|kA|$ has no fixed components for $k = 2$ and 3.

5.2. Apply the construction (4.2) with $\mathscr{M} = |3A|$. In a neighborhood of the point of index r ($r = 5$ or 12) we have $\mathscr{M} \sim -tK_X$, where

$$t = \begin{cases} 4 & \text{if } r = 5, \\ 9 & \text{if } r = 12. \end{cases}$$

Then by Lemma 4.2

$$\beta_3 \geq t\alpha. \tag{5.2}$$

The relation (4.4) for $k = 3$ has the form

$$3\hat{q} = 7s_3 + (7\beta_3 - 3\alpha)e \geq 7s_3 + (7t - 3)\alpha e, \tag{5.3}$$

where $\hat{q} \leq 13$ by Proposition 2.4. If the contraction $\bar{f}$ is not birational, then $\hat{q} = 1$ by Lemma 4.7. Hence, $\alpha \leq 3/(7t-3)$. On the other hand,

$$\alpha \geq 1/r > 3/(7t-3).$$

The contradiction shows that the contraction $\bar{f}$ must be birational. In particular, the movable linear system $\mathscr{M}$ is not contracted, that is, $s_3 \geq 1$.

5.3. If $\alpha \geq 1$, then inequality (5.3) and Proposition 2.4 give successively

$$3\hat{q} \geq 7s_3 + 25e, \qquad \hat{q} \geq 11, \qquad s_3 \geq 5, \qquad \hat{q} > 19,$$

a contradiction. Taking (5.1) into account we see that $P := f(E)$ is a non-Gorenstein point of X and f is the weighted blowup as in Proposition 2.51 (so-called *Kawamata blowup*). In particular, $\alpha = 1/\mathrm{r}(X, P)$. In this case by Lemma 4.4 we have

$$\mathrm{g}(\hat{X}) \geq \mathrm{g}(X) \geq 11.$$

Since $\hat{X}$ is not rational, according to Proposition 2.4 we have $\hat{q} \leq 7$. Note that $(7t-3)\alpha e \geq 5$. Then (5.3) implies $s_3 \leq 2$.

5.4 ($\mathrm{r}(X, P) = 2$). Then $\alpha = 1/2$ and $\beta_3 = 1/2 + m_3$, where $m_3 \geq 2$ by (5.2). We can rewrite (5.3) in the following form

$$3\hat{q} = 7s_3 + (7\beta_3 - 3\alpha)e = 2e + 7(s_3 + m_3 e).$$

Since $\hat{q} \leq 7$, this equation has no solutions.

5.5 ($\mathrm{r}(X, P) = 3$). Then, as above, $\alpha = 1/3$, $\beta_3 \geq 2$ is an integer, and (5.3) has the following form, which again has no solutions:

$$3\hat{q} = 7s_3 + (7\beta_3 - 3\alpha)e = -e + 7(s_3 + \beta_3 e).$$

5.6 ($\mathrm{r}(X, P) = r$, $r = 5$ or 12). Then $\beta_1 = t'/r + m_1$, where $m_1 \geq 0$, and $t' = 3$ if $r = 5$ and $t' = 7$ if $r = 12$. Then (4.4) for $k = 1$ has form

$$7 \geq \hat{q} = 7s_1 + (7\beta_1 - \alpha)e = 4e + 7(s_1 + m_1 e).$$

From this we obtain $s_1 = 0$ and $\hat{q} = 4$. Then from (5.3) we obtain $s_3 = 1$. Since $s_1 = 0$, the group $\mathrm{Cl}(\hat{X})$ is torsion free by Lemma 4.5. Thus $\hat{\mathscr{M}} \sim 0$ and so $\dim |\Theta| \geq 2$. This contradicts Proposition 2.4. □

Corollary 5.7. *Let X be a $\mathbb{Q}$-Fano threefold with $\mathrm{q}_{\mathbb{Q}}(X) = 7$ and let A be a Weil divisor such that $-K_X \sim_{\mathbb{Q}} 7A$ (here we do not claim that $-K_X \sim 7A$). Assume that* $\dim |2A| \geq 1$. *Then X is rational.*

Proof By Corollary 3.6, the group $\mathrm{Cl}(X)$ is torsion free. Then a computer search gives us $\mathrm{g}(X) \geq 11$. □

6 $\mathbb{Q}$-Fano Threefolds of Fano Index 13

Proposition 6.1. *Let X be a $\mathbb{Q}$-Fano threefold with $\mathrm{q}_{\mathbb{Q}}(X) = 13$. Then X is rational.*

Proof By Proposition 3.4, the group $\mathrm{Cl}(X)$ is torsion free. Assume X is not rational. According to [Pro10], we have to consider only one case:

$$A^3 = \tfrac{1}{210}, \qquad \mathbf{B} = (2, 3, 3, 5, 7). \tag{6.1}$$

One can expect, although it is not known, that all the varieties of this type are hypersurfaces $X_{12} \subset \mathbb{P}(3, 4, 5, 6, 7)$ (see [BS07]).

By orbifold Riemann–Roch, (6.1) implies that $|A| = |2A| = \emptyset$, the linear system $|kA|$ for $k = 3, 4, 5$ contains a unique irreducible surface M_k and for $k = 6, 7, 8$ the linear system $|kA|$ is a pencil $\mathscr{M}_k$ without fixed components [Pro10, Proposition 3.6].

6.2. Apply the construction (4.2) with $\mathscr{M} = |8A|$. Then near the point of index 7 we have $\mathscr{M} \sim -6K_X$. By Lemma 4.2

$$\beta_8 \geq 6\alpha. \tag{6.2}$$

The relation (4.4) for $k = 8$ has the form

$$8\hat{q} = 13s_8 + (13\beta_8 - 8\alpha)e \geq 13s_8 + 70e\alpha, \tag{6.3}$$

where $\hat{q} \leq 13$ by Proposition 2.4. Since $\alpha \geq 1/7$, we see that $\hat{q} > 1$. By Lemma 4.7, this implies that the contraction $\bar{f}$ is birational and so $s_8 > 0$. We also have

$$\tilde{\mathscr{M}} \sim_{\mathbb{Q}} \tilde{M}_3 + \tilde{M}_5 + (\beta_3 + \beta_5 - \beta_8)E \sim_{\mathbb{Q}} 2\tilde{M}_4 + (2\beta_4 - \beta_8)E,$$

where $\beta_3 + \beta_5 \geq \beta_8$ and $2\beta_4 \geq \beta_8$. Pushing forward this relation to $\hat{X}$ we obtain

$$s_8 = s_3 + s_5 + (\beta_3 + \beta_5 - \beta_8)e = 2s_4 + (2\beta_4 - \beta_8)e.$$

Since the $\bar{f}$-exceptional divisor is irreducible, only one of the numbers s_3, s_4, s_5 can be equal to 0. Therefore $s_8 \geq 2$.

6.3. If $\alpha \geq 2/3$, then the relation (6.3) gives us $\hat{q} \geq 10$. Then $\mathrm{Cl}(\hat{X})$ is torsion free by Proposition 3.4 and $\dim |k\Theta| \leq 0$ for $k = 1, 2, 3$ by Proposition 2.4. Hence, $s_8 \geq 4$. Then $\hat{q} \geq 13$ and so $s_8 \geq 6$, $\hat{q} > 13$, a contradiction. Therefore $P := f(E)$ is a non-Gorenstein point of X and f is the Kawamata blowup of P by Proposition 2.51. In particular, $\alpha = 1/\mathrm{r}(X, P)$, where $\mathrm{r}(X, P) = 2$, 3, 5 or 7.

6.4 ($\mathrm{r}(X,P)=2$). Then β_8 is an integer ≥ 3 by (6.2). The relation (6.3) has the form

$$8\hat{q} = -4e + 13(s_8 + \beta_8 e).$$

It has no solutions satisfying the inequalities $s_8 \geq 2$, $\beta_8 \geq 3$, $\hat{q} \leq 13$.

6.5 ($\mathrm{r}(X,P)=3$). Assume $\mathrm{r}(X,P)=3$. Then as above $\beta_8 = 2/3 + m_8$, $m_8 \geq 2$, and

$$8\hat{q} = 6e + 13(s_8 + m_8 e).$$

Again the equation has no suitable solutions.

6.6 ($\mathrm{r}(X,P)=5$). Then near the point of index 5 we have $-K_X \sim \mathscr{M}_8$. Hence $\beta_8 = 1/5 + m_8$, where $m_8 \geq 1$. The relation (6.3) has the form

$$8\hat{q} = e + 13(s_8 + m_8 e).$$

We get only one solution: $\hat{q}=5$, $e=1$, $s_8=2$. Since $e=1$, we have $d=n$ by Lemma 4.5. Since $|A| = |2A| = \emptyset$, we have $d \geq 3$ and so $n=d=3$ by Proposition 3.4. Thus $\mathrm{Cl}(\hat{X})_{\mathrm{t}} \simeq \mathbb{Z}/3\mathbb{Z}$. Then the image $\bar{f}(\bar{F})$ is a non-Gorenstein point according to Corollary 4.10. For $k=8$ the relation (4.5) yields $b \geq 7/3$. This contradicts Corollary 3.7.

6.7 ($\mathrm{r}(X,P)=7$). Finally suppose $\mathrm{r}(X,P)=7$. Then $\beta_8 = 6/7 + m_8$, where $m_8 \geq 0$. Hence,

$$8\hat{q} = 10e + 13(s_8 + m_8 e). \tag{6.4}$$

If $e \leq 2$ then the torsion part of $\mathrm{Cl}(\hat{X})$ is non-trivial Lemma 4.5 because $|2A| = \emptyset$. By Proposition 3.4, we have $\hat{q} \leq 7$ and then (6.4) has no solutions. Thus $e \geq 3$ and then there is only one possibility: $\hat{q}=7$, $s_8=2$. Then $\hat{X}$ is rational by Corollary 5.7. □

7 $\mathbb{Q}$-Fano Threefolds of Fano Index 11

Proposition 7.1. *Let X be a $\mathbb{Q}$-Fano threefold with $\mathrm{q}_{\mathbb{Q}}(X) = 9$. Then X is rational.*

Proof By Proposition 3.4, the group $\mathrm{Cl}(X)$ is torsion free. According to Proposition 2.4 and [Pro10], we have to consider only two cases:

			dim $\lvert kA\rvert$						
	B	A^3	$\lvert A\rvert$	$\lvert 2A\rvert$	$\lvert 3A\rvert$	$\lvert 4A\rvert$	$\lvert 5A\rvert$	$\lvert 6A\rvert$	$\mathrm{g}(X)$
1°	$(2,5,7)$	$1/70$	0	0	0	1	2	3	9
2°	$(2,2,3,4,7)$	$1/84$	-1	0	0	1	1	2	7

There are examples of varieties of these types: they are hypersurfaces $X_{12} \subset \mathbb{P}(1,4,5,6,7)$ and $X_{10} \subset \mathbb{P}(2,3,4,5,7)$ in cases 1° and 2°, respectively [BS07].

7.2. From the table above one can see that in both cases the linear systems $|kA|$ have no fixed components for $k = 4,\ 5,\ 6$. Apply the construction (4.2) with $\mathscr{M} = |5A|$. Then near the point of index 7 we have $A \sim -2K_X$, $\mathscr{M} \sim -3K_X$. By Lemma 4.2

$$\beta_5 \geq 3\alpha. \tag{7.1}$$

The relation (4.4) for $k = 5$ has the form

$$5\hat{q} = 11s_5 + (11\beta_5 - 5\alpha)e = -5\alpha e + 11(s_5 + \beta_5 e) \geq 11s_5 + 28\alpha e.$$

Assume that X is not rational. Then $\hat{q} \leq 11$ by Propositions 2.4 and 6.1.

7.3. Assume that $\alpha \geq 1$. Then $\hat{q} \geq 6$ and α is an integer by Proposition 2.5. Moreover, $\alpha = e = 1$ and $s_5 + \beta_5 e$ is also an integer. Hence, $\hat{q} \equiv -1 \mod 11$. This contradicts (2.1). Therefore, $\alpha < 1$ by Proposition 2.5*1*. In particular,

$$\alpha = 1/r, \qquad r := \mathrm{r}(X, P) = 2,\ 3,\ 4,\ 5 \text{ or } 7. \tag{7.2}$$

7.4. Assume that $\bar{f}$ is not birational. Since $\bar{X}$ is not rational by our assumptions, $\hat{q} = 1$ (see Lemma 4.7). Then $s_5 = 0$ and $5 = (11\beta_5 - 5\alpha)e$, where $11\beta_5 - 5\alpha \in \mathbb{Z}$. Then $\beta_5 = l/r$, $l \in \mathbb{Z}$ and $l \geq 3$ by (7.1). Thus we can write $5r = (11l - 5)e$. But this equation has no solutions satisfying (7.2). Therefore, the contraction $\bar{f}$ is birational. In particular, $s_5 > 0$.

7.5 (Cases 1° and 2° with $\mathrm{r}(X, P) = 2$)**.** Then $\beta_5 = 1/2 + m_5$, $m_5 \geq 1$. Thus (4.4) for $k = 5$ has the form

$$5\hat{q} = 3e + 11(s_5 + m_5 e).$$

We get one possibility: $\hat{q} = 5$, $e = 1$, $s_5 = 1$.

In the case 1°, the linear system $|A|$ contains a unique member M_1. Then (4.4) for $k = 1$ has a similar form

$$5 = \hat{q} = 5e + 11(s_1 + m_1 e), \quad m_1 \geq 0.$$

We obtain $s_1 = 0$. So, $\mathrm{Cl}(\hat{X})$ is torsion free by Lemma 4.5. Since $\dim|\Theta| = 2$, the variety $\hat{X}$ is rational by Proposition 2.4.

In the case 2° the map $\bar{f} \circ \chi \circ f^{-1}$ contracts a divisor $F \sim dA$ with $d > 1$ (because $|A| = \emptyset$). Since $e = 1$, by Lemma 4.5 we have $\mathrm{Cl}(\hat{X})_{\mathrm{t}} \simeq \mathbb{Z}/n\mathbb{Z}$

with $n = d > 1$. Apply (4.5)-(4.6). Recall that $n \le 3$ (see Proposition 3.4). In particular, $\gcd(n, 11) = 1$. Then the image $\bar{f}(\bar{F})$ is a non-Gorenstein point according to Corollary 4.10. For $k = 5$ the relation (4.5) yields $14 = n(b - 5\gamma_5)$. According to Corollary 3.7 this is impossible.

7.6 (Cases 1° and 2° with $\mathrm{r}(X, P) = 7$). Then $\beta_5 = 3/7 + m_5$ and $\beta_6 = 5/7 + m_6$, where $m_5, m_6 \ge 0$. Then (4.4) for $k = 5, 6$ has form

$$\begin{aligned} 5\hat{q} &= 4e + 11(s_5 + m_5 e) \\ 6\hat{q} &= 7e + 11(s_6 + m_6 e). \end{aligned} \tag{7.3}$$

Here $s_5 \le 3$ because $\hat{q} \le 11$. By Proposition 2.4, we have $\hat{q} \ne 9$ because $\mathrm{g}(\hat{X}) \ge \mathrm{g}(X) \ge 7$. Then in (7.3) one has $\hat{q} = 3e$, $s_5 = s_6 = e = 1$ or 2.

Assume that $\hat{q} = 6$ (and $e = s_5 = s_6 = 2$). In the case 1°, we have

$$6\bar{M}_1 + (6\beta_1 - \beta_6)\bar{E} \sim_{\mathbb{Q}} \bar{\mathscr{M}}_6 \sim_{\mathbb{Q}} 2\Theta, \quad 6\beta_1 \ge \beta_6.$$

Hence the divisor $\bar{M}_1$ is contracted (otherwise the class of Θ in the group $\mathrm{Cl}(\hat{X})/\mathrm{Cl}(\hat{X})_{\mathrm{t}}$ would be divisible). Since $e = 2$, this contradicts Lemma 4.5. In the case 2°, from the relation

$$3\bar{M}_2 + (3\beta_2 - \beta_6)\bar{E} \sim_{\mathbb{Q}} \bar{\mathscr{M}}_6 \sim_{\mathbb{Q}} 2\Theta,$$

we see that the divisor $\bar{M}_2$ must be contracted. Since $e = 2$, the group $\mathrm{Cl}(\hat{X})$ is torsion free by Lemma 4.5. Since $s_6 = 2$ and $\dim \mathscr{M}_6 = 2$, we have $\dim |2\Theta| \ge 2$. This contradicts Proposition 2.4.

Finally, assume that $\hat{q} = 3$ (and $e = s_5 = s_6 = 1$). In case 1°, we have

$$6\bar{M}_1 + (6\beta_1 - \beta_6)\bar{E} \sim_{\mathbb{Q}} \bar{\mathscr{M}}_6 \sim_{\mathbb{Q}} \Theta, \quad 6\beta_1 \ge \beta_6.$$

As above, the divisor $\bar{M}_1$ must be contracted and the group $\mathrm{Cl}(\hat{X})$ is torsion free. Since $s_6 = 1$ and $\dim \mathscr{M}_6 = 3$, we have $\dim |\Theta| \ge 3$. This contradicts Proposition 2.4.

In the case 2°, we have

$$3\bar{M}_2 + (3\beta_2 - \beta_6)\bar{E} \sim_{\mathbb{Q}} 2\bar{M}_3 + (2\beta_3 - \beta_6)\bar{E} \sim_{\mathbb{Q}} \bar{\mathscr{M}}_6 \sim_{\mathbb{Q}} \Theta,$$

where $3\beta_2 \ge \beta_6$, $2\beta_3 \ge \beta_6$. Since both $\bar{M}_2$ and $\bar{M}_3$ cannot be contracted simultaneously, this gives a contradiction.

7.7 (Case 2° with $\mathrm{r}(X, P) = 3$). Then $\beta_5 = 1/3 + m_5$, $m_5 \ge 1$. Thus

$$5\hat{q} = 2e + 11(s_5 + m_5 e)$$

and we obtain $\hat{q} = 7$ and $s_5 \le 2$. Then $\hat{X}$ is rational by Corollary 5.7.

7.8 (Case 2° with $\mathrm{r}(X, P) = 4$). Then $\beta_5 = 3/4 + m_5$, $m_5 \geq 0$. If $m_5 = 0$, then $\mathrm{ct}(X, \mathscr{M}) = 1/3$. In this case, $(X, \frac{1}{3}\mathscr{M})$ is canonical and points of indices 4 and 7 are canonical centers. Then we can apply our construction (4.2) starting with the point of index 7, as in 7.6. This gives a rationality construction.

Thus we assume that $m_5 \geq 1$. The relation (4.4) for has the form

$$5\hat{q} = 11(s_5 + m_5 e) + 7e$$

and then $\hat{q} = 8$, $s_5 \leq 2$. By Proposition 2.4, the variety $\hat{X}$ is rational.

7.9 (Case 1° with $\mathrm{r}(X, P) = 5$). Then $\mathscr{M}$ is a Cartier at P and so β_5 must be a positive integer. The relations (4.4) has the form $5\hat{q} = 11(s_5 + \beta_5 e) - e$. Since $\hat{q} \leq 11$, this equation has no solutions. □

8 $\mathbb{Q}$-Fano Threefolds of Fano Index 9

Proposition 8.1. *Let X be a $\mathbb{Q}$-Fano threefold with $\mathrm{q}_{\mathbb{Q}}(X) = 9$. Then X is rational.*

Proof By Proposition 3.4, the group $\mathrm{Cl}(X)$ is torsion free. Assume that X is not rational. According to [Pro10, Proposition 3.6], we have to consider only one case:

$$\mathbf{B} = (2, 2, 2, 5, 7), \qquad A^3 = 1/70. \tag{8.1}$$

By orbifold Riemann–Roch, (8.1) implies that

$$|A| = \emptyset, \quad \dim |2A| = \dim |3A| = 0, \quad \dim |4A| = \dim |5A| = 1.$$

Thus the linear system $|kA|$ contains a unique irreducible surface M_k for $k = 2, 3$ and $|kA|$ for $k = 4, 5$ is a pencil without fixed components.

8.2. Apply the construction (4.2) with $\mathscr{M} = |5A|$. Then near the point of index 7 we have $\mathscr{M} \sim -6K_X$. By Lemma 4.2, $\beta_5 \geq 6\alpha$. The relation (4.4) for $k = 5$ has the form $5\hat{q} = 9s_5 + (9\beta_5 - 5\alpha)e \geq 7s_5 + 49\alpha e$.

8.3. By Propositions 2.4, 6.1, and 7.1, we have $\hat{q} \leq 9$. Then, obviously, $\alpha < 1$. Therefore $P := f(E)$ is a non-Gorenstein point of X by Proposition 2.51 and $\alpha = 1/\mathrm{r}(X, P)$, where $\mathrm{r}(X, P) = 2$, 5 or 7.

8.4. If $\bar{f}$ is not birational, then $\hat{q} = 1$ by Lemma 4.7 and so $s_5 = 0$, that is, $\bar{\mathscr{M}}$ is $\bar{f}$-vertical. Note that $9\beta_5 - 5\alpha$ is an integer (since $9\mathscr{M}_5 + 5K_X$ is Cartier). Hence $9\beta_5 - 5\alpha = 1$ or 5. Let $r := \mathrm{r}(X, P)$. Then $\beta_5 = l/r$

for some l and $9l = r + 5$ or $5(r + 1)$. For $r = 2, 5, 7$ this equation has no solutions. The contradiction shows that $\bar{f}$ is birational. Thus $s_5 > 0$.

8.5 ($\mathrm{r}(X, P) = 2$). Then $\beta_5 = 1/2 + m_5$, $m_5 \geq 3$ and the relation (4.4) for $k = 5$ is $5\hat{q} = 2e + 9(s_5 + m_5 e)$, which is impossible since $\hat{q} \leq 9$.

8.6 ($\mathrm{r}(X, P) = 5$). Then β_5 is an integer ≥ 2 and, as above, $5\hat{q} = -e + 9(s_5 + \beta_5 e)$. We get one possibility: $\hat{q} = 7$, $e = 1$, $s_5 + \beta_5 = 4$. Since $|A| = \emptyset$, the group $\mathrm{Cl}(\hat{X})_{\mathrm{t}}$ is non-trivial by by Lemma 4.5. By Proposition 3.4, we have $\mathrm{Cl}(\hat{X})_{\mathrm{t}} \simeq \mathbb{Z}/2\mathbb{Z}$. By Corollary 4.10, the image $\bar{f}(\bar{F})$ is a point of even index. The relation (4.5) for $k = 5$ has the form

$$35 - 9s_5 = 2(bs_5 - 7\gamma_5), \qquad b \geq (35 - 9s_5)/2s_5 \geq 17/4.$$

Then we obtain a contradiction by Corollary 3.7.

8.7 ($\mathrm{r}(X, P) = 7$). Then $\beta_5 = 6/7 + m_5$, $m_5 \geq 0$,

$$5\hat{q} = 9s_5 + (9\beta_5 - 5\alpha)e = 7e + 9(s_5 + m_5 e).$$

We get the following possibilities $(\hat{q}, e) = (5, 1)$ or $(6, 3)$.

If $\hat{q} = 6$ then the group $\mathrm{Cl}(\hat{X})$ is torsion free by Proposition 3.4. Since $s_5 + 3m_5 = 1$, we have $s_5 = 1$. Hence $\dim |\Theta| \geq 1$. This contradicts Proposition 2.4.

Consider the case $\hat{q} = 5$. Then $s_5 \leq 2$ Since $e = 1$ and $|A| = \emptyset$, by Lemma 4.5 we have $\mathrm{Cl}(\hat{X})_{\mathrm{t}} \simeq \mathbb{Z}/n\mathbb{Z}$ with $n = d > 1$. Apply (4.5) with $k = 5$. We obtain $25 - 9s_5 \leq ns_5 b$ and so $b \geq 7/2n$. Since $n \leq 3$, we get a contradiction by Corollary 3.7. □

9 $\mathbb{Q}$-Fano Threefolds of Fano Index 8

Proposition 9.1. *Let X be a $\mathbb{Q}$-Fano threefold with $\mathrm{q}_{\mathbb{Q}}(X) = 8$. Then X is rational.*

Proof By Proposition 3.4, the group $\mathrm{Cl}(X)$ is torsion free. Assume that X is not rational. Using a computer search and taking Proposition 2.4 into account we obtain the following possibilities:

B	A^3	$\dim \lvert kA\rvert$					$\mathrm{g}(X)$
		$\lvert A\rvert$	$\lvert 2A\rvert$	$\lvert 3A\rvert$	$\lvert 4A\rvert$	$\lvert 5A\rvert$	
$(7, 13)$	$4/91$	0	0	1	2	3	11
$(5, 7)$	$1/35$	0	0	1	2	3	8
$(3, 5, 11)$	$4/165$	-1	0	0	1	2	6

Note that existence of varieties with $\mathbf{B}(X) = (7, 13)$ and $(5, 7)$ is not known. Varieties with $\mathbf{B}(X) = (3, 5, 11)$ can be realized as hypersurfaces $X_{12} \subset \mathbb{P}(1, 3, 4, 5, 7)$ which are rational. But again we do not know if this is the only family with corresponding invariants.

Apply the construction (4.2) with $\mathscr{M} = |4A|$. Since X is not rational by our assumption, we have $\hat{q} \le 8$ (see Propositions 2.4, 6.1, 7.1 and 8.1).

9.2 ($\mathbf{B}(X) = (5, 7)$). In a neighborhood of the point of index 7 we have $\mathscr{M} \sim -4K_X$. Thus by Lemma 4.2, $\beta_4 \ge 4\alpha$. The relation (4.4) for $k = 4$ has the form

$$8 \ge \hat{q} = 2s_4 + (2\beta_4 - \alpha)e \ge 2s_4 + 7\alpha e. \tag{9.1}$$

We claim that the contraction $\bar{f}$ is birational. Indeed, otherwise $\hat{q} = 1$ by Lemma 4.7 and so $s_4 = 0$, that is, $\bar{\mathscr{M}}$ is the pull-back of some linear system on $\hat{X}$. Since $\dim \bar{\mathscr{M}} = 2$, $\dim \hat{X} \ne 1$ (otherwise $\bar{\mathscr{M}} = \bar{f}^*|2p|$, where p is a point on $\hat{X} \simeq \mathbb{P}^1$, and then $\bar{M}_2 = \bar{f}^*p$ must be movable). Further, $4\bar{M}_1 \sim \bar{\mathscr{M}}$ and so $\bar{M}_1$ is also the pull-back of some divisor, say Λ, on the surface $\hat{X}$. Thus $\bar{M}_1 = \bar{f}^*\Lambda$ and $\bar{\mathscr{M}} = \bar{f}^*|4\Lambda|$. Clearly, Λ is a generator of the group $\mathrm{Cl}(\hat{X})$. Recall that $\hat{X}$ is a del Pezzo surface with at worst Du Val singularities of type $\mathrm{A_n}$ [MP08, Theorem 1.2.7]. According to the classification (see, for example, [MZ88, Lemmas 3 & 7]) for $\hat{X}$ there are only four possibilities:

$$\mathbb{P}^2, \qquad \mathbb{P}(1,1,2), \qquad \mathbb{P}(1,2,3) \quad \text{or} \quad \mathrm{DP}_5^{\mathrm{A}_4},$$

where $\mathrm{DP}_5^{\mathrm{A}_4}$ is a del Pezzo surface of degree 5 whose singular locus consists of one point of type A_4. Since $\dim|\bar{M}_1| = \dim|\bar{M}_2| = 0$, the divisors Λ and 2Λ are not movable. But one can easily check that $\dim|2\Lambda| > 0$ in all cases. The contradiction shows that the contraction $\bar{f}$ is birational. In particular, $s_4 \ge 1$. Then from (9.1) we immediately see that $\alpha < 1$. Therefore, $P := f(E)$ is a non-Gorenstein point of X and $\alpha = 1/\mathrm{r}(X, P)$, where $\mathrm{r}(X, P) = 5$ or 7 (see Proposition 2.5*1*).

9.3 (Subcase $\mathrm{r}(X, P) = 7$). Then we can write $\beta_1 = 1/7 + m_1$ and $\beta_4 = 4/7 + m_4$, where m_1 and m_4 are non-negative integers. We can rewrite the relation (4.4) for $k = 1$ and 4 as follows

$$8 \ge \hat{q} = 8(s_1 + m_1 e) + e = 2(s_4 + m_4 e) + e.$$

This yields $\hat{q} = e$ and $s_4 = 0$, a contradiction.

9.4 (Subcase $\mathrm{r}(X, P) = 5$). As above, $\beta_4 = 3/5 + m_4$, $\beta_1 = 2/5 + m_1$, where $m_1 \ge 0$ and $m_4 \ge 1$. Therefore,

$$8 \geq \hat{q} = 2(s_4 + m_4 e) + e = 8(s_1 + m_1 e) + 3e \geq 3.$$

This yields $s_1 = 0$ and $\hat{q} = 3e$. Since $s_1 = 0$, by Lemma 4.5 we have $e = 1$, $\hat{q} = 3$, and $s_4 = 0$, a contradiction.

9.5 ($\mathbf{B}(X) = (7, 13)$). Near the point of index 13 we have $\mathscr{M} \sim -7K_X$. Thus by Lemma 4.2

$$\beta_4 \geq 7\alpha, \qquad \beta_1 \geq \tfrac{7}{4}\alpha.$$

The relation (4.4) for $k = 4$ has the form

$$8 \geq \hat{q} = 2s_4 + (2\beta_4 - \alpha)e \geq 2s_4 + 13e\alpha.$$

From this, one immediately sees that $\alpha < 1$. Therefore, $P := f(E)$ is a non-Gorenstein point of X and f is the Kawamata blowup of P by Proposition 2.51. In particular, $\alpha = 1/\mathrm{r}(X, P)$, where $\mathrm{r}(X, P) = 7$ or 13.

9.6 (Subcase $\mathrm{r}(X, P) = 13$). Then we can write $\beta_1 = 5/13 + m_1$, where m_1 is a non-negative integer. Therefore,

$$8 \geq \hat{q} = 8(s_1 + m_1 e) + 3e \geq 3.$$

This gives us $s_1 = 0$ and $\hat{q} = 3e$. Since $s_1 = 0$, Lemma 4.5 we have $e = 1$, $\hat{q} = 3$, and $\mathrm{Cl}(\hat{X})$ is torsion free. Similarly, we can compute

$$\beta_5 = 12/13 + m_5, \qquad 5\hat{q} = 8(s_5 + m_5 e) + 7e, \qquad s_5 = 1.$$

Therefore, $\dim|\Theta| \geq \dim \mathscr{M}_5 = 3$. This contradicts Proposition 2.4.

9.7 (Subcase $\mathrm{r}(X, P) = 7$). Then we can write $\beta_1 = 1/7 + m_1$, where m_1 is a positive integer. Then

$$8 \geq \hat{q} = 8(s_1 + m_1 e) + e \geq 9,$$

which is a contradiction.

9.8 ($\mathbf{B}(X) = (3, 5, 11)$). Near the point of index 11, $\mathscr{M} \sim -6K_X$. Thus by Lemma 4.2, $\beta_4 \geq 6\alpha$. The relation (4.4) for $k = 4$ has the form

$$8 \geq \hat{q} = 2s_4 + (2\beta_4 - \alpha)e \geq 2s_4 + 11e\alpha.$$

From this we immediately see that $\alpha < 1$. Therefore $\alpha = 1/\mathrm{r}(X, P)$, where $\mathrm{r}(X, P) = 3$, 5 or 11 (see Proposition 2.51).

9.9 (Subcase $\mathrm{r}(X, P) = 3$). Then we can write $\beta_4 = 2/3 + m_4$, where $m_4 \geq 2$. Therefore

$$8 \geq \hat{q} = 2(s_4 + m_4 e) + e \geq 5.$$

In particular, $\bar{f}$ is birational and $s_4 > 0$. We get only one solution: $\hat{q} = 7$, $e = s_4 = 1$. By Corollary 5.7, the variety $\hat{X}$ is rational.

9.10 (Subcase $\mathrm{r}(X, P) = 5$). Then we can write $\beta_2 = 4/5 + m_2$ and $\beta_4 = 3/5 + m_4$, where $m_2 \geq 0$ and $m_4 > 0$. Therefore

$$8 \geq \hat{q} = 2(s_4 + m_4 e) + e = 4(s_2 + m_2 e) + 3e \geq 3.$$

In particular, $\bar{f}$ is birational and $s_4 > 0$. We obtain $\hat{q} = 7$ and $s_4 \leq 2$. Then $\hat{X}$ is rational again by Corollary 5.7.

9.11 (Subcase $\mathrm{r}(X, P) = 11$). Then we can write $\beta_4 = 6/11 + m_4$, where m_4 is a non-negative integer. Therefore

$$8 \geq \hat{q} = 2(s_4 + m_4 e) + e. \tag{9.2}$$

Similarly, the relation (4.4) for $k = 3$ has the form

$$3\hat{q} = 8(s_3 + m_3 e) + 7e, \qquad m_3 \geq 0. \tag{9.3}$$

One can see that there are only two solutions:

$$(\hat{q}, e) = (5, 1) \quad \text{or} \quad (7, 3).$$

If $\hat{q} = 7$, then by (9.2) we have $s_4 = 2$. This contradicts Corollary 5.7. Therefore $\hat{q} = 5$ and $e = 1$. Since $|A| = \emptyset$ and $e = 1$, we have an isomorphism $\mathrm{Cl}(\hat{X})_{\mathrm{t}} \simeq \mathbb{Z}/n\mathbb{Z}$ with $1 < n \leq 3$ by Proposition 3.4. If $n = 3$, then $s_3 = 0$ by Lemma 4.5. Then $\bar{f}(\bar{F})$ is a non-Gorenstein point by Corollary 4.10. The relation (4.5) gives us

$$3 s_4 b \geq 20 - 8 s_4,$$

where $s_4 \leq 2$ by (9.2). Hence, $b \geq 2/3$. This contradicts Corollary 3.7.

Assume that $n = 2$. Then $s_2 = 0$ by Lemma 4.5. The relation (4.5) for $k = 3$ has the form

$$15 - 8 s_3 = 2(b s_3 - 5\gamma_3),$$

where $s_3 = 1$ by (9.3). We see that $\bar{f}(\bar{F})$ is a non-Gorenstein point and $b \geq 7/2$. Again, this contradicts Corollary 3.7. □

Now Theorem 1.1 follows from Propositions 6.1, 7.1, 8.1, and 9.1.

Acknowledgements

The author was partially supported by the Higher School of Economics University Basic Research Program, Russian Academic Excellence Project 5-100.

References

[Ale94] Valery Alexeev. General elephants of **Q**-Fano 3-folds. *Compositio Math.*, 91(1):91–116, 1994.

[Bir21] Caucher Birkar. Singularities of linear systems and boundedness of Fano varieties. *Ann. of Math. (2)*, 193(2):347–405, 2021.

[BK] Gavin Brown and Alexander M. Kasprzyk. The graded ring database. http://www.grdb.co.uk/.

[BKQ18] Gavin Brown, Alexander M. Kasprzyk, and Muhammad Imran Qureshi. Fano 3-folds in $\mathbb{P}^2 \times \mathbb{P}^2$ format, Tom and Jerry. *Eur. J. Math.*, 4(1):51–72, 2018.

[BS07] Gavin Brown and Kaori Suzuki. Computing certain Fano 3-folds. *Japan J. Indust. Appl. Math.*, 24(3):241–250, 2007.

[Car08] Jorge Caravantes. Low codimension Fano-Enriques threefolds. *Note Mat.*, 28(2):117–147 (2010), 2008.

[CD20] Stephen Coughlan and Tom Ducat. Constructing Fano 3-folds from cluster varieties of rank 2. *Compos. Math.*, 156(9):1873–1914, 2020.

[Cor95] Alessio Corti. Factoring birational maps of threefolds after Sarkisov. *J. Algebraic Geom.*, 4(2):223–254, 1995.

[Kaw88] Yujiro Kawamata. Crepant blowing-up of 3-dimensional canonical singularities and its application to degenerations of surfaces. *Ann. of Math. (2)*, 127(1):93–163, 1988.

[Kaw92] Yujiro Kawamata. Boundedness of **Q**-Fano threefolds. In *Proceedings of the International Conference on Algebra, Part 3 (Novosibirsk, 1989)*, volume 131 of *Contemp. Math.*, pages 439–445. Amer. Math. Soc., Providence, RI, 1992.

[Kaw96] Yujiro Kawamata. Divisorial contractions to 3-dimensional terminal quotient singularities. In *Higher-dimensional complex varieties (Trento, 1994)*, pages 241–246. de Gruyter, Berlin, 1996.

[Kaw05] Masayuki Kawakita. Three-fold divisorial contractions to singularities of higher indices. *Duke Math. J.*, 130(1):57–126, 2005.

[MP08] Shigefumi Mori and Yuri Prokhorov. On $\mathbb{Q}$-conic bundles. *Publ. Res. Inst. Math. Sci.*, 44(2):315–369, 2008.

[MZ88] M. Miyanishi and D.-Q. Zhang. Gorenstein log del Pezzo surfaces of rank one. *J. Algebra*, 118(1):63–84, 1988.

[Oka19] Takuzo Okada. Stable rationality of orbifold Fano 3-fold hypersurfaces. *J. Algebraic Geom.*, 28(1):99–138, 2019.

[PR16] Yuri Prokhorov and Miles Reid. On $\mathbb{Q}$-Fano 3-folds of Fano index 2. *Minimal models and extremal rays (Kyoto, 2011)*, volume 70 of *Adv. Stud. Pure Math.*, pages 397–420. Math. Soc. Japan, 2016.

[Pro07] Yu. G. Prokhorov. The degree of $\mathbb{Q}$-Fano threefolds. *Mat. Sb.*, 198(11):153–174, 2007.

[Pro10] Yuri Prokhorov. $\mathbb{Q}$-Fano threefolds of large Fano index, I. *Doc. Math.*, 15:843–872, 2010.

[Pro13] Yu. G. Prokhorov. On Fano threefolds of large Fano index and large degree. *Mat. Sb.*, 204(3):43–78, 2013.

[Pro16] Yu. G. Prokhorov. $\mathbb{Q}$-Fano threefolds of index 7. *Tr. Mat. Inst. Steklova*, 294(Sovremennye Problemy Matematiki, Mekhaniki i Matematicheskoĭ Fiziki. II):152–166, 2016. English version published in Proc. Steklov Inst. Math. **294** (2016), no. 1, 139–153.

[Rei87] Miles Reid. Young person's guide to canonical singularities. *Algebraic geometry, Bowdoin, 1985 (Brunswick, Maine, 1985)*, volume 46 of *Proc. Sympos. Pure Math.*, pages 345–414. Amer. Math. Soc., Providence, RI, 1987.

[San96] Takeshi Sano. Classification of non-Gorenstein **Q**-Fano d-folds of Fano index greater than $d-2$. *Nagoya Math. J.*, 142:133–143, 1996.

[Suz04] Kaori Suzuki. On Fano indices of $\mathbb{Q}$-Fano 3-folds. *Manuscripta Math.*, 114(2):229–246, 2004.

[Tak06] Hiromichi Takagi. Classification of primary $\mathbb{Q}$-Fano threefolds with anti-canonical Du Val $K3$ surfaces. I. *J. Algebraic Geom.*, 15(1):31–85, 2006.

[TPG08] Bordeaux The PARI Group, 2008. *PARI/GP, version* 2.7.5.

An Exceptional Locus in the Perfect Compactification of A_g

N. I. Shepherd-Barron
with an Appendix with John Armstrong

To Miles, friend and teacher

1 Introduction

This paper is a sequel to [SB06]. The main result was a description of the cone $\overline{NE}(A_g^P)$ of curves on the perfect compactification of the coarse moduli space A_g (the notation will be explained below). In particular, the $\mathbb{Q}$-divisor class $L_g = 12H_g - D_g$ is nef but not ample. Here we make this result more explicit in two ways: we identify the exceptional locus $\mathrm{Exc}(L_g)$ as the locus $A_{1,g-1}^P$ inside A_g^P that is the closure of the locus in A_g that parametrizes abelian varieties with an elliptic factor, and we show that L_g is semi-ample in positive characteristic. We do not know whether L_g is semi-ample in characteristic zero.

Recall that A_g is known to be of general type if $g \geq 7$ and the characteristic is zero, and that the canonical class is given by $K_{A_g^P} \sim (g+1)H_g - D_g$. It follows (Corollary 4.7) that, if $7 \leq g \leq 10$, then $K_{A_g^P}$ is not nef and therefore the first step in running the characteristic zero Minimal Model Program on A_g^P is a flipping contraction that, in particular, crushes the locus that parametrizes products of g elliptic curves to a point. In particular, each non-trivial fibre of this contraction penetrates A_g, which we regard as the interior of A_g^P. Therefore to run the MMP on A_g^P (which has terminal singularities if $g \geq 6$; the appendix below completes the argument in [SB06]) while maintaining a modular interpretation would necessitate changing the definition of 'principally polarized abelian variety' in the presence of elliptic factors. In contrast, there is a prospect that the MMP for the coarse moduli space $\overline{M}_g$ of stable curves of genus g will run in a way that does not touch the interior of $\overline{M}_g$, where in this context 'interior' refers to the locus M_g of smooth curves [HH13].

Here is some of the notation used in this paper. We let $\mathcal{A}_g$ denote the stack, and $A_g = [\mathcal{A}_g]$ the coarse moduli space, of principally polarized abelian g-folds and $\mathcal{A}_g^P$, A_g^P their perfect compactifications. These are particular toroidal compactifications, and dominate the Satake compactification A_g^{Sat} of A_g. Stacks that classify abelian varieties with level n structures will be denoted by $\mathcal{A}_{g;n}$, etc. In general, the geometric quotient of an algebraic stack $\mathcal{X}$ will, when it exists, be denoted by $[\mathcal{X}]$.

We let H_g denote the line bundle of weight 1 modular forms on $\mathcal{A}_g^P$. It gives a $\mathbb{Q}$-line bundle on A_g^P that can, in turn, be identified with the pullback of a $\mathbb{Q}$-line bundle on A_g^{Sat}. The boundary $D_g = \mathcal{A}_g^P \setminus \mathcal{A}_g$ is a geometrically irreducible $\mathbb{Q}$-Cartier divisor, but is not Cartier; even at the stack level there are denominators.

The objects $\mathcal{A}_g^P$ and A_g^P exist over $\operatorname{Spec}\mathbb{Z}$ and there is a contraction $\pi\colon \mathcal{A}_g^P \to A_g^{Sat}$ that factors through A_g^P; see [FC90] for this. As mentioned above, the main result of [SB06] is a description of the cone of curves $\overline{NE}$ (in Mori's sense) of A_g^P over any field: $\overline{NE}(A_g^P)$ is the rational cone spanned by certain curves C_1 and C_2. Here C_1 is the closure of the locus of products $E\times B$, where E is a varying elliptic curve and $B \in A_{g-1}$ is fixed (so that C_1 is a copy of A_1^P, the compactified j-line) and C_2 is any curve in the boundary divisor $D_g = A_g^P \setminus A_g$ such that $\pi(C_2)$ is a point.

From [SB06], $aH_g - D_g$ is ample on A_g^P (or on $\mathcal{A}_g^P$) if and only if $a > 12$, and is nef if and only if $a \geq 12$. (I am grateful to Alexeev for pointing out that the notion of an ample invertible sheaf makes sense on any proper stack $\mathcal{X}$ with finite inertia and geometric quotient $X = [\mathcal{X}]$: the invertible sheaf $\mathcal{L}$ on $\mathcal{X}$ is ample if and only if, for all coherent sheaves $\mathcal{F}$ on $\mathcal{X}$, $H^i(\mathcal{X}, \mathcal{L}^{\otimes n} \otimes \mathcal{F}) = 0$ for all $i > 0$ and for all $n \gg 0$. If $\mathcal{L}$ corresponds to the $\mathbb{Q}$-line bundle L on X, then $\mathcal{L}$ is ample if and only if L is so. Note also that, even more obviously, the notions of nef and semi-ample are insensitive to the distinction between $\mathcal{X}$ and X, and $\operatorname{Exc}(\mathcal{L})$ is exactly the inverse image in $\mathcal{X}$ of $\operatorname{Exc}(L)$.)

Recall [Kee99] that if L is a nef divisor class on a projective variety X then the *exceptional locus* $\operatorname{Exc}(L)$ of L is the union of all the subvarieties Z of X with $L^{\dim Z}.Z = 0$, and that L is semi-ample on X provided that it is semi-ample on $\operatorname{Exc}(L)$ and the ground field k has $\operatorname{char} k > 0$. Notice that Keel's result holds without change when X is allowed to be any proper stack over a field whose inertia stack is finite (so that a geometric quotient of X exists as a proper algebraic space by [KM97]) and whose geometric quotient is projective.

Here is the main result of this paper. I am very grateful to Stefan Schröer who asked whether $12H_g - D_g$ is semi-ample in positive characteristic.

Theorem 1.1 (= Theorem 4.4). *Suppose that the base is a field k and set $L_g = 12H_g - D_g$.*

(1) $\mathrm{Exc}(L_g)$ *is the subvariety* $A_{1,g-1}^P$ *that is the image of the stack* $\mathcal{A}_{1,g-1}^P$ *defined in Corollary 3.2(3).*
(2) Suppose either *that* $\mathrm{char}\, k = 0$ *and that* $g \leq 11$ or *that* $\mathrm{char}\, k > 0$ *and that g is arbitrary. Then L_g is semi-ample. For a sufficiently divisible integer n the linear system $|mL_g|$ defines a birational morphism* $\phi_{|mL_g|} = \mathrm{contr}_{R_1} : A_g^P \to V_g$ *onto a normal projective k-variety V_g that contracts the ray R_1 generated by the curve C_1.*
(3) The normalization of the image $\phi_{|mL_g|}(\mathrm{Exc}(L_g))$ *is isomorphic to V_{g-1}.*

Observe that V_g can be regarded as being obtained from A_g^P by 'crushing elliptic factors to points'. From this point of view it is not clear *a priori* that V_g exists as an algebraic variety; indeed, we are unable to prove such a statement when $g \geq 12$ and the characteristic is zero.

2 A Remark and an Example

Given a principally polarized abelian variety (A, λ) there is not usually a theta divisor θ on A, although there is a natural line bundle $\mathcal{L}_{2\lambda}$ associated to 2λ. However, the stack $\mathcal{A}_g$ is naturally isomorphic to the stack $\mathcal{X}_g$ of pairs (X, θ) where X is a symmetric torsor under A and θ is an ample divisor on X that defines λ on $A = \mathrm{Aut}_X^0$. Moreover, (X, θ) has a 1-dimensional factor if and only if (A, λ) has an elliptic factor.

It is sometimes convenient to identify $\mathcal{X}_g$ with $\mathcal{A}_g$ and to consider pairs (X, θ) instead of ppavs (A, λ).

Now we recall [Hud90, DO88] some of the classical geometry of the moduli space A_2 over $\mathbb{C}$.

A level 2 structure on a principally polarized abelian surface (A, λ) is a symplectic isomorphism $\psi\colon A[2] \to G := (\mathbb{Z}/2)^2 \times \mu_2^2$, where G has its standard symplectic pairing and $A[2]$ has the Weil pairing defined by the principal polarization λ. There is a standard projective action of G

on $\mathbb{P}^3$; the linear system $|2\theta|$ on X then gives a G-equivariant morphism $X \to \mathbb{P}^3$ that factors through the Kummer surface $\mathrm{Km}(X) = X/(-1)$. The image of $\mathrm{Km}(X)$ lies in a unique G-invariant quartic; taking the coefficients of this quartic then determines a point on Σ, the Segre cubic threefold [Hud90]. This is the unique cubic threefold with 10 nodes and lies in $\mathbb{P}^5$ with equations $e_1 = e_3 = 0$, where e_i is the ith elementary symmetric function in 6 variables. If (X, θ) is irreducible then $|2\theta|$ is very ample on $\mathrm{Km}(X)$, so (X, θ) is determined by the point on Σ, while if $X = E_1 \times E_2$, where E_1, E_2 are curves of genus 1, then the image of $\mathrm{Km}(X)$ is $\mathbb{P}^1 \times \mathbb{P}^1$ which does not determine X.

There are also 15 planes on Σ. Let $\widetilde{\Sigma} \to \Sigma$ be the blow-up of the nodes, $\Pi = \sum \Pi_i$ the exceptional divisor of the blow-up and $D = \sum D_i$ the strict transform of the sum of the planes.

If G is a finite group acting on a variety V then we let V/G denote the geometric quotient.

Theorem 2.1. *(1) $\widetilde{\Sigma} = A^P_{2;2}$, the perfect compactification of the level 2 moduli space $A_{2;2}$. D is the toroidal boundary and Π the locus of products $E_1 \times E_2$ of elliptic curves.*

(2) A^P_2 has two distinct birational contractions: one is the standard contraction $\pi\colon A^P_2 \to A^{Sat}_2$ and the other is a birational contraction $\rho\colon A^P_2 \to \Sigma/\mathfrak{S}_6$, where $\mathfrak{S}_6$ is the symmetric group on 6 letters and is isomorphic to $\mathrm{Sp}_4(\mathbb{F}_2)$.

(3) $\Sigma/\mathfrak{S}_6$ is isomorphic to a weighted projective space $\mathbb{P}(2,4,5,6)$; $A^{Sat}_{2;2}$ is isomorphic to the Igusa quartic $e_1 = e_4 - e_2^2 = 0$ in $\mathbb{P}^5$ (this is also the projective dual of Σ); and A^{Sat}_2 is isomorphic to $\mathbb{P}(2,3,5,6)$.

From one point of view, the point of Theorem 1.1 is to extend this picture, albeit in a more abstract way, to all values of g.

3 The Structure of the Perfect Boundary

For any g let Λ_g be a fixed copy of $\mathbb{Z}^g$ and $B(\Lambda_g)$ be the $\mathbb{Z}$-module of symmetric bilinear $\mathbb{Z}$-valued bilinear forms on Λ_g. Let C_g denote the cone of real positive definite symmetric bilinear forms in g variables and $\overline{C}_g$ the cone of real positive semi-definite symmetric bilinear forms in g variables. We identify both of these as subsets of $B(\Lambda_g) \otimes_{\mathbb{Z}} \mathbb{R}$. Identify C_g with the interior $\overline{C}_g^0$ of $\overline{C}_g$. A toroidal compactification of $\mathcal{A}_g$ or A_g corresponds to a choice of $\mathrm{GL}_g(\mathbb{Z})$-*admissible* decomposition of $\overline{C}_g$, as defined in [AMRT10]. We consider here a particular admissible decomposition Σ_g^P of $\overline{C}_g$, namely, that defined by the perfect quadratic forms. It is called the *perfect decomposition* and the cones appearing in it (that is, the facets) are the *perfect cones.* It is described in Theorem 3.1 below. Since our results are particular to this specific toroidal compactification, the arguments of this paper depend upon knowing something about perfect forms and the combinatorics of the decomposition of $\overline{C}_g$ that they define. The cones of maximal dimension $g(g+1)/2$ correspond to perfect quadratic forms (this can be taken as the definition of a perfect quadratic form).

Theorem 3.1. *(1) (Voronoi) [AMRT10] The convex hull of the positive semi-definite integral forms in $\overline{C}_g$ defines a $\mathrm{GL}_g(\mathbb{Z})$-admissible decomposition Σ_g^P of $\overline{C}_g$.*

(2) (Barnes-Cohn) [BC76] This convex hull coincides with the convex hull of the primitive rank 1 forms. Moreover, a form of rank at least 2 lies in the interior of the hull.

(3) If $p = p(x_1, \dots, x_m)$ and $q = q(y_1, \dots, y_n)$ are perfect forms of equal minimal norm, then $p + q = r = r(x_1, \dots, x_m, y_1, \dots, y_n)$ is a form that defines a perfect cone in $\overline{C}_{m+n}$.

(4) Suppose that τ is a perfect cone in $\overline{C}_r$ that meets C_r. Then its closure contains perfect cones in some copy of $\overline{C}_{r-1}$. In particular, if τ is a minimal perfect cone in $\overline{C}_r$ that meets C_r, then $\tau \cap \overline{C}_{r-1}$ is a union of minimal perfect cones in $\overline{C}_{r-1}$ that meet C_{r-1}.

(5) Suppose that σ is a maximal perfect cone in $\overline{C}_r$, so that σ is defined by a perfect form q in r variables. Suppose also that τ_1, τ_2 are closed cones in $\overline{C}_{r+1}$ such that both contain σ in their boundary and $\dim \tau_i = \dim \sigma + 1$. That is, both cones τ_i are minimal with respect to containing σ in their boundary. Then each τ_i can be defined by a quadratic form $\lambda q + l_i^2$, where l_i is a primitive linear form and $\lambda \in \mathbb{Q}$ is the inverse of the minimal norm of q. In particular, τ_1 and τ_2 are conjugate under the parabolic subgroup of $\mathrm{GL}_{r+1}(\mathbb{Z})$ that preserves each of the first r variables.

Proof For *(1)* and *(2)* see the references. *(3)* is trivial. For *(4)*, we can suppose that σ is minimal with respect to meeting C_r. Suppose that $l_1, \dots, l_n$ are primitive elements of $\Lambda_r^\vee$ such that $l_1^2, \dots, l_n^2 \in B(\Lambda_r)$ span the 1-dimensional faces of σ. So there are $\lambda_1, \dots, \lambda_n > 0$ such that $\sum \lambda_i l_i^2 \in C_r$. Then, for some t, $\sum_1^t \lambda_i l_i^2$ is of rank at most $r-1$ for every $\lambda_1, \dots, \lambda_t > 0$, while $\sum_1^{t+1} \lambda_i l_i^2$ is of rank r for some $\lambda_1, \dots, \lambda_{t+1} > 0$.

Take t to be maximal subject to this, then the cone τ generated by $l_1^2,\dots,l_t^2$ is a face of σ with the stated properties. (*5*) is even easier. □

We denote by $\mathcal{A}_g^P$ the toroidal compactification of $\mathcal{A}_g$ that corresponds, as in [FC90], to the perfect decomposition of $\overline{C}_g$. The boundary $D = \mathcal{A}_g^P\setminus\mathcal{A}_g$ is a divisor and is the inverse image $\pi^{-1}(A_{g-1}^{Sat})$, where A_{g-1}^{Sat} is identified with the boundary $A_g^{Sat}\setminus A_g$. We also have the partial compactification $\mathcal{A}_g^{part}$, which is open in $\mathcal{A}_g^{tor}$ and is, by definition, the inverse image $\pi^{-1}(A_g\coprod A_{g-1})$. The universal abelian scheme $\mathcal{U}_{g-r}\to\mathcal{A}_{g-r}$ has an extension to a semi-abelian scheme $\mathcal{U}_{g-r}^{part}\to\mathcal{A}_{g-r}^{part}$ whose degenerate fibres have torus rank 1. Taking r-fold fibre products gives a semi-abelian scheme $\delta\colon\mathcal{U}_{g-r}^{r,part}\to\mathcal{A}_{g-r}^{part}$.

In higher codimension, the boundary is described as follows. There is a stratified scheme $F = F_r$, locally of finite type over the base, the closures of whose strata are projective toric varieties with an action of $\mathrm{GL}_r(\mathbb{Z})$ on F_r, and an F_r-bundle $\mathcal{F}_r\to\mathcal{U}_{g-r}^r$ with an equivariant action of $\mathrm{GL}_r(\mathbb{Z})$, such that $\pi^{-1}(A_{g-r}) = \mathcal{F}_r/\,\mathrm{GL}_r(\mathbb{Z})$, the stack quotient. The closures $F_{r,\tau}$ of the various strata of F_r correspond to the perfect cones τ in the perfect decomposition of $\overline{C}_r$ that meet the interior C_r. In particular, the irreducible components of F_r correspond to the minimal such cones.

Here is a more precise description of F_r. Denote by T_r the torus with character group $\mathbb{X}^*(T_r) = B(\Lambda_r)$. There is a locally finite $\mathrm{GL}_r(\mathbb{Z})$-equivariant torus embedding $T_r\hookrightarrow Y_r$ such that F_r is a $T_r\rtimes\mathrm{GL}_r(\mathbb{Z})$-equivariant closed subscheme of the boundary $Y_r\setminus T_r$. The closure $F_{r,\tau}$ of a stratum in F_r is then a torus embedding under a quotient T of T_r and gives rise to the closure $\mathcal{F}_{r,\tau}$ of a stratum in $\mathcal{F}_r$. This closure is a proper $F_{r,\tau}$-bundle $\mathcal{F}_{r,\tau}\to\mathcal{U}_{g-r}^r$ that is a relative T-equivariant compactification of a T-bundle $\mathcal{T}\to\mathcal{U}_{g-r}^r$.

In turn, the image of $\mathcal{F}_{r,\tau}$ in $\mathcal{A}_g^P$ is an irreducible closed substack $\mathcal{X}_{r,\tau}$ of $\pi^{-1}(A_{g-r})$. If $n\geq 3$ and is invertible in the base, then in the stack $\mathcal{A}_{g;n}^P$ the image $\mathcal{X}_{r,\tau}$ can be identified with $\mathcal{F}_{r,\tau}$. In this case (that is, at level n), T_r acts on $\mathcal{X}_{r,\tau}$ via the quotient $T_r\to T$, and this action extends to an action on the closure $\overline{\mathcal{X}}_{r,\tau}$ of $\mathcal{X}_{r,\tau}$ in $\mathcal{A}_{g;n}^P$.

Each such τ lies in the closure of finitely many maximal perfect cones σ in $\overline{C}_r$. Such a cone σ corresponds to the choice, up to scalars, of a perfect form q in r variables. The closure $\overline{\mathcal{U}}_{g-r,\sigma}^r$ of $\mathcal{U}_{g-r}^r$ is just $\overline{\mathcal{X}}_{r,\sigma}$. We let $\overline{\mathcal{U}}_{g-r,\sigma}^{r,norm}$ denote the normalization of $\overline{\mathcal{U}}_{g-r,\sigma}^r$.

Translating Theorem 3.1(*1–5*), respectively, into algebraic geometry yields the following statement.

Corollary 3.2. *(1) There exist toroidal compactifications* $\mathcal{A}_g^P$ *and* A_g^P *corresponding to this decomposition.*

(2) As a Deligne-Mumford stack, $\mathcal{A}_g^P$ *has terminal singularities and the boundary* D *is absolutely irreducible.*

(3) The product morphism $\mathcal{A}_g \times \mathcal{A}_h \to \mathcal{A}_{g+h}$ *extends to a morphism* $\mathcal{A}_g^P \times \mathcal{A}_h^P \to \mathcal{A}_{g+h}^P$ *whose image is denoted by* $\mathcal{A}_{g,h}^P$.

(4) $\pi^{-1}(A_{g-r}^{Sat})$ *lies in the closure of* $\pi^{-1}(A_{g-r+1})$ *and* $\pi^{-1}(A_{g-r+1}^{Sat})$ *is the closure of* $\pi^{-1}(A_{g-r+1})$.

(5) A maximal perfect cone σ *in* $\overline{C}_r$ *corresponds to an irreducible closed substack* $\overline{\mathcal{U}}_{g-r,\sigma}^r$ *of* $\pi^{-1}(A_{g-r}^{Sat})$ *that contains an open substack isomorphic to* $\mathcal{U}_{g-r}^{r,part}$. *The complement* $\overline{\mathcal{U}}_{g-r,\sigma}^r \setminus \mathcal{U}_{g-r}^{r,part}$ *has codimension at least 2 in* $\overline{\mathcal{U}}_{g-r,\sigma}^r$.

Proof (1), (2) and (3) are immediate. For (4), we use the fact that the irreducible components $\mathcal{Z}$ of $\pi^{-1}(A_{g-r})$ correspond to the (equivalence classes of the) minimal perfect cones τ in $\overline{C}_r$ that meet C_r. Since, by Theorem 3.1(4), $\tau \cap \partial\overline{C}_r$ is a union of minimal perfect cones in $\overline{C}_{r-1}$ that meet C_{r-1}, the closure $\overline{\mathcal{Z}}$ of $\mathcal{Z}$ lies in the closure of an irreducible component of $\pi^{-1}(A_{g-r+1})$. So $\pi^{-1}(A_{g-r}) \subseteq \overline{\pi^{-1}(A_{g-r+1})}$. Then, by induction, $\pi^{-1}(A_{g-r-m}) \subseteq \overline{\pi^{-1}(A_{g-r+1})}$, so that $\pi^{-1}(A_{g-r}^{Sat}) \subseteq \overline{\pi^{-1}(A_{g-r+1})}$. Therefore $\pi^{-1}(A_{g-r+1}) = \overline{\pi^{-1}(A_{g-r+1})}$.

For (5), note that the irreducible components of $\overline{\mathcal{U}}_{g-r,\sigma}^r \setminus \mathcal{U}_{g-r}^{r,part}$ correspond to the minimal perfect cones τ in $\overline{C}_{r+1}$ that contain σ. These are equivalent and we are done. □

Definition 3.3. An open substack $\mathcal{U}$ of an algebraic stack $\mathcal{X}$ is *nearly equal* to $\mathcal{X}$ if its complement has codimension at least 2 everywhere. A *near equality* is an open embedding $\mathcal{U} \hookrightarrow \mathcal{X}$ of stacks whose image is nearly equal to its target.

In particular, $\mathcal{A}_g^{part}$ is nearly equal to $\mathcal{A}_g^P$.

4 The Exceptional Locus of $12H_g - D_g$

Set $L_g = 12H_g - D_g$.

Proposition 4.1. *For sufficiently divisible* m, *the linear system* $|mL_g|$ *has no base points in* $\mathcal{A}_g^{part}$ *and contracts all curves (that is, complete 1-dimensional substacks) of the form* $\mathcal{A}_1^P \times \{B\}$ *where* B *is a point in* $\mathcal{A}_{g-1}$. *This morphism separates points except along* $A_{1,g-1}^P \cap A_g^{part}$.

Proof First, work over $\mathbb{Z}[1/2]$ and impose a level 2 structure. We then consider the morphism defined by the 2θ linear system: there is a universal family $f\colon V \to \mathcal{A}_{g;2}^{part}$ of projective schemes with level 2 structure and the 2θ linear system defines, by taking the cycle-theoretic image of each fibre of f, a morphism Φ from $\mathcal{A}_{g;2}^{part}$ to the Chow scheme of $\mathbb{P}^{2^g-1}$. After dividing by the finite group $\mathrm{Sp}_{2g}(\mathbb{Z}/2)$ we then get a morphism ϕ from $\mathcal{A}_g^{part}$ to some scheme, and ϕ contracts every curve of the form $A_1^P \times \{B\}$. Since the Kummer variety of an elliptic curve E is independent of E, so the j-line collapses to a point. It follows that ϕ is defined by some linear system $|mL_g|$, since for $g = 1$ the bundle H_g has degree $1/12$ and D_g has degree 1.

Over $\mathbb{Z}[1/3]$ we work at level 3. There is a normalized projective space $\mathbb{P}^{3^g-1}$ such that given (X, θ), X is embedded in $\mathbb{P}^{3^g-1}$ via $|3\theta|$. Let Gr denote the Grassmannian of quadrics in $\mathbb{P}^{3^g-1}$. Then to (X, θ) associate the point $P_{(X,\theta)}$ in Gr that corresponds to the space of quadrics that pass through the image of X in $\mathbb{P}^{3^g-1}$. Since a curve of genus 1 and degree 3 cannot be recovered from the quadrics that contain it (there are no such quadrics), this association defines a morphism whose source is $\mathcal{A}_{g;3}^{part}$ that performs a similar function of 'losing elliptic factors'. □

So pick $r \geq 2$ and consider L_g on the inverse image $\pi^{-1}(A_{g-r})$, where A_{g-r} is regarded as a stratum in A_g^{Sat}. The closure $\mathcal{Z}_r$ of $\pi^{-1}(A_{g-r})$ in $\mathcal{A}_g^P$ is a finite union of irreducible components $\overline{\mathcal{X}}_{r,\tau}$ as above. Note that $\mathcal{Z}_r = \pi^{-1}(A_{g-r}^{Sat})$, as already pointed out.

Fix a maximal perfect cone σ in $\overline{C}_r$ and a face τ of σ that meets C_r. There is, up to scalars, a unique perfect quadratic form q which defines σ and can be written, in many ways, as a linear combination $q = \sum_1^r \lambda_i x_i^2$ where x_i is a primitive integral linear form and $\lambda_i \in \mathbb{Q}_+$. Each rank 1 form x_i^2 corresponds, in $\mathrm{End}(\mathcal{U}_{g-r}^r) \otimes \mathbb{Q}$, to a projection $\xi_i\colon \mathcal{U}_{g-r}^r \to \mathcal{U}_{g-r}$ over $\mathcal{A}_{g-r}$. So (since homomorphisms of abelian schemes extend uniquely to homomorphisms of semi-abelian schemes) there is a commutative diagram

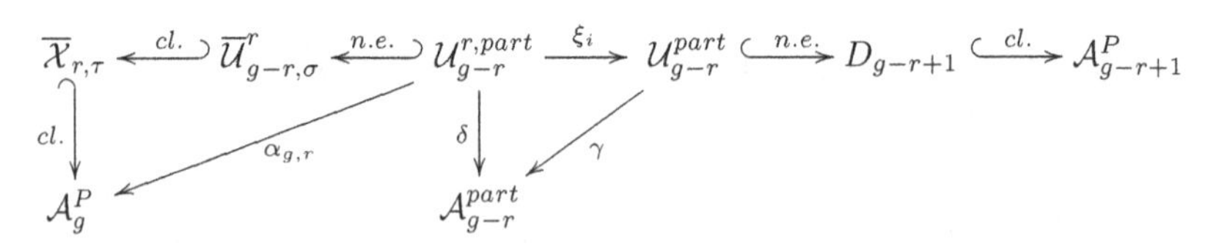

(Here the labels *cl.* and *n.e.* refer, respectively, to closed embeddings and near equalities.)

Proposition 4.2. *Let s denote the zero-section of δ. Then the restrictions $\alpha_{g,r}^* D_g|_s$ and $\delta^* D_{g-r}|_s$ are linearly equivalent.*

Proof Recall that the multiplication $\mathcal{A}_{g-r} \times \mathcal{A}_r \to \mathcal{A}_g$ extends to $\mathcal{A}_{g-r}^P \times \mathcal{A}_r^P \to \mathcal{A}_g^P$. The perfect form q corresponds to a maximally degenerate boundary point $x \in \mathcal{A}_r^P$, and then the image of $\{x\} \times \mathcal{A}_{g-r}^P$ in $\mathcal{A}_g^P$ is exactly the closure in $\mathcal{A}_g^P$ of the zero-section of $\mathcal{U}_{g-r}^r \to \mathcal{A}_{g-r}$.

Identify s with $\{x\} \times \mathcal{A}_{g-r}^{part}$ as above. Then there is a chain of rational curves in $\mathcal{A}_r^P$ leading from x to an interior point $y = E^r$, for any elliptic curve E (let E degenerate, and then take a chain of rational curves leading from this maximally degenerate boundary point to y).

So $D_g|_s$ is rationally equivalent to $D_g|_{\{y\} \times \mathcal{A}_{g-r}^{part}}$, and now the result is obvious. □

Set $\Lambda = -\alpha_{g,r}^* D_g + \delta^* D_{g-r}$, which is a divisor class on $\mathcal{U}_{g-r}^{r,part}$. Since $\mathcal{U}_{g-r}^{r,part} \to \mathcal{A}_{g-r}^{part}$ is semi-abelian and Λ is trivial on the zero-section, Λ is determined by the polarization that it defines on the generic fibre $\mathcal{U}_{g-r,\eta}^r$. (At this point we also use the fact that line bundles on the semi-abelian scheme $\mathcal{U}_{g-r}^{r,part} \to \mathcal{A}_{g-r}^{part}$ that are trivial on the zero-section are determined by the polarization that they define on the generic fibre. This is because the only global point of the self-dual abelian scheme $\mathcal{U}_{g-r}^r \to \mathcal{A}_{g-r}$ is zero.)

Corollary 4.3.

$$\Lambda \sim \sum_1^r \lambda_i \xi_i^* \left(-D_{g-r+1}|_{\mathcal{U}_{g-r}^{part}} + \gamma^* D_{g-r} \right).$$

Proof From its definition, and knowledge of the polarization defined by $-D_g$ on $\mathcal{U}_{g-r,\eta}^r$, the polarization defined by Λ on $\mathcal{U}_{g-r,\eta}^r$ is the quadratic form q. The polarization defined by $-D_{g-r+1}$ on the generic fibre of $\mathcal{U}_{g-r}$ is that given by the primitive rank one form x_i^2, and we are done. □

For $i = 1, \ldots, r$, choose a large positive integer n_i and set

$$\widetilde{\xi}_i = [n_i] \circ \xi_i \colon \mathcal{U}_{g-r}^{r,part} \to \mathcal{U}_{g-r}^{part} \quad \text{and}$$
$$\rho_i = \alpha_{g-r+1,1} \circ \widetilde{\xi}_i \colon \mathcal{U}_{g-r}^{r,part} \to \mathcal{A}_{g-r+1}^P.$$

Then

$$\alpha_{g,r}^* \Lambda \sim \sum_1^r \frac{\lambda_i}{n_i^2} \widetilde{\xi}_i^* \left(-D_{g-r+1}|_{\mathcal{U}_{g-r}^{part}} + \gamma^* D_{g-r} \right),$$

which can be re-written as

$$L_g|_{\mathcal{U}^{r,part}_{g-r}} \sim \left(1 - \sum_i \frac{\lambda_i}{n_i^2}\right)\delta^* L_{g-r} + \sum_i \frac{\lambda_i}{n_i^2}\rho_i^* L_{g-r+1}.$$

We abbreviate this to

$$L_g|_{\mathcal{U}^{r,part}_{g-r}} \sim a\delta^* L_{g-r} + \sum b_i \rho_i^* L_{g-r+1}. \tag{4.1}$$

Note that $a, b_i > 0$.

Theorem 4.4 (= Theorem 1.1). *(1) [SB06] L_g is nef.*
(2) $\mathrm{Exc}(L_g) = \mathcal{A}^P_{1,g-1}$.
Now suppose also that either the characteristic is zero and $g \leq 11$ or the characteristic is positive.

(3) Then L_g is semi-ample and there is a normal projective k-variety V_g and a birational contraction $A_g^P \to V_g$ of the ray R_1 whose exceptional locus is $A^P_{1,g-1}$.
(4) The normalization of the image of $A^P_{1,g-1}$ is isomorphic to V_{g-1}.

Proof (*1*) was proved in [SB06]. However, in order to prove (*2*) it is simplest to give here another proof of (*1*) whose techniques can be extended to prove (*2*) also.

Assume that L_g is not nef, and that $\mathcal{C}$ is a complete curve in $\mathcal{A}_g^P$ with $L_g.\mathcal{C} < 0$. By Proposition 4.1 and since $-D_g$ is π-ample, there is some minimal $r \geq 2$ such that $\mathcal{C}$ maps to a curve in A_{g-r}^{Sat} that does not lie in A_{g-r-1}^{Sat}.

Fix an integer $n \geq 3$ that is prime to $\mathrm{char}\, k$ and consider the perfect compactification $\mathcal{A}^P_{g;n}$ of the level n stack $\mathcal{A}_{g;n}$.

Consider L_g on the inverse image $\pi^{-1}(A_{g-r})$, where A_{g-r} is regarded as a locally closed subvariety of A_g^{Sat}. The closure $\mathcal{Z}^r$ of $\pi^{-1}(A_{g-r})$ in $\mathcal{A}_g^P$ is a finite union of irreducible components $\overline{\mathcal{X}}_{r,\tau}$, each of which is the image of an equivariant closure $\overline{\mathcal{X}}_{r,\tau;n}$ in $\mathcal{A}^P_{g;n}$ of a T-bundle over the universal level n abelian scheme $\mathcal{U}^r_{g-r;n}$, where T is a quotient of the torus T_r whose cocharacter group is B_r.

We can use the T-action at level n to construct a specialization (that is, a rational equivalence) $\mathcal{C} \sim \mathcal{C}_0 + \mathcal{F}$, where $\mathcal{C}_0$ is contained in a minimal stratum of $\mathcal{Z}_r$ and $\mathcal{F}$ is the image in $\mathcal{A}_g^P$ of a closed subvariety $\mathcal{F}_n$ in $\mathcal{A}^P_{g;n}$ such that $\mathcal{F}_n$ is preserved by T but no component of $\mathcal{F}_n$ consists of T-fixed points. Then $\mathcal{F}$ is π-vertical, so that $(-D_g).\mathcal{F} > 0$ and $H_g.\mathcal{F} = 0$, and then $\mathcal{C}_0$ is L_g-negative. So, we can assume that $\mathcal{C}$ lies in a minimal stratum of $\mathcal{Z}_r$. Recall that each such stratum is one of the closed substacks $\overline{\mathcal{U}}^r_{g-r,\sigma}$ considered previously and so $\mathcal{C} \subseteq \overline{\mathcal{U}}^r_{g-r,\sigma}$.

Note that, for any sufficiently divisible integer m, there are equalities

$$\begin{aligned} H^0(\overline{\mathcal{U}}^{r,norm}_{g-r,\sigma}, \mathcal{O}(mL_g)|_{\overline{\mathcal{U}}^{r,norm}_{g-r,\sigma}}) &= H^0(\mathcal{U}^{r,part}_{g-r}, \mathcal{O}(mL_g)|_{\mathcal{U}^{r,part}_{g-r}}) \\ &= H^0(\mathcal{U}^{r,part}_{g-r}, \mathcal{O}(ma\delta^* L_{g-r}) \otimes H^0(\mathcal{U}^{r,part}_{g-r}, \mathcal{O}(\sum mb_i\rho_i^* L_{g-r+1})). \end{aligned}$$

The first equality follows from the facts that $\mathcal{U}^{r,part}_{g-r}$ is nearly equal to $\overline{\mathcal{U}}^{r,norm}_{g-r,\sigma}$ and that $\overline{\mathcal{U}}^{r,norm}_{g-r,\sigma}$ is normal. The second follows from the linear equivalence (4.1) and the fact that, by Proposition 4.1, L_h has (stably) no base points on $\mathcal{A}^{part}_h$ for all values of h, and in particular for $h = g - r$ and $h = g - r + 1$.

Therefore the stable base locus $\mathbf{B}(L_g|_{\overline{\mathcal{U}}^{r,norm}_{g-r,\sigma}})$ is contained in the boundary $\overline{\mathcal{U}}^{r,norm}_{g-r,\sigma} \setminus \mathcal{U}^{r,part}_{g-r}$ of $\overline{\mathcal{U}}^{r,norm}_{g-r,\sigma}$.

The next lemma is well known but we include a proof for lack of a convenient reference.

Lemma 4.5. *If $\nu \colon X \to Y$ is a finite dominant morphism of integral proper algebraic spaces and $L \in \mathrm{Pic}_Y$, then $\nu^{-1}\mathbf{B}(L) = \mathbf{B}(\nu^* L)$.*

Proof The homomorphism

$$R(Y, L) = \oplus H^0(Y, L^{\otimes n}) \to R(X, \nu^* L) = \oplus H^0(X, \nu^* L^{\otimes n})$$

is injective and finite.

Suppose that $x \in \mathbf{B}(\nu^* L)$, so that $s(x) = 0$ for all $s \in R(X, \nu^* L)_{\geq 1}$. In particular, $t(x) = 0$ for all $t \in R(Y, L)_{\geq 1}$ and so $x \in \nu^{-1}\mathbf{B}(L)$. That is, $\mathbf{B}(\nu^* L) \subseteq \nu^{-1}\mathbf{B}(L)$.

Now suppose that $x \in \nu^{-1}\mathbf{B}(L) \setminus \mathbf{B}(\nu^* L)$. Choose $s \in H^0(X, \nu^* L^{\otimes n})$ with $s(x) \neq 0$. For some $a_i \in H^0(Y, L^{\otimes n(N-i)})$, there is an equation

$$s^N + a_{N-1}s^{N-1} + \cdots + a_0 = 0.$$

Then $a_i(x) = 0$ for all i, which is absurd. □

It follows from this that $\mathbf{B}(L_g|_{\overline{\mathcal{U}}^r_{g-r,\sigma}})$ is contained in the boundary $\overline{\mathcal{U}}^r_{g-r,\sigma} \setminus \mathcal{U}^{r,part}_{g-r}$ of $\overline{\mathcal{U}}^r_{g-r,\sigma}$. Therefore $\mathcal{C}$ is disjoint from the open substack $\mathcal{U}^{r,part}_{g-r}$. This contradicts the assumption that $\pi(\mathcal{C})$ does not lie in A^{Sat}_{g-r-1} and (*1*) is proved.

For (*2*), assume first that we are in positive characteristic so that Keel's theorem [Kee99] is available. Assume also, as an induction hypothesis, that L_h is semi-ample on $\mathcal{A}^P_h$ for all $h < g$.

Let $\mathcal{Z}$ be an irreducible closed substack of $\mathcal{A}^P_g$ with $\mathcal{Z}.L_g^{\dim \mathcal{Z}} = 0$, so that $\mathcal{Z}$ lies in $\mathrm{Exc}(L_g)$. If $\mathcal{Z}\cap\mathcal{A}_g^{part}$ is not empty, then $\mathcal{Z}$ lies in $\mathcal{A}_{1,g-1}^P$, as desired. So we can suppose that $\mathcal{Z}$ lies over A_{g-r}^{Sat}, but not over A_{g-r-1}^{Sat}, where $r\geq 2$. Then $\mathcal{Z}$ lies in some $\overline{\mathcal{X}}_r$.

Note first that $r<g$, since L_g is ample on the fibre $\pi^{-1}(A_0^{Sat})$ of π. (This is the statement that $-D_g$ is π-ample which holds because, as a toroidal compactification, $\mathcal{A}_g^P$ is defined by taking a convex hull.)

Once again we can use a torus action to construct a rational equivalence $\mathcal{Z}\sim\mathcal{Z}_0=\mathcal{Y}+\mathcal{W}$ where $\mathcal{Y}$ lies in $\overline{\mathcal{U}}_{g-r,\sigma}^r$ and the fibres of $\mathcal{W}\to A_g^{Sat}$ are of strictly positive dimension.

Suppose first that $\mathcal{Z}=\mathcal{Y}$, that is, that $\mathcal{Z}$ lies in some $\overline{\mathcal{U}}_{g-r,\sigma}^r$. Since L_g is nef, we have $\mathcal{Z}.L_g^{\dim\mathcal{Z}}=0$. Put $\mathcal{Z}^0=\mathcal{Z}\cap\mathcal{U}_{g-r}^{r,part}$; this is open and dense in $\mathcal{Z}$.

As in the proof of (*1*), it follows from the linear equivalence (4.1) that the stable base locus of the restriction $L_g|_{\overline{\mathcal{U}}_{g-r,\sigma}^r}$ lies in the boundary $\overline{\mathcal{U}}_{g-r,\sigma}^r\setminus\mathcal{U}_{g-r}^{r,part}$. Therefore by Kodaira's lemma [Kol96, VI.2.15, VI.2.16], the restriction L' of L_g to $\mathcal{U}_{g-r}^{r,part}$ is semi-ample but not big, so that $\mathcal{Z}^0$ is covered by open curves $\mathcal{C}^0$ on which the morphism defined by the linear system $|mL'|$ for some suitable integer m, is constant.

Then, for each i, the linear system $|mL_{g-r+1}|$ defines a constant morphism on each curve $\rho_i(\mathcal{C}^0)$. By induction, $|mL_{g-r+1}|$ has no base points on $\mathcal{A}_{g-r+1}^P$, so the closure of $\rho_i(\mathcal{C}^0)$ lies in $\mathrm{Exc}(L_{g-r+1})$.

By the induction hypothesis, $\mathrm{Exc}(L_{g-r+1})=\mathcal{A}_{1,g-r}^P$. Moreover, if $\mathcal{U}_{g-r}^{part}$ is identified with an open substack of the boundary D_{g-r+1} of $\mathcal{A}_{g-r+1}^P$, then taking the j-invariant of the elliptic factor to be ∞ shows that $\mathrm{Exc}(L_{g-r+1})\cap\mathcal{U}_{g-r}^{part}$ contains the closure of the zero-section of the semi-abelian scheme $\mathcal{U}_{g-r}^{part}\to\mathcal{A}_{g-r}^{part}$.

Consider the intersection $\mathcal{I}=\mathcal{A}_{1,g-r}^P\cap\mathcal{U}_{g-r}^{part}$, taken inside D_{g-r+1}.

Lemma 4.6. *$\mathcal{I}$ has just two irreducible components. One is the locus of points of the form*

$$(\infty,B,0_B),$$

where $B\in\mathcal{A}_{g-r}$ and ∞ is the point at infinity on $\mathcal{A}_1^P$; this is a copy of $\mathcal{A}_{g-r}$. The other is the locus of points of the form

$$(E\times V,(0_E,v)),$$

where $E\in\mathcal{A}_1$, $V\in\mathcal{A}_{g-r-1}$ and $v\in V$ is arbitrary; this is a copy of $\mathcal{A}_1^P\times\mathcal{U}_{g-r-1}$.

Proof The only thing to notice is that on $\mathcal{A}_{g-r+1}^{part}$, the exceptional locus $\mathrm{Exc}(L_{g-r+1})$ includes the image of $\mathcal{A}_1\times\mathcal{A}_{g-r}^{part}$, the locus where the cycle-theoretic image of the Kummer variety under the 2θ linear system does not determine the abelian variety. The pair (V, v) corresponds to a compactification $\widetilde{V}$ of some $\mathbb{G}_m$-bundle over V and then taking the Kummer variety of $E \times \widetilde{V}$ has the effect of 'losing the isomorphism class of E'. □

That is, for every $\widetilde{\xi}_i : \mathcal{U}_{g-r}^{r,part} \to \mathcal{U}_{g-r}^{part}$, the image $\widetilde{\xi}_i(\mathcal{Z})$ is contained in the union of these two loci. Now consider the summand $a\delta^* L_{g-r}$ that appears as a contribution to $L_g|_{\mathcal{U}_{g-r}^{r,part}}$ in (4.1); since, by induction, L_{g-r} is semi-ample and $\mathrm{Exc}(L_{g-r}) = \mathcal{A}_{1,g-r-1}^P$, this consideration shows that

$$\mathcal{Z}^0 \cap \mathcal{U}_{g-r}^r \subset \mathcal{A}_1 \times \mathcal{U}_{g-r-1}^r.$$

Now $\mathcal{Z}$ lies in the closure of $\mathcal{Z}^0 \cap \mathcal{U}_{g-r}^r$ in $\overline{\mathcal{U}}_{g-r,\sigma}^r$, so that $\mathcal{Z}$ is in (the image of) $\mathcal{A}_1^P \times \overline{\mathcal{U}}_{g-r-1,\sigma}^r$ in $\mathcal{A}_{1,g-1}^P$. In particular, $\mathcal{Z}$ lies in $\mathcal{A}_{1,g-1}^P$.

Now drop the assumption that $\mathcal{Z}$ lies in some $\overline{\mathcal{U}}_{g-r,\sigma}^r$. Then $\mathcal{Z}$ specializes as above to $\mathcal{Z}_0 = \mathcal{W} + \mathcal{Y}$ where $\mathcal{Y}$ lies in some $\overline{\mathcal{U}}_{g-r,\sigma}^r$. Since L_g is nef, both $\mathcal{W}$ and $\mathcal{Y}$ lie in $\mathrm{Exc}(L_g)$, so that, by what we have already proved, $\mathcal{Y}$ lies in $\mathcal{A}_{1,g-1}^P$.

Recall that $\mathcal{Z}$ lies in $\overline{\mathcal{X}}_{r,\tau}$, the image of the closure of a T-bundle $\mathcal{T} \to \mathcal{U}_{g-r}^r$. The specialization $\mathcal{Z} \sim \mathcal{Z}_0$ and the fact that $\mathcal{Y}$ lies in $\mathcal{A}_{1,g-1}^P$ show that $\mathcal{Z}$ lies in the image of the closure of the restriction of $\mathcal{T}$ to the closed substack $\mathcal{A}_1 \times \mathcal{U}_{g-r-1}^r$ of $\mathcal{U}_{g-r}^r$. But this restriction is of the form $\mathcal{A}_1 \times \mathcal{T}_1$, where $\mathcal{T}_1$ is a T-bundle over $\mathcal{U}_{g-r-1}^r$. So $\mathcal{Z}$ lies in $\mathcal{A}_1^P \times \overline{\mathcal{X}}'_{r,\tau}$, where $\overline{\mathcal{X}}'_{r,\tau}$ is the image of the closure of $\mathcal{T}_1$. However, $\overline{\mathcal{X}}'_{r,\tau}$ lies in $\mathcal{A}_{g-1}^P$, so that $\mathcal{Z}$ lies in $\mathcal{A}_{1,g-1}^P$, as required.

We have shown that in characteristic $p > 0$, $\mathrm{Exc}(L_g) \subset \mathcal{A}_{1,g-1}^P$; the other inclusion is an immediate consequence of the fact that L_1 is trivial. From Keel's theorem we deduce that L_g is semi-ample. It follows at once that $\mathrm{Exc}(L_g) = \mathcal{A}_{1,g-1}^P$ in characteristic zero, and (*2*) is proved.

(*3*) We know now that L_g is semi-ample, and it remains to show that the varieties V_g behave as stated.

The multiplication morphism $\phi \colon \mathcal{A}_{g-h}^P \times \mathcal{A}_h^P \to \mathcal{A}_g^P$ has the property that $\phi^*\mathcal{O}(H_g) \cong \mathcal{O}(H_{g-h}) \boxtimes \mathcal{O}(H_h)$ and $\phi^*\mathcal{O}(D_g) \cong \mathcal{O}(D_{g-h}) \boxtimes \mathcal{O}(D_h)$. So $\phi^*\mathcal{O}(L_g) \cong \mathcal{O}(L_{g-h}) \boxtimes \mathcal{O}(L_h)$.

Let $\psi_g \colon A_g^P \to V_g$ be the morphism defined by the linear system $|mL_g|$ for sufficiently divisible m and take $h = 1$. Since L_1 is trivial,

$$A_{g-1}^P \times A_1^P \to A_{g-1,1}^P \to A_g^P \to V_g$$

factors through $pr_1 \colon A^P_{g-1} \times A_1 \to A^P_{g-1}$, say via $A^P_{g-1} \to V_g$. Denote the Stein factorization of this by $A^P_{g-1} \to W_{g-1}$. Since $A_{g-1} \to V_g$ is defined by some system $|mL_{g-1}|$, so W_{g-1} is identified with V_{g-1}. □

Corollary 4.7. *Suppose that* $7 \leq g \leq 10$ *and* $\operatorname{char} k = 0$. *Then the first step in running the MMP on* A^P_g *is the contraction* $A^P_g \to V_g$. *Its fibres are the fibres of* $A^P_{g-1,1} \to V_{g-1}$ *and they meet the interior* A_g *of* A^P_g.

Proof This follows at once from the results above since A^P_g has only terminal singularities in this range by the following appendix. □

5 Appendix with John Armstrong

In this appendix, we complete the proof in [SB06] of a result that is suggested by Tai's paper [Tai82], that in characteristic zero the singularities of A^P_g are canonical if $g \geq 5$ and terminal if $g \geq 6$. However, Hulek and Sankaran have pointed out that the argument given in [SB06] is incomplete because [SB06, Proposition 3.2] is not strong enough. We complete it by replacing Proposition 3.2 in [SB06] with Proposition 5.8 below.

Recall the RST criterion for the geometric quotient $[\mathcal{X}]$ of a smooth Deligne–Mumford stack $\mathcal{X}$ over a field of characteristic zero to have canonical or terminal singularities. We phrase it in terms of the action of a finite group G acting on a smooth variety Y so as to fix a point P on Y.

For any $s \in G - \{1\}$, define $\lambda_{Y,P}(s) = \sum \arg(\zeta)/2\pi$ where the sum is over the eigenvalues ζ of s on the tangent space $T_{Y,P}$. This extends in an obvious way to the relative context and then, if $\mathrm{g} \colon Y \to Z$ is a smooth G-morphism, $\lambda_{Y,P}(s) = \lambda_{Y/Z,P}(s) + \lambda_{Z,\mathrm{g}(P)}(s)$. We omit P if Y is a vector space and P its origin.

(RST) G acts freely in codimension one in a neighbourhood of P and $[Y/G]$ has terminal (resp. canonical) singularities at the image of P if and only if $\lambda_{Y,P}(s) > 1$ (resp. $\lambda_{Y,P}(s) \geq 1$) for every $s \in G - \{1\}$.

It is well known that the local geometry of $\mathcal{A}^P_g$ and A^P_g reduces to that described in the next paragraph.

Suppose that $h, r \geq 0$, $g = h + r$, (C, μ) is a principally polarized abelian h-fold, $W = H^0(C, T_C)$, $\Lambda = \mathbb{Z}^r$, $V = \operatorname{Sym}^2 W \oplus (W \otimes \Lambda)$ and $\mathbb{T}$ is the torus with $\mathbb{X}_*(\mathbb{T}) = B(\Lambda)$, the group of symmetric $\mathbb{Z}$-valued bilinear forms on Λ. There is a torus embedding $\mathbb{T} \hookrightarrow X$ with terminal singularities and a $\mathbb{T}$-bundle $\pi \colon \mathcal{T} \to V$ such that locally $\mathcal{A}^P_g = \mathcal{X}/G$ for a finite group $G = \mathrm{Stab}_{\mathcal{X}}(P)$, where $\mathcal{X} = \mathcal{T} \times^{\mathbb{T}} X \to V$ is a G-equivariant X-bundle and $P \in \mathcal{X}$. This is compatible with a $\mathbb{Z}$-linear action of G on Λ and an action of G on (C, μ), so that $\mathbb{T} \hookrightarrow X$ is G-equivariant and G acts linearly on Λ and on W, and so on V. Moreover, G acts effectively on V.

In terms of co-ordinates, choose a $\mathbb{Z}$-basis $(x_1, \ldots, x_r)$ of Λ and write $Z_{pq} = \exp 2\pi i(x_p \odot x_q)$. Then locally $\mathcal{O}_{\mathcal{X}} = \mathcal{O}_V[Y_m]$ for some monomials Y_m in the Z_{pq} and the action of an element $s \in G$ on $\mathcal{O}_{\mathcal{X}}$ is given by $s^* Y_m = \Phi_{s,m} W_{s(m)}$, where $\Phi_{s,m} \in \mathcal{O}_V^{hol}$, $\Phi_{s,m}$ does not vanish at 0 and $W_{s(m)}$ is a monomial in the Z_{pq}.

Lemma 5.1. *If $h \geq 1$, then G acts effectively on V.*

Proof Note that the only elements of $\mathrm{GL}(W) \times \mathrm{GL}(\Lambda)$ that act trivially on V are ± 1. □

Fix $s \in G - \{1\}$, so that $s|_V \neq 1$. Let ϕ be the Euler function and $\rho = \frac{1}{2}\phi$. Say that n *appears in* a given complex linear transformation σ if some primitive nth root of unity is an eigenvalue of σ.

Lemma 5.2. *If n appears in $s|_W$ then $\lambda_{\mathrm{Sym}^2 W}(s) > \phi(n)^3/32n$.*

Proof Note that $W \oplus \overline{W} = H_1(C, \mathbb{Q})$ and so is defined over $\mathbb{Q}$.

Let $W_n \subseteq W$ be the direct sum of eigenlines of $s|_W$ belonging to eigenvalues that are primitive nth roots of unity. Then $W_n \neq 0$ and $W = \oplus W_n$. Moreover, $W_n \oplus \overline{W}_n$ is the subspace of $W \oplus \overline{W}$ spanned by the eigenvectors belonging to eigenvalues of exact order n and is therefore defined over $\mathbb{Q}$. So, all primitive nth roots ω of unity arise as eigenvalues of s acting on $W_n \oplus \overline{W}_n$. So either ω or $\overline{\omega}$ is an eigenvalue of $s|_{W_n}$. So, there are at least $\rho(n)$ distinct eigenvalues ω of $s|_{W_n}$ such that $\overline{\omega}$ is not an eigenvalue of $s|_{W_n}$. Then W_n contains a subspace on which s acts with at least $\rho(n)$ distinct eigenvalues all of which are of the form $\zeta_n^{t_i}$ with $t_i \in [1, n-1]$, $(t_i, n) = 1$, $t_i \not\equiv \pm t_j \pmod{n}$ for $i \neq j$. Note that

$$\lambda_{\mathrm{Sym}^2 W}(s) \geq \lambda_{\mathrm{Sym}^2 W_n}(s) = \sum_{i \geq j} \left\{ \frac{t_i + t_j}{n} \right\} > \frac{1}{2} \sum_{i,j} \left\{ \frac{t_i + t_j}{n} \right\}$$

where $\{x\}$ denotes the fractional part of the real number x. For a fixed value of i, the fractions $\{\frac{t_i + t_j}{n}\}$ are distinct and non-zero, so that

$$\sum_j \left\{ \frac{t_i + t_j}{n} \right\} \geq \sum_{l=1}^{\rho(n)} l/n = \rho(n)(\rho(n)+1)/2n$$

and therefore

$$\frac{1}{2}\sum_{i,j}\left\{\frac{t_i+t_j}{n}\right\} \geq \rho(n)^2(\rho(n)+1)/4n$$

$$> \rho(n)^3/4n = \phi(n)^3/32n. \qquad \square$$

Lemma 5.3. *If $\phi(n)^3 < 32n$ then n divides 2520 and $n \leq 30$.*

Proof Consider the prime factorization $n = \prod p^{e_p}$, and notice that $(p-1)^2 > p(p-2)$. If n is even this leads to

$$64 > 2\phi(n)^3/n \geq 2^{2(e_2-1)} \prod_{p>2}(p-1)(p-2)p^{2(e_p-1)},$$

so that $e_2 \leq 3$ and if n is odd then

$$32 > \phi(n)^3/n \geq \prod(p-1)(p-2)p^{2(e_p-1)}.$$

So $(p-1)(p-2)p^{2(e_p-1)} < 64$ for all odd p whether n is even or odd. Then $e_3 \leq 2$, $e_5, e_7 \leq 1$ and $e_p = 0$ for $p \geq 11$. So, n divides $2^3 \cdot 3^2 \cdot 5 \cdot 7 = 2520$ and we verify using Mathematica [Mat] that $n \leq 30$. $\square$

For any integer $N \geq 1$ let S_N denote the set of primitive Nth roots of unity. For $\eta \in \mathbb{C}^*$ define

$$\psi_N(\eta) = \sum_{\zeta \in S_N} \arg(\zeta/\eta)/2\pi.$$

Lemma 5.4. *$\psi_N(\eta) \geq 1/3, 1/2, 1/3$ for $N = 3, 4, 6$ and $\psi_N(\eta) > 1$ if $N \neq 1, 2, 3, 4, 6$.*

Proof Clearly $\psi_N(\eta)$ is minimized by some $\eta \in S_N$, for otherwise we can rotate the whole picture clockwise by a small amount so as to decrease $\psi_N(\eta)$. For $N = 3, 4, 6$ the result is now immediate, so assume that $N \neq 1, 2, 3, 4, 6$.

The angle between any two members of S_N that are adjacent on the unit circle is at least $2\pi/N$, while if N is even then this angle is at least $4\pi/N$. So

$$\psi_N(\eta) \geq \sum_{i=0}^{\phi(N)-1} \frac{i}{N} = \frac{1}{N}\binom{\phi(N)}{2} > 1$$

if N is odd and

$$\psi_N(\eta) \geq \sum_{i=0}^{\phi(N)-1} \frac{2i}{N} = \frac{2}{N}\binom{\phi(N)}{2} > 1$$

if N is even and $N \neq 12$. It is easy to see that $\min_\eta \psi_{12}(\eta) = 4/3$. $\square$

Lemma 5.5. *Suppose that $h, r \geq 1$ and that $\lambda_{W\otimes\Lambda}(s) \leq 1$. Then only $1, 2, 3, 4, 6$ can appear in $s|_{\Lambda_\mathbb{C}}$.*

Proof Since $1 \geq \lambda_{W\otimes\Lambda}(s) = \sum_{i=1}^{h} \sum_N \psi_N(\zeta_i)$, where N runs over the numbers appearing in $s|_{\Lambda_\mathbb{C}}$ and ζ_i runs over the eigenvalues of $s|_W$, this follows from Lemma 5.4. $\square$

The next two lemmas deal with involutions. We let $\sim$ denote conjugacy of matrices and $(a_i^{h_i})$ or $(a_1^{h_1}, \ldots)$ the diagonal matrix whose diagonal entries are a_i with multiplicity h_i.

Lemma 5.6. *If $s|_\Lambda \neq 1$ and $\lambda_{\Lambda_\mathbb{C}}(s) < 1$ then $s|_{\Lambda_\mathbb{C}} \sim (1^{r-1}, -1)$, $s|_\Lambda^2 = 1$ and $\lambda_{\Lambda_\mathbb{C}}(s) = 1/2$.*

Proof If some $q \geq 3$ appears in $s|_{\Lambda_\mathbb{C}}$ then

$$\lambda_{\Lambda_\mathbb{C}}(s) \geq \sum_{a \in (\mathbb{Z}/q\mathbb{Z})^*} \frac{a}{q} \geq \frac{1}{q} + \frac{q-1}{q} = 1.$$

Otherwise $s|_{\Lambda_\mathbb{C}} \sim (1^a, -1^b)$, $b \geq 1$ and $\lambda_{\Lambda_\mathbb{C}}(s) \geq b/2$. $\square$

Lemma 5.7. *Assume that $h + r \geq 3$, $s|_W^2 = 1$ and $\lambda_V(s) < 1$. Then $h = 1$, $(s|_W;\ s|_{\Lambda_\mathbb{C}}) \sim \pm(1;\ (1^{r-1}, -1))$, $s|_V^2 = 1$ and $\lambda_V(s) = 1/2$.*

Proof If $s|_W \neq \pm 1$ then $s|_W \sim (1^{h_0}, -1^{h_1})$ with $h_0, h_1 > 0$. Therefore $\lambda_{\mathrm{Sym}^2 W}(s) \geq \frac{1}{2} h_0 h_1$, so $h_0 = h_1 = 1$. Then

$$\begin{aligned}\lambda_{W\otimes\Lambda}(s) &= \lambda_{\Lambda_\mathbb{C}}(s) + \lambda_{\Lambda(-1)_\mathbb{C}}(s) \\ &= \sum_\zeta \left(\frac{\arg \zeta}{2\pi} + \frac{\arg(-\zeta)}{2\pi} \right) \geq r/2 \geq 1/2,\end{aligned}$$

where ζ runs over the eigenvalues of $s|_{\Lambda_\mathbb{C}}$ and so $\lambda_V(s) \geq 1$.

If $s|_W = 1$ then, by Lemma 5.6, $\lambda_V(s) = h\lambda_{\Lambda_\mathbb{C}}(s) \geq 1$ unless $h = 1$ and $s|_{\Lambda_\mathbb{C}} \sim (1^{r-1}, -1)$. If $s|_W = -1$ then replace W by the twist $W(-1)$, and then $\lambda_V(s) \geq 1$ unless $s|_{\Lambda_\mathbb{C}} \sim (1, -1^{r-1})$. $\square$

Proposition 5.8. *Assume that $g \geq 5$ and that $h \geq 1$.*

(1) If $s|_W \neq \pm 1$, then $\lambda_{\mathrm{Sym}^2 W}(s) \geq 1$ if $h = 5$ and $\lambda_{\mathrm{Sym}^2 W}(s) > 1$ if $h \geq 6$.
(2) If $\lambda_V(s) < 1$ then $s|_V^2 = 1$, $h = 1$, $(s|_W, s|_{\Lambda_\mathbb{C}}) \sim \pm(1, (1^{r-1}, -1))$ and $\lambda_V(s) = 1/2$.

Proof We can assume that $\lambda_{\mathrm{Sym}^2 W}(s) \leq 1$ and that

(1) if n appears in $s|_W$ then n divides 2520 and $n \leq 30$ (by Lemmas 5.2 and 5.3);
(2) $s|_W^2 \neq 1$, so that some $n \geq 3$ appears in $s|_W$ (by Lemma 5.7);
(3) if q appears in $s|_{\Lambda_\mathbb{C}}$ then $q = 1, 2, 3, 4, 6$ (by Lemma 5.5).

Case (A): some $n \neq 1, 2, 3, 4, 6$ appears in $s|_W$.

For $I \subseteq (\mathbb{Z}/n\mathbb{Z})^*$ define $\overline{I} = \{n - a | a \in I\}$. For all $I \subseteq (\mathbb{Z}/n\mathbb{Z})^*$ such that $\#I = \rho(n)$ and $I \cap \overline{I} = \emptyset$ define

$$\lambda(n, I) = \sum_{a,b \in I, a \leq b} \left\{ \frac{a+b}{n} \right\}.$$

Note that $\lambda_{\mathrm{Sym}^2 W}(s) \geq \lambda(n, I)$. We verify using Mathematica that $\lambda(n, I) > 1$ for all n in the range (*1*) except when $(n, I) = (12, \{1, 7\})$. In this case, $\lambda(n, I) = 1$ and $h = 2$. Then $r \geq 3$ and $\lambda_{W \otimes \Lambda}(s) > 0$ so that $\lambda_V(s) > 1$.

Case (B): only $1, 2, 3, 4, 6$ appear in $s|_W$.

So $s|_W \sim ((\zeta^i)^{h_i})$ and $s|_{\Lambda_\mathbb{C}} \sim ((\zeta^k)^{r_k})$ where $\zeta = \exp(2\pi i/12)$,

$$\sum h_i = h, \ \sum_{i \neq 0,6} h_i \geq 1, \ h_i \geq 0, \ h_1 = h_5 = h_7 = h_{11} = 0 \quad \text{and}$$

$$\sum r_k = r, r_k \geq 0, \ r_1 = r_5 = r_7 = r_{11} = 0, \ r_2 = r_{10}, \ r_3 = r_9, \ r_4 = r_8.$$

Then

$$\lambda_V(s) = \sum_{0 \leq i \leq j \leq 11} \left\{ \frac{i+j}{12} \right\} h_i h_j + \sum_{i,k=0}^{11} \left\{ \frac{i+k}{12} \right\} h_i r_k.$$

Decreasing any h_i or r_k can only decrease $\lambda_V(s)$. However, the two sets of constraints just given imply that each shortest step in the process of decrease will diminish $h + r$ by either 1 or 2, and so we are reduced to considering the cases where $h + r = 5$ or 6. We verify using Mathematica that $\lambda_V(s) \geq 1$ for all $(h_i), (r_k)$ in this range. □

Theorem 5.9. *A_g^P has terminal (resp. canonical) singularities if $g \geq 6$ (resp. $g = 5$).*

Proof Proposition 5.8 shows, in particular, that the result holds in A_g.

We use notation introduced at the beginning of this appendix. Choose a $\mathbb{T} \rtimes G$-equivariant resolution $\widetilde{X} \to X$ and put $\widetilde{\mathcal{X}} = \mathcal{T} \times^{\mathbb{T}} \widetilde{X}$. If $s \in G - \{1\}$ acts trivially on some divisor E in $\widetilde{\mathcal{X}}$, then E is in the boundary $\partial \widetilde{\mathcal{X}} = \widetilde{\mathcal{X}} - \mathcal{T}$. Each boundary divisor dominates V, since $\widetilde{\mathcal{X}} \to V$ is a fibre bundle, so s acts trivially on V. This is false, so G acts freely in codimension one on $\widetilde{\mathcal{X}}$.

For sufficiently divisible $m > 0$ there is a G-invariant generator σ of the sheaf $\mathcal{O}(mK_{\mathcal{X}})$. Then σ is also a generator of $\mathcal{O}(mK_{[\mathcal{X}/G]})$. Since $\mathcal{X}$ has terminal singularities, σ vanishes along all exceptional divisors in $\widetilde{\mathcal{X}}$, so that σ also vanishes along all exceptional divisors in $[\widetilde{\mathcal{X}}/G]$. So, it is enough to show, via (RST), that the singularities of $[\widetilde{\mathcal{X}}/G]$ are terminal (respectively canonical).

Suppose $P \in \widetilde{\mathcal{X}}$ is fixed by G.

Case (1): $h \geq 1$, $g \geq 6$ (resp. $g = 5$) and $\lambda_{\widetilde{\mathcal{X}},P}(s) \leq 1$ (resp. < 1). So $P \in \partial\widetilde{\mathcal{X}}$.

If $\lambda_{\widetilde{\mathcal{X}}/V,P}(s) = 0$ then s acts trivially on the relative tangent space $T_{\widetilde{\mathcal{X}}/V,P}$ so that a suitable point $Q \in \mathcal{T}$ near to P would give $Q \in \mathcal{A}_g$ with $\lambda_{\mathcal{A}_g,Q}(s) = \lambda_{\widetilde{\mathcal{X}},P}(s) \leq 1$ (respectively < 1), which is impossible. Since $\lambda_{\widetilde{\mathcal{X}},P}(s) = \lambda_{\widetilde{\mathcal{X}}/V,P}(s) + \lambda_V(s)$, it follows that $\lambda_{\widetilde{\mathcal{X}}/V,P}(s) > 0$, $\lambda_V(s) < 1$, and then, by Proposition 5.8, $s|_V^2 = 1$, $\lambda_V(s) = 1/2$ and $h = 1$. Since G acts effectively on V, $s^2 = 1$ and also $\lambda_{\widetilde{\mathcal{X}}/V,P}(s) = 1/2$ and $\lambda_{\widetilde{\mathcal{X}},P}(s) = 1$.

There is a G-equivariant factorization $\widetilde{\mathcal{X}} \overset{\alpha}{\to} \mathcal{Y} \overset{\beta}{\to} V$ of $\widetilde{\mathcal{X}} \to V$, where α is a vector bundle and β is a torus bundle. The $\mathcal{O}_V$-algebra $\mathcal{O}_{\mathcal{Y}}$ is generated by those monomials Y_m that are invertible at P and the rank n of α is the number of irreducible boundary divisors D_i of $\widetilde{\mathcal{X}}$ that pass through P. The D_i are permuted by G, so by s. Since we are in characteristic zero, each D_i has a defining equation $z_i = 0$ such that either $s^*z_i = u_i z_i$ with $u_i^2 = 1$ or $s^*z_i = z_j$ for some j.

If $\lambda_{\widetilde{\mathcal{X}}/\mathcal{Y},P}(s) = 0$ then, as before, we find $Q \in \mathcal{A}_g$ with $\lambda_{\mathcal{A}_g,Q}(s) = 1$. So $\lambda_{\widetilde{\mathcal{X}}/\mathcal{Y},P}(s) = 1/2$ and $\lambda_{\mathcal{Y}/V,\alpha(P)}(s) = 0$. It follows that $s^*z_i = z_i$ for $i \leq n-2$ and either $s^*z_{n-1} = z_n$ and $s^*z_n = z_{n-1}$ or $s^*z_{n-1} = z_{n-1}$ and $s^*z_n = -z_n$. The log cotangent space $\Omega^1_{\widetilde{\mathcal{X}}/\mathcal{Y}}(\log\partial\widetilde{\mathcal{X}})(P)$ is based by $\{\frac{dz_i}{z_i}(P)\}$, so that

$$\operatorname{tr}\left(s|_{T_{\widetilde{\mathcal{X}}/\mathcal{Y}}(-\log\partial\widetilde{\mathcal{X}})(P)}\right) = n-2 \text{ or } n.$$

The middle term of the G-equivariant log tangent bundle sequence

$$0 \to T_{\widetilde{\mathcal{X}}/\mathcal{Y}}(-\log\partial\widetilde{\mathcal{X}}) \to T_{\widetilde{\mathcal{X}}/V}(-\log\partial\widetilde{\mathcal{X}}) \to \alpha^*T_{\mathcal{Y}/V} \to 0$$

is constant and its fibre is $\operatorname{Lie}\mathbb{T}$. This is naturally isomorphic to $B(\Lambda_{\mathbb{C}})$. Since $\lambda_{\mathcal{Y}/V,\alpha(P)}(s) = 0$, $\operatorname{tr}(s|_{\operatorname{Lie}(\mathbb{T})}) = \dim\mathbb{T}-2$ or $\dim\mathbb{T}$. However, $s|_{\Lambda_{\mathbb{C}}} \sim (1, -1^{r-1})$, so that the invariant subspace $B(\Lambda_{\mathbb{C}})^s$ has codimension $r-1$, while $r \geq 4$.

Case (2) : $h = 0$. We use an unpublished theorem of Snurnikov [Snu02] for which we include a proof.

Proposition 5.10 ([Snu02]). *Suppose $\mathbb{T} \hookrightarrow X$ is a G-equivariant torus embedding, $U \subset X$ a G-invariant neighbourhood of $X - \mathbb{T}$, G acts freely in codimension one on $U \cap \mathbb{T}$ and that U and $[(U \cap \mathbb{T})/G]$ have terminal (resp. canonical) singularities. Then G acts freely in codimension one on U and $[U/G]$ has terminal (resp. canonical) singularities.*

Proof Let $\pi\colon \widetilde{X} \to X$ be a $\mathbb{T} \rtimes G$-equivariant resolution. As before, it is enough to show that the G-action on $\widetilde{U} = \pi^{-1}(U)$ satisfies (RST).

Near P, $\widetilde{X}$ is an s-equivariant vector bundle

$$q\colon \widetilde{X} = \operatorname{Spec} \mathbb{C}[x, y_j^{\pm}] \to \mathbb{T}_1 = \operatorname{Spec} \mathbb{C}[y_j^{\pm}];$$

the y_j are those characters of $\mathbb{T}$ invertible at P and $s^* x_i = \phi_i x_{s(i)}$, where ϕ_i is a monomial in the y_j. We can identify P with the origin of the vector space $F = q^{-1}(q(P))$.

We can assume $\widetilde{X} \neq \mathbb{T}_1$, for else $\widetilde{X} = \mathbb{T}$. Then the locus $x_i = t$, $y_j^{\pm} = 1$ for all i, j is a line L of s-fixed points in $\widetilde{X}$ that passes through P and meets $\widetilde{U} \cap \mathbb{T} = U \cap \mathbb{T}$. Take any point $Q \in U \cap L$; then $\lambda_{\widetilde{X},P}(s) = \lambda_{U \cap \mathbb{T}, Q}(s)$. □

If we take U (respectively $U \cap \mathbb{T}$) to be a chart of $\mathcal{A}_g^P$ (respectively $\mathcal{A}_g$), then Theorem 5.9 is proved when $h = 0$. □

References

[AMRT10] Avner Ash, David Mumford, Michael Rapoport and Yung-Sheng Tai. *Smooth compactifications of locally symmetric varieties.* Cambridge Mathematical Library. Cambridge University Press, Cambridge, second edition, 2010. With the collaboration of Peter Scholze.

[BC76] E. S. Barnes and M. J. Cohn. On the inner product of positive quadratic forms. *J. London Math. Soc. (2)*, 12(1):32–36, 1975/76.

[DO88] Igor Dolgachev and David Ortland. Point sets in projective spaces and theta functions. *Astérisque*, (165):210 pp. (1989), 1988.

[FC90] Gerd Faltings and Ching-Li Chai. *Degeneration of abelian varieties*, volume 22 of *Ergebnisse der Mathematik und ihrer Grenzgebiete (3).* Springer-Verlag, Berlin, 1990. With an appendix by David Mumford.

[HH13] Brendan Hassett and Donghoon Hyeon. Log minimal model program for the moduli space of stable curves: the first flip. *Ann. of Math. (2)*, 177(3):911–968, 2013.

[Hud90] R. W. H. T. Hudson. *Kummer's quartic surface.* Cambridge Mathematical Library. Cambridge University Press, Cambridge, 1990. With a foreword by W. Barth, Revised reprint of the 1905 original.

[Mat] Wolfram Research, Inc. Mathematica. Champaign, IL, 2021.

[Kee99] Seán Keel. Basepoint freeness for nef and big line bundles in positive characteristic. *Ann. of Math. (2)*, 149(1):253–286, 1999.

[KM97] Seán Keel and Shigefumi Mori. Quotients by groupoids. *Ann. of Math. (2)*, 145(1):193–213, 1997.

[Kol96] János Kollár. *Rational curves on algebraic varieties*, volume 32 of *Ergebnisse der Mathematik und ihrer Grenzgebiete. 3. Folge. A Series of Modern Surveys in Mathematics.* Springer-Verlag, Berlin, 1996.

[SB06] N. I. Shepherd-Barron. Perfect forms and the moduli space of abelian varieties. *Invent. Math.*, 163(1):25–45, 2006.

[Snu02] V. S. Snurnikov. *Quotients of canonical toric singularities.* PhD thesis, University of Cambridge, 2002.

[Tai82] Yung-Sheng Tai. On the Kodaira dimension of the moduli space of abelian varieties. *Invent. Math.*, 68(3):425–439, 1982.

Variation of Stable Birational Types of Hypersurfaces

Evgeny Shinder
with an Appendix by Claire Voisin

To Miles Reid

Abstract

We introduce and study the question how can stable birational types vary in a smooth proper family. Our starting point is the specialization for stable birational types of Nicaise and the author, and our emphasis is on stable birational types of hypersurfaces. Building up on the work of Totaro and Schreieder on stable irrationality of hypersurfaces of high degree, we show that smooth Fano hypersurfaces of large degree over a field of characteristic zero are in general not stably birational to each other. In the appendix, Claire Voisin proves a similar result in a different setting using the Chow decomposition of diagonal and unramified cohomology.

1 Introduction

Let k be an uncountable algebraically closed field of characteristic zero. Recall that k-varieties X, Y of the same dimension are called *stably birational* if $X \times \mathbb{P}^m$ and $Y \times \mathbb{P}^m$ are birational for some $m \geq 0$. If in the above definition Y is a projective space, then X is called *stably rational.* There has been recently a lot of progress in showing that for large classes of varieties, including Fano hypersurfaces of high degree, very general members are stably irrational [Voi15, CTP16b, Tot16, Sch19b]. In this paper, we introduce and study the following more general question:

Question 1.1. Given a family of smooth projective varieties, how can we decide if all members are stably birational to each other?

We answer this question for Fano hypersurfaces of sufficiently high degree. Our main result is the following:

Theorem 1.2 (See Theorem 3.4). *If there exists a stably irrational smooth projective hypersurface of dimension n and degree $d \leq n+1$, then very general hypersurfaces of dimension n and degree d are not stably birational to each other.*

Here by very general hypersurfaces we mean pairs of hypersurfaces corresponding to points in the parameter space $\mathbb{P}(H^0(\mathbb{P}^{n+1}, \mathcal{O}(d)))^2$ lying in the complement of a countable union of divisors.

Thus the only case when smooth hypersurfaces of given degree and dimension are stably birational to each other is when they are all stably rational. This happens in degrees one and two, and for cubic surfaces, and it is widely expected that no other such cases exist.

It has been proved by Totaro [Tot16] that in every dimension $n \geq 3$ very general Fano hypersurfaces of degree $d \geq 2\lceil \frac{n+2}{3} \rceil$ are stably irrational. Schreieder improved Totaro's bound to $d \geq \log_2(n) + 2$ [Sch19b]. Using [Sch19b, Corollary 1.2] and the Theorem above we deduce the following.

Corollary 1.3. *For $n \geq 3$ and $d \geq \log_2(n) + 2$, very general hypersurfaces of dimension n and degree d are not stably birational to each other.*

In particular, we see that there are uncountably many stable birational types of such hypersurfaces. The first interesting case when Corollary applies is that of quartic threefolds ($n = 3, d = 4$, here stable irrationality of the very general member follows from [CTP16b]).

Under the assumptions of the Theorem every stable birational type is attained at a countable union of Zariski closed subsets in the parameter space of smooth hypersurfaces. A more explicit description of which hypersurfaces of fixed dimension and degree would be stably birational to the given one, seems completely out of reach.

Our approach to stable birational types relies on the Grothendieck ring of varieties, the Larsen—Lunts Theorem [AM72] and the specialization map [NS19, KT19]. Firstly, we reformulate results of [NS19] by introducing the idea of a variation of stable birational types and show that if stable birational type in a family is not constant, then it has to vary in a strong sense (Theorem 3.2). Then by constructing an appropriate degeneration of smooth hypersurfaces to a hyperplane arrangement, with desingularized total space (Lemma 3.6) and showing that the class of this hyperplane arrangement in the Grothendieck ring is congruent to 1 modulo $\mathbb{L}$ (Lemma 2.1) we deduce that under the conditions of the theorem, stable birational types of hypersurfaces can not be constant (Theorem 3.4). The same method would apply to any family that has a smooth stably irrational member alongside a smooth stably rational member, or more generally, a member with mild singularities and whose class in the Grothendieck modulo $\mathbb{L}$ is equal to one and provided that the total space of the degeneration is smooth or has mild singularities.

In addition to using the Grothendieck ring of varieties and the specialization map, one novelty of this work is making use of degeneration of a hypersurface to a hyperplane arrangement. Such degenerations are ubiquitous in algebraic geometry, starting from computing the genus of a plane curve and all the way to the modern Gross—Siebert program. These degenerations also played their role in rationality problems [CTO89]. Our contribution, however, is the direct link between having a semistable fiber in a family and variation of the stable birational types of the smooth fibers. One familiar example of this behaviour is that an isotrivial elliptic surface can not have semistable fibers. This well-known fact is an easy corollary of Proposition 3.3.

In the Appendix to this paper Claire Voisin proves a similar result regarding variation of stable birational types in a slighly different setting using decomposition of diagonal and unramified cohomology. Very soon after appearance of this work, Stefan Schreieder gave a different proof of Corollary 1.3 using degeneration to hyperplane arrangement and decomposition of the diagonal, also relying on [Sch19c]; in fact Schreieder's proof does not use resolution of singularities and thus generalizes the statement to a field of an arbitrary characteristic.

Finally, we note that unlike in Hodge theory, where the term 'variation' can be understood using the period map between moduli spaces, our term 'variation of stable birational types' has a very naive meaning; it is not at all clear how one could introduce a reasonable moduli space of stable birational types.

Notation

By a variety we mean a separated irreducible and reduced scheme of finite type over k. By a point of a variety we a mean a closed point. We say that a property holds for very general points of a variety if it holds away from a countable union of divisors.

2 Preliminary Results

2.1 Grothendieck Ring of Varieties

Recall that the Grothendieck ring of varieties $K_0(Var/k)$ is generated as an abelian group by isomorphism classes $[X]$ of schemes of finite type X/k modulo the scissor relations

$$[X] = [U] + [Z]$$

for every closed $Z \subset X$ with open complement $U \subset X$. The product structure on $K_0(Var/k)$ is induced by product of schemes. We write $\mathbb{L} \in K_0(Var/k)$ for the class of the affine line $[\mathbb{A}^1]$.

The following lemma is useful when degenerating smooth varieties to hyperplane arrangements.

Lemma 2.1. *Let $H_1, \ldots, H_r \subset \mathbb{P}^{n+1}$ be a collection of distinct hyperplanes in $\mathbb{P}^{n+1}$ such that $\bigcup_{i=1}^{r} H_i$ is a simple normal crossing divisor, that is, we assume that any intersection of k hyperplanes is either empty or of codimension k. Then we have*

$$[H_1 \cup \cdots \cup H_r] = \sum_{j=0}^{n} (-1)^j \binom{r}{j+1} [\mathbb{P}^{n-j}]$$

and if $r \leq n+1$, then $[H_1 \cup \cdots \cup H_r] \equiv 1 \mod \mathbb{L}$.

Proof Let $P_{r,n} \in K_0(Var/k)$ be the class of a simple normal crossing hyperplane arrangement of r hyperplanes in $\mathbb{P}^{n+1}$ in the Grothendieck ring of varieties. It follows from the inductive argument below that the class $P_{r,n}$ only depends on r and n and not on the relative positions of the hyperplanes.

We prove the formula for $P_{r,n}$ for $r \geq 1$, $n \geq 0$ using induction. For the induction base we have for all $r \geq 1$, $P_{r,0} = r$ (r points in $\mathbb{P}^1$) and for all $n \geq 0$ we have $P_{1,n} = [\mathbb{P}^n]$. We assume that the formula is true for $P_{r,n-1}$ and $P_{r-1,n-1}$. Given $r \geq 2$ hyperplanes in $\mathbb{P}^{n+1}$, intersecting the first $r-1$ of them with the last one, gives rise to an arrangement of $r-1$ hyperplanes in $\mathbb{P}^n$, which is still simple normal crossing. Using inclusion-exclusion we obtain

$$P_{r,n} = P_{r-1,n} + [\mathbb{P}^n] - P_{r-1,n-1},$$

which by induction hypothesis can be rewritten as

$$P_{r,n} = \sum_{j=0}^{n}(-1)^j\binom{r-1}{j+1}[\mathbb{P}^{n-j}] + [\mathbb{P}^n] - \sum_{i=0}^{n-1}(-1)^i\binom{r-1}{i+1}[\mathbb{P}^{n-i-1}]$$

which easily gives the desired result.

Finally, if $r \leq n+1$, then

$$P_{r,n} \equiv \sum_{j=0}^{n}(-1)^j\binom{r}{j+1} = \sum_{i=1}^{r}(-1)^{i-1}\binom{r}{i} = 1 \mod \mathbb{L}. \qquad \square$$

2.2 Resolution of One Toric Singularity

When constructing resolutions of singularities for the total space of a degeneration of hypersurfaces the following result is useful. We refer to [Ful93] for standard facts and constructions from toric geometry.

Lemma 2.2. *Let be $\mathcal{X}$ be a hypersurface in $\mathbb{A}^{n+2}$ defined by equation*

$$t \cdot y = z_1 \cdots z_n$$

and let $\pi\colon \mathcal{X} \to \mathbb{A}^1$ be the morphism given by the t coordinate.

(1) Let $N = \mathbb{Z}^{n+1}$ with standard basis $e_1, \ldots, e_{n+1}$ and let $N_\mathbb{R} = N \otimes \mathbb{R}$. For every $1 \leq i \leq n$, let $f_i = e_i + e_{n+1} \in N$. Let $\sigma \subset N_\mathbb{R} = \mathbb{R}^{n+1}$ be the cone generated by the vectors $e_1, \ldots, e_n, f_1, \ldots, f_n$. Then $\mathcal{X}$ is the toric variety corresponding to the cone σ, that is, $\mathcal{X} = \mathrm{Spec}(k[N^\vee \cap \sigma^\vee])$.

(2) Subdivision of σ into n cones

$$\sigma_k := \mathbb{R}_{\geq 0}f_1 + \cdots + \mathbb{R}_{\geq 0}f_k + \mathbb{R}_{\geq 0}e_k + \cdots + \mathbb{R}_{\geq 0}e_n \subset N_\mathbb{R}, \quad 1 \leq k \leq n$$

provides a resolution of singularities $\tau\colon \tilde{\mathcal{X}} \to \mathcal{X}$. The composition $\tilde{\pi} := \pi \circ \tau$ has a reduced simple normal crossing fiber over $0 \in \mathbb{A}^1$.

(3) A desingularization τ is obtained by a sequence of blow ups of proper preimages of $n-1$ Weil divisors $V(t, z_1), \ldots, V(t, z_{n-1}) \subset \mathcal{X}$.

Proof The proof is a standard computation in toric geometry.

(*1*) Let $M = N^\vee$ be the dual lattice with the dual basis $e_1^*, \ldots, e_{n+1}^*$. The dual cone $\sigma^\vee \subset M_\mathbb{R}$ is described by the system of inequalities for $(a_1, \ldots, a_{n+1}) \in M_\mathbb{R}$:

$$a_1 \geq 0, \; \ldots, \; a_n \geq 0,$$

$$a_1 + a_{n+1} \geq 0, \; \ldots, \; a_n + a_{n+1} \geq 0.$$

It is clear that the $n+2$ vectors

$$e_1^*, \ \ldots, \ e_n^*, \ e_{n+1}^*, \ e_1^* + \cdots + e_n^* - e_{n+1}^* \in M$$

all satisfy these inequalities and that every integral point in $\sigma^\vee$ can be written as a non-negative integer combination of these vectors (indeed, if $a_{n+1} \geq 0$, then we are done, while if $a_{n+1} < 0$, all other coordinates must be positive and a multiple of $e_1^* + \cdots + e_n^* - e_{n+1}^*$ can be subtracted).

If we set

$$z_1, \ldots, z_n, t, y$$

to be the monomials corresponding to the vectors above, they satisfy a single relation $ty = z_1 \cdots z_n$.

(*2*) The cone σ combinatorially is a cone over a prism $\Delta^{n-1} \times [0,1]$ (where Δ is a simplex), and the subdivision we consider corresponds to a standard subdivision of this prism into n simplices. To describe this construction in detail note that we have seen that the dual cone $\sigma^\vee$ is generated by $e_1^*, \ \ldots, \ e_n^*, \ e_{n+1}^*, \ e_1^* + \cdots + e_n^* - e_{n+1}^* \in M$, so that the cone $\sigma \subset N_{\mathbb{R}}$ is the set of solutions of

$$x_1 \geq 0, \ \ldots, \ x_{n+1} \geq 0, \quad x_{n+1} \leq x_1 + \cdots + x_n.$$

For every $1 \leq k \leq n$ let us consider the cones given by

$$x_1 \geq 0, \ \ldots, \ x_{n+1} \geq 0, \quad x_1 + \cdots + x_{k-1} \leq x_{n+1} \leq x_1 + \cdots + x_k.$$

These cones obviously form a partition of σ and it is easy to see that these are precisely the cones σ_k with boundary rays generated by $f_1, \ldots, f_k$, $e_k, \ldots, e_n$.

The new fan, consisting of the cones σ_k and all their faces has its cones generated by basis vectors of the lattice N, hence the corresponding morphism $\tau \colon \tilde{\mathcal{X}} \to \mathcal{X}$ is a resolution of singularities.

To check the fiber $\tilde{\pi}$ is reduced simple normal crossing over $0 \in \mathbb{A}^1$, we consider each affine toric chart U_k, corresponding to the cone σ_k. By our choice of coordinates, the restriction of $\tilde{\pi}$ to U_k corresponds to the projection onto the last coordinate $N_{\mathbb{R}} = \mathbb{R}^{n+1} \to \mathbb{R}$. The fiber over $0 \in \mathbb{A}^1$ being a reduced simple normal crossing divisor in U_k translates into the fact that every vector $f_1, \ldots, f_k, e_k, \ldots, e_n$ has zero or one as its last coordinate.

(*3*) Let us describe the effect of the blow up of $V(t, z_1) \subset \mathcal{X}$ on our toric model. We have two open charts, on the first open chart which we call U_1 we have $z_1 = tz_1'$ so the coordinates are $t, y, z_1', z_2, \ldots, z_n$ and the equation is

$$y = z_1' z_2 \cdots z_n.$$

On the other open chart which we call $\mathcal{X}'$, we have $t = z_1 t'$. So, coordinates are $t', y, z_1, \ldots, z_n$ and the equation is

$$t'y = z_2 \cdots z_n.$$

The gluing between the two open charts is $t' = \frac{1}{z_1'}$. We see that U_1 is the affine space with coordinates $z_1', z_2, \ldots, z_n$ which are monomials corresponding to vectors

$$e_1^* - e_{n+1}^*,\ e_2^*,\ \ldots,\ e_n^*,\ e_{n+1}^*,$$

that are precisely the generators for the dual cone to σ_1, while $\mathcal{X}'$ has coordinates corresponding to the monomials

$$e_{n+1}^* - e_1^*,\ e_1^*,\ \ldots,\ e_n^*,\ e_1^* + \cdots + e_n^* - e_{n+1}^*$$

and it follows easily that $\mathcal{X}'$ is the toric variety corresponding to the cone

$$\sigma' := \sigma_2 \cup \cdots \cup \sigma_n = \sum_{i=2}^{n} \mathbb{R}_{\geq 0} e_i + \sum_{i=1}^{n} \mathbb{R}_{\geq 0} f_i \subset N_{\mathbb{R}}.$$

Furthermore $\mathcal{X}'$ is the product of $\mathbb{A}^1$ (z_1 coordinate) with the same model in $n-1$ variables.

Since U_1 is already smooth, it will not be affected by further blow ups of divisors, while the proper preimage of $V(t, z_k)$ (for $k \geq 2$) in $\mathcal{X}'$ is $V(t', z_k)$, and the same argument can be applied to $\mathcal{X}'$ to get open charts $U_2, \ldots, U_n$, corresponding to the subdivision of σ'.

The process terminates after $n-1$ steps, when two smooth charts are produced. □

3 Stable Birational Types of Hypersurfaces

3.1 Variation of Stable Birational Types

We recall the following result of Larsen and Lunts which holds over arbitrary fields of characteristic zero and which provides the link between birational geometry and the Grothendieck ring of varieties.

Theorem 3.1. *[AM72] If X and Y are smooth projective varieties with classes $[X], [Y] \in K_0(Var/k)$, then X and Y are stably birational if and only if*

$$[X] \equiv [Y] \mod \mathbb{L}.$$

Thus the element $[X] \in K_0(Var/k)/(\mathbb{L})$ encodes the stable birational class of X, for a smooth projective variety X.

We now introduce the idea of the variation of stable birational types in the smooth and simple normal crossing settings. These rely on the results of [NS19].

Theorem 3.2. *Let S be a variety and let $\pi\colon \mathcal{X} \to S$ be a smooth proper morphism with connected fibers. Then one of the following is true:*

(1) **Constant stable birational type:** *All fibers $\pi^{-1}(t)$, $t \in S$ are stably birational.*
(2) **Variation of stable birational type:** *For very general points $(t, t') \in S \times S$ the fibers $\pi^{-1}(t)$ and $\pi^{-1}(t')$ are not stably birational to each other.*

Proof For $i = 1, 2$, let us write $p_i\colon S \times S \to S$ for the two projections. Let $\pi_i\colon \mathcal{X}_i \to S \times S$ denote the base change of π by p_i. Thus π_1, π_2 are smooth proper morphisms. Let $Z \subset S \times S$ be the set of points where the fibers of π_1 and π_2 are stably birational. In other words, Z consists of points (t_1, t_2) such that $\pi^{-1}(t_1)$ and $\pi^{-1}(t_2)$ are stably birational.

By [NS19, Theorem 4.1.4], Z is a countable union of Zariski closed subsets of $S \times S$. Thus either $Z = S \times S$, which corresponds to the case (*1*), or $Z \subsetneq S \times S$, so that points in $S \times S \setminus Z$ are very general which corresponds to (*2*). $\square$

The next proposition provides a generalization of the Theorem above to simple normal crossing singularities. Instead of stable birational types, we work with classes in $K_0(Var/k)/(\mathbb{L})$.

Proposition 3.3. *Let C be a smooth connected curve and let $\pi\colon \mathcal{X} \to C$ be a flat proper morphism with connected fibers and smooth total space $\mathcal{X}$. Let $0 \in C$, and assume that the restriction of π to $C \setminus 0$ is smooth, and that $\pi^{-1}(0)$ is reduced simple normal crossing.*

If all fibers $\pi^{-1}(t)$ for $t \neq 0$ are stably birational to a smooth projective variety X, then the class of the central fiber satisfies

$$[\pi^{-1}(0)] \equiv [X] \mod \mathbb{L}.$$

Proof Let us form a constant family $\pi'\colon X \times C \to C$. By assumption, the two morphisms π, π' have stably birational fibers for $t \neq 0$.

Thus using [NS19, Proposition 4.1.1], we deduce that fibers over $t = 0$ satisfy

$$[\pi^{-1}(0)] \equiv [\pi'^{-1}(0)] = [X] \mod \mathbb{L}. \qquad \square$$

3.2 Application to Hypersurfaces

We study stably birational types of hypersurfaces $X \subset \mathbb{P}^{n+1}$. The interesting case is the Fano case, that is, the case when the degree of X satisfies $d \leq n+1$.

Theorem 3.4. *Assume that there exists a smooth projective hypersurface of dimension n and degree $d \leq n+1$ which is stably irrational. Then smooth projective hypersurfaces of dimension n and degree d admit a variation of stable birational types, that is, two very general such hypersurfaces are not stably birational to each other.*

Remark 3.5. By the main result of [NS19], existence of a single stably irrational smooth projective hypersurface of dimension n and degree d is equivalent to very general such hypersurfaces being stably irrational.

Before we prove the theorem, we need the following lemma, which provides a convenient degeneration of smooth hypersurfaces.

Lemma 3.6. *For every $n \geq 2$, $1 \leq d \leq n+1$, there exists a smooth connected curve C with a point $0 \in C$ and a flat proper morphism*

$$\pi\colon \mathcal{X} \to C$$

with smooth $\mathcal{X}$ such that

(1) All fibers $\pi^{-1}(t)$, for $t \neq 0$ are smooth projective hypersurfaces of dimension n and degree d,
(2) The fiber $\pi^{-1}(0)$ is reduced simple normal crossing and satisfies

$$[\pi^{-1}(0)] \equiv 1 \mod \mathbb{L}.$$

Proof We consider two sections $F_0, F_1 \in H^0(\mathbb{P}^{n+1}, \mathcal{O}(d))$. We take F_0 to be a product of d linearly independent linear forms, and F_1 to be a general section. In particular, the hypersurface $V(F_1) \subset \mathbb{P}^{n+1}$ is smooth, and it intersects all the strata of the hyperplane arrangement $V(F_0)$ transversally.

We set $\mathcal{X}$ to denote the zero locus of $F_0 + tF_1$ in $\mathbb{P}^{n+1} \times \mathbb{A}^1$. After restricting to an open subset $C \subset \mathbb{A}^1$, we may assume that the fibers of $\pi\colon \mathcal{X}' \to C$ for $t \neq 0$ are smooth. The fiber $\pi^{-1}(0)$ is the hyperplane arrangement $V(F_0)$. Since $d \leq n+1$, Lemma 2.1 implies that the fiber $\pi^{-1}(0)$ satisfies

$$[\pi^{-1}(0)] \equiv 1 \mod \mathbb{L}.$$

Thus the morphism $\pi\colon \mathcal{X} \to C$ satisfies all the requirements of the Lemma except for smoothness of the total space $\mathcal{X}$.

We provide an explicit desingularization of $\mathcal{X}$. Let $E_1, \ldots, E_d$ be the components of $\pi^{-1}(0)$. We claim that blowing up the Weil divisors $E_1, \ldots, E_n$ (in any order) produces a model satisfying all the required properties.

Locally at every point $P \in \mathcal{X}$ in the central fiber $t = 0$, the model $\mathcal{X}$ is given by equations of the form

$$t \cdot f + l_1 \cdots l_d = 0,$$

where l_i are linear polynomials and f is a polynomial of degree d, in $n+1$ variables.

We change coordinates so that $P = 0$, and by our transversality assumptions the equation can be written as

$$t \cdot (x_{n+1} + \text{terms of deg.} \geq 2) + x_1 \cdots x_k \cdot g(x_1, \ldots, x_{n+1}) = 0,$$

where $g(0) \neq 0$ and $k \leq n$. Taking formal completion of $\mathcal{X}$ at P we can change the coordinates again to rewrite the defining local equation as

$$t \cdot x_{n+1} = x_1 \cdots x_k.$$

According to Lemma 2.2 such singularities are resolved by a sequence of blows up of Weil divisors $V(t, x_i)$ (which are precisely the components of the central fiber containing the point P) and this new model is semistable over $0 \in \mathbb{A}^1$ (and the rest of the fibers are unchanged, so they are smooth hypersurfaces).

Since each open chart of the blow up is a hypersurface in $\mathbb{A}^{n+2}$, the resulting blow ups only glue in Zariski locally trivial $\mathbb{P}^1$-fibrations. In particular, the class of the central fiber in $K_0(Var/k)/(\mathbb{L})$ does not change at each blow up. □

Proof of Theorem 3.4 Let $U \subset \mathbb{P}(H^0(\mathbb{P}^{n+1}, \mathcal{O}(d)))$ be the open subset parametrizing smooth hypersurfaces. By Theorem 3.2, if stable birational types of hypersurfaces of dimension n and degree d, does NOT vary, it has to be constant, that is, all such smooth hypersurfaces are stably birational to a smooth projective variety X.

We now consider the family $\pi \colon \mathcal{X} \to C$ given by Lemma 3.6. From what we explained above, all fibers $\pi^{-1}(t)$, for $t \neq 0$ have to be stably birational to X. By Proposition 3.3, the special fiber has to satisfy

$$1 \equiv [\pi^{-1}(0)] \equiv [X] \mod \mathbb{L}.$$

This is a contradiction, since Larsen-Lunts Theorem 3.1 implies that X is stably rational, contrary to our assumptions. □

Remark 3.7. Using motivic volume expressed in terms of log smooth models [NS19, Appendix A] the result of Theorem 3.4 can be obtained without explicit resolution of singularities of the model by applying [NS19, Theorem A.3.9] to the appropriate log scheme. However, the explicit resolution obtained in Lemma 3.6 can be useful for other purposes, such as in the proof of the same result in positive characteristic by Schreieder [Sch19c, Theorem 5.1].

4 Appendix: Stable Birational Equivalence and Decomposition of the Diagonal, by Claire Voisin

We prove in this appendix that if a family of projective varieties has a mildly singular member with a nonzero unramified cohomology class with given coefficients, while the very general member Y is smooth and has no such class, the stable birational equivalence class of the fibers Y_t is not constant. In particular, quartic and sextic double covers of $\mathbb{P}^3$ do not have a constant stable birational type. This result is inspired by the main theorem of Shinder in this paper. Note however that the assumptions and range of applications in both statements are different. We will work over any algebraically closed field k of infinite transcendence degree over the prime field but the main application (Theorem 4.8) will assume characteristic zero. We refer to Schreieder's recent note [Sch19c] for generalizations and a similar statement in nonzero characteristic.

We start with the following decomposition of the diagonal result for stable birational equivalence.

Proposition 4.1. *Let X, Y be two smooth projective varieties of dimension n. Assume X and Y are stably birational. Then there exist codimension n cycles*

$$\Gamma \in \mathrm{CH}^n(X \times Y), \qquad \Gamma' \in \mathrm{CH}^n(Y \times X)$$

such that

$$\begin{aligned} \Gamma' \circ \Gamma &= \Delta_X + Z_X \text{ in } \mathrm{CH}^n(X \times X), \\ \Gamma \circ \Gamma' &= \Delta_Y + Z_Y \text{ in } \mathrm{CH}^n(Y \times Y), \end{aligned} \tag{4.1}$$

where Z_X is supported on $D_X \times X$ for some proper closed algebraic subset $D_X \subset X$, and Z_Y is supported on $D_Y \times Y$ for some proper closed algebraic subset $D_Y \subset Y$.

Proof When X and Y are actually birational, this statement is proved in [CTV12]. In this case, we simply take for Γ the graph of a birational map $\phi\colon X \dashrightarrow Y$ and for Γ' the graph of ϕ^{-1}. The equality $\Gamma' \circ \Gamma = \Delta_X$ (respectively $\Gamma \circ \Gamma' = \Delta_Y$) is in this case satisfied at the level of cycles on $U \times X$ (respectively $V \times Y$) where $U \cong V$ is a Zariski open set of X on which ϕ is an isomorphism onto its image $V \subset Y$. Assume now that

$$\phi\colon X \times \mathbb{P}^r \dashrightarrow Y \times \mathbb{P}^r$$

is a birational map for some r. Then by the previous step, there exist

$$\Gamma_\phi \in \mathrm{CH}^{n+r}(X \times \mathbb{P}^r \times Y \times \mathbb{P}^r), \quad \Gamma'_\phi \in \mathrm{CH}^{n+r}(Y \times \mathbb{P}^r \times X \times \mathbb{P}^r)$$

such that (4.1) hold for some proper closed algebraic subsets $D \subset X \times \mathbb{P}^r$ (respectively $D' \subset Y \times \mathbb{P}^r$). For any point $O \in \mathbb{P}^r$, define

$$\Gamma := p_{XY*}(\Gamma_{\phi|X\times O\times Y\times\mathbb{P}^r}), \quad \Gamma' := p_{YX*}(\Gamma'_{\phi|Y\times O\times X\times\mathbb{P}^r}),$$

where p_{XY} is the projection from $X \times O \times Y \times \mathbb{P}^r$ to $X \times Y$ and p_{YX} is the projection from $Y \times O \times X \times \mathbb{P}^r$ to $Y \times X$. We have to show that (4.1) holds.

Let us decompose $\mathrm{CH}(X \times \mathbb{P}^r \times Y \times \mathbb{P}^r)$ as polynomials in h_1, h_2 with coefficients in $\mathrm{CH}(X \times Y)$, where $h_1 = \mathrm{pr}_2^* c_1(\mathcal{O}_{\mathbb{P}^r}(1))$, $h_2 = \mathrm{pr}_4^* c_1(\mathcal{O}_{\mathbb{P}^r}(1))$

$$\mathrm{CH}(X \times \mathbb{P}^r \times Y \times \mathbb{P}^r) = \bigoplus_{0\le i\le r,\, 0\le j\le r} h_1^i h_2^j \mathrm{CH}(X \times Y),$$

which gives, in particular,

$$\Gamma_\phi = \sum_{i,j} h_1^i h_2^j \Gamma_{\phi,i,j}, \qquad \Gamma'_\phi = \sum_{i,j} h_1^i h_2^j \Gamma'_{\phi,i,j}, \tag{4.2}$$

with $\Gamma_{\phi,i,j} \in \mathrm{CH}(X \times Y)$ and $\Gamma'_{\phi,i,j} \in \mathrm{CH}(Y \times X)$. We obviously have $\Gamma = \Gamma_{\phi,0,r}$, $\Gamma' = \Gamma'_{\phi,0,r}$. With the notation (4.2), we have

$$\Gamma'_\phi \circ \Gamma_\phi = \sum_{i,j,j'} h_1^i h_2^{j'} \Gamma'_{\phi,r-j,j'} \circ \Gamma_{\phi,i,j} \text{ in } \mathrm{CH}(X \times \mathbb{P}^r \times X \times \mathbb{P}^r) \tag{4.3}$$

while $\Delta_{X\times\mathbb{P}^r} = \sum_{i+j=r} h_1^i h_2^j \Delta_X$ in $\mathrm{CH}(X \times \mathbb{P}^r \times X \times \mathbb{P}^r)$. The fact that $\Gamma'_\phi \circ \Gamma_\phi - \Delta_{X\times\mathbb{P}^r}$ is rationally equivalent to a cycle supported via the first projection over a proper closed algebraic subset of $X \times \mathbb{P}^r$ then implies (by taking $i = 0$, $j' = r$ in (4.3)) that the cycle

$$\sum_j \Gamma'_{\phi,r-j,r} \circ \Gamma_{\phi,0,j} - \Delta_X \tag{4.4}$$

is supported via the first projection over a proper closed algebraic subset of X. We observe now that $\dim \Gamma = n + r$, $n = \dim X$, so that $\Gamma_{\phi,0,j}$

for $j < r$ has dimension $< n$, hence does not dominate X via the first projection. It follows that $\sum_{j<r}\Gamma'_{\phi,r-j,r}\circ\Gamma_{\phi,0,j}$ does not dominate X via the first projection, so the remaining term in (4.4) with $j = r$, namely

$$\Gamma'_{\phi,0,r}\circ\Gamma_{\phi,0,r} - \Delta_X$$

is rationally equivalent to a cycle supported over a proper closed algebraic subset of X via the first projection. Exchanging X and Y concludes the proof. □

Remark 4.2. In the sequel, we will use a weaker version of Proposition 4.1, stating only the first decomposition in (4.1). There is, in this case, no need to assume that X and Y are of the same dimension. Furthermore, as noticed by Shinder, the proof can be then made simpler by observing that the stated existence property for Γ, Γ' holds for pairs of birational varieties and also for the pair $(X, X\times\mathbb{P}^r)$.

We now prove the following version of the specialization theorem for decomposition of the diagonal fist proved in [Voi15] and later improved in [CTP16b]. We will say that a variety Z has mild singularities if there exists a desingularization morphism $\tau\colon \widetilde{Z}\to Z$ which is CH_0-universally trivial in the sense of [CTP16b]. This means that τ_* is an isomorphism on CH_0 over any field K containing k. The easiest way to make this condition satisfied is to ask that τ has the following property: for each (irreducible) subvariety $M\subset Z$, the induced morphism $\widetilde{Z}_M\to M$ has generic fiber smooth rational over $k(M)$. For example, ordinary quadratic singularities in dimension ≥ 2 are mild. We refer to [CTP16a] for a more general geometric interpretation of the mildness condition.

Theorem 4.3. *(1) Let $\mathcal{Y}\to B$ be a projective flat morphism of relative dimension n, where B is smooth. Assume the general fiber $\mathcal{Y}_b$ is smooth and stably birational to a fixed smooth projective variety Y of dimension n. Then, for any desingularization $\widetilde{\mathcal{Y}_0}$ of $\mathcal{Y}_0$, there exist codimension n cycles $\Gamma\in\mathrm{CH}^n(\widetilde{\mathcal{Y}_0}\times Y)$, $\Gamma'\in\mathrm{CH}^n(Y\times\widetilde{\mathcal{Y}_0})$ such that*

$$\Gamma'\circ\Gamma = \Delta_{\widetilde{\mathcal{Y}_0}} + Z + Z' \text{ in } \mathrm{CH}^n(\widetilde{\mathcal{Y}_0}\times\widetilde{\mathcal{Y}_0}), \tag{4.5}$$

where Z is supported on $D\times\widetilde{\mathcal{Y}_0}$ for some proper closed algebraic subset D of $\widetilde{\mathcal{Y}_0}$ and Z' is supported over $\widetilde{\mathcal{Y}_0}\times E$, where E is the exceptional locus of τ.

(2) If the special fiber $\mathcal{Y}_0$ has mild singularities, one can achieve for an adequate choice of desingularization $\widetilde{\mathcal{Y}_0}$ that $Z' = 0$ in (4.5).

Proof (*1*) We can assume B is a smooth curve. By assumption and using Proposition 4.1, there exist for a general point $t \in B$ a divisor $D_t \subset \mathcal{Y}_t$ and codimension n cycles $\Gamma_t \in \mathrm{CH}^n(\mathcal{Y}_t \times Y)$, $\Gamma'_t \in \mathrm{CH}^n(Y \times \mathcal{Y}_t)$, such that

$$\Gamma'_t \circ \Gamma_t = \Delta_{\mathcal{Y}_t} + Z_t \text{ in } \mathrm{CH}^n(\mathcal{Y}_t \times \mathcal{Y}_t), \tag{4.6}$$

where Z_t is supported on $D_t \times \mathcal{Y}_t$. By a countability argument for the Chow varieties parameterizing cycles in fibers, D_t and the cycles Z_t, Γ_t, Γ'_t can be constructed in families after a base change $B' \to B$. We will denote $\mathcal{Y}' := \mathcal{Y} \times_B B'$. This provides us with varieties and cycles

$$\mathcal{D} \subset \mathcal{Y}, \quad \mathcal{Z} \in \mathrm{CH}(\mathcal{D} \times_{B'} \mathcal{Y}'), \quad \Gamma \in \mathrm{CH}(\mathcal{Y}' \times Y), \quad \Gamma' \in \mathrm{CH}(Y \times \mathcal{Y}')$$

whose fiber at the general point $t \in B'$ satisfies (4.6) (see [Voi15] for the details). Restricting to the regular locus of the morphism $\mathcal{Y}' \to B'$, the composition in (4.6) still makes sense as a relative composition because for $\Gamma \in \mathrm{CH}^n(U \times Y)$, $\Gamma' \in \mathrm{CH}^n(Y \times U')$, the composition $\Gamma' \circ \Gamma$ is well-defined whenever U and Y are smooth, and Y is projective. Furthermore, by specialization of rational equivalence, (4.6) holds in $\mathrm{CH}^n(\mathcal{Y}_{0,reg} \times \mathcal{Y}_0)$ for any $0 \in B$. Here, as we assumed B (hence B') is a curve, the divisor $\mathcal{D}$ can be assumed not to contain any component of the fiber $\mathcal{Y}_0$, hence to restrict to a proper divisor $D_0 \subset \mathcal{Y}_0$. Identifying $\mathcal{Y}_{0,reg}$ with $\widetilde{\mathcal{Y}}_0 \setminus E$, we get cycles

$$\widetilde{Z}_0, \quad \widetilde{\Gamma}_0 \in \mathrm{CH}^n(\widetilde{\mathcal{Y}}_0 \times Y), \quad \widetilde{\Gamma}'_0 \in \mathrm{CH}^n(Y \times \widetilde{\mathcal{Y}}_0)$$

with $\widetilde{Z}_0$ supported on $\widetilde{D}_0$ such that the equality

$$\widetilde{\Gamma}'_0 \circ \widetilde{\Gamma}_0 = \Delta_{\widetilde{\mathcal{Y}}_0} + \widetilde{Z}_0$$

holds in $\mathrm{CH}^n((\widetilde{\mathcal{Y}_0} \setminus E) \times (\widetilde{\mathcal{Y}_0} \setminus E))$. It follows from the localization exact sequence that the cycle $\widetilde{\Gamma}'_0 \circ \widetilde{\Gamma}_0 - \Delta_{\widetilde{\mathcal{Y}}_0} - \widetilde{Z}_0 \in \mathrm{CH}^n(\widetilde{\mathcal{Y}}_0 \times \widetilde{\mathcal{Y}}_0)$ is rationally equivalent to a cycle supported on $E \times \widetilde{\mathcal{Y}}_0 \cup \widetilde{\mathcal{Y}}_0 \times E$. This last cycle is the sum of a cycle Z_1 supported on $E \times \widetilde{\mathcal{Y}}_0$ and a cycle Z_2 supported on $\widetilde{\mathcal{Y}}_0 \times E$. We thus proved (4.5) with $\Gamma = \widetilde{\Gamma}_0$, $\Gamma' = \widetilde{\Gamma}'_0$, $Z = \widetilde{Z}_0 + Z_1$ and $Z' = Z_2$.

(*2*) As in [CTP16b], using the fact that (4.6) holds in $\mathrm{CH}^n(\mathcal{Y}_{0,reg} \times \mathcal{Y}_0)$, we observe that the cycle $Z' \in \mathrm{CH}_n(\widetilde{\mathcal{Y}}_0 \times E)$ vanishes by construction in $\mathrm{CH}_n(U \times \mathcal{Y}_0)$ for some dense Zariski open subset U of $\widetilde{\mathcal{Y}}_0$. On the other hand, we can work by assumption with the resolution $\tau \colon \widetilde{\mathcal{Y}}_0 \to \mathcal{Y}_0$ for which the morphism τ is universally CH_0-trivial. It follows that the cycle Z', seen over the generic point of $\widetilde{\mathcal{Y}}_0$ as a 0-cycle of $\widetilde{\mathcal{Y}}_0$ defined on the field $k(\mathcal{Y}_0)$, vanishes in $\mathrm{CH}_0((\widetilde{\mathcal{Y}_0})_{k(\mathcal{Y}_0)})$. Hence Z' vanishes in $\mathrm{CH}^n(U \times \widetilde{\mathcal{Y}_0})$ for some dense Zariski open set U of $\widetilde{\mathcal{Y}_0}$. By the localization exact sequence, it is thus supported on $D \times \widetilde{\mathcal{Y}_0}$, where $D = \widetilde{\mathcal{Y}_0} \setminus U$ and thus can be absorbed in the term Z. □

Corollary 4.4. *Under the same assumptions as in Theorem 4.3(2), assume that $H^i_{nr}(Y, A) = 0$ for some integer i and abelian group A. Then $H^i_{nr}(\widetilde{\mathcal{Y}_0}, A) = 0$.*

Proof We let both sides of (4.5) with $Z' = 0$ act on $H^i_{nr}(\widetilde{\mathcal{Y}_0}, A)$ (see [CTV12] for a construction of the action). The action of $\Gamma' \circ \Gamma$ factors through $H^i_{nr}(Y, A)$, hence it is 0. Moreover the diagonal acts by the identity map. We thus conclude that for any $\alpha \in H^i_{nr}(\widetilde{\mathcal{Y}_0}, A)$,

$$\alpha = Z^* \alpha.$$

On the other hand, as Z is supported on $D \times \widetilde{\mathcal{Y}_0}$, the class $Z^*\alpha$ vanishes on $U \times X$, where $U := X \setminus D$. Hence $\alpha_{|U} = 0$, which implies $\alpha = 0$ by [BO74]. □

We are now in position to prove the following result.

Theorem 4.5. *Let $\mathcal{Y} \to B$ be a projective flat morphism of relative dimension n, where B is smooth, the generic fiber is smooth, and the special fiber $\mathcal{Y}_0$ has mild singularities. Assume*

(1) The very general fiber $\mathcal{Y}_b$ satisfies $H^i_{nr}(\mathcal{Y}_b, A) = 0$.
(2) $H^i_{nr}(\widetilde{\mathcal{Y}_0}, A) \neq 0$ for some (equivalently, any) desingularization $\widetilde{\mathcal{Y}_0}$ of $\mathcal{Y}_0$.

Then two very general fibers $\mathcal{Y}_b$, $\mathcal{Y}_{b'}$ are not stably birational.

Proof Fix one very general fiber $\mathcal{Y}_{b'}$ and denote it by Y. We want to show that the general fiber $\mathcal{Y}_b$ is not stably birational to Y. If it is, Corollary 4.4 and the vanishing $H^i_{nr}(\mathcal{Y}_{b'}, A) = 0$ given by (*1*) imply that $H^i_{nr}(\widetilde{\mathcal{Y}_0}, A) = 0$, contradicting assumption (*2*). □

The following variant of Theorem 4.5 is proved as above, using Corollary 4.7 below instead of Corollary 4.4.

Theorem 4.6. *Let $\mathcal{Y} \to B$ be a projective flat morphism of relative dimension n, where B is smooth. Assume*

(1) The very general fiber $\mathcal{Y}_b$ satisfies $H^i_{nr}(\mathcal{Y}_b, A) = 0$.

(2) *The central fiber admits a desingularization $\widetilde{\mathcal{Y}_0}$ with exceptional divisor $E = \cup_j E_j$ with E_j smooth, and $\widetilde{\mathcal{Y}_0}$ has a nonzero class $\alpha \in H^i_{nr}(\widetilde{\mathcal{Y}_0}, A)$ which vanishes on all the divisors E_j.*

Then two very general fibers $\mathcal{Y}_b$, $\mathcal{Y}_{b'}$ are not stably birational.

The proof uses the following variant of Corollary 4.4, which is based on Schreieder's criterion [Sch19a].

Corollary 4.7. *Under the same assumptions as in Theorem 4.3(1), assume that $H^i_{nr}(Y, A) = 0$ for some integer i and abelian group A, and that $\mathcal{Y}_0$ has a desingularization $\widetilde{\mathcal{Y}_0}$ with exceptional divisor $E = \cup_j E_j$ with E_j smooth. Then any unramified cohomology class $\alpha \in H^i_{nr}(\widetilde{\mathcal{Y}_0}, A)$ which vanishes on each component E_j of E is identically 0.*

Proof We let both sides of (4.5) act on $H^i_{nr}(\widetilde{\mathcal{Y}_0}, A)$ (see [CTV12] for a construction of the action). The action of $\Gamma' \circ \Gamma$ factors through $H^i_{nr}(Y, A)$ hence it is 0 since this group is assumed to be 0. We conclude as before that for any $\alpha \in H^i_{nr}(\widetilde{\mathcal{Y}_0}, A)$,

$$\alpha = Z^*\alpha + Z'^*\alpha.$$

If α vanishes on all the components E_j of the exceptional divisor E, we have $Z'^*\alpha = 0$. Thus $\alpha = Z^*\alpha$ and we conclude as before that $\alpha = 0$. □

The families to which Theorem 4.5 applies are essentiall, in characteristic 0, all the families of weighted Fano hypersurfaces for which the stable irrationality has been proved by a degeneration argument to a mildly singular member having a nonzero unramified cohomology class of degree ≤ 3. For example, we have

Theorem 4.8. (1) *Two very general quartic or sextic double solids or quartic hypersurfaces of dimension 3 or 4 over $\mathbb{C}$ are not stably birational.*

(2) *Two very general hypersurfaces over $\mathbb{C}$ of degree ≥ 5 and dimension n with $5 \leq n \leq 9$ are not stably birational.*

Proof The case (*1*) uses Theorem 4.5. We know by [AM72] in case of quartic double solids, by Beauville [Bea16] in case of sextic double solids, by Colliot–Thélène–Pirutka [CTP16b] in case of quartic threefolds and Schreieder [Sch19b] in the case of quartic fourfolds, that they admit degenerations with mild singularities having a nonzero unramified cohomology class of degree 2, which is given by a nonzero torsion class in $H^3_B(\widetilde{\mathcal{X}}_0, \mathbb{Z})$. On the other hand, for all these classes of varieties, the smooth member X does not have torsion in $H^3_B(X, \mathbb{Z})$. Theorem 4.5 thus applies.

For case (*2*), we use Schreieder's degeneration, which in the numerical range above produces a desingularized central fiber with a nonzero unramified cohomology class of degree 3 with torsion coefficients on the desingularized central fiber, vanishing on the exceptional divisor. The very general hypersurface X on the other hand has trivial unramified cohomology of degree 3. Indeed, it is proved in [CTV12] that such a class measures the defect of the Hodge conjecture for degree 4 integral Hodge classes on X. But the smooth hypersurface of degree ≥ 5 in $\mathbb{P}^{n+1}$ for $5 \leq n \leq 9$ has no integral Hodge class of degree 4 not coming from $\mathbb{P}^n$ by the Lefschetz theorem on hyperplane sections. □

The reason we can not a priori extend Theorem 4.5 to all hypersurfaces shown by Schreieder [Sch19b] not to be stably rational is the fact that we do not know how to compute unramified cohomology of degree ≥ 4 for general hypersurfaces, so we are not able to check Condition (*1*) in Theorem 4.5. Note that the case of smooth hypersurfaces is covered by Shinder's main theorem.

Acknowledgements

The author would like to thank Adel Betina, Christian Böhning, Jean-Louis Colliot-Thélène, Sergey Galkin, Alexander Kuznetsov, Johannes Nicaise, Alexander Pukhlikov, Claire Voisin, Stefan Schreieder, Konstantin Shramov for discussions and encouragement. The idea of using degeneration to a hyperplane arrangement, as opposed to a nodal hypersurface is due to an e-mail correspondence with Sergey Galkin. A suggested simplification in the proof of Lemma 3.6, as explained in Remark 3.7 is due to the referee.

The author was partially supported by Laboratory of Mirror Symmetry NRU HSE, RF government grant, ag. N 14.641.31.0001.

The paper is dedicated to Miles Reid who inspired and educated several generations of algebraic geometers through direct contact as well as with his books and papers, which often seem to be intentionally written to inspire and to educate. In 2014, Miles told me that at some point young people should stop dreaming and do something useful instead and I try to follow this advice ever since.

References

[AM72] M. Artin and D. Mumford. Some elementary examples of unirational varieties which are not rational. *Proc. London Math. Soc. (3)*, 25:75–95, 1972.

[Bea16] Arnaud Beauville. A very general sextic double solid is not stably rational. *Bull. Lond. Math. Soc.*, 48(2):321–324, 2016.

[BO74] Spencer Bloch and Arthur Ogus. Gersten's conjecture and the homology of schemes. *Ann. Sci. École Norm. Sup. (4)*, 7:181–201 (1975), 1974.

[CTO89] Jean-Louis Colliot-Thélène and Manuel Ojanguren. Variétés unirationnelles non rationnelles: au-delà de l'exemple d'Artin et Mumford. *Invent. Math.*, 97(1):141–158, 1989.

[CTP16a] Jean-Louis Colliot-Thélène and Alena Pirutka. Cyclic covers that are not stably rational. *Izv. Ross. Akad. Nauk Ser. Mat.*, 80(4):665–677, 2016.

[CTP16b] Jean-Louis Colliot-Thélène and Alena Pirutka. Hypersurfaces quartiques de dimension 3: non-rationalité stable. *Ann. Sci. Éc. Norm. Supér. (4)*, 49(2):371–397, 2016.

[CTV12] Jean-Louis Colliot-Thélène and Claire Voisin. Cohomologie non ramifiée et conjecture de Hodge entière. *Duke Math. J.*, 161(5):735–801, 2012.

[Ful93] William Fulton. *Introduction to toric varieties*, volume 131 of *Annals of Mathematics Studies*. Princeton University Press, Princeton, NJ, 1993. The William H. Roever Lectures in Geometry.

[KT19] Maxim Kontsevich and Yuri Tschinkel. Specialization of birational types. *Invent. Math.*, 217(2):415–432, 2019.

[NS19] Johannes Nicaise and Evgeny Shinder. The motivic nearby fiber and degeneration of stable rationality. *Invent. Math.*, 217(2):377–413, 2019.

[Sch19a] Stefan Schreieder. On the rationality problem for quadric bundles. *Duke Math. J.*, 168(2):187–223, 2019.

[Sch19b] Stefan Schreieder. Stably irrational hypersurfaces of small slopes. *J. Amer. Math. Soc.*, 32(4):1171–1199, 2019.

[Sch19c] Stefan Schreieder. Variation of stable birational types in positive characteristic. `arXiv:1903.06143`, 2019.

[Tot16] Burt Totaro. Hypersurfaces that are not stably rational. *J. Amer. Math. Soc.*, 29(3):883–891, 2016.

[Voi15] Claire Voisin. Unirational threefolds with no universal codimension 2 cycle. *Invent. Math.*, 201(1):207–237, 2015.

Triangle Varieties and Surface Decomposition of Hyper-Kähler Manifolds

Claire Voisin

Pour Miles Reid, avec estime et amitié

Abstract

We introduce and study the notion of 'surface decomposable' variety, and discuss the possibility that any projective hyper-Kähler manifold is surface decomposable, which would produce new evidence for Beauville's weak splitting conjecture. We show that surface decomposability relates to the Beauville–Fujiki relation, a constraint on the cohomology ring of the variety, and that general varieties with $h^{2,0} \neq 0$ are not surface decomposable. We also formalize the notion of triangle variety that is useful to produce surface decomposition. We show the existence of these geometric structures on most explicitly constructed classes of projective hyper-Kähler manifolds of Picard number 1.

1 Introduction

Let X be a complex manifold of dimension $2n$ equipped with a holomorphic 2-form σ_X which is everywhere of maximal rank $2n$. Locally for the Euclidean topology on X, Darboux's theorem tells that one can write, for an adequate choice of holomorphic coordinates,

$$\sigma_X = \sum_{i=1}^{n} dz_i \wedge dz_{n+i},$$

that is, σ_X is the sum of n closed holomorphic 2-forms of rank 2.

A natural question is whether this statement can be made more global, particularly in the projective case:

Question 1.1. Does there exist a generically finite cover $\phi\colon Y \to X$, such that $\phi^*\sigma_X$ is the sum of n closed holomorphic 2-forms of rank 2 on Y?

Our goal in this paper is to study a geometric variant of this question, namely the possibility that any projective hyper-Kähler manifold is *surface decomposable* (or admits a *surface decomposition*) in the following sense:

Definition 1.2. A smooth projective variety X of dimension $2n$ will be said to be surface decomposable if there exist a smooth variety Γ, smooth projective surfaces $S_1, \ldots, S_n$, and generically finite surjective morphisms $\phi\colon \Gamma \to X$, $\psi\colon \Gamma \to S_1 \times \ldots \times S_n$ such that for any holomorphic 2-form $\sigma \in H^0(X, \Omega_X^2)$,

$$\phi^*\sigma = \psi^*\left(\sum_i \mathrm{pr}_i^*\sigma_i\right) \tag{1.1}$$

for some holomorphic 2-forms σ_i on S_i.

We will show that surface decomposability is restrictive for general projective varieties of dimension $2n \geq 4$ (see Theorem 2.6). The reason is that it essentially implies the Beauville–Fujiki formulas describing the top self-intersection on cohomology of degree 2 as the power of a quadratic form, or at least a product of quadratic forms (see Proposition 2.5).

Remark 1.3. Surface decompositions are natural in the hyper-Kähler context which is the one of this paper, but one can of course introduce, in general, decompositions into summands of other dimensions. For example, we can consider *curve decompositions* of any variety X of dimension n, given by the data of generically finite surjective morphisms

$$\phi\colon \Gamma \to X, \qquad \psi\colon \Gamma \to C_1 \times \ldots \times C_n$$

such that for any 1-form $\alpha \in H^{1,0}(X)$,

$$\phi^*\alpha = \psi^*\left(\sum_i \mathrm{pr}_i^*\alpha_i\right)$$

for some forms $\alpha_i \in H^{1,0}(C_i)$. Abelian varieties are clearly curve decomposable. One can show similarly that this notion puts strong restrictions on the structure of the intersection pairing $\bigwedge^{2n} H^1(X,\mathbb{C}) \to H^{2n}(X,\mathbb{C}) = \mathbb{C}$, when $h^{1,0}(X)$ is large compared to n.

The first examples of hyper-Kähler manifolds were constructed by Beauville [Bea83] and Fujiki [Fuj87] as punctual Hilbert schemes of $K3$ surfaces or abelian surfaces and hence were rationally dominated by products of surfaces. They were thus obviously surface decomposable. However, it follows from deformation theory that these $K3$ or abelian surfaces disappear under a general deformation to a projective hyper-Kähler manifold with Picard number 1. Indeed, the parameter space for $K3$ surfaces is too small to parameterize also general deformations with Picard number 1 of their punctual Hilbert schemes. The starting point of this paper is the observation that on many explicitly described general deformations as above, a surface decomposition still exists.

Let us make several remarks concerning Definition 1.2. First of all, the condition (1.1) has been asked only for holomorphic 2-forms, but by an elementary argument involving Hodge structures (see Section 2), it then follows that it is satisfied for any transcendental class $\eta \in H^2(X,\mathbb{Q})_{tr}$, the latter space being defined as the smallest Hodge substructure of $H^2(X,\mathbb{Q})$ whose complexification contains $H^{2,0}(X)$.

Next, if we allow an arbitrarily large number N of summands S_i and only ask that ϕ is surjective and ψ is generically finite, then the definition is (at least conjecturally) not restrictive since the property is satisfied by any smooth projective variety X satisfying the Lefschetz standard conjecture in degree 2 (see Proposition 2.2). Similarly, we could consider, instead of (1.1), the weaker condition that ϕ is surjective generically finite and

$$\sigma = \phi_*\left(\psi^*\left(\sum_i \mathrm{pr}_i^*\sigma_i\right)\right) \text{ in } H^{2,0}(X), \tag{1.2}$$

but, as before, this is implied by Lefschetz standard conjecture, and we even can take $N = 1$ to achieve (1.2), as shows the case of the $(2,0)$-forms on the symmetric product or rather punctual Hilbert scheme $X = S^{[k]}$ of a simply connected surface S: they are all obtained starting from a $(2,0)$-form on S by a formula like (1.2).

The reason why (1.2) is much weaker than (1.1) is the fact that pull-back maps are compatible with cup-products, while push-forward maps are not. More precisely, we will show (see Proposition 2.5) that the surface decomposability implies and provides a geometric explanation for

Beauville–Fujiki's famous formula for the self-intersection of degree 2 cohomology on a hyper-Kähler manifold:

$$\int_X \alpha^{2n} = \lambda q(\alpha)^n,$$

where q is the Beauville–Bogomolov quadratic form on $H^2(X, \mathbb{Q})_{tr}$ (see [Bea83]). To prove this implication, we have to assume that the Mumfort–Tate group of the Hodge structure on $H^2(X, \mathbb{Q})_{tr}$ is large enough to guarantee that all the quadratic forms on $H^2(X, \mathbb{Q})_{tr}$ induced from $(\, , \,)_{S_i}$ via the morphism of Hodge structures $\psi_i \colon \eta \mapsto \eta_i$ (see Section 2) are proportional, but this is automatic if X is the general member of a family of polarized hyper-Kähler manifolds.

This observation suggests that surface decomposability could be a way to approach the weak splitting property of hyper-Kähler manifolds (see Conjecture 2.7) conjectured by Beauville in [Bea07]. It says that cohomological polynomial relations between divisors on hyper-Kähler manifolds X are satisfied on the Chow level. A weaker version asks that X has a canonical 0-cycle $o_X \in \mathrm{CH}_0(X)$ such that D^{2n} is proportional to o_X in $\mathrm{CH}_0(X)$ for any divisor D on X.

In this direction, we prove Theorem 2.8 which has the following consequence:

Theorem 1.4. *Assume that the general member X_t of a family $(X_t)_{t \in B}$ of hyper-Kähler manifolds with given Picard lattice Λ has a surface decomposition with a simply connected Γ. Then the Beauville weak splitting property holds for the divisor classes belonging to $\mathrm{NS}(X_t)^{\perp \Lambda}$ if and only if there exists a 0-cycle $o_{X_t} \in \mathrm{CH}_0(X_t)$ such that D^{2n} is proportional to o_{X_t} in $\mathrm{CH}_0(X_t)$ for any divisor $D \in \mathrm{NS}(X_t)^{\perp \Lambda}$.*

The importance of Theorem 1.4 is that it reduces the weak splitting property to checking the weak version, namely in top degree. The weakness of the result is that it applies only to divisors of class perpendicular to Λ, which means, in practice, primitive. One has to understand separately what happens with the powers h^k of the polarizing class. In all the geometric examples we have, the natural surface decomposition that we exhibit provides (1.1) only on primitive cohomology.

Remark 1.5. This raises the question whether hyper-Kähler manifolds could admit a surface decomposition such that formula (1.1) holds on the whole cohomology H^2, instead of only primitive or transcendental cohomology. This question is very interesting in the case of the Hilbert scheme $S^{[n]}$ of a $K3$ surface S, where the natural surface decomposition provides (1.1) only on the part of H^2 which is orthogonal to the class of the exceptional divisor over the diagonal. Combined with the theorem above, this could be a way to reproving the weak splitting conjecture for $S^{[n]}$ (proved by Maulik and Neguţ [MN19]), as the weak version is known for them by [Voi08].

Remark 1.6. The statement above is empty for the very general member of the family since it has $\mathrm{NS}(X_t)^{\perp\Lambda} = 0$. The statement above is interesting for special hyper-Kähler manifolds with higher Picard rank $\rho \geq \rho_{gen} + 2$, which are parameterized by a countable union of closed algebraic subsets in the base B, which is dense in B if $\dim B \geq 2$.

The second geometric notion that will play an important role in this paper is the following.

Definition 1.7 ([Voi16]). An algebraically coisotropic subvariety of a hyper-Kähler manifold X of dimension $2n$ is a subvariety $Z \subset X$ of codimension $k \leq n$ which admits a rational fibration $\phi\colon Z \dashrightarrow W$, where W is smooth and $\dim W = 2n - 2k$, such that

$$\sigma_{X|Z_{\mathrm{reg}}} = \phi_{\mathrm{reg}}^* \sigma_W,$$

where $\phi_{\mathrm{reg}}\colon Z_{\mathrm{reg}} \dashrightarrow W$ is the restriction of ϕ to the regular locus of Z and σ_W is a holomorphic 2-form on W.

This notion is to be distinguished from the notion of coisotropic subvariety, which just asks that the restriction $\sigma_{X|Z_{\mathrm{reg}}}$ has rank $2n - 2k$ at any point, or equivalently that

$$T_{Z_{\mathrm{reg}},x}^{\perp\sigma_X} \subset T_{Z_{\mathrm{reg}},x} \tag{1.3}$$

at any point x of Z_{reg}. Equation (1.3) defines then a foliation on Z_{reg} and Z is algebraically coisotropic when the leaves of this foliation are algebraic. The two notions coincide in the case $n = k$ of Lagrangian varieties. For $k = 1$, any divisor is coisotropic but smooth ample divisors are not algebraically isotropic (see [AC17]). It is not easy a priori to construct algebraically coisotropic divisors in a projective hyper-Kähler manifolds. Examples are given by uniruled (singular) divisors. Indeed, starting from a singular uniruled divisor $D \subset X$, consider a desingularization $\widetilde{D} \to D$ with induced morphism $j\colon \widetilde{D} \to X$. Then by assumption $\widetilde{D}$ is uniruled, hence its maximal rationally connected fibration (or MRC fibration, see [KMM92]) is nontrivial, producing, possibly after changing the birational model of $\widetilde{D}$, a morphism

$$f\colon \widetilde{D} \to B$$

with rationally connected fibers of positive dimension. Any holomorphic form on $\widetilde{D}$ is pulled-back via f from a holomorphic form on B. We apply this to the pull-back $j^*\sigma_X \in H^{2,0}(\widetilde{D})$ and conclude that

$$j^*\sigma_X = f^*\sigma_B \text{ in } H^{2,0}(\widetilde{D}). \tag{1.4}$$

Formula (1.4) implies that the fibers of f are tangent to the kernel of $j^*\sigma_X$. As the generic rank of $j^*\sigma_X$ is $2n-1$, $n = \dim X$, we conclude that the fibers of f are 1-dimensional (so in fact D is ruled), and D is algebraically coisotropic. Unfortunately, the uniruled divisors in a hyper-Kähler manifold X are rigid, because rational curves cannot cover X (indeed, the MRC fibration of X is trivial as shows the argument above). One open question is whether a projective hyper-Kähler manifold can always be swept out by algebraically coisotropic divisor. This is certainly true if X has a surface decomposition (see Proposition 1.11 below). In the other direction, we show in Proposition 2.10 that the existence of a 1-parameter family of algebraically coisotropic divisors for X implies a decomposition, on a generically finite cover of X, of the holomorphic 2-form of X as a sum of one rank 2 and one rank $2n-2$ holomorphic forms.

The theory of coisotropic subvarieties of higher codimension is more complicated. In the paper [Voi16], we discussed the constraints on the cohomology classes of coisotropic subvarieties of higher codimension and asked whether the space of coisotropic classes, namely those satisfying these constraints, are generated by classes of algebraically co-isotropic subvarieties. We also proposed the construction of algebraically coisotropic subvarieties as total spaces of $2(n-k)$-dimensional families of constant cycles varieties (in the sense of Huybrechts [Huy14]) of dimension k.

The third notion that will be introduced and studied in this paper is that of *triangle variety*:

Definition 1.8. A triangle variety for X (equipped with a holomorphic 2-form σ_X) is a subvariety of $X \times X \times X$ which dominates X by the three projections, maps in a generically finite way to its image in $X \times X$ via the three projections and is Lagrangian for the holomorphic form $\sigma_1+\sigma_2+\sigma_3$ on X^3, where $\sigma_i = \mathrm{pr}_i^*\sigma_X$.

The following example will be generalized in Section 3.5.

Example 1.9. Let $S \to B$ be an elliptic surface with a section. Then the graph of minus the relative sum map $S \times_B S \dashrightarrow S$, which is naturally contained in S^3, is a triangle variety for any holomorphic 2-form σ on S.

Triangle varieties seem to exist for most explicitly constructed classes of projective hyper-Kähler manifolds. In fact, the simplest example of them, namely actual triangles in the Fano variety $F_1(Y)$ of lines of a smooth cubic fourfold Y, is studied with detail in [SV16] by Shen and Vial, who use them to study a decomposition (Beauville splitting) of the Chow groups of $F_1(Y)$. The main geometric examples, including this one, will be presented in Section 3.

Remark 1.10. More generally, we can also introduce (and we will use) $(k+1)$-angle subvarieties $T_{k+1} \subset X^{k+1}$ which are Lagrangian for the 2-form $\sum_i \pm \mathrm{pr}_i^* \sigma_X$. They are easily constructed by iteration starting from a triangle variety.

The first link between triangle varieties, surface decompositions and algebraically coisotropic subvarieties is the following obvious implication:

Proposition 1.11. *If X has a surface decomposition, then it has mobile algebraically coisotropic subvarieties of any codimension $k \leq n$. If the surfaces appearing in a surface decomposition of X have triangle varieties, (for example, if they are elliptic) then so does X.*

Proof Indeed, let $\phi\colon \Gamma \to X$, $\psi\colon \Gamma \to S_1 \times \ldots \times S_n$ be surjective generically finite maps such that

$$\phi^* \sigma_X = \psi^*(\mathrm{pr}_1^* \sigma_{S_1} + \ldots + \mathrm{pr}_n^* \sigma_{S_n}) \text{ in } H^{2,0}(\Gamma).$$

For any integer $k \leq n$, let $C_i \subset S_i$, $i = 1, \ldots, k$, be very ample curves in general position. Then $\phi(\psi^{-1}(C_1 \times \ldots \times C_k \times S_{k+1} \times \ldots \times S_n))$ is an algebraically coisotropic subvariety of X of codimension k. If $T_i \subset S_i^3$ are triangle varieties, then $\phi^3(T_1 \times \ldots \times T_n)$ is a triangle variety for X. □

We will show in the paper how conversely triangle (or $(n+1)$-angle) subvarieties for X and algebraically coisotropic subvarieties of X of codimension $n-1$, where dimension $= 2n$, can be used to construct surface decompositions and algebraically coisotropic varieties of X of any codimension $1 \leq k \leq n$. We prove the following result (see Theorem 4.1).

Theorem 1.12. *Let X be a projective hyper-Kähler variety of dimension $2n$. Assume X has an $(n+1)$-angle variety (see Remark 1.10) $T_{n+1} \subset X^{n+1}$ and an algebraically coisotropic subvariety $\tau\colon Z \dashrightarrow \Sigma$ of dimension $n+1$. Let $F \subset X$ be the general fiber of τ. Then if the intersection number $F^n \cdot p_{1\ldots n}(T_{n+1})$ of cycles in X^n is nonzero, X admits a surface decomposition.*

Finally, we will prove (see Theorem 4.3), as an application of Theorem 1.12 or variants of it, that most hyper-Kähler manifolds that have been explicitly constructed from algebraic geometry admit surface decompositions.

2 The Decomposition Problem for Hyper-Kähler Varieties

Let X be a smooth projective manifold of dimension $2n$ (we will later on focus on the hyper-Kähler case). We wish to study the existence and consequences of a surface decomposition of the form described in the introduction (see Definition 1.2), namely the existence of smooth projective surfaces S_i, $i = 1, \ldots, n$, and an effective correspondence Γ (which can be assumed to be smooth and projective)

$$\phi \colon \Gamma \to X, \qquad \psi \colon \Gamma \to S_1 \times \ldots \times S_n,$$

with ϕ, ψ dominant generically finite, such that for any $\sigma_X \in H^{2,0}(X)$

$$\phi^* \sigma_X = \psi^* \left(\sum_{i=1}^{n} \mathrm{pr}_i^* \sigma_{S_i} \right) \text{ in } H^{2,0}(\Gamma) \tag{2.1}$$

for some $\sigma_{S_i} \in H^{2,0}(S_i)$, where the $\mathrm{pr}_i \colon S_1 \times \ldots \times S_n \to S_i$ are the various projections. Let us spell out the proof of the following Lemma.

Lemma 2.1. *Condition* (2.1) *implies more generally that, for any* $\eta \in H^2(X, \mathbb{Q})_{tr}$,

$$\phi^* \eta = \psi^* (\mathrm{pr}_1^* \eta_1 + \ldots + \mathrm{pr}_n^* \eta_n) \text{ in } H^2(\Gamma, \mathbb{Q}), \tag{2.2}$$

for some $\eta_i \in H^2(S_i, \mathbb{Q})_{tr}$.

Proof Indeed, the form $\sigma_{S_i} \in H^{2,0}(S_i)$ in (2.1) can be reconstructed from σ_X by the action of the morphism of Hodge structures

$$\psi_i \colon H^2(X, \mathbb{Q})_{tr} \to H^2(S_i, \mathbb{Q}), \qquad \eta \mapsto \eta_i$$

$$\psi_i(\eta) = \frac{1}{NN_i} \mathrm{pr}_{i*} \left(d^{2n-2} \cup \psi_*(\phi^* \eta) \right), \tag{2.3}$$

where N is the degree of ψ and $d = \sum_i \mathrm{pr}_i^* d_i$ is the first Chern class of an ample divisor on $\prod_i S_i$ with the property that $\mathrm{pr}_{i*}(d^{2n-2}) = N_i 1_{S_i}$ in $H^0(S_i, \mathbb{Q})$ for all i. Indeed the last condition guarantees that

$$\frac{1}{NN_i}\mathrm{pr}_{i*}\left(d^{2n-2}\cdot\psi_*\psi^*\left(\sum_j \mathrm{pr}_j^*\eta_j\right)\right)=\eta_i$$

for any cohomology classes η_j on S_j such that $\eta_j \cup d_j = 0$ for all j. The morphisms of Hodge structures ψ_i being defined as in (2.3), condition (2.1) then rewrites as

$$\phi^*\sigma_X = \psi^*\left(\sum_i \mathrm{pr}_i^*(\psi_i(\sigma_X))\right) \text{ in } H^{2,0}(\Gamma).$$

This equality defines a Hodge substructure of $H^2(X,\mathbb{Q})$. Hence, once it is satisfied on $H^{2,0}(X)$, it is satisfied on $H^2(X,\mathbb{Q})_{tr}$. □

A variant of this definition assumes that $S_1 = \ldots = S_n$ and the correspondence Γ is symmetric with respect to the symmetric group action on S^n, but this is not essential. Another variant asks that condition (2.2) is satisfied for any $\eta \in H^2(X,\mathbb{Q})_{prim}$, where the subscript "prim" refers to the choice of an ample line bundle L on X and primitive cohomology is primitive with respect to $l = c_1(L)$. If we work with very general hyper-Kähler manifolds of Picard number 1, the two notions coincide. In the hyper-Kähler case, (2.1) decomposes the smooth projective manifold X in the sense that the rank $2n$ holomorphic 2-form on X gets decomposed as the sum of n (generically) rank 2 holomorphic 2-forms $\psi^*(\mathrm{pr}_i^*\sigma_{S_i})$ on the generically finite cover Γ.

If we now relax the conditions on ϕ, ψ in Definition 1.2 and just ask that ϕ is surjective and $\phi^*\sigma_X = \psi^*(\sum_{i=1}^N \mathrm{pr}_i^*\sigma_{S_i})$ for any holomorphic 2-form on X, allowing an arbitrarily large number of summands, then a decomposition as in (2.1) should always exist, for any smooth projective variety X. More precisely:

Proposition 2.2. *Let X be a smooth projective variety. Assume X satisfies the Lefschetz standard conjecture for degree 2 cohomology. Then there is a generically finite cover $\phi\colon \Gamma \to X$, surfaces $S_1, \ldots, S_N$, and a morphism $\psi\colon \Gamma \to S_1 \times \ldots \times S_N$, such that any $(2,0)$-form σ on X satisfies*

$$\phi^*\sigma = \psi^*\left(\sum_i \mathrm{pr}_i^*\sigma_i\right) \text{ in } H^{2,0}(\Gamma) \tag{2.4}$$

for some $(2,0)$-forms σ_i on S_i.

Remark 2.3. The proof will even show that we can take $S_1 = \ldots = S_N$ and Γ symmetric.

Proof of Proposition 2.2 The Lefschetz standard conjecture for degree 2 cohomology on X provides a codimension 2-cycle Z on $X \times X$ such that, if $n = \dim X$, $Z^*\colon H^{2n-2}(X, \mathbb{Q}) \to H^2(X, \mathbb{Q})$ is the inverse of the Lefschetz isomorphism $l^{n-2}\colon H^2(X, \mathbb{Q}) \cong H^{2n-2}(X, \mathbb{Q})$ induced by the first Chern class l of a very ample line bundle L on X. Let $j\colon S \to X$ be a smooth surface which is the complete intersection of $n-2$ ample hypersurfaces in $|L|$. Then by the Lefschetz theorem on hyperplane sections, the Gysin morphism $j_*\colon H^2(S, \mathbb{Q}) \to H^{2n-2}(X, \mathbb{Q})$ is surjective, so that, denoting by Z_S the restriction of Z to $X \times S$, we find that

$$Z_S^* = Z^* \circ j_*\colon H^2(S, \mathbb{Q}) \to H^2(X, \mathbb{Q})$$

is also surjective. We can make Z_S is effective by replacing if necessary its negative components $-Z_{S,i}$ by effective residual cycles Z_i' of class $H^2 - Z_{S,i}$, where $H = \mathrm{pr}_1^* H_1 + \mathrm{pr}_2^* H_2$ is a sufficiently ample line bundle on $X \times S$. The cycle H^2 acts trivially on transcendental cohomology, so this change does not affect $Z_S^*\colon H^2(S, \mathbb{Q})_{tr} \to H^2(X, \mathbb{Q})_{tr}$. Because Z_S is effective, it is given by a rational map

$$\phi_{Z_S}\colon X \dashrightarrow S^{(N)},$$

so that

$$Z_S^* \sigma_S = \phi_{Z_S}^* \sigma_{S^{(N)}} \ \text{ in } H^{2,0}(X),$$

for any holomorphic 2-form σ_S on S, where $\sigma_{S^{(N)}}$ denotes the induced 2-form on $S^{(N)}$. Recall that, denoting by $\mu\colon S^N \to S^{(N)}$ the quotient map,

$$\mu^* \sigma_{S^{(N)}} = \sum_{i=1}^{N} \mathrm{pr}_i^* \sigma_S \ \text{ in } H^{2,0}(S^N). \tag{2.5}$$

The finite cover μ induces a finite cover $\phi\colon \Gamma := X \times_{S^{(N)}} S^N \to X$, $\psi\colon \Gamma \to S^N$ and we have a commutative diagram

$$\begin{array}{ccc} \Gamma & \xrightarrow{\psi} & S^N \\ \Big\downarrow{\scriptstyle\phi} & & \Big\downarrow{\scriptstyle\mu} \\ X & \xrightarrow{\phi_{Z_S}} & S^{(N)} \end{array} \tag{2.6}$$

From the commutativity of (2.6), we deduce the equality of holomorphic 2-forms on Γ

$$\phi^*(\phi_{Z_S}^* \sigma_{S^{(N)}}) = \psi^*(\mu^* \sigma_{S^{(N)}}). \tag{2.7}$$

Combining (2.5) and (2.7), we get (2.4) with $\sigma_X = Z_S^*\sigma_S$, $\sigma_i = \sigma_S$ for all i. □

Remark 2.4. To be fully rigorous in the above proof, we should introduce desingularizations of $S^{(n)}$ and Γ to write the equalities above. This is done in [Mum68].

2.1 Surface Decomposition and Cohomology Ring

It is a well-known and fundamental result (see [Bog96, Fuj87]) that for a hyper-Kähler manifold X of dimension $2n$, there exist a quadratic form q on $H^2(X,\mathbb{Q})$ and a positive rational number λ such that for any $\eta \in H^2(X,\mathbb{Q})$

$$\int_X \eta^{2n} = \lambda q(\eta)^n. \tag{2.8}$$

Let us show how this property follows, at least on transcendental cohomology, from the existence of a surface decomposition for a smooth projective variety X, assuming it satisfies the following property $(*)$. Recall first that a quadratic form q on a rational weight 2 Hodge structure, consisting of a $\mathbb{Q}$-vector space H and a Hodge decomposition of $H_\mathbb{C} := H \otimes \mathbb{C}$

$$H_\mathbb{C} = H^{2,0} \oplus H^{1,1} \oplus H^{0,2}, \qquad \overline{H^{p,q}} = H^{q,p},$$

is said to satisfy the first Hodge–Riemann relations if the Hodge decomposition is orthogonal for the Hermitian pairing $h(\alpha,\beta) = q(\alpha,\overline{\beta})$ on $H_\mathbb{C}$ or, equivalently, $q(H^{2,0}, H^{2,0} \oplus H^{1,1}) = 0$. We will say that it satisfies the weak second Hodge–Riemann relations if $q(\alpha,\overline{\alpha}) \geq 0$ for $\alpha \in H^{2,0}$ and $q(\alpha,\overline{\alpha}) \leq 0$ for $\alpha \in H^{1,1}$. Consider the condition

$(*)$ *There exists up to a coefficient a unique quadratic form q satisfying the first Hodge–Riemann relations on $H^2(X,\mathbb{Q})_{tr}$.*

Property $(*)$ is well-known to be satisfied by a very general lattice polarized projective hyper-Kähler manifold. Note that we need in any case to use transcendental cohomology, namely $H^2(X,\mathbb{Q})^{\perp NS(X)}$, instead of primitive cohomology, as $(*)$ is never satisfied on $H^2(X,\mathbb{Q})_{prim}$ if it is different from $H^2(X,\mathbb{Q})_{tr}$, that is, if it contains rational classes of type $(1,1)$. We have the following.

Proposition 2.5. *(1) If a smooth projective variety X of dimension $2n$ admits a surface decomposition (Definition 1.2) there exist quadratic*

forms $q_1, \ldots, q_n$ satisfying the first and weak second Hodge–Rie-mann relations on $H^2(X,\mathbb{Q})_{tr}$, such that, for any $\eta \in H^2(X,\mathbb{Q})_{tr}$

$$\int_X \eta^{2n} = q_1(\eta)\ldots q_n(\eta). \tag{2.9}$$

(2) If furthermore X satisfies property $(*)$, there exists a rational number λ and a quadratic form q satisfying the first and weak second Hodge–Riemann relations on $H^2(X,\mathbb{Q})_{tr}$, such that, for any $\eta \in H^2(X,\mathbb{Q})_{tr}$

$$\int_X \eta^{2n} = \lambda q(\eta)^n.$$

Proof We have by assumption, for any $\eta \in H^2(X,\mathbb{Q})_{tr}$, an equality

$$\phi^*\eta = \psi^*\left(\sum_{i=1}^{n} \mathrm{pr}_i^*\eta_i\right), \tag{2.10}$$

where ψ, ϕ are as in (2.1). For each surface S_i, we have the Poincaré pairing $(\,,\,)_{S_i}$ on $H^2(S_i,\mathbb{Q})_{tr}$ which satisfies the first and second Hodge–Riemann relations, and, as the morphism ψ_i which maps η to η_i is a morphism of Hodge structures (see (2.3)), it provides an intersection form $q_i(\eta) := (\eta_i,\eta_i)_{S_i}$ on $H^2(X,\mathbb{Q})_{tr}$, which satisfies the first and weak second Hodge–Riemann relations.

Let now N, M be the respective degrees of the maps ϕ, ψ. We deduce from (2.10) the following equality:

$$\begin{aligned} N\int_X \eta^{2n} &= M\int_{S_1\times\ldots\times S_n} \sum_{i=1}^{n} (\mathrm{pr}_i^*\eta_i)^{2n} \\ &= M\frac{(2n)!}{2^n n!}(\eta_1,\eta_1)_{S_1}\ldots(\eta_n,\eta_n)_{S_n}. \end{aligned} \tag{2.11}$$

Let $q_1(\eta) := (\eta_1,\eta_1)_{S_1}, \ldots, q_n(\eta) := (\eta_n,\eta_n)_{S_n}$, where the η_i's are defined by (2.10). Then (2.11) gives (2.9) up to a multiplicative coefficient, which proves (*1*).

We next assume property $(*)$ which implies that $(\eta_i,\eta_i)_{S_i} = \mu_i q(\eta)$ for some rational numbers μ_i, since q_i satisfies the first Hodge–Riemann relations. Equation (2.11) then gives:

$$N\int_X \eta^{2n} = M\frac{(2n)!}{2^n n!}\mu_1\ldots\mu_n q(\eta)^n,$$

proving (*2*). □

Proposition 2.5(*1*) now implies the following result, showing that having a surface decomposition is a restrictive condition:

Theorem 2.6. *Let S_1, S_2, S_3 be three smooth projective surfaces with $h^{1,0}(S_i) = 0$, $h^{2,0}(S_i) \neq 0$ for all i, and let $H = \mathrm{pr}_1^* H_1 + \mathrm{pr}_2^* H_2 + \mathrm{pr}_3^* H_3 \in \mathrm{Pic}\,(S_1 \times S_2 \times S_3)$ be a very ample divisor on $S_1 \times S_2 \times S_3$. Let $Y \subset S_1 \times S_2 \times S_3$ be the smooth complete intersection of two general members of $|H|$. Then Y is not surface decomposable.*

Proof As $h^{1,0}(S_i) = 0$, we have

$$H^2(S_1 \times S_2 \times S_3, \mathbb{Q}) = H^2(S_1, \mathbb{Q}) \oplus H^2(S_2, \mathbb{Q}) \oplus H^2(S_3, \mathbb{Q})$$

and similarly for transcendental cohomology. By the Lefschetz hyperplane section theorem, we get, as $\dim Y = 4$:

$$H^2(Y, \mathbb{Q})_{tr} = H^2(S_1, \mathbb{Q})_{tr} \oplus H^2(S_2, \mathbb{Q})_{tr} \oplus H^2(S_3, \mathbb{Q})_{tr}.$$

We now compute $\int_Y \alpha^4$ for $\alpha \in H^2(Y, \mathbb{Q})_{tr}$. For $\alpha = \alpha_1 + \alpha_2 + \alpha_3$, using $\int_{S_i} \alpha_i \cup h_i = 0$, where $h_i := c_1(H_i)$, we get

$$\begin{aligned} \int_Y \alpha^4 &= \int_{S_1 \times S_2 \times S_3} (\mathrm{pr}_1^*\alpha_1 + \mathrm{pr}_2^*\alpha_2 + \mathrm{pr}_3^*\alpha_3)^4 (\mathrm{pr}_1^* h_1 + \mathrm{pr}_2^* h_2 + \mathrm{pr}_3^* h_3)^2 \\ &= \lambda_1 q_2(\alpha_2) q_3(\alpha_3) + \lambda_2 q_1(\alpha_1) q_3(\alpha_3) + \lambda_3 q_1(\alpha_1) q_2(\alpha_2), \qquad (2.12) \end{aligned}$$

where $q_i(\alpha_i) := \int_{S_i} \alpha_i^2$ and the constants λ_i are nonzero rational numbers. It is immediate to see that (2.12) is not of the form (2.9), namely the product of two quadrics in $\alpha = \alpha_1 + \alpha_2 + \alpha_3$. Indeed, the hypersurface in $\mathbb{P}(H^2(Y, \mathbb{C})_{tr})$ defined by (2.12) is irreducible, being fibered with irreducible fibers over the smooth conic in $\mathbb{P}^2_{\mathbb{C}}$ with equation $\lambda_1 y_2 y_3 + \lambda_2 y_1 y_3 + \lambda_3 y_1 y_2 = 0$, via the rational map

$$\begin{aligned} \mathbb{P}(H^2(Y, \mathbb{C})_{tr}) &\dashrightarrow \mathbb{P}^2_{\mathbb{C}}, \\ \alpha = \alpha_1 + \alpha_2 + \alpha_3 &\mapsto (q_1(\alpha_1), q_2(\alpha_2), q_3(\alpha_3)). \end{aligned}$$

□

2.2 Application to Beauville's Weak Splitting Conjecture

In the paper [BV04], it was observed that a projective $K3$ surface has the following property: there is a canonical 0-cycle $o_S \in \mathrm{CH}_0(S)$ of degree 1 (in fact, it can be defined as $c_2(S)/24$) such that for any divisor $D \in \mathrm{Pic}\, S = \mathrm{CH}^1(S) = \mathrm{NS}(S)$, one has

$$D^2 = q(D) o_S \ \text{ in } \mathrm{CH}_0(S), \qquad (2.13)$$

where $q(D) = ([D], [D])_S$. One can rephrase this result by saying:

(*) *Any cohomological polynomial relation*

$$Q([D_1],\dots,[D_k])=0 \text{ in } H^*(S,\mathbb{Q})$$

involving only divisor classes is already satisfied in $\mathrm{CH}(S)_{\mathbb{Q}}$.

In [Bea07], Beauville made the following conjecture, generalizing the result above:

Conjecture 2.7. *Let X be a projective hyper-Kähler manifold. Then the cycle class map is injective on the subalgebra of* $\mathrm{CH}^*(X)_{\mathbb{Q}}$ *generated by divisor classes.*

This conjecture is called Beauville's weak splitting conjecture. It is equivalent to the fact that property (*), that we will call the weak splitting property, is satisfied for any divisor classes on X. Let us discuss Conjecture 2.7 in relation with the notion of surface decomposition. Let X be a projective hyper-Kähler manifold, and let $\Lambda\subset \mathrm{NS}(X)$ be a lattice polarization (which means that Λ contains an ample class). The very general Λ-polarized deformation X_t of X is the very general fiber of a family $\mathcal{X}\to B$ parameterized by a quasiprojective base B and it satisfies $H^2(X_t,\mathbb{Q})_{tr}=H^2(X_t,\mathbb{Q})^{\perp\Lambda}$. We assume that the general (or very general) Λ-polarized deformation X_t of X has a surface decomposition. Then, by standard spreading arguments involving relative Chow varieties, after passing to a generically finite cover B' of B, we have projective morphisms $\Gamma\to B'$, $\mathcal{S}_i\to B'$, with $\dim \mathcal{S}_i/B'=2$, and morphisms over B'

$$\phi\colon \Gamma\to\mathcal{X},\qquad \psi\colon\Gamma\to\mathcal{S}_1\times_{B'}\ldots\times_{B'}\mathcal{S}_n$$

inducing a surface decomposition at the general point $t\in B'$. After shrinking B', by desingularization of the general fiber, one can assume that the fibers Γ_t and $S_{i,t}$ are smooth and we get by specialization a diagram

$$\phi_t\colon\Gamma_t\to X_t,\qquad \psi_t\colon\Gamma_t\to S_{1,t}\times\ldots\times S_{n,t}$$

such that

$$\phi_t^*\sigma_{X_t}=\psi_t^*\left(\sum_i \mathrm{pr}_i^*\sigma_{S_{i,t}}\right) \text{ in } H^{2,0}(\Gamma_t) \tag{2.14}$$

for some $(2,0)$-forms $\sigma_{S_{i,t}}$ on $S_{i,t}$. We proved in Lemma 2.1 that the relation (2.14) then holds for any class $\alpha\in H^2(X_t,\mathbb{Q})_{tr}=H^2(X_t,\mathbb{Q})^{\perp\Lambda}$

and that there is for each i a (locally constant) morphism of Hodge structures

$$\psi_{i,t}\colon H^2(X_t,\mathbb{Q})^{\perp\Lambda} \to H^2(S_{i,t},\mathbb{Q})$$

given by (2.3) such that

$$\phi_t^*\alpha = \psi_t^*\left(\sum_i \mathrm{pr}_i^*(\psi_{i,t}(\alpha))\right) \text{ in } H^2(\Gamma_t,\mathbb{Q}). \tag{2.15}$$

Let us now assume furthermore that $H^1(\Gamma_t,\mathbb{Z})=0$, or equivalently

$$\mathrm{NS}\,(\Gamma_t) = \mathrm{Pic}\,(\Gamma_t). \tag{2.16}$$

Note that, since Γ_t dominates each $S_{i,t}$, this implies the same equality for each $S_{i,t}$. In the situation described above, we have the following result.

Theorem 2.8. *For any $t \in B'$, the weak splitting property holds for divisor classes on X_t which are in $H^2(X_t,\mathbb{Q})^{\perp\Lambda}$ if and only if, for each surface $S_{i,t}$, the Beauville–Voisin relation* (2.13) *holds on* $\mathrm{Im}\,\psi_{i,t}$ *for an adequate 0-cycle $o_{S_{i,t}} \in \mathrm{CH}_0(S_{i,t})$.*

Proof Using (2.16), we conclude that (2.15) holds in $\mathrm{Pic}\,(\Gamma_t)_{\mathbb{Q}}$ for $\alpha \in \mathrm{Pic}\,(X_t)^{\perp\Lambda} = \mathrm{NS}\,(X_t)^{\perp\Lambda}$ (where the point t is now special in B', being in a Noether–Lefschetz locus), and more precisely, that the morphism of Hodge structures $\psi_{i,t}$ induces for any $t \in B'$ a $\mathbb{Q}$-linear map

$$\psi_{i,t}\colon \mathrm{Pic}\,(X_t)_{\mathbb{Q}}^{\perp\Lambda} \to \mathrm{Pic}\,(S_{i,t})_{\mathbb{Q}}$$

such that, for any $D \in \mathrm{Pic}\,(X_t)_{\mathbb{Q}}^{\perp\Lambda}$:

$$\phi_t^* D = \psi_t^*\left(\sum_i \mathrm{pr}_i^*(\psi_{i,t}(D))\right) \text{ in } \mathrm{Pic}(\Gamma_t)_{\mathbb{Q}} = \mathrm{CH}^1(\Gamma_t)_{\mathbb{Q}}. \tag{2.17}$$

As in the cohomological setting which has been studied in the previous section, the important point here is the fact that the pull-back maps appearing on both sides are compatible with intersection product. Note also that they are injective since the maps ϕ_t and ψ_t are dominant. For any point $t \in B$, let $D_1,\ldots, D_k \in \mathrm{CH}^1(X)_{\mathbb{Q}}$ and let Q be a degree l homogeneous polynomial with $\mathbb{Q}$-coefficients in k variables. Then we get from (2.17):

$$\phi_t^* Q(D_1,\ldots,D_k) = \psi_t^*(Q(D_1',\ldots, D_k')) \text{ in } \mathrm{CH}^l(\Gamma_t)_{\mathbb{Q}}, \tag{2.18}$$

where $D_j' := \sum_i \mathrm{pr}_i^*(\psi_{i,t}(D_j))$. Assume that X_t satisfies the weak splitting property, at least for divisor classes $D \in \mathrm{CH}^1(X_t)^{\perp\Lambda}$. There is then a 0-cycle $o_X \in \mathrm{CH}_0(X)$ of degree 1 such that

$$D^{2n} = (\deg D^n) o_X \text{ in } \mathrm{CH}_0(X)$$

for any $D \in \mathrm{CH}^1(X_t)^{\perp\Lambda}$. Pulling-back to Γ_t and using (2.18), we have

$$\phi_t^*(D^{2n}) = \frac{(2n)!}{2^n n!} \psi_t^* \left(\prod_{j=1}^{n} \mathrm{pr}_j^*(\psi_{j,t}(D)^2) \right) \text{ in } \mathrm{CH}_0(\Gamma_t). \tag{2.19}$$

Note that any $D \in \mathrm{CH}^1(X_t)^{\perp\Lambda}$ satisfies $q(D) \neq 0$ by the Hodge index theorem, where q is the Beauville–Bogomolov quadratic form on $H^2(X_t, \mathbb{Q})$, which can also be defined as the Lefschetz intersection pairing on $H^2(X_t, \mathbb{Q})^{\perp\Lambda}$ (see [Bea83]). As we have $\deg D^{2n} = \lambda q([D])^n$ with $\lambda \neq 0$ by (2.8), we conclude that $\deg D^{2n} \neq 0$. Let

$$o_{S_{j,t}} := \mathrm{pr}'_{j*} \left(\frac{1}{\deg \phi_t} (\psi_{t*}(\phi_t^* o_X)) \right) \in \mathrm{CH}_0(S_{j,t})_{\mathbb{Q}}.$$

This cycle has degree 1 and so from (2.19) by pushing-forward to $S_{j,t}$ via $\mathrm{pr}_j \circ \psi$, we get $\psi_{j,t}(D)^2$ is proportional to $o_{S_{j,t}}$. Indeed, $(\mathrm{pr}_j \circ \psi)_*(\phi_t^*(D^{2n}))$ is a 0-cycle of degree different from 0 on $S_{j,t}$, which by (2.19) is proportional to both $o_{S_{j,t}}$ and $\psi_{j,t}(D)^2$. This proves the 'only if' direction.

Conversely, assume each surface $S_{i,t}$ has a 0-cycle $o_{S_{i,t}}$ of degree 1 with the property that divisors D_i in $\mathrm{Im}\,\psi_{i,t} \subset \mathrm{NS}(S_{i,t})_{\mathbb{Q}} = \mathrm{Pic}\,(S_{i,t})_{\mathbb{Q}}$ satisfy $D_i^2 = (D_i, D_i)_{S_{i,t}} o_{S_{i,t}}$ in $\mathrm{CH}_0(S_{i,t})$ or equivalently that for any D_i, $D_i' \in \mathrm{Im}\,\psi_{i,t}$

$$D_i \cdot D_i' = (D_i, D_i')_{S_{i,t}} o_{S_{i,t}} \text{ in } \mathrm{CH}_0(S_{i,t}). \tag{2.20}$$

We now use the fact that, at the very general point of B', the Mumford–Tate group of the Hodge structure on $H^2(X_t, \mathbb{Q})^{\perp\Lambda}$ is the orthogonal group, and thus the intersection form $\psi_{i,t}^*((\,,\,)_{S_{i,t}})$ equals $\mu_i q$ on $H^2(X_t, \mathbb{Q})^{\perp\Lambda}$, for some coefficient μ_i. It then follows from (2.20) that a numerical relation $q(D) = 0$ for $D \in \mathrm{Pic}\,(X_t)_{\mathbb{C}}$ produces relations

$$D_i^2 = 0 \text{ in } \mathrm{CH}_0(S_{i,t})_{\mathbb{C}}, \tag{2.21}$$

for any $i = 1, \ldots, n$, where $D_i := \psi_{i,t}(D)$.

By [Bog96] (or rather, the same arguments as in [Bog96] using the fact that the Beauville–Bogomolov pairing q remains nondegenerate on $\mathrm{NS}(X_t)_{\mathbb{Q}}$), we know that the relations in the subalgebra of $H^*(X_t, \mathbb{C})$ generated by $\mathrm{NS}\,(X_t)_{\mathbb{C}}$ are generated by the Bogomolov–Verbitsky relations

$$d^{n+1} = 0 \text{ if } q(d) = 0. \tag{2.22}$$

This is true as well (for the same reasons) if we restrict to the subalgebra generated by $\mathrm{NS}\,(X_t)_{\mathbb{C}}^{\perp\Lambda} = \mathrm{Pic}\,(X_t)_{\mathbb{C}}^{\perp\Lambda}$. Next, (2.18) provides for any $D \in \mathrm{Pic}\,(X_t)_{\mathbb{C}}^{\perp\Lambda}$

$$\phi_t^*(D^{n+1}) = \psi_t^*\left(\left(\sum_{i=1}^{n} \mathrm{pr}_i^* D_i\right)^{n+1}\right) \tag{2.23}$$

$$= \sum_i (\mathrm{pr}_1^* D_1 \cdot \ldots \cdot \mathrm{pr}_i^* D_i^2 \cdot \ldots \cdot \mathrm{pr}_n^* D_n) + \ldots \text{ in } \mathrm{CH}(\Gamma_t)_{\mathbb{C}},$$

where the remaining term '...' involves products $\mathrm{pr}_i^* D_i^2 \cdot \mathrm{pr}_j^* D_j^2$ of two squares, then three squares $\mathrm{pr}_i^* D_i^2 \cdot \mathrm{pr}_j^* D_j^2 \cdot \mathrm{pr}_k^* D_k^2$, etc. Using (2.21) and (2.23), we get $\phi_t^*(D^{n+1}) = 0$ in $\mathrm{CH}^{n+1}(\Gamma_t)_{\mathbb{C}}$, hence $D^{n+1} = 0$ in $\mathrm{CH}^{n+1}(X_t)_{\mathbb{C}}$, whenever $q(D) = 0$. In other words, the Bogomolov–Verbitsky relations (2.22) are satisfied in $\mathrm{CH}^{n+1}(X_t)_{\mathbb{C}}$, which concludes the proof. □

We get the following corollary:

Corollary 2.9. *(See Theorem 1.4) Under the same assumptions as in Theorem 2.8, the weak splitting property holds for divisor classes on X_t which are in $H^2(X_t, \mathbb{Q})^{\perp\Lambda}$ if and only if they hold in top degree, that is,*

(∗) *there exists a canonical 0-cycle $o_{X_t} \in \mathrm{CH}_0(X_t)$ such that for any $D \in \mathrm{NS}(X_t)^{\perp\Lambda}$, D^{2n} is proportional to o_{X_t} in $\mathrm{CH}_0(X_t)$.*

Proof The 'only if' is clear. In the other direction, examining the proof of Theorem 2.8, we observe that we only used relations (2.19) in top degree $2n$ to conclude that, if (∗) holds, defining $o_{S_{i,t}} \in \mathrm{CH}_0(S_{i,t})$ by $o_{S_{i,t}} := \frac{1}{\deg \phi_t} \mathrm{pr}_{i*}(\psi_{t*}(\phi_t^* o_{X_t}))$, the 0-cycle $D_{i,t}^2$ is proportional to $o_{S_{i,t}}$ in $\mathrm{CH}_0(S_{i,t})$, for any $D_t \in \mathrm{NS}(X_t)^{\perp\Lambda}$, where $D_{i,t} := \psi_{i,t}(D_t)$. Hence by Theorem 2.8, (∗) implies the weak splitting property for $\mathrm{NS}(X_t)^{\perp\Lambda}$. □

2.3 Decomposition from Families of Algebraically Coisotropic Divisors

We study in this section a weaker notion of decomposition for a holomorphic 2-form into forms of smaller rank (see Question 1.1). The following is a weak converse to Proposition 1.11.

Proposition 2.10. *Let X be smooth projective variety of dimension $2n$ equipped with a generically nondegenerate holomorphic 2-form σ_X. Assume X is swept-out by (possibly singular) algebraically coisotropic* *divisors. Then there exists a generically finite cover* $\Phi\colon \mathcal{D}' \to X$ *such that*

$$\Phi^*\sigma_X = \eta_1 + \eta_2 \text{ in } H^{2,0}(\mathcal{D}'),$$

where $\operatorname{rank}\eta_1 = 2$, *and* $\operatorname{rank}\eta_2 = 2n-2$. *More precisely,* η_2 *is the pull-back of a holomorphic* 2*-form on a variety of dimension* $\leq 2n-1$.

Here by the rank we mean the generic rank of the considered forms.

Proof of Proposition 2.10 By assumption, there exists a 1-parameter family

$$\mathcal{D} \to C, \qquad \mathcal{D} \to X$$

of divisors $D_t \subset X$ whose characteristic foliation (on the regular locus of D_t) is algebraically integrable, that is, there exists a rational map

$$\phi_t\colon D_t \dashrightarrow B_t$$

with $\dim B_t = 2n-2$, such that the equality $\sigma_{X|D_t} = \phi_t^*\sigma_{B_t}$, for some holomorphic 2-form σ_{B_t} on the regular locus of B_t, holds on the regular locus of D_t. Note that, by desingularisation, we can assume D_t and B_t smooth, at least for general t. Indeed, the 2-form σ_{B_t} extends holomorphically on any smooth projective model $\widetilde{B}_t$ of B_t, because it can be constructed as

$$\tilde{\phi}_{t*}(\tilde{j}_t^*\sigma_X \wedge \omega)$$

where $\tilde{j}_t\colon \widetilde{D}_t \to X$ is a smooth model of D_t such that $\tilde{\phi}_t\colon \widetilde{D}_t \to \widetilde{B}_t$ is a morphism, and ω is a closed $(1,1)$-form on $\widetilde{D}_t$ whose integral over the fibers of $\tilde{\phi}_t$ is 1.

As usual, the data above (namely the family of varieties B_t and morphisms ϕ_t) can be put in a family, possibly after base change from the original family $\mathcal{D} \to C$ of divisors on X and birational transformations. We thus get the following diagram

$$\begin{array}{ccc} \mathcal{D}' & \xrightarrow{J} & X \\ \downarrow{\scriptstyle \Phi} & & \\ B & & \end{array}$$

where all the varieties are smooth and projective, the morphism J is surjective generically finite, $\dim B = 2n-1$ and B admits a morphism $f\colon B \to C$ such that, considering the induced diagram of fibers over a general point $t \in C$

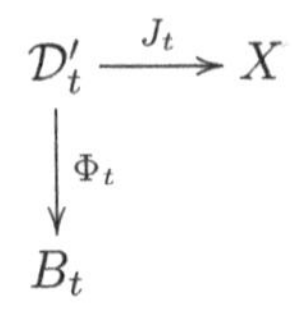

one has

$$J_t^*\sigma_X = \Phi_t^*\sigma_{B_t} \text{ in } H^{2,0}(B_t). \tag{2.24}$$

We deduce from this last equality that the forms σ_{B_t}, $t \in C$, form a locally constant section of the bundle $\mathcal{H}^{2,0} \subset R^2 f_*\mathbb{C} \otimes \mathcal{O}_C$ on the open set of C of regular values of f. By the global invariant cycles theorem [Del71], [Voi07, 4.3.3], there exists a holomorphic 2-form $\sigma_B \in H^{2,0}(B)$ such that

$$\sigma_{B|B_t} = \sigma_{B_t}. \tag{2.25}$$

We conclude from (2.24) and (2.25) that the 2-form $\Phi^*\sigma_B - J^*\sigma_X$ vanishes on the divisors $\mathcal{D}'_t = \Phi^{-1}(B_t)$ which cover $\mathcal{D}'$. This form thus has rank ≤ 2 on $\mathcal{D}'$. Finally, as the rank of $\Phi^*\sigma_B - J^*\sigma_X$ is ≤ 2 and the rank of $\Phi^*\sigma_B$ is $\leq 2n-2$, while $\operatorname{rank} J^*\sigma_X = 2n$, one concludes that $\operatorname{rank} \Phi^*\sigma_B - J^*\sigma_X = 2$ and $\operatorname{rank} \Phi^*\sigma_B = 2n-2$. □

This statement raises the following question.

Question 2.11. Is any projective hyper-Kähler manifold swept out by algebraically coisotropic divisors?

The following question was asked by G. Pacienza.

Question 2.12. Is any projective hyper-Kähler manifold swept out by elliptic curves?

The following proposition relates Question 2.11 and Question 2.12.

Proposition 2.13. *If a very general polarized hyper-Kähler manifold with $b_2 \geq 5$ is swept out by elliptic curves, then it is swept out by algebraically coisotropic divisors.*

Here, "very general" means that X is the very general member of a complete family of polarized hyper-Kähler manifolds.

Proof There exists by assumption a covering family of elliptic curves

$$\phi\colon \mathcal{E} \to X, \qquad \psi\colon \mathcal{E} \to B$$

with ϕ surjective generically finite and $\dim B = 2n-1$. If these elliptic curves have constant moduli, after passing to a generically finite cover of B, $\mathcal{E}$ becomes birational to a product $E \times B$ and we conclude that there is an injective morphism of Hodge structures

$$H^2(X,\mathbb{Q})_{tr} \to H^1(E,\mathbb{Q}) \otimes H^1(B,\mathbb{Q}).$$

Indeed, $\phi^*\sigma_X$ is not in the image of ψ^* because $\psi^* H^{2,0}(B)$ consists of holomorphic forms of generic rank $< \dim X$, while $\phi^*\sigma_X$ has generic rank equal to $\dim X$. Hence $\phi^*\sigma_X$ has a nontrivial image in $H^1(E,\mathbb{C}) \otimes H^1(B,\mathbb{C})$. The natural morphism $H^2(X,\mathbb{Q})_{tr} \to H^1(E,\mathbb{Q}) \otimes H^1(B,\mathbb{Q})$ given by pull-back and projection to a Leray summand is thus nonzero on $H^{2,0}(X)$, hence injective on $H^2(X,\mathbb{Q})_{tr}$. When $b_2 \geq 5$ and X is very general, the existence of such injective morphism contradicts the result of [vGV16]. Hence the elliptic curves E must have variable modulus. For a fixed $t \in \mathbb{P}^1$, consider the divisor $B_t \subset$ parameterizing elliptic curves E_b with fixed j-invariant determined by t. Over B_t, the family $\mathcal{E}_t = \psi^{-1}(B_t)$ is birational (possibly after after base change) to $E_t \times B_t$. Let

$$\phi_t \colon \mathcal{E}_t \to X, \qquad \psi_t \colon \mathcal{E}_t \to B_t$$

be the restricted family. The same argument as above shows that $\phi_t^*\sigma_X$ has to vanish in $H^1(E_t,\mathbb{C}) \otimes H^1(B_t,\mathbb{C})$. This is exactly saying that $\phi(\mathcal{E}_t)$ is an algebraically coisotropic divisor in X, as this implies that $\phi_t^*\sigma_X$ is pulled-back from B_t. □

It seems plausible that Question 2.12 has a negative answer while Question 2.11 has a positive answer.

3 Triangle Varieties: Examples

Recall from Definition 1.8 in the introduction that a triangle variety T for a hyper-Kähler manifold X of dimension $2n$ is a subvariety of $X \times X \times X$ which has dimension $3n$, maps surjectively onto the various summands and maps in a generically finite way on its image in the product of two summands, and is such that

$$(\mathrm{pr}_1^*\sigma_X + \mathrm{pr}_2^*\sigma_X + \mathrm{pr}_3^*\sigma_X)_{|T_{\mathrm{reg}}} = 0, \tag{3.1}$$

where σ_X is the holomorphic 2-form of X. Note that (3.1) says that T is Lagrangian for the everywhere nondegenerate holomorphic 2-form $\mathrm{pr}_1^*\sigma_X + \mathrm{pr}_2^*\sigma_X + \mathrm{pr}_3^*\sigma_X$ on X^3. A variant of the main deformation invariance theorem of [Voi92] says now the following:

Theorem 3.1. *Let X be hyper-Kähler and let $j\colon L \hookrightarrow X \times X \times X$ be a smooth triangle subvariety (hence L is Lagrangian for the 2-form $\mathrm{pr}_1^*\sigma_X + \mathrm{pr}_2^*\sigma_X + \mathrm{pr}_3^*\sigma_X$). Then for a small deformation X_t of X with constant Picard group, there is a deformation $j_t\colon L_t \hookrightarrow X_t \times X_t \times X_t$ of L, and L_t is a triangle variety for X_t.*

The last statement follows from the fact that, denoting $\Lambda = \mathrm{NS}(X)$, the subgroup $H^2(X_t, \mathbb{Q})^{\perp\Lambda}$ and the restriction map

$$j_t^*\colon H^2(X_t \times X_t \times X_t, \mathbb{Q}) \to H^2(L_t, \mathbb{Q})$$

are locally constant on the base B of deformations of X with fixed Picard number. Hence the diagonal image of the subgroup $H^2(X_t, \mathbb{Q})^{\perp\Lambda}$ in $H^2(X_t \times X_t \times X_t, \mathbb{Q}) = H^2(X_t, \mathbb{Q})^3$ is annihilated by j_t^*, since it is annihilated by j^* (note here that $H^2(X, \mathbb{Q})^{\perp\Lambda} = H^2(X, \mathbb{Q})_{tr}$).

Remark 3.2. A smooth triangle subvariety $L \subset X \times X \times X$ cannot deform in products $X_t \times X_{t'} \times X_{t''}$ unless $t = t' = t''$. Indeed, the kernel H of $j^*\colon H^2(X, \mathbb{Q})^3_{tr} \to H^2(L, \mathbb{Q})$ is exactly the diagonal image of $H^2(X, \mathbb{Q})_{tr}$, as it follows from the fact that L maps to a subvariety of dimension $3n$ in the three products $X \times X$. If there is a deformation $L_{t,t',t''}$ of L in $X_t \times X_{t'} \times X_{t''}$, there is a Hodge substructure

$$H_{t,t',t''} \subset H^2(X_t, \mathbb{Q})^{\perp\Lambda} \oplus H^2(X_{t'}, \mathbb{Q})^{\perp\Lambda} \oplus H^2(X_{t''}, \mathbb{Q})^{\perp\Lambda}$$

deforming H. But then $H_{t,t',t''}$ is isomorphic by projections to the three Hodge structures $H^2(X_t, \mathbb{Q})^{\perp\Lambda}$, $H^2(X_{t'}, \mathbb{Q})^{\perp\Lambda}$, $H^2(X_{t''}, \mathbb{Q})^{\perp\Lambda}$. By the local Torelli theorem, we then have $t = t' = t''$.

Theorem 3.1 suggests possibly that triangle subvarieties tend to be stable under deformations with constant Picard number, but in the examples we will describe below, the triangle subvarieties are never smooth, so in fact Theorem 3.1 does not apply.

Considering the conjectures made in [Bea07, Voi16], it would be very nice if the triangle varieties T satisfied a cycle-theoretic variant of (3.1), asking the following: for any $t = (t_1, t_2, t_3) \in T \subset X^3$

$$t_1 + t_2 + t_3 = c \text{ in } \mathrm{CH}_0(X), \tag{3.2}$$

for some fixed zero-cycle c of X. Note that (3.2) implies indeed (3.1) by Mumford's theorem [Mum68]. Let us explain why it is not possible to achieve (3.2) starting from dimension 4.

Proposition 3.3. *Let X be a projective hyper-Kähler manifold of dimension $2n \geq 4$. Let T be a triangle subvariety of $X \times X \times X$. Then the cycle $t_1 + t_2 + t_3 \in \mathrm{CH}_0(X)$ for $t = (t_1, t_2, t_3) \in T$ is not constant along T.*

Proof Indeed, if (3.2) holds, then Mumford's theorem [Mum68] says that for any power σ_X^l, $l > 0$, of σ_X,

$$(\mathrm{pr}_1^*\sigma_X^l + \mathrm{pr}_2^*\sigma_X^l + \mathrm{pr}_3^*\sigma_X^l)_{|T_{\mathrm{reg}}} = 0 \text{ in } H^0(T_{\mathrm{reg}}, \Omega^{2l}_{T_{\mathrm{reg}}}).$$

We now set $l = 2$. We then have the two equations

$$(\mathrm{pr}_1^*\sigma_X)_{|T_{\mathrm{reg}}} = -(\mathrm{pr}_2^*\sigma_X + \mathrm{pr}_3^*\sigma_X)_{|T_{\mathrm{reg}}} \text{ in } H^0(T_{\mathrm{reg}}, \Omega^2_{T_{\mathrm{reg}}}),$$
$$(\mathrm{pr}_1^*\sigma_X^2)_{|T_{\mathrm{reg}}} = -(\mathrm{pr}_2^*\sigma_X^2 + \mathrm{pr}_3^*\sigma_X^2)_{|T_{\mathrm{reg}}} \text{ in } H^0(T_{\mathrm{reg}}, \Omega^4_{T_{\mathrm{reg}}}).$$

It follows that

$$-(\mathrm{pr}_2^*\sigma_X^2 + \mathrm{pr}_3^*\sigma_X^2)_{|T_{\mathrm{reg}}} = (\mathrm{pr}_2^*\sigma_X + \mathrm{pr}_3^*\sigma_X)^2_{|T_{\mathrm{reg}}} \tag{3.3}$$

in $H^0(T_{\mathrm{reg}}, \Omega^4_{T_{\mathrm{reg}}})$. Let us write the above equation as

$$\omega_{|T_{\mathrm{reg}}} \wedge \omega'_{|T_{\mathrm{reg}}} = 0 \text{ in } H^0(T_{\mathrm{reg}}, \Omega^4_{T_{\mathrm{reg}}}),$$

where

$$\omega := \mathrm{pr}_2^*\sigma_X - \frac{-1 + i\sqrt{3}}{2}\mathrm{pr}_3^*\sigma_X, \quad \omega' := \mathrm{pr}_2^*\sigma_X - \frac{-1 - i\sqrt{3}}{2}\mathrm{pr}_3^*\sigma_X.$$

We next have the following

Lemma 3.4. *Let V be a vector space and let $\omega, \omega' \in \bigwedge^2 V^*$ such that $\omega \neq 0$, $\omega' \neq 0$ and $\omega \wedge \omega' = 0$ in $\bigwedge^4 V^*$. Then there exists a quotient $V \to V'$ with $\dim V' \leq 4$ such that both ω and ω' are pulled back from 2-forms on V'.*

Proof This follows from the fact that for a 2-form ω of rank 6 on a 6-dimensional vector space V, the wedge product map $\omega\wedge\colon \bigwedge^2 V^* \to \bigwedge^4 V^*$ is an isomorphism. This fact already implies that if $\omega \wedge \omega' = 0$ in $\bigwedge^4 V^*$, with $\omega \neq 0$, $\omega' \neq 0$, the rank of ω is at most 4 and similarly for ω'. If both forms ω and ω' are of rank 2, the conclusion of the lemma holds. If $\omega = e_1^* \wedge e_2^* + e_3^* \wedge e_4^*$ is of rank 4, let $V' = \langle e_1^*, \ldots, e_4^* \rangle$. Choosing a decomposition $V^* = V'^* \oplus W^*$, we can write $\omega' = \alpha + \beta + \gamma$ with

$$\alpha \in \bigwedge^2 V'^*, \qquad \beta \in V'^* \otimes W^*, \qquad \gamma \in \bigwedge^2 W^*,$$

and we must have $\omega\wedge\beta=0$, $\omega\wedge\gamma=0$, which clearly implies that $\beta=0$ and $\gamma=0$ because ω has rank 4 so $\omega\wedge$ is injective on V'^*. Thus ω' belongs to $\bigwedge^2 V'^*$. □

The contradiction now comes from (3.3) and Lemma 3.4 which imply that either $\omega_{|T_{\rm reg}}=0$ or $\omega'_{|T_{\rm reg}}=0$, or both forms $\omega_{|T_{\rm reg}}$ and $\omega'_{|T_{\rm reg}}$ have rank ≤ 4 at any point t of T and more precisely, at any point $t\in T_{\rm reg}$, are pulled-back via a quotient map $T_{T,t}\to T'$, with $\dim T'\leq 4$. The form $\omega_{|T}$ cannot be 0 because this would imply that the projection of T in $X\times X$ via $({\rm pr}_2,{\rm pr}_3)$ is Lagrangian for a form which has rank $4n$ everywhere on $X\times X$ while by assumption $\dim({\rm pr}_2,{\rm pr}_3)(T)=3n$. The same argument also works for ω'. We thus conclude that the last possibility should hold. In that case, the restrictions to T of ${\rm pr}_2^*\sigma_X$ and ${\rm pr}_3^*\sigma_X$ are also pulled-back via the quotient map $T_{T,t}\to T'$ hence have rank ≤ 4. The form σ_X on X is everywhere nondegenerate and the projections ${\rm pr}_2$, ${\rm pr}_3$ restricted to T are dominant, so we conclude that the forms ${\rm pr}_2^*\sigma_X$, ${\rm pr}_3^*\sigma_X$ restricted to T have rank equal to $\dim X$. As they are of rank ≤ 4 at a general point $t\in T$, we get a contradiction if $n\geq 3$. If $n=2$, these forms pulled-back to T have rank 4 but they do not have the same kernel, because their respective kernels at a point $t\in T_{\rm reg}$ are the spaces ${\rm Ker}\,({\rm pr}_{2|T})_*$ and ${\rm Ker}\,({\rm pr}_{3|T})_*$ which are different generically on T by the assumption that $({\rm pr}_2,{\rm pr}_3)$ is generically of maximal rank. This contradiction concludes the proof. □

We construct in the next subsections triangle varieties for the main "known" classes of hyper-Kähler manifolds, for which we have an explicit projective model.

3.1 Hilbert Schemes of $K3$ Surfaces

Recall from [BV04] (see also Section 2.2) that a projective $K3$ surface S has a canonical 0-cycle o_S of degree 1 satisfying many properties, including the following: for any integer $k\geq 1$, the degree-k 0-cycle ko_S on S has a k-dimensional orbit

$$O_{ko_S}=\{z\in S^{(k)},\ z=ko_S \text{ in } {\rm CH}_0(S)\}$$

in $S^{(k)}$ for rational equivalence on S. An explicit example of a k-dimensional orbit component of ko_S and of a triangle variety for $S^{[n]}$ is as follows. Assume S has a very ample polarization $L\in{\rm Pic}\,S$ with $\deg L^2=2g-2$. Let $k=2g-2$. One component of the orbit $O_{L^2}\subset S^{(2g-2)}$ of the 0-cycle L^2 is birational to the Grassmannian $G(2,H^0(S,L))$ and is made of complete intersections $H_1 \cap H_2$, with $H_1,\ H_2 \in |L|$, or rather of their supports. Note that $\dim G(2, H^0(S,L)) = 2g - 2$ as we want. Assume furthermore that $2g - 2 = 3n$ is divisible by 3 and consider

$$T := \{(z_1,\ z_2,\ z_3) \in (S^{[n]})^3,\ c(z_1) + c(z_2) + c(z_3) \in O_{L^2} \subset S^{(3n)}\},$$

where $c\colon S^{[l]} \to S^{(l)}$ denotes the Hilbert–Chow morphism.

Proposition 3.5. *T is a triangle variety for $S^{[n]}$.*

Proof The relation $(\mathrm{pr}_1^*\sigma_{S^{[n]}} + \mathrm{pr}_2^*\sigma_{S^{[n]}} + \mathrm{pr}_3^*\sigma_{S^{[n]}})_{|T_{\mathrm{reg}}}$ follows from the fact that the 0-cycle $c(z_1) + c(z_2) + c(z_3)$ is constant in $\mathrm{CH}_0(S)$ along T and from Mumford's theorem [Mum68] because the holomorphic 2-form $\sigma_{S^{[n]}}$ is induced by the holomorphic 2-form σ_S via the incidence correspondence. The fact that the dimension of T is $3n$ follows from the fact that T is birational to a generically finite cover of O_{L^2} which has dimension $2g - 2 = 3n$. It remains to see that T dominates the three summands and that it maps in a generically finite way to its images in the three products $S^{[n]} \times S^{[n]}$. The first statement follows from the fact that L is very ample with $h^0(S,L) = g + 1$, where $3n = 2g - 2$. This implies that for a general set $z_1 = \{x_1, \ldots,\ x_n\}$ of n points of S, there is a reduced complete intersection Z of two members of $|L|$ containing all the x_i. Then writing Z as the union $z_1 \sqcup z_2 \sqcup z_3$ of three sets of cardinality n, we have $(z_1, z_2, z_3) \in T$.

For the second statement, we observe that for such a general reduced 0-dimensional complete intersection

$$Z = H_1 \cap H_2 = \{x_1, \ldots, x_{2g-2}\},$$

with $2g - 2 = 3n$, the first $2n$ points $x_1, \ldots,\ x_{2n}$ already impose $g - 1$ conditions on $|L|$, hence the space of hypersurfaces in $|L|$ containing these $2n$ points is the projective line $\langle H_1,\ H_2\rangle$. Setting

$$z_1 = \{x_1, \ldots, x_n\}, \quad z_2 = \{x_{n+1}, \ldots, x_{2n}\}, \quad z_3 = \{x_{2n+1}, \ldots, x_{3n}\},$$

we have $(z_1, z_2, z_3) \in T$. The fiber of the projection $p_{12}\colon T \to S^{[n]} \times S^{[n]}$ over (z_1, z_2) consists by definition of the single element z_3. $\square$

The numerical condition $3n = 2g - 2$ used for the construction above is not important, as there are variants of this construction, starting from other Lagrangian subvarieties of $S^{[3n]}$, also obtained as components of dimension $3n$ of the orbit of $3no_S$ in $S^{(3n)}$.

3.2 Fano Variety of lines in a Cubic Fourfold

The Fano variety $F_1(Y)$ of lines in a smooth cubic fourfold Y is a hyper-Kähler fourfold (see [BD85]). In this case, the triangles are just triangles in a usual sense, namely the plane sections of Y which are the unions of three lines (plus an ordering of these lines). They form a 6-dimensional subvariety of $F_1(Y)^3$. Indeed, for each line $l \subset Y$, consider the $\mathbb{P}^3_l$ of planes containing l. Each of these planes cuts Y along the union of l and a conic, and when the conic is degenerate, that is along a surface in $\mathbb{P}^3_l$, the conic becomes the union of two lines, which together with l form a triangle. In this case, the fact that the family of these triangles is a Lagrangian subvariety of $F_1(Y)^3$ is a consequence of Mumford's theorem [Mum68]. Indeed we know that, via the incidence correspondence $p\colon P \to F_1(Y)$, $q\colon P \to Y$ given by the universal family of lines in Y, one has $\sigma_{F_1(Y)} = P^*\eta_Y$ for some class $\eta_Y \in H^1(Y, \Omega_Y^3)$, and furthermore, for any triangle $([l], [l'], [l'']) \in F_1(Y)^3$

$$P_*l + P_*l' + P_*l'' = h^3 \text{ in } \mathrm{CH}_1(Y),$$

where $h = c_1(\mathcal{O}_Y(1))$.

3.3 Debarre–Voisin Hyper-Kähler Fourfolds

Let V_{10} be a 10-dimensional vector space and let $\lambda \in \bigwedge^3 V_{10}^*$. The associated Debarre–Voisin fourfold $F_\lambda \subset G(6, V_{10})$ is the set of 6-dimensional vector subspaces $W_6 \subset V_{10}$ such that $\lambda_{|W_6} = 0$. This is a hyper-Kähler fourfold for a general parameter λ (see [DV10]). In [Baz19], Bazhov considered the subvariety $T \subset F_\lambda \times F_\lambda \times F_\lambda$ parameterizing triples $([W], [W'], [W'']) \in F_\lambda^3$ such that the three subspaces W, W' and W'' of V_{10} generate only a $V_9 \subset V_{10}$. He proved the following:

Theorem 3.6. *(1) T has dimension 6. It is birationally equivalent via the projection p_{12} to the incidence subvariety $I \subset F_\lambda \times F_\lambda$ defined as the set of couples $([W], [W'])$ such that W and W' generate only a $V_9 \subset V_{10}$.*

(2) I dominates F_λ by the first projection.

(3) T is a Lagrangian subvariety of $F_\lambda \times F_\lambda \times F_\lambda$.

These three facts together say that T is a triangle variety for F_λ.

3.4 Double EPW Sextics

The double EPW sextics X constructed by O'Grady [O'G06] are quasi-étale double covers of sextic hypersurfaces in $\mathbb{P}^5$ singular along a surface discovered by Eisenbud–Pospescu–Walter [EPW01]. We will follow the description given by Iliev–Manivel [IM11], which is very convenient to study subvarieties and relations between 0-cycles of X. More precisely, the Iliev–Manivel description provides a Fano fourfold Y, such that X parameterizes 1-cycles in Y and the $(2,0)$-form on X is induced via the incidence relation from a cohomology class of Hodge type $(3,1)$ in Y. By Proposition 3.3, we cannot obtain enough relations (3.1) in $\mathrm{CH}_0(X)$ to construct, using Mumford's theorem, triangle varieties in X, that is, Lagrangian subvarieties of X^3. In the present case, as exploited in the case of the Fano variety of lines of a cubic fourfold, it suffices to exhibit relations between the corresponding 1-cycles in $\mathrm{CH}_1(Y)$.

The Iliev–Manivel construction is as follows. Let V_5 be a 5-dimensional vector space and let $\mathbb{G} := G(2,5) \subset \mathbb{P}^9$. Let $Y \subset \mathbb{G}$ be the generic complete intersection of a linear Plücker section $H \subset \mathbb{G}$ and a quadratic Plücker section Q of $\mathbb{G}$. The fourfold Y is Fano of index 2 with Picard number 1 and its variety of conics $\mathcal{H}_{2,0}$ is 5-dimensional. It is fibered into $\mathbb{P}^1$'s, because if $C \subset \mathbb{G}$ is a conic, there exists a hyperplane $V_4 \subset V_5$ such that $C \subset G(2, V_4)$. (Indeed, the surface in $\mathbb{P}(V_5)$ swept-out by lines parameterized by C has degree 2, so is contained in some $\mathbb{P}(V_4) \subset \mathbb{P}(V_5)$.) Thus C is contained in the del Pezzo surface $\Sigma = H \cap Q \cap G(2, V_4)$ which has index 1 and degree 4. But then C moves in a pencil in Σ. Next Iliev and Manivel show that Y has a $(3,1)$-form $\eta_Y \in H^{3,1}(Y)$ and considering the incidence diagram given by the universal conic

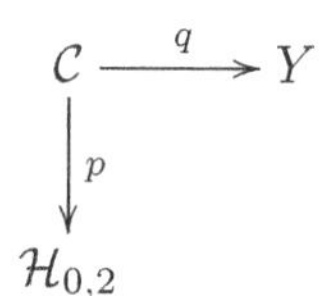

They show that the $(2,0)$-form $p_*q^*\eta$ has generic rank 4 on $\mathcal{H}_{0,2}$. It follows that the base of the MRC fibration of $\mathcal{H}_{0,2}$ is 4-dimensional, with fibers given by the $\mathbb{P}^1$'s described above. Finally, it is shown in [IM11] that this base is birational to a general double EPW sextic X.

This construction is very convenient to exhibit Lagrangian subvarieties in X^l and, for $l = 1$, this is already done in [IM11]. For example, the variety of conics contained in a general hyperplane section $Y' \subset Y$ is 3-dimensional and its image in X is a Lagrangian surface constructed in [IM11]. This follows from the fact that the class η vanishes on Y' and that the pull-back of σ_X to $\mathcal{H}_{0,2}$ is defined as $p_*(q^*\eta)$. We now explain how to use this description of X to produce a triangle variety for X.

First of all, we observe that nondegenerate rational curves of degree 4 on Y are parameterized by a 9-dimensional variety $\mathcal{H}_{0,4}$, while nondegenerate elliptic curves of degree 6 are parameterized by a 12-dimensional variety $\mathcal{H}_{1,6}$. Furthermore, there is a dominant rational map

$$\Phi\colon \mathcal{H}_{1,6} \dashrightarrow \mathcal{H}_{0,4}$$

with general fiber $\mathbb{P}^3$. This map is obtained by liaison. Indeed, a nondegenerate rational curve C of degree 4 on Y has $h^0(C, \mathcal{O}_C(1)) = 5$ and the restriction map $H^0(Y, \mathcal{O}_Y(1)) \to H^0(C, \mathcal{O}_C(1))$ is surjective, hence has a 4-dimensional kernel. As C is general, C is defined in Y by linear Plücker equations. Thus, taking three general equations $\sigma_1, \sigma_2, \sigma_3$ vanishing on C, the locus defined by these three equations is a curve of degree 10 that contains C and is the union of C and an elliptic curve of degree 6. Conversely, starting from a nondegenerate elliptic curve E of degree 6, we have $h^0(E, \mathcal{O}_E(1)) = 6$, and the restriction map $H^0(Y, \mathcal{O}_Y(1)) \to H^0(E, \mathcal{O}_E(1))$ is surjective, hence has a 3-dimensional kernel. The locus defined by this 3-dimensional set of linear Plücker equations is a curve of degree 10 containing E and is in fact the union of E and a residual rational curve of degree 4.

There is a 4-dimensional (or codimension 1) family $\Gamma_4 \subset \mathcal{H}_{0,2}$ of conics in Y (which must be contracted to a surface in X), which is constructed as follows. Consider the variety $Z := H \cap \mathbb{G}$ and its variety of planes $P \subset Z$. The equation defining H is a 2-form $\omega \in \bigwedge^2 V_5^*$. It is well-known that a plane in $\mathbb{G}$ corresponds to a point $x \in \mathbb{P}(V_5)$ together with a $\mathbb{P}(V_4) \subset \mathbb{P}(V_5)$ passing through x and defining the plane P of lines in $\mathbb{P}(V_4)$ passing through x. This plane is contained in Z when V_4 is contained in $x^{\perp\omega}$, which provides the desired 4-dimensional family (parameterized birationally by the choice of $x \in \mathbb{P}(V_5)$). Any such plane P determines a conic $C = P \cap Y$ in Y (or is contained in Y, but this does not happen for generic Y). This provides us with a rational 4-dimensional subvariety

$$\Gamma_4 \subset \mathcal{H}_{0,2}.$$

It is obvious that the subvariety of X we get this way is Lagrangian for σ_X, because it is dominated by the rational variety Γ_4.

We now make the following construction. Inside $\Gamma_4 \times \Gamma_4$, there is a 6-dimensional subvariety Γ_6 consisting of pairs of intersecting conics. We observe that Γ_6 maps naturally to $\mathcal{H}_{0,4}$, via the 2 to 1 map which associates to a pair of intersecting conics the rational curve of degree 4 which is the union of the two conics. This way we get a 6-dimensional variety parameterizing degree 4 rational curves in Y, and applying the residual construction explained previously, we get a 9-dimensional subvariety Γ_9^1 of $\mathcal{H}_{1,6}$.

Let now $\mathcal{T} \subset \mathcal{H}_{0,2} \times \mathcal{H}_{0,2} \times \mathcal{H}_{0,2}$ be the set of triples of conics (C_1, C_2, C_3) in Y, intersecting each other (a triangle of conics), and such that the singular elliptic curve $E = C_1 \cup C_2 \cup C_3$ is a member of the family parameterized by Γ_9^1.

Theorem 3.7. *For general Y, the image T of $\mathcal{T}$ in X^3 is a triangle variety.*

Proof The triples (C_1, C_2, C_3) of conics in Y parameterized by $\mathcal{T}$ have the property that the singular elliptic curve $E = C_1 \cup C_2 \cup C_3 \subset Y$ is residual in Y to a rational curve of degree 4 which is the union of two conics C_4, C_5 meeting at one point, where C_4 and C_5 are cut on Y by planes in Z. All the planes contained in Z are rationally equivalent in Z, so we conclude that the elliptic curves E parameterized by Γ_9^1 are all rationally equivalent. By [IM11], the holomorphic 2-form on X pulls-back to a holomorphic 2-form $\widetilde{\sigma}_X$ on $\mathcal{H}_{0,2}$ which is induced from a cohomology class of type $(3,1)$ on Y via the incidence correspondence. Mumford's theorem [Mum68] implies that $\mathrm{pr}_1^*\widetilde{\sigma}_X + \mathrm{pr}_2^*\widetilde{\sigma}_X + \mathrm{pr}_3^*\widetilde{\sigma}_X$ vanishes on $\mathcal{T}$, hence equivalently that $\mathrm{pr}_1^*\sigma_X + \mathrm{pr}_2^*\sigma_X + \mathrm{pr}_3^*\sigma_X$ vanishes on T. We leave to the reader checking the dimension count for general Y and the fact that T dominates X by the various projections and is generically finite on its image in $X \times X$ by the various projections. □

Remark 3.8. The method described in the next section and the existence of a covering of X by a family of Lagrangian surfaces given in [IM11] can also be used to construct triangle varieties for X, see Theorem 3.11.

3.5 Lagrangian Fibrations and Lagrangian Coverings

Let $\phi\colon X \to B$ be a projective Lagrangian fibration on a hyper-Kähler manifold of dimension $2n$. Recall from Lin [Lin20] that ϕ has a Lagrangian constant cycle multisection $\widetilde{B}$. By base change from B to $\widetilde{B}$, we get (possibly after desingularization) an induced fibration $\widetilde{X} \to \widetilde{B}$ which has a section, hence is (over a dense open set of $\widetilde{B}$) a family of abelian varieties. Let $\widetilde{I} := \widetilde{X} \times_{\widetilde{B}} \widetilde{X} \subset \widetilde{X} \times \widetilde{X}$. Using the relative addition map, we get a rational map $\mu \colon \widetilde{I} \dashrightarrow \widetilde{X}$. Finally, we define T as the image of $\widetilde{I}$ in $X \times X \times X$ under the rational map $(r \circ \mathrm{pr}_1, r \circ \mathrm{pr}_2, r \circ -\mu)$ where $r \colon \widetilde{X} \to X$ is the natural map and the pr_i's are the projections from $\widetilde{X} \times \widetilde{X}$ to $\widetilde{X}$, restricted to $\widetilde{I}$.

Proposition 3.9. *The variety T is a triangle variety.*

Proof As T is the union over $t \in \widetilde{B}$ of the graphs of the sum map in the fibers X_b, it is clear that T dominates X by the three projections and maps in a generically finite way to the products $X \times X$ of any two factors (the image is $X \times_B X$ but the map is not birational because of the base change $\widetilde{B} \to B$). We want to prove that $\mathrm{pr}_1^* \sigma_X + \mathrm{pr}_2^* \sigma_X + \mathrm{pr}_3^* \sigma_X = 0$ on T_{reg}, or, equivalently

$$\mathrm{pr}_1^* \sigma_{\widetilde{X}} + \mathrm{pr}_2^* \sigma_{\widetilde{X}} = \mu^* \sigma_{\widetilde{X}} \tag{3.4}$$

on $\widetilde{I}$, where $\sigma_{\widetilde{X}} := r^* \sigma_X$. As $\tilde{\phi} \colon \widetilde{X}_{\mathrm{reg}} \to \widetilde{B}_{\mathrm{reg}}$ is a Lagrangian fibration with respect to $\sigma_{\widetilde{X}}$, we have

$$\sigma_{\widetilde{X} | \widetilde{X}_{\mathrm{reg}}} \in H^0(\widetilde{X}_{\mathrm{reg}}, F^1 \Omega^2_{\widetilde{X}_{\mathrm{reg}}}), \tag{3.5}$$

where $F^1 \Omega_{\widetilde{X}^2_{\mathrm{reg}}} := \tilde{\phi}^* \Omega_{\widetilde{B}} \wedge \Omega_{\widetilde{X}_{\mathrm{reg}}}$. Let $F^2 \Omega^2_{\widetilde{X}_{\mathrm{reg}}} = \tilde{\phi}^* \Omega^2_{\widetilde{B}}$. The quotient bundle $F^1 \Omega_{\widetilde{X}^2_{\mathrm{reg}}} / F^2 \Omega_{\widetilde{X}^2_{\mathrm{reg}}}$ is isomorphic to $\tilde{\phi}^* \Omega_{\widetilde{B}} \otimes \Omega_{\widetilde{X}_{\mathrm{reg}}/\widetilde{B}}$. We have

$$\mathrm{pr}_1^* \sigma_{\widetilde{X}} + \mathrm{pr}_2^* \sigma_{\widetilde{X}} = \mu^* \sigma_{\widetilde{X}} \text{ in } H^0(\widetilde{I}_{\mathrm{reg}}, F^1 \Omega_{\widetilde{I}^2_{\mathrm{reg}}} / F^2 \Omega_{\widetilde{I}^2_{\mathrm{reg}}}) \tag{3.6}$$

by (3.5) and because on the fibers $\widetilde{X}_b$, we have $\mu^* \alpha = \mathrm{pr}_1^* \alpha + \mathrm{pr}_2^* \alpha$ for any $\alpha \in H^0(\widetilde{X}_b, \Omega_{\widetilde{X}_b})$, so that

$$\mu^* = \mathrm{pr}_1^* + \mathrm{pr}_2^* \colon \mu^* \Omega_{\widetilde{X}_{\mathrm{reg}}/\widetilde{B}} \to \Omega_{\widetilde{I}_{\mathrm{reg}}/\widetilde{B}}.$$

It follows from (3.6) that

$$\mathrm{pr}_1^* \sigma_{\widetilde{X}} + \mathrm{pr}_2^* \sigma_{\widetilde{X}} - \mu^* \sigma_{\widetilde{X}} \in H^0(\widetilde{I}_{\mathrm{reg}}, \tilde{\phi}^* \Omega^2_{\widetilde{B}_{\mathrm{reg}}}) \subset H^{2,0}(\widetilde{I}_{\mathrm{reg}}),$$

which gives an equality of 2-forms on $\widetilde{I}_{\mathrm{reg}}$

$$\mathrm{pr}_1^* \sigma_{\widetilde{X}} + \mathrm{pr}_2^* \sigma_{\widetilde{X}} - \mu^* \sigma_{\widetilde{X}} = \tilde{\phi}^* \eta \tag{3.7}$$

for some $\eta \in H^0(\widetilde{B}, \Omega^2_{\widetilde{B}_{\mathrm{reg}}})$. On the other hand, recall that the multisection $\widetilde{B}$ of ϕ, or 0-section $\widetilde{B}$ of $\tilde{\phi}$, was chosen to be Lagrangian for σ_X (or equivalently $\sigma_{\widetilde{X}}$). Restricting (3.7) to the 0-section $\widetilde{B}$, we then conclude that $\eta = 0$, which proves (3.4). □

Let us say that a hyper-Kähler manifold X has a Lagrangian covering if there exists a diagram

$$\begin{array}{ccc} \mathcal{L} & \xrightarrow{\Phi} & X \\ \downarrow{\scriptstyle \pi} & & \\ B & & \end{array}$$

where $\mathcal{L}$ and B are smooth projective varieties, the morphism Φ is surjective and maps birationally the general fiber L_t, $t \in B$, of π to a (possibly singular) Lagrangian subvariety of X, and furthermore, the following condition holds. As $\Phi(L_t)$ is Lagrangian, one has a natural morphism of coherent sheaves

$$\lrcorner\sigma_X \colon N_{L_t/X} \to \Omega_{L_t}$$

which is a generic isomorphism, and induces a morphism at the level of global sections

$$H^0(L_t, N_{L_t/X}) \to H^0(L_t, \Omega_{L_t}).$$

We ask that for $t \in B$ generic, the composite map

$$T_{B,t} \to H^0(L_t, N_{L_t/X}) \stackrel{\lrcorner\sigma_X}{\to} H^0(L_t, \Omega_{L_t})$$

where the first map is the classifying map, is an isomorphism. In particular $\dim B = h^{1,0}(L_t) =: g$. This condition is satisfied by unobstructedness results for deformations of Lagrangian submanifolds (see [Voi92]) if, for general $t \in B$, the fiber L_t is isomorphic via Φ to a smooth Lagrangian subvariety of X, and $\mathcal{L} \to B$ identifies near t to the universal family of deformations of L_t in X. For singular Lagrangian subvarieties, the deformation theory is not well understood. Note that, with the hypotheses above, the surjectivity of Φ has the following interpretation.

Lemma 3.10. *The surjectivity of Φ is equivalent to the fact that the Albanese map $\mathrm{alb}_{L_t} \colon L_t \to \mathrm{Alb}\, L_t$ is generically finite on its image for general t.*

Proof The second property is equivalent to the fact that, for general $t \in B$, alb_{L_t} has a generically injective differential, or equivalently, that the evaluation map

$$\mathrm{ev} \colon H^0(L_t, \Omega_{L_t}) \otimes \mathcal{O}_{L_t} \to \Omega_{L_t}$$

is generically surjective on L_t. The surjectivity of Φ is equivalent to the fact that Φ is submersive generically along L_t for general t. As L_t imbeds generically into X, this is well-known to be equivalent to the fact that the map

$$\mathrm{ev}\colon T_{B,t} \otimes \mathcal{O}_{L_t} \to N_{L_t/X}, \tag{3.8}$$

which is the composition of the evaluation map and of the classifying map $T_{B,t} \to H^0(L_t, N_{L_t/X})$ is generically surjective. We now use the map

$$\lrcorner\sigma_X\colon N_{L_t/X} \to \Omega_{L_t}$$

which is a generic isomorphism on L_t and induces a morphism at the level of global sections which composed with the classifying map makes the following diagram commutative

$$\begin{array}{ccc} T_{B,t} \otimes \mathcal{O}_{L_t} & \xrightarrow{\mathrm{ev}} & N_{L_t/X} \\ \downarrow{\scriptstyle \lrcorner\sigma_X} & & \downarrow{\scriptstyle \lrcorner\sigma_X} \\ H^0(L_t, \Omega_{L_t}) \otimes \mathcal{O}_{L_t} & \xrightarrow{\mathrm{ev}} & \Omega_{L_t} \end{array}$$

As the first vertical map is by assumption an isomorphism, it follows that the generic surjectivity of the evaluation map (3.8) is equivalent to the generic surjectivity of the evaluation map (3.8) □

We show the following variant of Proposition 3.9.

Theorem 3.11. *Let X be a projective hyper-Kähler manifold admitting a Lagrangian covering $\Phi\colon \mathcal{L} \to X$. Assume that the general fibers L_t have the property that the sum map $L_t \times L_t \to \operatorname{Alb} L_t$ is surjective (in particular $2n \geq g$). Assume that there exists a Lagrangian subvariety $K \subset X$ such that the general fiber $\Phi(L_t)$ intersects K in a finite (nonzero) number of points. Then X admits a triangle variety.*

Note that by the same arguments as above, the assumption on K will be satisfied by taking $K = L_s$, for general s, assuming that the fibers $L_t \subset X$ are smooth Lagrangian, and a general form $\alpha \in H^0(L_t, \Omega_{L_t})$ has finitely many zeroes.

Proof of Theorem 3.11 Consider, over the open set B_{reg} of regular values of π, the Albanese fibration $\mathcal{A} \to B_{\mathrm{reg}}$ with fiber $\operatorname{Alb} L_b$ over $b \in B$. By assumption, a general variety $\Phi(L_b) \subset X$ intersects K in finitely many points, which provides a generically finite cover

$$B_K = \Phi^{-1}(K) \to B_{\mathrm{reg}}$$

parameterizing the pairs (b, k), where $k \in L_b$ is such that $\Phi(k) \in K$. Denoting by $\mathcal{L}_K$ the fibered product $\mathcal{L} \times_B B_K$, there is a natural section

$$\sigma \colon B_K \to \mathcal{L}_K = \mathcal{L} \times_B B_K,$$

$$(b, k) \mapsto k.$$

We denote by $\mathrm{alb}_K \colon \mathcal{L}_K \to \mathcal{A}_K = \mathcal{A} \times_{B_{\mathrm{reg}}} B_K$ the relative Albanese map defined by the section σ, so that

$$\mathrm{alb}_K(x) = \mathrm{alb}_{L_b}(x - \sigma(b)),$$

where $b = \pi(x) \in B_K$. Let now, for any integer $N \neq 0$, $\widetilde{T}_N \subset \mathcal{L}_K \times_{B_K} \mathcal{L}_K \times_{B_K} \mathcal{L}_K$ be defined as

$$\begin{aligned} \widetilde{T}_N := \{(x, y, z) \in &\mathcal{L}_K \times_{B_K} \mathcal{L}_K \times_{B_K} \mathcal{L}_K, \\ &N(\mathrm{alb}_K(x) + \mathrm{alb}_K(y) + \mathrm{alb}_K(z)) = 0 \\ &\qquad \text{in } \mathrm{Alb}\, L_b,\ b := \pi(x) = \pi(y) = \pi(z)\}. \end{aligned} \tag{3.9}$$

The variety $\mathcal{L}_K$ has a morphism $\Phi_K \colon \mathcal{L}_K \to X$ composed from Φ and the natural map $\mathcal{L}_K \to \mathcal{L}$. We define T_N as the Zariski closure in $X \times X \times X$ of $(\Phi_K, \Phi_K, \Phi_K)(\widetilde{T}_N^0)$, where $\widetilde{T}_N^0$ is the union of the irreducible components of $\widetilde{T}_N$ dominating B_K, and where the point $\mathrm{alb}_K(x) + \mathrm{alb}_K(y) + \mathrm{alb}_K(z)$ is of order exactly N. It remains to show that T_N has the required properties for large N. First of all, the proof given for Proposition 3.9 works as well to show

Lemma 3.12. *One has* $(\mathrm{pr}_1^*\sigma_X + \mathrm{pr}_2^*\sigma_X + \mathrm{pr}_3^*\sigma_X)_{|T_{N,\mathrm{reg}}} = 0$. *Equivalently,* $(\mathrm{pr}_1^*(\Phi_K^*\sigma_X) + \mathrm{pr}_2^*(\Phi_K^*\sigma_X) + \mathrm{pr}_3^*(\Phi_K^*\sigma_X))_{|\widetilde{T}_{N,\mathrm{reg}}} = 0$.

We next observe that, if X is of dimension $2n$, $\widetilde{T}_N$ has expected dimension $3n$, which is the dimension of a triangle variety. Indeed, let $g := \dim B = \dim \mathrm{Alb}\, L_b$. Then, as $\dim L_b = n$,

$$\dim \mathcal{L}_K \times_{B_K} \mathcal{L}_K \times_{B_K} \mathcal{L}_K = g + 3n,$$

while from (3.9), we see that $\widetilde{T}_N^0$ is the inverse image of the N-torsion multisection of $\mathcal{A}_K \to B_K$ via the sum morphism $\mathrm{alb}_K \circ \mathrm{pr}_1 + \mathrm{alb}_K \circ \mathrm{pr}_2 + \mathrm{alb}_K \circ \mathrm{pr}_3$, over the regular locus B_K^0 of $\mathcal{L}_K \to B_K$. Hence the expected codimension of $\widetilde{T}_N$ is g and the expected dimension of $\widetilde{T}_N$ is $3n$. The proof of the theorem concludes with

Lemma 3.13. *Under the assumptions of the theorem, $\widetilde{T}_N^0$ is actually of dimension $3n$, the projections $\widetilde{T}_N^0 \to \mathcal{L}_K$ are dominant and the projections $\widetilde{T}_N^0 \to \mathcal{L}_K \times_{B_K} \mathcal{L}_K$ are generically finite on their images.*

Proof By assumption, the sum map $L_t \times L_t \to \operatorname{Alb} L_t$ is surjective for general t, while by Lemma 3.10, the Albanese map $L_t \to \operatorname{Alb} L_t$ is generically finite on its image. (Here the Albanese map of L_t is computed using one of the finitely many points of $L_t \cap K$, in other words, t is taken in B_K rather than B.) This implies that for general $x \in L_t$, there is a solution to the equation

$$N(\mathrm{alb}_{L_t} x + \mathrm{alb}_{L_t} y + \mathrm{alb}_{L_t} z) = 0, \tag{3.10}$$

with $y, z \in L_t$. This is saying that the three projections $\widetilde{T} \to \mathcal{L}_K$ are surjective. Finally, using (3.10), we find that the projections $\widetilde{T}_N \to \mathcal{L}_K \times_{B_K} \mathcal{L}_K$ are generically finite on their image because the Albanese map of L_t is generically finite on its image by Lemma 3.10. □

It remains to see that the same properties hold for $T_N \subset X \times X \times X \times X$. This follows from the following lemma which is proved exactly as Lemma 3.10.

Lemma 3.14. *The assumptions that the sum map $L_t \times L_t \to \operatorname{Alb} L_t$ is surjective is equivalent to the fact that the natural map*

$$(\Phi, \Phi) \colon \mathcal{L}_K \times_{B_K} \mathcal{L}_K \to X \times X$$

is generically finite on its image.

As Φ is surjective, the fact that the projections $\mathrm{pr}_i \colon \widetilde{T}_N^0 \to \mathcal{L}_K$ are dominant for $i = 1, 2, 3$ implies the same property for the projections $\mathrm{pr}_i \colon T_N \to X$. As $(\Phi, \Phi) \colon \mathcal{L}_K \times_{B_K} \mathcal{L}_K \to X \times X$ is generically finite on its image, the fact that the projections $\mathrm{pr}_{ij} \colon \widetilde{T}_N^0 \to \mathcal{L}_K \times_{B_K} \mathcal{L}_K$ are dominant for $i = 1, 2, 3$ does not necessarily imply the same property for the projections $\mathrm{pr}_{ij} \colon T_N \to X \times X$, but it will imply it if N is large, using the Zariski density of torsion points. □

Example 3.15. In the case of the variety of lines $X = F_1(Y)$ of a smooth cubic fourfold, we get by applying Theorem 3.11 constructions of triangle varieties for X, different from the one constructed in Section 3.2, by using its Lagrangian covering by Fano surfaces $S_H := F_1(Y_H)$, or rather their desingularizations, where $Y_H \subset Y$ is a singular (generically 1-nodal) hyperplane section $Y \cap H$ of Y, and S_H is its surface of lines. The construction depends on the choice of a Lagrangian surface $K \subset X$.

A similar construction can be done for the double covers of EPW sextics, using again the constructions of Iliev and Manivel.

4 Surface Decompositions from Triangle Varieties

Let X be a smooth projective variety of dimension $2n$ and $\sigma_X \in H^{2,0}(X)$ a holomorphic 2-form on X. First of all, note that from a triangle variety $T \subset X \times X \times X$, we can construct for each $k \geq 3$ a subvariety T_k of X^k of dimension kn satisfying the following property: the holomorphic 2-form $\sum_{i=1}^{k} \epsilon_i \mathrm{pr}_i^* \sigma_X$ vanishes on T_k, with $\epsilon_i = \pm 1$. The k-angle variety T_k is defined inductively by composition in the sense of correspondences. For $k = 4$, let

$$T'_4 = \mathrm{pr}_{1245*}\left(\mathrm{pr}_{123}^{-1}(T) \cap \mathrm{pr}_{345}^{-1}(T)\right) \subset X^4,$$

where the projections are defined on X^5, pr_{1245} takes value in X^4 and pr_{123}, pr_{345} take value in X^3. On $\mathrm{pr}_{123}^{-1}(T)$, one has $\mathrm{pr}_1^*\sigma_X + \mathrm{pr}_2^*\sigma_X + \mathrm{pr}_3^*\sigma_X = 0$ and on $\mathrm{pr}_{345}^{-1}(T)$, one has $\mathrm{pr}_3^*\sigma_X + \mathrm{pr}_4^*\sigma_X + \mathrm{pr}_5^*\sigma_X = 0$ so that, by subtracting, one has on the regular locus of $\mathrm{pr}_{123}^{-1}(T) \cap \mathrm{pr}_{345}^{-1}(T)$, hence also on $T'_{4,\mathrm{reg}}$:

$$\mathrm{pr}_1^*\sigma_X + \mathrm{pr}_2^*\sigma_X - \mathrm{pr}_4^*\sigma_X - \mathrm{pr}_5^*\sigma_X = 0, \tag{4.1}$$

where now the projections are defined on X^4 with factors indexed by $1, 2, 4, 5$. As T dominates X by the projections, the variety T'_4 so defined also dominates X by the various projections. As the fibers of the projection $T \to X$ have dimension n, T'_4 has at least one component which is of dimension $\geq 4n$. We take for T_4 the union of the irreducible components of dimension $4n$ of T_4. Note that, if X is hyper-Kähler, the 2-form σ_X is everywhere nondegenerate, so T_4 does not have components of dimension $> 4n$, because we already know by (4.1) that the components are Lagrangian for the holomorphic symplectic form $\mathrm{pr}_1^*\sigma_X + \mathrm{pr}_2^*\sigma_X - \mathrm{pr}_4^*\sigma_X - \mathrm{pr}_5^*\sigma_X$ on X^4. The variety T_k is similarly defined inductively by composing T_{k-1} and T.

Recall that for X as above, an algebraically coisotropic subvariety $Z \subset X$ of dimension $n+1$ admits a rational map $\tau \colon Z \dashrightarrow \Sigma$, where Σ is a surface and, denoting $\tau_{\mathrm{reg}} := \tau_{|Z_{\mathrm{reg}}}$,

$$\sigma_{X|Z_{\mathrm{reg}}} = \tau_{\mathrm{reg}}^* \sigma_\Sigma$$

for some holomorphic 2-form σ_Σ on Σ.

Theorem 4.1. *Let X be a projective hyper-Kähler variety of dimension $2n$. Assume X has a triangle variety $T \subset X^3$ and an algebraically coisotropic subvariety $\tau \colon Z \dashrightarrow \Sigma$ of dimension $n+1$. Let $F \subset X$ be the general fiber of τ. Then if the intersection of $F^n \subset X^n$ and $\mathrm{pr}_{1\ldots n}(T_{n+1}) \subset X^n$ satisfies*

$$F^n \cdot \mathrm{pr}_{1\dots n}(T_{n+1}) \neq 0, \tag{4.2}$$

X admits a surface decomposition. In particular it admits mobile algebraically coisotropic subvarieties of any codimension $\leq n$.

In (4.2), we have $\dim F = n-1$, and so $\dim F^n = n(n-1)$ and $\dim T_{n+1} = n(n+1)$, while the intersection takes place in X^n which has dimension $2n^2 = n(n-1) + n(n+1)$.

Proof We construct $\phi\colon \Gamma_0 \to X$, $\psi\colon \Gamma_0 \to \Sigma^n$ by the formulas

$$\Gamma_0 = \mathrm{pr}_{1\dots n}^{-1}(Z^n) \cap T_{n+1} \subset T_{n+1} \subset X^{n+1},$$
$$\phi := \mathrm{pr}_{n+1}\colon \Gamma_0 \to X, \qquad \psi = \tau^n \circ \mathrm{pr}_{1\dots n}\colon \Gamma_0 \to \Sigma^n.$$

As $\Gamma_0 \subset T_{n+1}$, the form $\sum_{i=1}^{n+1} \epsilon_i \mathrm{pr}_i^* \sigma_X$ vanishes on $\Gamma_{0,\mathrm{reg}}$, where the ϵ_i are the signs introduced in the construction of T_{n+1}. In other words, using $\phi = \mathrm{pr}_{n+1}$

$$\phi^* \sigma_X = \sum_{i=1}^{n} \epsilon_i' \mathrm{pr}_i^* \sigma_{X|\Gamma_{0,\mathrm{reg}}}$$

in $H^0(\Omega^2_{\Gamma_{0,\mathrm{reg}}})$, where $\epsilon_i' = \pm\epsilon_i$. We next use the fact that $\mathrm{pr}_i(\Gamma) \subset Z$ and that $\sigma_{X|Z} = \tau^*(\sigma_\Sigma)$. We then get the desired formula characterizing a surface decomposition

$$\phi^* \sigma_X = \psi^* \left(\sum_{i=1}^{n} \epsilon_i' \mathrm{pr}_i^* \sigma_\Sigma \right) \text{ in } H^{2,0}(\Gamma_{0,\mathrm{reg}}). \tag{4.3}$$

We need to show that ϕ and ψ are dominant, and that we can assume that they are generically finite. The fact that ψ is dominant is a consequence of (4.2), which can be seen as saying that $\mathrm{pr}_{1\dots n}^{-1}(Z^n) \cap T_{n+1}$ intersects nontrivially the fibers of $\tau^n\colon \mathrm{pr}_{1\dots n}^{-1}(Z^n) \to \Sigma^n$. Knowing that ψ is dominant, we conclude that the form $\psi^*(\sum_{i=1}^n \epsilon_i' \mathrm{pr}_i^* \sigma_\Sigma)$ has generic rank $2n$ on $\Gamma_{0,\mathrm{reg}}$. It thus follows from (4.3) that $\phi^*\sigma_X$ has generic rank $2n$ on Γ, hence that ϕ is also dominant. The last argument applies to any irreducible component Γ_0' of Γ_0 dominating Σ^n, which thus also has to dominate X. Finally, by cutting Γ_0' by hyperplane sections and reapplying the same arguments if necessary, we get a Γ which is generically finite onto both Σ^n and X, and still satisfies (4.3). □

We also have the following result, whose proof is a variant of that of Theorem 4.1, and shows how to construct new algebraically coisotropic subvarieties out of old ones, using a triangle variety:

Let X be smooth projective variety of dimension $2n$ with an everywhere nondegenerate holomorphic 2-form σ_X. Denote by $I_r \subset X$ an algebraically coisotropic subvariety of X of codimension $n-r$. Hence there exists a rational map

$$\phi_r \colon I_r \dashrightarrow B_r$$

to a smooth projective variety B_r of dimension $2r$, with general fiber F_r of dimension $n-r$, such that

$$\sigma_{X|I_r} = \phi_r^* \sigma_{B_r}, \tag{4.4}$$

for some holomorphic 2-form σ_{B_r} on B_r which is generically of maximal rank $2r = \dim B_r$.

Theorem 4.2. *Assume that X has a triangle variety T relative to σ_X and let I_r, $I_{r'}$ be two algebraically coisotropic varieties of X of respective codimensions r, r'. Assume that*

($*$) *the class* $\mathrm{pr}_{3*}([T] \cup \mathrm{pr}_{12}^*([I_r \times I_{r'}]))$ *is nonzero in* $H^{2n-2r-2r'}(X, \mathbb{Q})$ *(so in particular $r + r' \leq n$).*

Then $\mathrm{pr}_3(T \cap \mathrm{pr}_{12}^{-1}(I_r \times I_{r'})) \subset X$ *contains an algebraically coisotropic subvariety $I_{r+r'}$ of codimension $n - r - r'$.*

As usual, pr_i and pr_{ij} denote the projections from $X \times X \times X$ to its factors, or products of two factors.

Proof The variety $Y := T \cap \mathrm{pr}_{12}^{-1}(I_r \times I_{r'}) \subset X \times X \times X$ maps to $B_r \times B_{r'}$ by the map $\phi_{r+r'} := (\phi_r, \phi_{r'}) \circ \mathrm{pr}_{12|Y}$. By the definition of a triangle variety and using (4.4), we get that

$$\mathrm{pr}_3^* \sigma_{X|Y} = -\phi_{r+r'}^* \left(\mathrm{pr}_1^* \sigma_{B_r} + \mathrm{pr}_2^* \sigma_{B_{r'}}\right) \text{ in } H^{2,0}(Y_{\mathrm{reg}}).$$

It thus follows that the rank of $\mathrm{pr}_3^* \sigma_X$ restricted to Y_{reg} is nowhere greater than $2r + 2r'$. On the other hand, Condition ($*$) implies that $\mathrm{pr}_3(Y)$ has at least one component of dimension $\geq n + r + r'$. This component thus must have dimension exactly $n + r + r'$ and is coisotropic. This is the desired variety $I_{r+r'}$, and it is in fact algebraically coisotropic, choosing a subvariety $Y' \subset Y$ mapping to $I_{r+r'}$ in a generically finite way and using the diagram

$$\begin{array}{ccc} Y' & \xrightarrow{\mathrm{pr}_3} & I_{r+r'} \\ \Big\downarrow{\scriptstyle (\phi_r, \phi_{r'}) \circ \mathrm{pr}_{12}} & & \\ B_r \times B_{r'} & & \end{array}$$

in which $\mathrm{pr}_3^*\sigma_{X|Y'} = \phi_r^*\sigma_{B_r} + \phi_{r'}^*\sigma_{B_{r'}}$. □

As a consequence of Theorem 4.1 (or using methods similar as above), we conclude now that many explicitly constructed projective hyper-Kähler manifolds admit a surface decomposition:

Theorem 4.3. *The following hyper-Kähler manifolds admit surface decompositions:*

(1) The Fano variety of lines $X = F_1(Y)$ of a cubic fourfold Y ([BD85]).
(2) The Debarre–Voisin hyper-Kähler fourfold ([DV10]).
(3) The double EPW sextics ([O'G06]).
(4) The LLSvS hyper-Kähler 8-fold ([LLSvS17]).
(5) The LSV compactification of the intermediate Jacobian fibration associated with a cubic fourfold ([LSV17]).

Proof of cases 1 *and* 2 The case of the Beauville–Donagi hyper-Kähler fourfold $X = F_1(Y)$ is done as follows: recall first that $F_1(Y)$ has an ample (singular) uniruled divisor D which can be constructed using the rational self-map of degree 16

$$\phi\colon X \dashrightarrow X$$

constructed in [Voi04]. This map associates to a general point $[l]$ parameterizing a line $l \subset Y$ the point $[l']$ parameterizing the line $l' \subset Y$ such that there is a unique plane $P \subset \mathbb{P}^5$ with $P \cap Y = 2l + l'$. It satisfies the property that

$$\phi^*\sigma_X = -2\sigma_X.$$

This map has indeterminacies when the plane P is not unique and this happens along a surface Σ which is studied in [Ame09]. After blowing-up Σ, the map ϕ becomes a morphism $\tilde{\phi}\colon \widetilde{F_1(Y)} \to F_1(Y)$ which is finite (see [Ame09]). The image of the exceptional divisor E under $\tilde{\phi}$ is thus a uniruled divisor E' in $F_1(Y)$ which has in fact normalization isomorphic to E. We thus have a diagram

$$\begin{array}{ccc} E & \xrightarrow{\tilde{\phi}_E} & E' \subset X = F_1(Y) \\ \downarrow{\scriptstyle \tau_E} & & \\ \Sigma & & \end{array} \tag{4.5}$$

where τ_E is the restriction of the blowing-up morphism to E, such that $\tilde{\phi}_E^*\sigma_X = \tau_E^*\sigma_\Sigma$ for some holomorphic 2-form σ_Σ on Σ.

On the other hand, we have the triangle variety $T \subset X \times X \times X$ described in Section 3.2. We thus have the ingredients needed to apply Theorem 4.1, but we have to check the condition (4.2). This is easy because $\mathrm{pr}_{12}(T) \subset X \times X$ has codimension 2, and the classes of the fibers F of $\tau_E \colon E' \to \Sigma$ must be proportional to h^3, where h is the first Chern class of an ample line bundle on X, because $\rho(X) = 1$. Hence the intersection number $(F \times F) \cdot \mathrm{pr}_{12}(T)$ is nonnegative. We claim that the intersection number is strictly positive. If the intersection number is 0, then for a general complete intersection curve $C \subset X$, the 3-fold $(C \times X) \cap \mathrm{pr}_{12}(T)$ does not map to a 3-fold by the second projection to X. This implies that $\mathrm{pr}_{12}(T)$ contains $C \times \Sigma$ for some surface $\Sigma \subset X$. Recalling that $X = F_1(Y)$ and that $\mathrm{pr}_{12}(T)$ is the set of pairs of intersecting lines in Y, we can easily contradict this conclusion, proving the claim. We thus get a surface decomposition given by

$$\Gamma = \mathrm{pr}_{12}^{-1}(E \times E) \cap T \subset X^3,$$
$$\psi = (\tau_E, \tau_E) \circ \mathrm{pr}_{12} \colon \Gamma \to \Sigma \times \Sigma, \qquad \phi = \mathrm{pr}_{3|\Gamma} \colon \Gamma \to X. \tag{4.6}$$

The proof in the case *2* works similarly. We use on the one hand the triangle variety T constructed by Bazhov (see [Baz19] or Section 3.3), and on the other hand the existence of a uniruled divisor $\tau \colon D \to \Sigma$, $D \to X$ that we can exhibit either by looking at the indeterminacies of Bazhov's construction, or by applying [CMP], using the fact that the Debarre–Voisin fourfold has the deformation type of $K3^{[2]}$. As the very general Debarre–Voisin varieties have Picard number 1, the fibers of a uniruled divisors have as before a class proportional to h^3, where h is an ample divisor class, hence the variety

$$\mathrm{pr}_{12}^{-1}(D \times D) \subset T \subset X \times X \times X$$

dominates $\Sigma \times \Sigma$ by $(\tau, \tau) \circ \mathrm{pr}_{12}$, hence X by the projection pr_3. The rest of the argument is identical. □

Proof of cases 4 *and* 5 The LLSvS manifold $F_3'(Y)$ is a hyper-Kähler 8-fold contructed in [LLSvS17] as a smooth hyper-Kähler model of the basis of the rationally connected fibration of the Hilbert scheme $F_3(Y)$ of degree 3 rational curves in a smooth cubic fourfold X not containing a plane. In the paper [Voi16], we constructed a dominating rational map

$$\mu \colon F_1(Y) \times F_1(Y) \dashrightarrow F_3'(Y).$$

Two non intersecting lines l, l' in Y generate a $\mathbb{P}^3$ which intersects Y along a cubic surface S. The $\mathbb{P}^2$ of degree 3 rational curves corresponding to $\mu(l,l')$ is the linear system $|h+l-l'|$ on S, where $h = \mathcal{O}_S(1)$. It follows from this formula, Mumford's theorem and the fact that the holomorphic 2-forms on the considered varieties come from a $(3,1)$-class on Y by the corresponding incidence correspondences, that

$$\mu^*\sigma_{F_3'(Y)} = \mathrm{pr}_1^*\sigma_{F_1(Y)} - \mathrm{pr}_2^*\sigma_{F_1(Y)}. \tag{4.7}$$

Together with Case *1*, this gives us a surface decomposition for $F_3'(Y)$. Indeed, we have the surface decomposition $\phi\colon \Gamma \to F_1(Y)$, $\psi\colon \Gamma \to \Sigma\times\Sigma$ for $F_1(Y)$ of (4.6). The maps satisfy

$$\phi^*\sigma_{F_1(Y)} = \psi^*\left(\mathrm{pr}_1^*\sigma_\Sigma + \mathrm{pr}_2^*\sigma_\Sigma\right). \tag{4.8}$$

Taking products, we get

$$\phi'\colon \Gamma\times\Gamma \to F_1(Y)\times F_1(Y), \qquad \psi'\colon \Gamma\times\Gamma \to \Sigma\times\Sigma\times\Sigma\times\Sigma.$$

Composing the first map with μ and desingularizing, we get

$$\phi''\colon \widetilde{\Gamma\times\Gamma} \to F_3(Y), \qquad \psi''\colon \widetilde{\Gamma\times\Gamma} \to \Sigma\times\Sigma\times\Sigma\times\Sigma. \tag{4.9}$$

By (4.8) and (4.7), the morphisms in (4.9) satisfy

$$\phi''^*\sigma_{F_3(Y)} = \psi''^*\left(\mathrm{pr}_1^*\sigma_\Sigma + \mathrm{pr}_2^*\sigma_\Sigma - \mathrm{pr}_3^*\sigma_\Sigma - \mathrm{pr}_4^*\sigma_\Sigma\right),$$

which gives the desired decomposition in case *4*.

We now turn to the LSV hyper-Kähler fourfold $J(Y)$, a 10-dimensional hyper-Kähler manifold associated to a general cubic fourfold Y. As it has a Lagrangian fibration, we will be able to use the triangle variety described in Section 3.5. Another ingredient we will use is the following:

Lemma 4.4. *There exists a codimension* 3 *algebraically coisotropic subvariety of* $J(Y)$ *which is birational to a* $\mathbb{P}^3$*-bundle over* $F_1(Y)$.

Proof For each line $l \subset Y \subset \mathbb{P}^5$, there is a $\mathbb{P}^3 \subset (\mathbb{P}^5)^*$ of hyperplane sections of Y containing l. This determines a $\mathbb{P}^3$-bundle $P \to F_1(Y)$. Each of these hyperplanes H determines a hyperplane section Y_H of Y. Then $l \subset Y_H$ and the 1-cycle $3l - h^2$ is homologous to 0 on Y_H, at least for Y_H smooth, which allows us to define a point $\Psi_{Y_H}(3l-h^2) \in J(Y_H)$ using the Abel–Jacobi map Ψ_{Y_H} of Y_H. As $J(Y)$ is fibered over $(\mathbb{P}^5)^*$ into intermediate Jacobians $J(Y_H)$, we thus constructed the desired rational map $P \dashrightarrow J(Y)$. It is not hard to see that this map is birational onto its image P' which thus has dimension 7. Since the only holomorphic 2-forms on P are those coming from $F_1(Y)$, we conclude that $P' \subset J(Y)$ is algebraically coisotropic. □

Corollary 4.5. *There exists an algebraically coisotropic subvariety Z of $J(Y)$ which has codimension 4 (dimension 6). This variety dominates $(\mathbb{P}^5)^*$.*

Proof We use for this the existence of the uniruled divisor $E' \subset F_1(Y)$ appearing in (4.5). Let $P_{E'}$ be the inverse image of E' in P and let Z be its image in $J(Y)$. We have to prove that Z dominates $(\mathbb{P}^5)^*$. This is saying that any hyperplane section Y_H of Y contains a line residual to a special line of Y, which is implied by the fact that no hyperplane section of Y contains a 2-dimensional family of lines residual to a special line of Y. The last statement is proved in [LSV17]. □

We now construct a surface decomposition for $J(Y)$: Let Z be as in Corollary 4.5, so Z is fibered into curves over B. We use the sum map on the fibers of the fibration $J(Y) \to B := (\mathbb{P}^5)^*$. We then get a morphism:

$$\mu_{Z,5}\colon Z \times_B \ldots \times_B Z \dashrightarrow J(Y)$$
$$(a_1, \ldots, a_5) \mapsto \sum_i a_i.$$

We first observe that μ_5 is dominant: the fibers of $J(Y) \to B$ are generically irreducible abelian varieties and the fibers of $Z \to B$ are curves Z_H which must generate $J(Y_H)$, which is 5-dimensional. Finally, it remains to prove that the construction above provides a surface decomposition for $J(Y)$. First of all, by Proposition 3.9, for the relative sum map

$$\mu_5\colon J(Y) \times_B \ldots \times_B J(Y) \dashrightarrow J(Y)$$
$$(a_1, \ldots, a_5) \mapsto \sum_i a_i,$$

one has

$$\mu_5^*(\sigma_{J(Y)}) = \sum_i \mathrm{pr}_i^* \sigma_{J(Y)}. \tag{4.10}$$

Next we use the rational map $f\colon Z \to \Sigma$ which is the composition of $Z \to E' \subset F_1(Y)$, with E' birational to E, and $\tau_E\colon E \to \Sigma$. We clearly have

$$f^*\sigma_\Sigma = \sigma_{J|Z}. \tag{4.11}$$

Next f induces a morphism

$$f_5\colon Z \times_B \ldots \times_B Z \to \Sigma^5$$

and combining (4.10) and (4.11), one concludes that the diagram

$$\begin{array}{ccc} Z\times_B\ldots\times_B Z & \xrightarrow{\mu_5} & J(Y) \\ \downarrow{\scriptstyle f_5} & & \\ \Sigma^5 & & \end{array}$$

provides a surface decomposition of $J(Y)$. □

References

[AC17] Ekaterina Amerik and Frédéric Campana. Characteristic foliation on non-uniruled smooth divisors on hyperkähler manifolds. *J. Lond. Math. Soc. (2)*, 95(1):115–127, 2017.

[Ame09] Ekaterina Amerik. A computation of invariants of a rational self-map. *Ann. Fac. Sci. Toulouse Math. (6)*, 18(3):445–457, 2009.

[Baz19] Ivan Bazhov. On the variety of triangles for a hyper-Kähler fourfold constructed by Debarre and Voisin. *J. Pure Appl. Algebra*, 223(6):2530–2542, 2019.

[BD85] Arnaud Beauville and Ron Donagi. La variété des droites d'une hypersurface cubique de dimension 4. *C. R. Acad. Sci. Paris Sér. I Math.*, 301(14):703–706, 1985.

[Bea83] Arnaud Beauville. Variétés Kähleriennes dont la première classe de Chern est nulle. *J. Differential Geom.*, 18(4):755–782 (1984), 1983.

[Bea07] Arnaud Beauville. On the splitting of the Bloch-Beilinson filtration. *Algebraic cycles and motives. Vol. 2*, volume 344 of *London Math. Soc. Lecture Note Ser.*, pages 38–53. Cambridge Univ. Press, Cambridge, 2007.

[Bog96] F. A. Bogomolov. On the cohomology ring of a simple hyper-Kähler manifold (on the results of Verbitsky). *Geom. Funct. Anal.*, 6(4):612–618, 1996.

[BV04] Arnaud Beauville and Claire Voisin. On the Chow ring of a $K3$ surface. *J. Algebraic Geom.*, 13(3):417–426, 2004.

[CMP] F. Charles, G. Mongardi and G. Pacienza. Families of rational curves on holomorphic symplectic varieties and applications to 0-cycles. A corrected version of arXiv:1401.4071.

[Del71] Pierre Deligne. Théorie de Hodge. II. *Inst. Hautes Études Sci. Publ. Math.*, (40):5–57, 1971.

[DV10] Olivier Debarre and Claire Voisin. Hyper-Kähler fourfolds and Grassmann geometry. *J. Reine Angew. Math.*, 649:63–87, 2010.

[EPW01] David Eisenbud, Sorin Popescu and Charles Walter. Lagrangian subbundles and codimension 3 subcanonical subschemes. *Duke Math. J.*, 107(3):427–467, 2001.

[Fuj87] Akira Fujiki. On the de Rham cohomology group of a compact Kähler symplectic manifold. In *Algebraic geometry, Sendai, 1985*, volume 10 of *Adv. Stud. Pure Math.*, pages 105–165. North-Holland, Amsterdam, 1987.

[Huy14] D. Huybrechts. Curves and cycles on K3 surfaces. *Algebr. Geom.*, 1(1):69–106, 2014. With an appendix by C. Voisin.

[IM11] Atanas Iliev and Laurent Manivel. Fano manifolds of degree ten and EPW sextics. *Ann. Sci. Éc. Norm. Supér. (4)*, 44(3):393–426, 2011.

[KMM92] János Kollár, Yoichi Miyaoka and Shigefumi Mori. Rationally connected varieties. *J. Algebraic Geom.*, 1(3):429–448, 1992.

[Lin20] Hsueh-Yung Lin. Lagrangian constant cycle subvarieties in Lagrangian fibrations. *Int. Math. Res. Not. IMRN*, (1):14–24, 2020.

[LLSvS17] Christian Lehn, Manfred Lehn, Christoph Sorger and Duco van Straten. Twisted cubics on cubic fourfolds. *J. Reine Angew. Math.*, 731:87–128, 2017.

[LSV17] Radu Laza, Giulia Saccà and Claire Voisin. A hyper-Kähler compactification of the intermediate Jacobian fibration associated with a cubic 4-fold. *Acta Math.*, 218(1):55–135, 2017.

[MN19] D. Maulik and A. Neguţ. Lehn's formula in Chow and conjectures of Beauville and Voisin. arXiv:1904.05262, 2019.

[Mum68] D. Mumford. Rational equivalence of 0-cycles on surfaces. *J. Math. Kyoto Univ.*, 9:195–204, 1968.

[O'G06] Kieran G. O'Grady. Irreducible symplectic 4-folds and Eisenbud-Popescu-Walter sextics. *Duke Math. J.*, 134(1):99–137, 2006.

[SV16] Mingmin Shen and Charles Vial. The Fourier transform for certain hyperkähler fourfolds. *Mem. Amer. Math. Soc.*, 240(1139):vii+163, 2016.

[vGV16] Bert van Geemen and Claire Voisin. On a conjecture of Matsushita. *Int. Math. Res. Not. IMRN*, (10):3111–3123, 2016.

[Voi92] Claire Voisin. Sur la stabilité des sous-variétés lagrangiennes des variétés symplectiques holomorphes. In *Complex projective geometry (Trieste, 1989/Bergen, 1989)*, volume 179 of *London Math. Soc. Lecture Note Ser.*, pages 294–303. Cambridge Univ. Press, Cambridge, 1992.

[Voi04] Claire Voisin. Intrinsic pseudo-volume forms and K-correspondences. In *The Fano Conference*, pages 761–792. Univ. Torino, Turin, 2004.

[Voi07] Claire Voisin. *Hodge theory and complex algebraic geometry. II*, volume 77 of *Cambridge Studies in Advanced Mathematics*. Cambridge University Press, Cambridge, english edition, 2007. Translated from the French by Leila Schneps.

[Voi08] Claire Voisin. On the Chow ring of certain algebraic hyper-Kähler manifolds. *Pure Appl. Math. Q.*, 4(3, Special Issue: In honor of Fedor Bogomolov. Part 2):613–649, 2008.

[Voi16] Claire Voisin. Remarks and questions on coisotropic subvarieties and 0-cycles of hyper-Kähler varieties. In *K3 surfaces and their moduli*, volume 315 of *Progr. Math.*, pages 365–399. Birkhäuser/Springer, [Cham], 2016.

Printed in the United States
by Baker & Taylor Publisher Services